# PROBABILITY AND STATISTICS

for Science and Engineering

# PROBABILITY AND STATISTICS

## for Science and Engineering

G Shanker Rao

Department of Mathematics
University College of Engineering
Osmania University, Hyderabad

Universities Press

**UNIVERSITIES PRESS (INDIA) PRIVATE LIMITED**

*Registered Office*
3-6-747/1/A & 3-6-754/1, Himayatnagar, Hyderabad 500 029 (A.P.), India
E-mail: info@universitiespress.com; Web: www.universitiespress.com

*Distributed by*
Orient Blackswan Private Limited

*Registered Office*
3-6-752 Himayatnagar, Hyderabad 500 029 (A.P.), India

*Other Offices*
Bangalore / Bhopal / Bhubaneshwar / Chandigarh / Chennai / Ernakulam / Guwahati
Hyderabad / Jaipur / Kolkata / Lucknow / Mumbai / New Delhi / Noida / Patna


First published 2011

ISBN 978 81 7371 744 4



*Typeset in* Times New Roman 10/12 *by*
Techastra Solutions Pvt. Ltd., Hyderabad 500 029

*Printed at*
Graphica Printers, Hyderabad 500 013

*Published by*
Universities Press (India) Private Limited
3-6-747/1/A & 3-6-754/1, Himayatnagar
Hyderabad 500 029 (A.P.), India

*To*

*My Mother*

# Contents

# Preface

*Probability and Statistics for Science and Engineering* is intended to provide students an adequate understanding of the basic concepts of probability theory and statistics for applying them to problem solving in science and engineering. The book grew out of my attempt to meet the continued and persistent demand of my students for materials that are easy to follow and yet are mathematically rigorous. The topics included in the book cover the syllabus of most undergraduate engineering programmes in India.

The basics of probability theory and statistical analysis methods dealt with in the book are spread over twelve chapters as follows.

Chapter 1 discusses the basic concepts of probability while Chapter 2 deals with random variables, one of the most significant concepts encountered in probability theory. In Chapter 3, a detailed discussion on probability distributions—binomial, Poisson, geometric and uniform—is provided with applications and examples.

Statistical inference, which includes sampling, estimation theory and hypothesis testing theory, are explored in the next few chapters. Chapter 4 deals with the fundamentals of sampling theory. Chapter 5 introduces the theory of estimation and some of the basic concepts essential for understanding hypothesis testing. Chapter 6 explains the meaning of hypothesis and delves into some of the basic concepts essential for the better understanding of hypothesis testing methods. Chapter 7 introduces chi-square distribution, which is used for testing the goodness of fit of a model under consideration. The chi-square test is also used for finding associations and relations between attributes. Chapter 8 discusses tests of significance for small samples. The important tests for small samples like the chi-square test, the $t$-test and the $F$-test are explained in detail.

Chapter 9 deals with the problem of fitting an equation or a curve to data involving paired values.

Chapter 10 discusses the concepts of correlation and regression, which help while dealing with two variables that may have a relationship or association between themselves.

Chapter 11 introduces the queueing theory. Queues arise whenever a shared facility needs to be accessed for service by a large number of jobs or customers, e.g, in telecommunications, traffic control, retail stores, banks, etc. This chapter describes the behaviour of waiting lines—the general characteristics of queues—and the mathematical solution of some basic queueing models.

Chapter 12 gives a brief introduction to stochastic processes inherent in dynamic systems of the real world. It looks at the various types of stochastic processes including Markov chains.

In all the above chapters, a large number of solved problems and sets of well-graded exercises have been included in order to give the student ample guidance on typical problems encountered in their course of study and help them to apply the mathematical tools and models in real life decision making.

I welcome feedback from readers and would appreciate any suggestion that would help me improve the book further in its future edition.

## CHAPTER ONE

# Probability

## 1.1 Introduction

This chapter is devoted to the study of probability—what it is and some of the basics concepts of the theory that surrounds it. Historically, the original purpose of the theory of probability was to describe the exceedingly narrow domain of experience connected with the games of chance. The main effort was directed to the calculation of certain probabilities. We know that statistics is concerned with inferences about population based on sample information. In this chapter we assume that population is known and calculate the chances of obtaining various samples from the population. Thus probability is the reverse of statistics. In probability we use population information to infer the probable nature of the sample. Many experiments do not yield exactly the same results when performed repeatedly. For example, if we toss a coin, we are not sure if we will get a head or a tail and if we toss a six-faced die, we have no way of knowing which of the six possible outcomes 1, 2, 3, 4, 5 or 6, will turn up. The outcomes of these experiments are associated with uncertainties, and experiments of this type are called probabilistic experiments. The outcome of a probabilistic experiment cannot be predicted in advance. Any prediction made involves uncertainty.

Probability is a concept which tends to measure the degree of this uncertainty. The general meaning of the term "probability" is chance or possibility. The term is used in two senses viz. in a qualitative sense and a quantitative sense. When we say that "most probably it will rain tomorrow" or "it is likely that Australia wins this tournament this time", then the probability is used in the qualitative sense. When we say that "there is a 45% chance of rain tomorrow", the chance is expressed as a percentage and the term probability is used in the quantitative sense. In general, probability is the chance that something will happen. Probabilities are expressed as fractions or decimals between 0 and 1. Although probability was used in games of chance, the theory of probability is now applied in many diverse fields. In this chapter, we shall calculate a few typical probabilities. Before we deal with the definition and the theory of probability, it is necessary to understand the following concepts.

## 1.2 Preliminary Concepts

### 1.2.1 Experiment

**Definition 1.1** An experiment is an act or process that leads to a single outcome that cannot be predicted with certainty.

**Remark** A result of an experiment is called an outcome. Our definition of an experiment is broader than that used in physical sciences. Experiments are usually denoted by uppercase letters of the English alphabet.

**Example 1** Tossing a coin is an experiment. It has two possible outcomes: "Head up" or "Tail up".

**Example 2** Rolling a fair die is an experiment. It has six possible outcomes: 1 or 2 or 3 or 4 or 5 or 6.

### 1.2.2 Random experiment

Each experiment may yield one or more outcomes. When these are not predetermined but subject to chance, we have a random experiment. Probability is concerned with the analysis of random experiments.

**Definition 1.2** An experiment in which all the outcomes can be enumerated, but it is not known which particular outcome will result is called a random experiment.

**Example 1** Counting the number of misprints in a page is a random experiment.

**Example 2** Drawing a card from a well-shuffled pack of cards is a random experiment.

**Example 3** Arranging 52 bridge cards such that no two aces are adjacent is a random experiment.

### 1.2.3 Event

**Definition 1.3** The results of experiments or observations are called events.
Events are usually denoted by the letters E, A, B, C etc.

### 1.2.4 Simple event

**Definition 1.4** A simple event is an experiment that cannot be decomposed into a simpler outcome. A simple event is a single outcome of an experiment. Getting a head or a tail when a coin is tossed is a simple event. A simple event is also called an elementary event.

### 1.2.5 Compound event

An event is a collection (aggregate) of one or more simple events. Events can be decomposed into simpler outcomes.

**Definition 1.5** The joint occurrence of two or more simple events is called a compound event.

**Example** Getting an even number when a fair die is rolled is a compound event. The possible even numbers are 2, 4 and 6.

### 1.2.6 Sample space

The sample space of an experiment is the collection (aggregate) of all its simple events. It is defined as follows.

**Definition 1.6** The set of all possible outcomes of a random experiment is called a sample space.

The sample space of an experiment is denoted by the letter $S$. Each simple event of an experiment is called a sample point of the sample space $S$.

For example, consider an experiment in which a coin is tossed. There are simple events associated with this experiment. If $S$ denotes the sample space of the experiment, each element of $S$ is a simple event. The total number of outcomes of an experiment may be finite, countable or uncountable. For instance, consider an experiment in which two coins are tossed simultaneously. There are four possible outcomes. If $S$ denotes the sample space, then

$$S = \{\text{HH, HT, TH, TT}\}, \text{H} = \text{head}, \text{T} = \text{tail}$$

where the first letter denotes the outcome of the first toss and the second letter denotes the outcome of the second toss. An event is a subset of a sample space.

### 1.2.7 Mutually exclusive events (incompatible events)

**Definition 1.7** Events are said to be mutually exclusive if the occurrence of one preludes, the occurrence of the others.

Two mutually exclusive events cannot occur in the same trial. In other words, two events $E_1$ and $E_2$ are mutually exclusive if there is no point in common in between the points belonging to $E_1$ and $E_2$.

**Example** In tossing a coin, the events Head (H) and Tail (T) are mutually exclusive.

### 1.2.8 Exhaustive events

**Definition 1.8** All possible events in any trial are known as exhaustive events.

**Example 1** In tossing a fair coin, there are two exhaustive events viz. head and tail.

**Example 2** In tossing two coins, there are four exhaustive events. They are

HH (two heads), HT (one head, one tail), TH (one tail, one head), TT (two tails i.e. no heads)

### 1.2.9 Impossible event

**Definition 1.9** An event which is certain not to occur is called an impossible event. It is denoted by $\phi$.

**Example** Consider an experiment in which two dice are rolled simultaneously. Getting a score of 15 is an impossible event.

### 1.2.10 Complement of an event

**Definition 1.10** Complement of any event $E$ is the event that $E$ does not occur. The complement of an event is denoted by "$E^c$" or "$\overline{E}$" or "$E'$".

Consider the experiment in which two coins are tossed. If $E$ denotes the event in which at least one head appears, then the complement of $E$ is $\overline{E} = \{TT\}$.

If $E$ is an element and $\overline{E}$ is its complement then $E \cap \overline{E} = \phi$, $E \cup \overline{E} = S$.

### 1.2.11 Certain event

**Definition 1.11** An event which is certain to occur is called a certain event.

### 1.2.12 Equally likely events

**Definition 1.12** Events are said to be equally likely when there is no reason to expect any one of them to occur rather than any one of the others.

**Definition 1.13 (Equally likely events)** Two or more events are said to be equally likely if one of them cannot be expected to occur in preference to the others.

### 1.2.13 Venn diagrams

In studying probability, the words "set" and "event" are interchangeable. If we say that $E_1$ and $E_2$ are mutually exclusive events, it means that $E_1 \cap E_2 = \phi$ (null set).

If $E_1$ and $E_2$ are exhaustive events then it means that $E_1 \cup E_2 = S$. These can be extended to $n$ events and stated as follows.

1. $E_1, E_2, \ldots, E_n$ are mutually exclusive events means that $E_i \cap E_j = \phi$, $(i \neq j)$

2. $E_1, E_2,\ldots, E_n$ are exhaustive events means that $E_1 \cup E_2 \cup \ldots \cup E_n = S$

3. If $E_1, E_2,\ldots, E_n$ are mutually exclusive and exhaustive events then,

$$E_i \cap E_j = \phi,\ (i \neq j) \text{ and } E_1 \cup E_2 \cup \ldots \cup E_n = S.$$

A graphical method called Venn diagram is useful for presenting sample space and its sample points. It helps in discovering and considering relationships that may exist between the subjects of a given universe.

The sample space is a closed curve labelled $S$. The figure $S$ contains a set of points called sample points, with each point representing a simple event. The sample space $S$ can also be represented by a rectangle, whereas events are represented by regions within the rectangle, usually by circles or parts of circles.

**Example 1** Consider an experiment in which a coin is tossed. If $S$ is the sample space, then $S = \{H,T\}$. The Venn diagram in Fig. 1.1 represents the sample space.

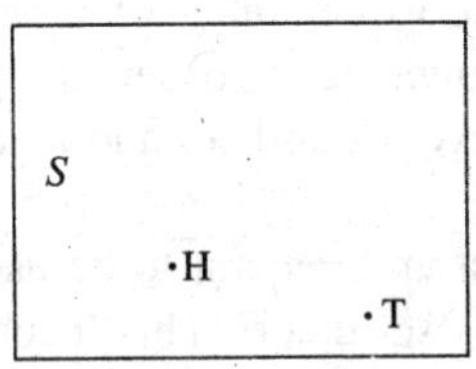


**Fig. 1.1(a)**

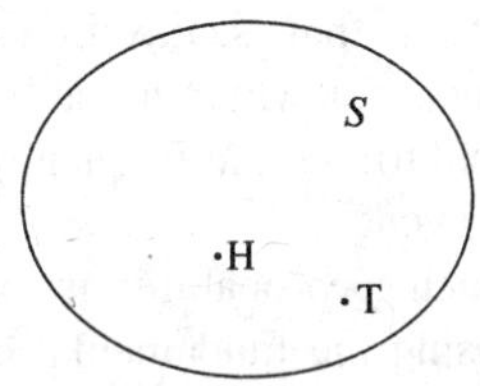


**Fig. 1.1(b)**

**Example 2** If a die is rolled and $S$ denotes the sample, the Venn diagram representing $S$ of the experiment is shown in Fig. 1.2.

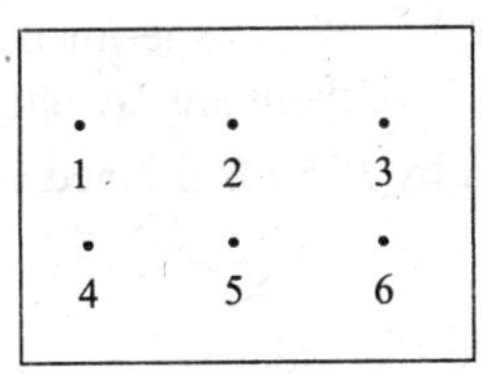


**Fig. 1.2(a)**

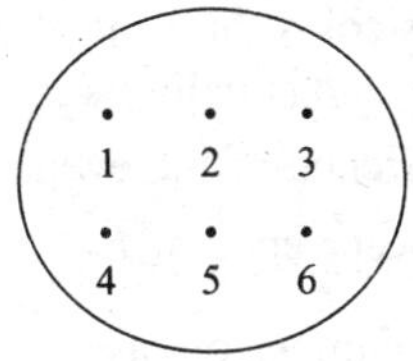


**Fig 1.2(b)**

**Example 3** If $E_1$ and $E_2$ are two events of a sample space $S$, such that $E_1$ is a subset of $E_2$, then the Venn diagram to represent it is shown in Fig. 1.3.

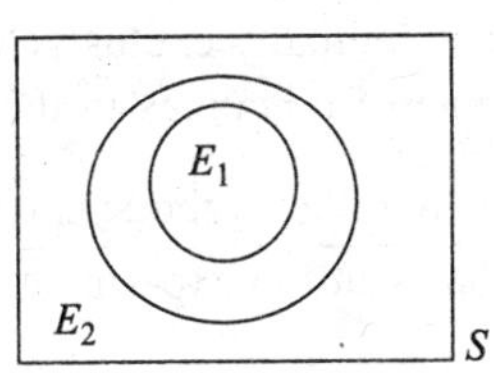


**Fig. 1.3**

## 1.3 Definition and Types of Probability

There are several approaches to the concept of probability. In this chapter we will consider three different approaches, namely

1. The classical approach
2. The frequency approach
3. The axiomatic approach

The classical approach is to calculate the number of favourable outcomes of a trial and divide it by the total number of possible outcomes. The classical definition of probability is also called the mathematical definition of probability and it fails when the number of possible outcomes of the experiment is infinite. The frequency approach is based on the frequency of an event in a long series of trials. The frequency of an event $E$ in a series of $N$ trials is the proportion of times the event $E$ would occur, that is, the limit of the ratio of number of times the event $E$ occurs. There are random events for which a stability of frequencies is observed. With an increase of $N$ (number of independent trials), the frequency of the event stabilises and tends to a quantity called the probability of the event.

The modern approach to probability is too abstract and general to be useful. It is based on an axiomatic approach using the fundamental concept of set theory This theory is known as the set theoretical approach.

## 1.4 Classical Probability

**Definition 1.14 (Classical definition of probability)** If an experiment can result in $n$ exhaustive, mutually exclusive and equally likely cases and $m$ of them are favourable to the event $E$, then the probability of occurrence of the event $E$ denoted by $P(E)$ is defined as the ratio $\frac{m}{n}$.

The probability of occurrence of $E$

$$P(E) = \frac{\text{The number of cases favourable to the event } E}{\text{The total number of exhaustive mutually exclusive and equally likely cases in which the experiment can result}}$$

$$= \frac{m}{n}$$

Since the total number of exhaustive mutually exclusive and equally likely cases in which the experiment can result = the total number of outcomes in the experiment, we can define,

$$P(E) = \frac{\text{Number of outcomes favourable to the event E}}{\text{Total number of outcomes in the experiment}}$$

$$= \frac{n(E)}{n(s)}$$

$$= \frac{m}{n}$$

where $S$ is sample space of the experiment.

The above definition is called the classical definition. It is also called the mathematical or Laplacian definition of probability.

If $m = 0$, then $P(E) = 0$, and the event $E$ is called an impossible event.

When $m = n$, we have $P(E) = 1$. In this case, the event $E$ is said to be a certain event.

The classical definition is known as a "a priori" probability, since the definition does not require the results of a certain number of repeated trials of the experiment.

Classical definition of probability is based on cases which are "equally likely" and as such cannot be applied to experiments where all outcomes are not likely. When the number of possible outcomes of the experiment is infinite, this method fails. In some cases, it may not be possible to enumerate all the possible outcomes of the experiment and in this case also, the method fails. The definition of classical probability depends on equal probability.

From the definition, it is clear that the number of favourable outcomes favourable to the event $E$ cannot be greater than the total number of outcomes $n$ in the experiment. Therefore, we have

$$0 \leq m \leq n$$

or $$0 \leq \frac{m}{n} \leq 1$$

or $$0 \leq P(E) \leq 1$$

## 1.4.1 Properties of classical probability

If $E$ is an event of a sample space $S$, then

1. $P(E)$ is a pure number.
2. $0 \leq P(E) \leq 1$
3. If $E_1, E_2, \ldots, E_k$ are $k$ mutually exclusive and exhaustive events in the sample space $S$ of an experiment, then $\sum_{i=1}^{k} P(E_i) = 1$.
4. The probability of an impossible event is zero.
5. If $E_1 \subseteq E_2$, then $P(E_1) \leq P(E_2)$

## 1.4.2 Uses and limitations

The advantage of the classical definition of probability is that it can be applied without experimentation. The method fails when the number of possible outcomes is infinite and the outcomes are not equally likely. The definition also restricts the value of probability to rational numbers only. In some cases, it may not be practically possible to enumerate all the equally likely cases.

### 1.4.3 Probability of failure

Suppose an event E can happen in $m$ ways out of $n$ possible equally likely outcomes. Then the number of times the event does not occur is $n-m$. The non-occurrence of the event is denoted by "$E^c$" or "$\overline{E}$" or "$E'$".

$P(\overline{E})$ denotes the probability of non-occurrence of the event $E$. $P(\overline{E})$ is given by

$$P(\overline{E}) = \frac{n-m}{n} = \frac{n}{n} - \frac{m}{n} = 1 - \frac{m}{n}$$

$$= 1 - P(E)$$

**Note:** 1. $P(E)+P(\overline{E}) = 1$

2. $P(E) = 1 - P(\overline{E})$

The probability of the occurrence of the event $E$ is called the probability of its success and is usually denoted by $p$. The probability of non-occurrence of the event $E$, is called the probability of its failure and is usually denoted by $q$. We have $p+q=1$.

## 1.5 Relative Frequency Approach

By introducing the relative frequency approach, some of the drawbacks of classical probability can be removed.

**Definition 1.15** Let an experiment be conducted a large number of times under homogeneous and identical conditions. If $n$ denotes the number of times the experiment is conducted out of which $m$ are favourable to an event $E$, then the probability of the event $E$ is $P(E) = \lim_{n\to\infty} \frac{m}{n}$.

This definition is also known as the statistical or empirical definition of probability and was developed by R. A. Fisher and R.Von Mises. The statistical definition requires the results of a certain number of repeated trials of the experiment and hence is called the posteriori definition of probability. This definition also removes all the objections and drawbacks of mathematical definition. The only objection of this definition is that it is difficult to prove the existence of a limit to the relative frequency.

## 1.6 Symbolic Notation

Let $S$ denote sample space (i.e., set of all possible outcomes of an experiment). If $A$ and $B$ are any two events of the experiment, $A$ and $B$ are subsets of $S$. Then

(1) The non-occurrence of $A$ is denoted by $\bar{A}$ (or by $A'$ or $A^c$).

(2) $A \cup B$ is an event which denotes the occurrence of "either $A$ or $B$" (or both).

(3) $A \cap B = AB$ denotes the occurrence of both $A$ and $B$.

(4) If $A \cap B = \phi$; i.e., if $A$ and $B$ are mutually exclusive, then $A \cup B$ is denoted by $A + B$.

(5) Neither $A$ or $B$ occurs is denoted by $\overline{A} \cap \overline{B} = \overline{A \cup B} = S - (A \cup B)$.

(6) $A$ occurs and $B$ does not occurs is denoted by $A \cap \overline{B}$ (or $A - B$).

(7) If $A$, $B$, $C$ are any three events,

(a) Neither $A$, $B$, $C$ occur is denoted by $\overline{A} \cap \overline{B} \cap \overline{C}$.

(b) At least one of $A$, $B$, $C$ occurs is denoted by $A \cup B \cup C$.

(c) All the three events $A$, $B$, $C$ occur is denoted by $A \cap B \cap C$.

## 1.7 Axiomatic Approach

In this section, we will state the following postulates (axioms) of probability which require no proof. The probabilities in axiomatic approach are defined as mathematical objects that behave according to certain well-defined rules.

### 1.7.1 Axioms of probability

Let $S$ be a sample space, $\xi$ be the class of events, and $P$ be a real-valued function on $\xi$. Then $P$ is called a probability function. If $A$ is an event of the sample space, then $P(A)$ is called the probability of event $A$ if the following axioms hold.

$P_1$: For every event $A : 0 \leq P(A) \leq 1$. (Axiom of non-negativity)

$P_2$: $P(S) = 1$. (Axiom of certainty)

$P_3$: If $A$ and $B$ are mutually exclusive events then $P(A \cup B) = P(A) + P(B)$. (Axiom of additivity)

$P_4$: If $A_1, A_2, A_3, \ldots, A_n, \ldots$ is a sequence of mutually exclusive events, then

$$P(A_1 \cup A_2 \cup A_3 \ldots\ldots \cup A_n \ldots.) = P(A_1) + P(A_2) + \ldots + P(A_n) + \ldots$$

**Solved examples**

**Example 1** A coin is tossed once. Find the probability of getting a head.

**Solution** A coin is tossed once, therefore the number of equally likely cases is 2 i.e., heads or trials.

We have $n = 2$

Number of favourable cases $= 1$

i.e., $m = 1$

$$\text{Probability of getting a head} = P\text{ (Head)} = \frac{m}{n} = \frac{1}{2}$$

**Example 2** Two coins are tossed once. Find the probability of getting (a) one head and (b) at least one head.

**Solution** The sample space of possible outcomes when two coins are tossed once is $S = \{$HH, HT, TH, TT$\}$ where HH denotes the case in which both coins turn up head, HT denotes the outcome that the first coin turns up head (H) and the second coin turns up tail (T) and so on. Therefore, the number of points in the sample space $S$ of the experiment $= n = 4$.

(a) Let $A$ be the event in which exactly one head turns up. Then

$A = \{$HT, TH$\}$

The number of favourable cases $= m = 2$.

Hence $P(A) = \dfrac{m}{n} = \dfrac{2}{4} = \dfrac{1}{2}$

(b) Let $B$ denote the event in which at least one head turns up. Then,

$B = \{$HH, HT, TH$\}$

Number of cases favourable to $B$ is 3 i.e., $m = 3$.

Hence, $P(B) = \dfrac{m}{n} = \dfrac{3}{4}$

**Example 3** A coin is tossed thrice. Find the probability of getting all tails.

**Solution** The sample space for the experiment is

$S = \{$HHH, HHT, HTH, THH, HTT, THT, TTH, TTT$\}$

Number of points in the sample space $S = 8$.

i.e., $n(S) = n = 8$

Let $E$ be the event of getting all tails. Then,

$E = \{$TTT$\}$

We have $m = 1$.

Hence, $P(E) = \dfrac{m}{n} = \dfrac{1}{8}$

**Example 4** A die is rolled. List the sample space and find the probability in which an even number appears on the die.

**Solution** A die has six faces, marked with the numbers 1, 2, 3, 4, 5 and 6.

The sample space is $S = \{1,2,3,4,5,6\}$

Number of elements in $S = n = 6$

Let $E$ be the event of getting an even number. Then,

$$E = \{2,4,6\}$$

We have $m = 3$.

The probability of getting an even number when a die is rolled $= P(E) = \frac{m}{n} = \frac{3}{6} = \frac{1}{2}$

**Example 5** A pair of dice is tossed and the two numbers appearing on the top are recorded
(1) Describe the sample space.
(2) Find the probability of the event in which the sum is 10 or more.

**Solution** There are six possible numbers 1, 2, 3, 4, 5 and 6 on each die. Therefore, the total number of possible outcomes is $6 \times 6 = 36$.

We have

(1) $$\begin{aligned} S = \{&(1,1),(1,2),(1,3),(1,4),(1,5),(1,6)\\ &(2,1),(2,2),(2,3),(2,4),(2,5),(2,6)\\ &(3,1),(3,2),(3,3),(3,4),(3,5),(3,6)\\ &(4,1),(4,2),(4,3),(4,4),(4,5),(4,6)\\ &(5,1),(5,2),(5,3),(5,4),(5,5),(5,6)\\ &(6,1),(6,2),(6,3),(6,4),(6,5),(6,6)\}\end{aligned}$$

Hence, $n(S) = n = 36$

(2) Let $E$ be the event in which the sum is 10 or more. Then,

$$E = \{(4,6),(5,5),(6,4),(5,6),(6,5),(6,6)\}$$

Number of cases favourable to $E = 6$.

We have, $m = 6$ and $n = 36$.

$$P(E) = \frac{m}{n} = \frac{6}{36} = \frac{1}{6}$$

**Example 6** What is the chance that a non-leap year selected at random will contain 53 Sundays?

**Solution** A non-leap year has 365 days i.e., 52 weeks and 1 day.

The one day i.e., the 365th day can be any one of the following days:

Sun, Mon, Tue, Wed, Thu, Fri, Sat

The total number of possibilities = 7

Number of favourable cases = 1

Required probability $= \frac{1}{7}$

**Example 7** There are $n$ particles, each of which can occupy each of $N(N > n)$ cells with the same probability $\frac{1}{N}$. Find the probability that (a) there will be one particle in each of $n$ definite cells, (b) there will be one particle in each of $n$ arbitrary cells.

**Solution** Each particle may be located in each of the $N$ cells; therefore, $n$ particles may be distributed in the $N$ cells in $N^n$ different ways.

(a) Let $A$ denote the event in which there will be one particle in each of $n$ definite cells. The number of favourable cases for $A = n!$

Hence, the probability that one particle will fall in each of $n$ definite cells is $P(A) = \frac{n!}{N^n}$

(b) Let $B$ denote the event in which there will be one particle in each of $n$ arbitrary cells. The number of favourable cases in this case is ${}^NC_n.n!$

Therefore, we get $P(B) = \frac{n!\,{}^NC_n}{N^n}$

$$= \frac{n!N!}{N^n n!(N-n)!}$$

$$= \frac{N!}{N^n(N-n)!}$$

**Example 8** What is the chance that a leap year, selected at random will contain 53 Sundays?

**Solution** A leap year has 366 days i.e., 52 weeks + 2 extra days.

The 356th and 366th days can be as follows:

(Sun, Mon), (Mon, Tue), (Tue, Wed), (Wed, Thu), (Thu, Fri), (Fri, Sat), (Sat, Sun)

Total number of cases $= n = 7$

Number of favourable cases in which there are 53 Sundays $= m = 2$

Let $E$ be the event of getting 53 Sundays. Then,

$$P(E) = \frac{m}{n} = \frac{2}{7}$$

**Example 9** A single card is drawn from an ordinary pack of 52 cards. Find the probability that

(a) The card drawn is a king.

(d) The card is a face card (jack, queen or king).

(c) The card is a heart.

**Solution** A card is drawn from a pack of 52 cards.

$\therefore$ Number of points in the sample space = 52

(a) Number of kings in a pack = 4

Number of ways a king can be drawn from 4 kings = 4

Probability that the card drawn is a king $= \frac{4}{52} = \frac{1}{13}$

(b) Number of face cards in a pack = 4 jacks + 4 queens + 4 kings = 12

Number of ways of drawing a face card = 12

Probability that the card drawn from the pack is a face card $= \frac{12}{52} = \frac{3}{13}$

(c) There are 13 hearts in a pack of cards. Hence the probability that the card drawn is a heart $= \frac{13}{52} = \frac{1}{4}$

**Example 10** A bag contains 4 red, 3 green and 5 black balls. If a ball is drawn at random from the bag, find the chance that it is a red ball.

**Solution** Number of balls in the bag = 4 red + 3 green + 5 black = 12

Number of ways of drawing a ball from the bag = 12

Number of red balls in the bag = 4

Number of ways of drawing red balls from the bag = 4

Let $E$ be the event of drawing a red ball from the bag

$$\therefore \quad P(E) = \frac{4}{12} = \frac{1}{3}$$

**Example 11** A group of 5 men and 4 woman agree that three should be chosen by lot to form a special committee. What is the probability that the committee consists of one man and two women?

**Solution** We have $n = 5$ men + 4 women = 9

The number of possible ways in which the committee consisting of 3 can be selected

$$= C(9,3) = {}^9C_3 = \frac{9.8.7}{1.2.3} = 84$$

If the committee is to include one man and two women, then the number of ways it can be done is

$$C(5,1)C(4,2) = {}^5C_1{}^4C_2 = \frac{5}{1}.\frac{4.3}{1.2} = 30$$

We have $n = 84$, $m = 30$

Let $E$ be the event.

$$\therefore P(E) = \frac{m}{n} = \frac{30}{84} = \frac{5}{14}$$

**Example 12** From a set of 17 cards, numbered 1, 2, 3,...,17, one card is drawn at random. Find the chance that the number on the card is divisible by 3 or 5.

**Solution** Number of cards = 17

Number of ways of drawing a card from 17 cards $= n = {}^{17}C_1 = 17$

Let $E$ be the event in which the card drawn is divisible by 3 or 5.

Favourable cases $= E = \{3, 5, 6, 9, 10, 12, 15\}$

Number of favourable cases = 7

We have $m = 7$

Therefore, $P(E) = \dfrac{m}{n} = \dfrac{7}{17}$

**Example 13** Find the probability that in a family of 5 children, there will be at least one boy.

**Solution** Number of children in a family = 5

Probability that at least one of them is a boy $= 1 - \dfrac{1}{2^n} = 1 - \dfrac{1}{2^5}$

$$= 1 - \frac{1}{32} = \frac{31}{32}$$

**Example 14** In a throw with a pair of dice, what is the chance of throwing a doublet?

**Solution** Number of dice thrown = 2

Number of points in the sample space $= n = 6^2 = 6 \times 6 = 36$

Let $E$ be the event of a doublet, then $E\{(1,1), (2,2), (3,3), (4,4), (5,5,), (6,6)\}$

Number of cases favourable to $E = 6$

$\therefore \quad m = 6$

Required probability $= P(E) = \frac{m}{n} = \frac{6}{36} = \frac{1}{6}$

**Example 15** A bag contains 7 white, 5 black and 4 red balls. If two balls are drawn at random, find the probability that

(i) both the balls are black.

(ii) one ball is black and the other is a red ball.

**Solution** Total number of balls in the bag = 7 white + 5 black + 4 red = 16
Total number of ways of drawing two balls from 16 balls $= n = {}^{16}C_2 = 120$

(i) Number of black balls in the bag = 5

Number of ways of drawing 2 balls from the 5 black balls $= m = {}^{5}C_2 = 10$

The probability of drawing 2 black balls from the bag $= \frac{10}{120} = \frac{1}{12}$

(ii) Number of black balls = 5

Number of red balls = 4

Number of ways of drawing a black ball and a red ball from the bag

$= {}^{5}C_1 \times {}^{4}C_1 = 20$

Probability of drawing one red ball and one black ball from the bag $= \frac{20}{120} = \frac{1}{6}$

**Example 16** $n$ letters, to each of which corresponds an addressed envelope, are placed in the envelopes at random. What is the probability that all the letters are placed in right envelopes?

**Solution** Number of ways $n$ letters can be placed in the envelopes is $n!$

Let $E$ be the event in which all the letters are placed correctly in the addressed envelopes.

Number of cases favourable to $E = m = 1$

$P(E) = \frac{m}{n} = \frac{1}{n!}$

**Example 17** A class consists of 6 girls and 10 boys. If a committee of 3 is to be chosen from the class. Find the probability that (i) 3 boys are selected, (ii) exactly 2 girls is selected.

(JNTU Feb, 2008)

**Solution** Total number of students = 6 girls + 10 boys = 16

Number of ways of choosing 3 from 16 is ${}^{16}C_3$

(i) Number of boys = 10

Number of ways of selecting 3 boys from 10 boys = ${}^{10}C_3$

Let $E$ be the event of selecting 3 boys

$\therefore \quad n(E) = m = {}^{10}C_3$

Hence $P(E) = \dfrac{m}{n} = \dfrac{{}^{10}C_3}{{}^{16}C_3} = \dfrac{3}{14}$

(ii) When 2 girls are selected, the committee of 3 includes one boy. This can be done in ${}^{6}C_2 \times {}^{10}C_1$ ways.

Let $E$ be the event. Then $P(E) = \dfrac{{}^{6}C_2 \times {}^{10}C_1}{{}^{16}C_3} = \dfrac{15}{56}$

## 1.8 Odds in Favour and Odds Against an Event

Let $S$ be a sample space and $A$ be an event of $S$. Then $P(A){:}P(\overline{A})$ is the odds in favour of the event $A$ and $P(\overline{A}){:}P(A)$ is the odds against the event $A$.

If $P(A) = \dfrac{m}{n}$, then $P(\overline{A}) = 1 - P(A) = \dfrac{n-m}{n}$

Therefore, the odds in favour of the event $A$ is $\dfrac{m}{n} : \dfrac{n-m}{n}$ i.e., $m : n-m$. and the odds against the event $A$ is $n-m : m$.

If the odds in favour of the event $A$ are given as $a$ to $b$ i.e., $a : b$, then we have

$$P(A) : P(\overline{A}) = a : b$$

i.e., $$\frac{P(A)}{P(\overline{A})} = \frac{a}{b}$$

$$\Rightarrow \quad bP(A) = aP(\overline{A}) = a(1 - P(A))$$

$$bP(A) = a - aP(A)$$

$$aP(A) + bP(A) = a$$

$$(a+b)P(A) = a$$

$$P(A) = \frac{a}{a+b}$$

Similarly, we can prove that $P(\overline{A}) = \dfrac{b}{a+b}$

## Solved examples

**Example 1** If the odds against a certain event are given as 7 to 3, what is the probability of the event?

**Solution** Let $A$ be the event. Then we have,

$$P(\overline{A}) : P(A) = 7 : 3$$

$$\therefore \quad P(\overline{A}) = \frac{7}{7+3} = \frac{7}{10}$$

and $P(A) = 1 - P(\overline{A})$

$$= 1 - \frac{7}{10} = \frac{3}{10}$$

**Example 2** Two dice are tossed. Find the odds in favour of throwing a sum of 8.

**Solution** Let $E$ be the event in which the sum obtained is 8.

Favourable cases: $(2,6), (3,5)(4,4), (5,3), (6,2)$

Number of favourable cases $= 5$

$\Rightarrow \quad m = 5$

Number of dice thrown = 2

Number of points in the sample space $= 6 \times 6 = 36$

We get $P(E) = \dfrac{5}{36}$ and $P(\overline{E}) = \dfrac{31}{36}$

$$P(E) : P(\overline{E}) = \frac{5}{36} : \frac{31}{36} = 5 : 31$$

Therefore, odds in favour of throwing a sum of 8 are 5 to 31.

**Example 3** From a bag containing 4 white, and 5 black balls, a man draws 3 balls at random. What are the odds in favour of they being all black?

**Solution** Number of balls in the bag = 4 white + 5 black = 9

Number of ways of drawing 3 balls from the bag = ${}^9C_3$

Number of black balls in the bag = 5

Number of ways of drawing three black balls from the 5 black balls = ${}^5C_3$

Let $E$ be the event of drawing three black balls from the bag.

Then, $P(E) = \dfrac{{}^5C_3}{{}^9C_3} = \dfrac{5}{42}$

$$P(\overline{E}) = 1 - P(E) = 1 - \frac{5}{42} = \frac{37}{42}$$

Odds in favour of the event $E$ is $P(E) : P(\overline{E})$

i.e., $\dfrac{5}{42} : \dfrac{37}{42}$

i.e., $5 : 37$

Hence $P(E) : P(\overline{E}) = 5 : 37$

## Exercise 1.1

1. Define the terms
   (a) Event (b) Simple event (c) Complementary event
   (d) Mutually exclusive event (e) Equally likely events.
   Give examples.
2. Define the terms
   (a) Probability (b) Impossible event (c) Certain event
3. Define: (a) Odd in favour of an event, (b) odds against an event.
4. From a well-shuffled deck of playing cards, if a card is drawn find the probability that it is an honours card.
5. What is the probability of getting a head on one toss of a coin.
6. A box contains 3 red and 4 yellow balls. Two balls are drawn from the box at random. What is the probability that the two balls are of different colours?
7. A cubical die is thrown. Find the probability of obtaining an odd number.
8. A coin is tossed twice. Find the probability of getting
   (a) one tail (b) two tails (c) at least one head (d) at least one tail

9. Three coins are tossed. Find the probability of getting at least one head.

10. What is the probability of getting a number greater than 3 when a die is rolled once?

11. Find the probability of obtaining a score of 8 with two dice in a single throw.

12. Two dice are tossed. Find the chance that the sum of the number obtained is 7.

13. If a die is thrown, what is the probability that the upper face shows

    (a) 5 dots (b) more that 5 dots (c) an even number of dots

14. Find the probability of three half-rupee coins falling all heads up when tossed simultaneously.

15. When two dice are rolled, find the probability that the total score on the two dice show even numbers.

16. Two cards are drawn from a pack of cards. Find the probability that the cards are (a) aces, (b) queens.

17. If a card is drawn from a pack, what is the probability of getting an ace?

18. A card is drawn from a pack. What is the probability of getting a spade?

19. When an unbiased coin is tossed five times, find the probability of getting "Heads" at least once.

20. From a set of 17 cards numbered 1, 2, 3, ..., 17, one card is drawn at random. Find the chance that its number is divisible by 3 or 7.

21. When two unbiased dice are rolled simultaneously, find the probability of getting a prime score.

22. Dialling a telephone number, an old man forgets the last two digits. He remembers only that they are different and dials at random. Find the probability that the number is dialed correctly.

23. Two balls are drawn at random from a bag containing 7 red and 11 white balls. Find the chance that both the balls are white.

24. Find the probability that in a family of four children, there will be at least one boy.

25. Three bulbs are chosen at random from 15 bulbs of which 5 are defective. Find the probability that none is defective.

26. What is the probability that 4 Ss appear consecutively in the word MISSISSIPPI, assuming that the letters are arranged at random.

27. Three cards are drawn at random from a pack of cards. Find the chance that they are a king, a queen and a jack.

28. An unbiased coin is tossed 5 times. Find the probability of getting at least one tail. Also find the odds in favour of getting at least one tail.

29. $n$ letters are to be placed in $n$ addressed envelopes. All the $n$ letters are placed at random. Find the probability that at least one letter may be placed in a wrongly addressed envelope.

30. If $P(A \cup B) = 0.65$, $P(A \cap B) = 0.15$, find $P(\overline{A}) + P(\overline{B})$.

31. Find the odds against drawing two red balls from a bag containing 3 red and 2 white balls.

32. If a pack of cards is distributed among four people A, B, C and D, what is the probability that A may get cards of the same colour?

33. Eight people sit indiscriminately at a round table. Find the probability that two particular persons will be next to each other.

34. From 30 tickets marked with the first 30 numerals, one ticket is drawn at random. Find the chance that it is a multiple of 5 or 7.

35. A man's pocket contains 5 fifty paisa coins, 4 twenty-five paisa coins and 4 ten paisa coins. A boy is asked to draw two coins at random. What is the probability of the boy drawing the maximum possible amount?

    [Hint: To get the maximum possible amount, the boy should draw 2 fifty paisa coins. Therefore, the required probability is $\frac{^5C_2}{^{13}C_2}$]

36. Find the probability of getting 4 when a die is rolled.

37. Find the probability of throwing a number greater than 4 when a die is rolled.

38. In a single throw with two dice, find the probability of throwing 11.

39. A card is drawn from a packet of 100 cards numbered 1 to 100. Find the probability of drawing a number which is a square.

40. A card is drawn from a packet of 100 cards numbered 1 to 100. Find the probability of drawing a number divisible by 13.

41. Two uniformed dice marked 1 to 6 are thrown together. Find the probability of getting a total count of 7 in a single throw.

42. When two unbiased dice are rolled, find the probability that the dice show up an even number.

43. Six boys and six girls sit in a row at random. What is the probability that

    (1) the six girls sit together?

    (2) the boys and girls sit alternatively?

44. Five digit numbers are formed with 0, 1, 2, 3, 4 (not allowing a digit to be repeated) in any place i.e., without repetition of digits. Find the probability of getting 2 in the ten's place and 0 in the unit's place always.

45. Show that in a single throw with two dice, the chance of throwing more than seven is equal to that of throwing a score less than seven.

46. Two different digits are chosen at random from the set 1, 2, 3, 4, 5, 6, 7, 8. Show that the probability that the sum of the digits will be equal to 5 is the same as the probability that their sum will exceed 13, each being $\frac{1}{14}$.

47. If a test consists of 12 true or false questions, in how many different ways can a student make the test paper with one answer to each question?

48. If three squares are chosen at random on a chess board, show that the chance that they should be in a diagonal is $\frac{7}{744}$.

49. Determine the probability of finding a non-defective bolt, if out of 600 bolts already examined, 12 were found defective.

50. In a sample of 446 cars stopped at a road block, only 67 of the drivers had their seat belts fastened. Estimate the probability that a driver stopped on that road will have his or her seat belt fastened. (JNTU 2003)

51. Give two definitions of probability. (Agra B.Sc 1968)

52. A party of $n$ persons sit at a round table. Find the odds against two specified individuals sitting next to each other.

53. There are $n$ letters and $n$ addressed envelopes. If the letters are placed in the envelopes at random, what is the probability that all the letters are not placed in the right envelopes?

## Answers 1.1

(4) $\frac{5}{13}$ (5) $\frac{1}{2}$ (6) $\frac{1}{7}$ (7) $\frac{1}{2}$

(8) (a) $\frac{1}{2}$ (b) $\frac{1}{4}$ (c) $\frac{3}{4}$ (d) $\frac{3}{4}$ (9) $\frac{7}{8}$ (10) $\frac{1}{2}$ (11) $\frac{5}{36}$

(12) $\frac{1}{6}$ (13) (a) $\frac{1}{6}$ (b) $\frac{1}{6}$ (c) $\frac{1}{2}$ (14) $\frac{1}{8}$ (15) $\frac{1}{2}$

(16) (a) $\frac{1}{221}$ (b) $\frac{1}{221}$ (17) $\frac{1}{13}$ (18) $\frac{1}{4}$ (19) $\frac{31}{32}$

(20) $\frac{7}{17}$ (21) $\frac{5}{12}$ (22) $\frac{1}{90}$ (23) $\frac{55}{153}$

(24) $\frac{31}{32}$ (25) $\frac{24}{91}$ (26) $\frac{8!\,4!}{1!}$ (27) $\frac{64}{^{50}C_3}$

(28) $\frac{31}{32}$ (29) $\frac{n!-1}{n!}$ (30) 1.2 (31) 7 : 3

(32) $\frac{4}{^{52}C_{13}}$ (33) $\frac{5!\,2!}{7!}$ (34) $\frac{1}{3}$ (36) $\frac{1}{6}$

(37) $\frac{1}{3}$ (38) $\frac{1}{18}$ (39) $\frac{1}{10}$ (40) $\frac{7}{100}$

(41) $\frac{1}{6}$ (42) $\frac{1}{2}$ (43) (1) $\frac{7!\,6!}{12!}$ (2) $\frac{2!\,6!\,6!}{12!}$

(44) $\frac{1}{10}$ (47) $2^{12}$ (49) $\frac{147}{150}$ (50) $\frac{67}{446}$

(52) $\frac{n-3}{2}$ i.e $n-3:2$ (53) $1-\frac{1}{1!}+\frac{1}{2!}-\ldots+(-1)^n\frac{1}{n!}$

**Theorem 1.1** If $S$ is a sample space of an experiment, then $p(\phi)=0$

*Proof* The sets $\phi$ and $S$ are mutually exclusive and we have $S\cup\phi=S$

It follows that $P(S)=P(S\cup\phi)$

$$=P(S)+P(\phi) \qquad \text{(by the axiom of additivity)}$$

Hence, $P(\phi)=0$

**Theorem 1.2** If events $A$ and $\overline{A}$ are complementary events in a sample space $S$, then

$$P(\overline{A})=1-P(A)$$

*Proof* The events $A$ and $\overline{A}$ are mutually exclusive and we have $A\cup\overline{A}=S$

Since $P(S)=1$, we get $P(A\cup\overline{A})=1$

or $P(A)+P(\overline{A})=1$

i.e., $P(\overline{A})=1-P(A)$

Hence proved.

**Theorem 1.3** If $A$ and $B$ are events in a sample space $S$ and $A \subseteq B$, then $P(A) \leq P(B)$.

*Proof* Since $A \subseteq B$, the event $B$ can be written as $B = A + (B - A)$

where $A$ and $B - A$ are mutually exclusive.

Therefore, we get $P(B) = P(A) + P(B - A)$

or $P(B) - P(A) = P(B - A) \geq 0$

or $P(B) - P(A) \geq 0$

or $P(B) \geq P(A)$

i.e., $P(A) \leq P(B)$

**Theorem 1.4** If $A$ and $B$ are any two events of sample space $S$, then

$$P(A - B) = P(A) - P(A \cap B)$$

*Proof* $A$ can be written as the union of two mutually exclusive events, $A - B$ and $A \cap B$ as follows

$$A = (A - B) \cup (A \cap B)$$

Therefore, $P(A) = P(A - B) + P(A \cap B)$

or $P(A - B) = P(A) - P(A \cap B)$

Hence proved.

## 1.9 Addition Theorem

**Theorem 1.5** If $A$ and $B$ are two arbitrary events in the sample space $S$, then

$$P(\text{A or B}) = P(A \cup B) = P(A) + P(B) - P(A \cap B)$$

*Proof* $A \cup B$ can be decomposed into two mutually exclusive events $A - B$ and $B$ as follows

$$A \cup B = (A - B) \cup B$$

Therefore, $P(A \cup B) = P(A - B) + P(B)$

or $P(A \cup B) = P(A) - P(A \cap B) + P(B)$

i.e., $P(A \cup B) = P(A) + P(B) - P(A \cap B)$

**Corollary 1.5.1 (Addition theorem for mutually exclusive events)**

If $A$ and $B$ are two mutually exclusive events of a sample space $S$, then

$$P(A \cup B) = P(A) + P(B)$$

*Proof* If $A$ and $B$ are mutually exclusive, then we have $A \cap B = \phi$

Substituting in $P(A \cup B) = P(A) + P(B) - P(A \cap B)$

We get $P(A \cup B) = P(A) + P(B)$

**Theorem 1.6** For any events $A$, $B$ and $C$ defined on $S$

$$P(A \cup B \cup C) = P(A) + P(B) + P(C) - P(A \cap B) - P(B \cap C) - P(C \cap A) + P(A \cap B \cap C)$$

*Proof* We have $P(A \cup B \cup C) = P(A \cup (B \cup C))$

$$= P(A) + P(B \cup C) - P(A \cap (B \cup C))$$

$$= P(A) + P(B) + P(C) - P(B \cap C) - P[(A \cap B) \cup (A \cap C)]$$

$$= P(A) + P(B) + P(C) - P(B \cap C) - [P(A \cap B) + P(A \cap C) - P[(A \cap B)(A \cap C)]$$

$$= P(A) + P(B) + P(C) - P(A \cap B) - P(B \cap C) - P(C \cap A) + P(A \cap B \cap C)$$

(Since $A \cap C = C \cap A$ and $(A \cap B) \cap (A \cap C) = A \cap B \cap C$)

Hence proved.

**Corollary 1.6.1** If $A$, $B$, $C$ are mutually exclusive events, then

$$P(A \cup B \cup C) = P(A) + P(B) + P(C)$$

**Corollary 1.6.2** The addition rule of probabilities can be generalised to an arbitrary number $n$ of incompatible (mutually exclusive) events.

If $A_i \cap A_j = \phi$ for $i \neq j$, then

$$P\left[\bigcup_{i=1}^{n} A_i\right] = P\left[\sum_{i=1}^{n} A_i\right] = \sum_{i=1}^{n} P(A_i)$$

**Corollary 1.6.3 (Generalisation of addition rule)**

For events $A_1, A_2, A_3, \ldots, A_n$ defined on the sample space $S$,

$$P\left[\bigcup_{i=1}^{n} A_i\right] = P\left[\sum_{i=1}^{n} A_i\right] - \sum_{i=1}^{n} P(A_i) - \sum_{i<j=1}^{n} P(A_i \cap A_j) + \sum_{i<j<k=1}^{n} P(A_i \cap A_j \cap A_k)$$

$$- \sum_{i<j<k<r=1}^{n} P(A_i \cap A_j \cap A_k \cap A_r) + \ldots + (-1)^n P(A_1 \cap A_2 \cap A_3 \cap \ldots \cap A_n), (n > 3)$$

**Remarks**

The addition rule can easily be generalised to the case of infinite (countable) number of events.

**Theorem 1.7** If A, B are any two events, then $P(A \cap B) \leq P(A) \leq P(A \cup B) \leq P(A) + P(B)$ (Boole's inequality)

*Proof* We have $A \cap B \subseteq A \subseteq A \cup B$

Therefore, we get $P(A \cap B) \leq P(A) \leq P(A \cup B)$ (1)

Since $P(A \cup B) = P(A) + P(B) - P(A \cap B)$,

we have $P(A \cup B) \leq P(A) + P(B)$ (2)

From (1) and (2)

$$P(A \cap B) \leq P(A) \leq P(A \cup B) \leq P(A) + P(B)$$

The above inequality can be extended to $n$ events and stated as follows

**Theorem 1.8** If $A_1, A_2, A_3, \ldots, A_n$ are $n$ events defined on a sample space $S$ of an experiment, then $P\left[\bigcup_{i=1}^{n} A_i\right] \leq P\left[\sum_{i=1}^{n} A_i\right]$.

*Proof* We prove the theorem by the method of mathematical induction.

For $n = 2$, we have

$$P\left[\bigcup_{i=1}^{2} A_i\right] = P(A_1) + P(A_2) - P(A_1 \cap A_2)$$ [By theorem 1.5]

Therefore, $P\left[\bigcup_{i=1}^{2} A_i\right] \leq P(A_1) + P(A_2)$

i.e., the result is true for $n = 2$.

Let us assume that the result holds for $n = k$ (say)

Now consider, $\left[\bigcup_{i=1}^{k+1} A_i\right] = \left[\bigcup_{i=1}^{k} A_i\right] \cup A_{k+1}$

Therefore $P\left[\bigcup_{i=1}^{k+1} A_i\right] = P\left[\left[\bigcup_{i=1}^{k} A_i\right] \cup A_{k+1}\right]$

$$\leq P\left[\bigcup_{i=1}^{k} A_i\right] + P(A_{k+1})$$

$$\leq \left[\sum_{i=1}^{k} P(A_i)\right] + P(A_{k+1})$$

Thus, $P\left[\bigcup_{i=1}^{k+1} A_i\right] \leq P\left[\sum_{i=1}^{k+1} A_i\right]$ i.e., the result holds for $n = k+1$ also.

Hence, by the principle of mathematical induction

$$P\left[\bigcup_{i=1}^{n} A_i\right] \leq P\left[\sum_{i=1}^{n} A_i\right], \ (n \geq 2)$$

Hence proved.

We now state the following.

**Theorem 1.9** If $A_1, A_2, A_3, \ldots, A_n$ are any $n$ events defined on a sample space $S$, then

$$P\left(\bigcap_{i=1}^{n} A_i\right) \geq \sum_{i=1}^{n} P(A_i) - (n-1)$$

*Proof* We can prove the theorem by mathematical induction.

We know that for any event $A_1$ of the sample space $S$, $0 \leq P(A_1) \leq 1$

When $n = 2$, we have $P(A_1 \cup A_2) \leq 1$

i.e., $P(A_1) + P(A_2) - P(A_1 \cap A_2) \leq 1$

or $P(A_1) + P(A_2) \leq 1 + P(A_1 \cap A_2)$

or $1 + P(A_1 \cap A_2) \geq P(A_1) + P(A_2)$

i.e. $P(A_1 \cap A_2) \geq P(A_1) + P(A_2) - 1$

i.e., $P(A_1 \cap A_2) \geq P(A_1) + P(A_2) - (2-1)$

The result is true for $n = 2$.

Let us assume that the result is true for $n = k$

Therefore, we have $P\left(\bigcap_{i=1}^{k} A_i\right) \geq \sum_{i=1}^{k} P(A_i) - (k-1)$

$$\text{Then, } P\left(\bigcap_{i=1}^{k+1} A_i\right) = P\left(\bigcap_{i=1}^{k} A_i \cap A_{k+1}\right)$$

$$\geq P\left(\bigcap_{i=1}^{k} A_i\right) + P(A_{k+1}) - 1$$

$$\geq \sum_{i=1}^{k} P(A_i) - (k-1) + P(A_{k+1}) - 1$$

i.e., $P\left(\bigcap_{i=1}^{k+1} A_i\right) \geq \sum_{i=1}^{k+1} P(A_i) - k + 1 - 1$

Hence, $P\left(\bigcap_{i=1}^{k+1} A_i\right) \geq \sum_{i=1}^{k+1} P(A_i) - k$

Proving the result for $n = k+1$ also.

By the principle of mathematical induction, $P\left(\bigcap_{i=1}^{n} A_i\right) \geq \sum_{i=1}^{n} P(A_i) - (n-1)$ for all integral values of $n$.

## Solved examples

**Example 1** The probability that at least one of the events $A$ and $B$ occur is 0.7 and that they occur simultaneously is 0.2. Find $P(\overline{A}) + P(\overline{B})$.

**Solution** We have $P(A \cup B) = 0.7$, $P(A \cap B) = 0.2$

By addition theorem, $P(A \cup B) = P(A) + P(B) - P(A \cap B)$

Therefore, $P(A \cup B) = 1 - P(\overline{A}) + 1 - P(\overline{B}) - P(A \cap B)$

or $0.7 = 2 - P(\overline{A}) - P(\overline{B}) - 0.2$

or $P(\overline{A}) + P(\overline{B}) = 2 - 0.2 - 0.7 = 1.1$

**Example 2** Two events $A$ and $B$ have probabilities 0.25 and 0.50 respectively. The probability that both $A$ and $B$ occur is 0.14. Find the probability that neither $A$ nor $B$ occurs.

**Solution** We have $P(A) = 0.25$, $P(B) = 0.50$ and $P(A \cap B) = 0.14$

Substituting in $P(A \cup B) = P(A) + P(B) - P(A \cap B)$

We get $P(A\cup B)=0.25+0.50-0.14=0.61$

Therefore, $P(\bar{A}\cap\bar{B})=P(\overline{A\cup B})$

$=1-P(A\cup B)$

$=1-0.61$

$=0.39$

**Example 3** If $P(A)=\frac{3}{8}$, $P(B)=\frac{1}{2}$ and $P(A\cap B)=\frac{1}{4}$, find (1) $P(A\cup B)$, (2) $P(\bar{A})$, (3) $P(\bar{B})$, (4) $P(\bar{A}\cap\bar{B})$.

**Solution**

(1) $P(A\cup B)=P(A)+P(B)P(A\cap B)$

$$=\frac{3}{8}+\frac{1}{2}-\frac{1}{4}=\frac{5}{8}$$

(2) $P(\bar{A})=1-P(A)=1-\frac{3}{8}=\frac{5}{8}$

(3) $P(\bar{B})=1-P(B)=1-\frac{1}{2}=\frac{1}{2}$

(4) $P(\bar{A}\cap\bar{B})=P(\overline{A\cup B})$

$=1-P(A\cup B)$

$=1-\frac{5}{8}=\frac{3}{8}$

**Example 4** Two dice are rolled. Find the probability of getting a sum of 10 or 11.

**Solution** Total number of possible outcomes when two dice are rolled $=6\times 6=36$

Let $A$ be the event of getting a sum 10 and $B$ be the event of getting a sum of 11.

$A$ and $B$ are mutually exclusive.

We have $A=\{(4,6),(5,5),(6,4)\}$

$B=\{(5,6),\ (6,5)\}$

$n(A)=3,\ n(B)=2$

We get $P(A) = \frac{3}{36} = \frac{1}{12}$

$$P(B) = \frac{2}{36} = \frac{1}{18}$$

$$P(A \cup B) = P(A) + P(B) = \frac{1}{12} + \frac{1}{18} = \frac{5}{36}$$

**Example 5** $A$ and $B$ are two mutually exclusive events such that $A \cup B = S$ and $P(B) = 2P(A)$. Find $P(A)$.

**Solution** $A$ and $B$ are mutually exclusive.

$P(B) = 2P(A)$ and $A \cup B = S$ (given)

Now $P(S) = P(A \cup B)$

or $1 = P(A) + P(B) = P(A) + 2P(A) = 3P(A)$

or $P(A) = \frac{1}{3}$

**Example 6** From a pack, two cards are drawn at random. What is the probability that either both are black or both are aces?

**Solution** Two cards can be drawn from a pack in ${}^{52}C_2$ ways. Therefore, the number of points in the sample space $= {}^{52}C_2$.

Let $A$ denote the event of drawing 2 black cards and $B$ denote the event of drawing 2 aces. There are 26 black cards in the pack. Therefore, the number of ways in which 2 black cards can be drawn $= {}^{26}C_2$.

There are 4 aces in the pack. Therefore, 2 aces can be drawn from the pack in ${}^{4}C_2$ ways.

There are 2 black aces in the pack. Hence, two black aces can be drawn from the pack in ${}^{2}C_2$ ways.

We have, $P(A) = \frac{{}^{26}C_2}{{}^{52}C_2}$, $P(B) = \frac{{}^{4}C_2}{{}^{52}C_2}$

$$P(A \cap B) = \frac{{}^{2}C_2}{{}^{52}C_2} = \frac{1}{{}^{52}C_2}$$

Therefore, $P(A \cup B) = P(A) + P(B) - P(A \cap B)$

$$= \frac{{}^{26}C_2}{{}^{52}C_2} + \frac{{}^{4}C_2}{{}^{52}C_2} - \frac{1}{{}^{52}C_2} = \frac{165}{663}$$

**Example 7** A bag contains 4 red , 6 black and 7 green balls. A ball is drawn at random. What is the probability that it is either a red or a black ball?

**Solution** Total number of balls in the bag = 4 red + 6 black + 7 green = 17

Number of ways of drawing a ball from 17 balls = ${}^{17}C_1$

$\therefore n(s) = 17$

Let $A$ denote the event of drawing a red ball and $B$ denote the event of drawing a black ball.

Number of ways of drawing a red ball from 4 red balls = ${}^4C_1 = 4$

Number of ways of drawing a black ball from 6 black balls = ${}^6C_1 = 6$

Therefore, we get $P(A) = \dfrac{4}{17}$ and $P(B) = \dfrac{6}{17}$

$A$ and $B$ are mutually exclusive.

$$P(A \text{ or } B) = P(A \cup B) = P(A) + P(B) = \frac{4}{17} + \frac{6}{17} = \frac{10}{17}$$

**Example 8** A card is drawn from a well-shuffled pack of cards. What is the probability that it is either a spade or an ace?

**Solution** Number of ways of drawing a card from the pack $= n(S) = {}^{52}C_1 = 52$

where $S$ is the sample space.

Let $A$ denote the event in which a spade is drawn and $B$ denote the event in which an ace is drawn.

Number of spades in a pack = 13

Number of ways of drawing a spade = ${}^{13}C_1 = 13$

Number of aces in a pack = 4

Number of ways of drawing an ace = ${}^4C_1 = 4$

Number of spade aces in a pack = 1

Number of ways of drawing an ace which is a spade = ${}^1C_1 = 1$

Therefore, we get $P(A) = \dfrac{13}{52}$, $P(B) = \dfrac{4}{52}$, $P(A \cap B) = \dfrac{1}{52}$

Substituting in $P(A \cup B) = P(A) + P(B) - P(A \cap B)$

$$= \frac{13}{52} + \frac{4}{52} - \frac{1}{52} = \frac{16}{52} = \frac{4}{13}$$

We get $P(A \cup B) = \dfrac{4}{13}$

**Example 9** Three students *A*, *B*, *C* are running a race. *A* and *B* have the same probability of winning, and each is twice as likely to win as *C*. Find the probability that *B* or *C* wins.

**Solution** We have $A \cup B \cup C = S$ (sample space of the race)

$$P(A) = P(B)$$

$$P(A) = 2P(C) = P(B)$$

Since $P(S) = 1$, we have $P(A) + P(B) + P(C) = 1$

i.e., $2P(C) + 2P(C) + P(C) = 1$

or $5P(C) = 1$

we get $P(C) = \dfrac{1}{5}$, $P(A) = \dfrac{2}{5}$, $P(B) = \dfrac{2}{5}$,

The probability that B or C wins the race $= P(B \cup C) = P(B) + P(C) - P(B \cap C)$

$$= \frac{2}{5} + \frac{1}{5} - 0 = \frac{3}{5}$$

**Example 10** If $P(A) = \dfrac{1}{2}$, $P(B) = \dfrac{1}{3}$, $P(A \cap B) = \dfrac{1}{5}$ then, find (1) $P(A \cup B)$, (2) $P(A^c \cap B)$, (3) $P(A \cap B^c)$, (4) $P(A^c \cap B^c)$

**Solution** We have $P(A \cup B) = P(A) + P(B) - P(A \cap B)$

$$= \frac{1}{2} + \frac{1}{3} - \frac{1}{5} = \frac{19}{30}$$

$$P(A^c \cap B) = P(B) - P(A \cap B) = \frac{1}{3} - \frac{1}{5} = \frac{2}{15}$$

$$P(A \cap B^c) = P(A) - P(A \cap B) = \frac{1}{2} - \frac{1}{5} = \frac{3}{15}$$

$$P(A^c \cap B^c) = 1 - P(A \cup B) = 1 - \frac{19}{30} = \frac{11}{30}$$

## Exercise 1.2

1. State the addition theorem on probability.
2. State and prove Boole's inequality.

3. Two dice are rolled. Find the probability that the sum of the numbers obtained is 8 or 9.

4. A bag contains 4 red, 3 black and 5 white balls. If a ball is drawn at random from the bag, what is the probability that it may be either red or black?

5. From a set of 17 cards numbered 1, 2, 3, ...,17, one card is drawn at random. Find the chance that it is a multiple of 3 or 8.

6. One ticket is drawn at random from a bag containing 25 tickets numbered 1 to 25. Find the probability that it is a multiple 5 or 7.

7. A bag contains 5 green, 8 red and 6 white balls. If a ball is drawn from the bag, find the chance that it is either a red or a white ball.

8. A bag contains 4 green, 6 black and 9 white balls. A ball is drawn at random from the bag. What is the probability that it is either a green or a black ball?

9. There are 5 green, 6 black and 7 white balls in a bag. A ball is drawn at random from the bag. Find the probability that it may be either a green or a black ball.

10. A card is drawn at random from a pack. Find the probability that the card drawn is either a spade or a diamond.

11. A card is drawn at random from a pack. Find the chance that it is either a black card or an ace.

12. An urn contains 13 balls numbered 1 to 13. If a ball is drawn at random from the urn, find the probability that the number on the ball is divisible by 3 or 4.

13. Suppose a student is selected at random from 100 students, where 30 students are taking mathematics, 20 are taking chemistry and 10 are taking mathematics and chemistry. Find the probability $P$ that the student is taking mathematics or chemistry.

14. A single card is drawn from an ordinary pack of 52 cards. Find the probability that the card is a face card or a heart.

15. In a certain college, 25% of the students failed in mathematics, 15% failed in chemistry and 10% failed in both mathematics and chemistry. A student is selected at random. What is the probability that he failed in mathematics or chemistry?

16. Let $A$ and $B$ be events with $P(A) = 0.6$, $P(B) = 0.3$, $P(A \cap B) = 0.2$. Find $P(A \cup B)$.

17. From a city, 3 newspapers $A$, $B$, $C$ are being published. $A$ is read by 20%, $B$ is read by 16%, $C$ is read by 14%, both $A$ and $B$ are read by 8%, both $A$ and $C$ are read by 5%, both $B$ and $C$ are read by 4%, and $A$, $B$, $C$ are read by 2%. What is the percentage of the population that read at least one paper?

18. $A$, $B$ and $C$ in order tossed a coin. The first one to toss a head wins the game. What are the probabilities of winning, assuming that the game may continue indefinitely?

19. A problem in statistics is given to the three students $A$, $B$, $C$, whose chances of solving it are $\frac{1}{2}, \frac{3}{4}$ and $\frac{1}{4}$ respectively. What is the probability that the problem is solved?
(JNTU 2007)

20. If $P(A \cup B) = 0.65$, $P(A \cap B) = 0.15$, find $P(\overline{A}) + P(\overline{B})$.

21. If $P(A) = 0.6$, $P(B) = 0.5$, $P(A \cap B) = 0.3$, find $P(\overline{A})$, $P(A \cup B)$, $P(\overline{A} \cap B)$, $P(\overline{A} \cap \overline{B})$ and $P(\overline{A} \cup \overline{B})$.

22. If $P(A) = \frac{1}{2}$, $P(B) = \frac{1}{3}$, $P(A \cap B) = \frac{1}{5}$, then find (1) $P(A \cup B)$, (2) $P(A^c \cap B)$, (3) $P(A \cap B^c)$, (4) $P(A^c \cap B^c)$
(JNTU, Aug 2007)

## Answers 1.2

(3) $\frac{1}{4}$ (4) $\frac{7}{12}$ (5) $\frac{7}{17}$ (6) $\frac{6}{25}$

(7) $\frac{14}{19}$ (8) $\frac{10}{19}$ (9) $\frac{11}{18}$ (10) $\frac{7}{18}$

(12) $\frac{6}{13}$ (13) $\frac{2}{5}$ (14) $\frac{11}{16}$ (15) 0.30

(16) 0.7 (17) 35 (18) $\frac{4}{7}$ (19) $\frac{4}{7}, \frac{1}{4}, \frac{1}{7}$

(20) 1.2 (21) 0.4, 0.8, 0.2, 0.2, 0.7 (22) $\frac{19}{30}, \frac{2}{15}, \frac{3}{15}, \frac{11}{30}$

## 1.10 Independent and Dependent Events

Consider an experiment in which a coin is tossed twice. The result in the second throw of the coin in this experiment is in no way affected by the result of the first throw and the result in the first throw of the coin is in no way affected by the result of the second throw. That is the result of the events has no effect on each other. Such events are called independent events.

**Definition 1.16** Two events $A$ and $B$ are independent if and only if

$$P(A \cap B) = P(A).P(B)$$

From the definition, it is clear that two events $A$ and $B$ are independent if the occurrence or non-occurrence of either one does not affect the probability of the occurrence of the other. In general, if $A$ and $B$ are any two events in a sample space $S$, we say that $A$ is independent of $B$ if and only if $P(A|B) = P(A)$. The event $B$ is independent of $A$ whenever $A$ is independent of $B$.

**Remark** If $A$ and $B$ are mutually exclusive events with $P(A) \neq 0, P(B) \neq 0$, then the events $A$ and $B$ cannot be independent.

**Example** Consider a trial in which two cards are drawn in succession from a pack of cards. If the first card drawn is not replaced, then the probability of the first drawn card to be a king is $\frac{4}{52}$. Since the card drawn is not replaced, the number of cards remaining after the first draw is $52 - 1 = 51$.

The probability that the second card drawn is a queen is $\frac{4}{51}$. Hence the probability is affected.

This is a case of events which are not independent. The events in this case are said to be dependent.

In the case of dependent events, the probability of occurrence of the event $B$ after the occurrence of the event $A$ is denoted by $P(B|A)$. In this case the occurrence of $B$ is affected by the occurrence of $A$.

In the case of independent events, the probability of occurrence of the event $B$ after the occurrence of the event $A$ is denoted by $P(B|A) = P(B)$. In this case we have $P(A) \neq 0, P(B) \neq 0$, B is independent of A.

The probability of occurrence of $B$ is not affected by the occurrence of $A$.

## 1.11 Conditional Probability

Let $A$ and $B$ denote two events associated with a random experiment. Then $P(B|A)$ represents the conditional probability of occurrence of $B$ given that $A$ has already occurred and is also referred to as the probability of occurrence of $B$ relative to $A$.

In particular, if $S$ is a finite equiprobable space, and $A$ and $B$ are two events of $S$, then

$$P(B \cap A) = \frac{n(B \cap A)}{n(S)}, P(B) = \frac{n(B)}{n(S)}$$

$$\Rightarrow P(B|A) = \frac{n(A \cap B)}{n(A)} \text{ (since } A \cap B = B \cap A)$$

$$= \frac{\text{Number of ways } A \text{ and } B \text{ occur}}{\text{Number of ways } A \text{ can occur}}$$

It can also be defined as $(P(B|A) = \frac{P(A \cap B)}{P(A)}$

Conditional probability satisfies all the axioms of unconditional probability:

If $A$ and $B$ are any two events of a sample space $S$ with $P(B) > 0$, then

(i) $0 \leq P(A|B) \leq 1$

(ii) $P(S|B) = 1$

(iii) $P(A \cup C|B) = P(A|B) + P(C|B)$
where $A$ and $C$ are mutually exclusive events defined on $S$.
Conditional probability of events on a sample space $S$ satisfies the following properties

(i) $P(\overline{A}|B) = 1 - P(A|B)$, $P(B) > 0$

(ii) If $A$, $B$, $C$ are any three events defined on $S$, with $P(B) > 0$, then
$P(A \cup C|B) = P(A|B) + P(C|B) - P(A \cap C|B)$

(iii) If $A$ and $B$ are mutually exclusive then, $P(A|B) = P(A)$

(iv) If $A \subset B$, then $P(A|B) = P(A)$

## 1.12 Multiplication Theorem

**Theorem 1.10** If $A$ and $B$ are two events, then $P(A \cap B) = P(A)\ P(B|A) = P(B)P(A|B)$

*Proof* From the definition, we have $P(B|A) = \dfrac{P(A \cap B)}{P(A)}$

Therefore, $P(B \cap A) = P(A)P(B|A)$

i.e., $$P(A \cap B) = P(A)P(B|A)$$

**Corollary** If $A_1, A_2, \ldots, A_n$ are $n$ events related to an experiment, then

$$P(A_1 \cap A_2 \ldots \cap A_n) = P(A_1)P(A_2|A_1)\ P(A_3|A_2 \cap A_1) \ldots P(A_n|A_1 \cap A_2 \cap \ldots \cap A_n)$$

## 1.13 Addition Theorem for Independent Events

$A_1, A_2, \ldots, A_n$ are $n$ independent events, then

$$P(A_1 \cup A_2 \cup \ldots \cup A_n) = 1 - P(\overline{A_1})P(\overline{A_2}) \ldots P(\overline{A_n})$$

## Solved examples

**Example 1** If $A$ and $B$ are two independent events in a sample space, then show that

(i) $P(A\cup B)=P(A)+P(B)-P(A)P(B)$

(ii) A and $\overline{B}$ are independent.

(iii) $\overline{A}$ and $B$ are independent.

(iv) $\overline{A}$ and $\overline{B}$ are independent.

**Solution** $A$ and $B$ are independent.

Therefore, $P(A\cap B)=P(A)P(B)$

(i) $$\begin{aligned}P(A\cup B)&=P(A)+P(B)-P(A\cap B)\\&=P(A)+P(B)-P(A)P(B)\end{aligned}$$

(ii) $$\begin{aligned}P(A\cap \overline{B})&=P(A)-P(A\cap B)\\&=P(A)-P(A)P(B)\\&=P(A)(1-P(B))\\&=P(A)P(\overline{B})\end{aligned}$$

Therefore, $A$ and $\overline{B}$ are independent.

(iii) $$\begin{aligned}P(\overline{A}\cap B)&=P(B)-P(A\cap B)\\&=P(B)-P(A)P(B)\\&=P(B)(1-P(A))\\&=P(B)P(\overline{A})\end{aligned}$$

Thus $B$ and $\overline{A}$ are independent.

(iv) $$\begin{aligned}P(\overline{A}\cap \overline{B})&=P(\overline{A\cup B})\\&=1-P(A\cup B)\\&=1-(P(A)+P(B)-P(A)P(B))\\&=1-P(A)-P(B)+P(A)P(B)\\&=[1-P(A)]-P(B)[1-P(A)]\\&=[1-P(A)][1-P(B)]\\&=P(\overline{A})P(\overline{B})\end{aligned}$$

Therefore, $\overline{A}$ and $\overline{B}$ are independent.

**Example 2** An urn contains 4 white and 3 green balls. 2 balls are drawn one after the other and the ball drawn is not replaced. Find the probability that the first drawn is white and the second ball is green.

**Solution** Number of balls in the urn $= 4W + 3G = 7$

Number of white balls = 4

Probability that the first ball drawn is white $= \frac{4}{7}$

When the ball is not replaced, the number of balls remaining in the urn $= 7 - 1 = 6$

The probability that the second ball drawn from the urn is green $= \frac{3}{6} = \frac{1}{2}$

The required probability $= \frac{4}{7} \cdot \frac{1}{2} = \frac{2}{7}$

**Example 3** A bag contains 6 red and 5 black balls. A ball is drawn at random from the bag and without replacing it, another ball is drawn from the bag. Find the probability that both the balls drawn are red.

**Solution** Total number of balls in the bag = 6 red + 5 black = 11

Let $A$ denote the event of drawing the first red ball and $B$ denote the event of drawing the second red ball (after the first red ball is drawn).

Number of ways of drawing one ball from 11 balls $= {}^{11}C_1 = 11$

The first red ball can be drawn from the 6 red balls in ${}^{6}C_1 = 6$ ways

Therefore, $P(A) = \frac{6}{11}$

After the event $A$ (i.e., when the first red ball is drawn), the number of balls remaining

$= 11 - 1 = 10$

Number of ways of drawing 1 ball from 10 balls $= {}^{10}C_1 = 10$

Number of red balls remaining in the bag (after the event $A$) = 5

Number of ways of drawing one red ball from the remaining 5 red balls $= {}^{5}C_1 = 5$

The events are dependent events

$$P(B|A) = \frac{5}{10} = \frac{1}{2}$$

Hence, $P(A \cap B) = P(A)P(A|B) = \frac{6}{11}\frac{1}{2} = \frac{3}{11}$

**Example 4** Find the probability of drawing a spade on each of consecutive two draws from a well-shuffled pack of 52 cards, without replacement.

**Solution** Let $A$ denote the event in which a spade is drawn in the first draw and $B$ denote the event in which a spade is drawn in the second draw.

Number of spades in a pack = 13

$$\therefore P(A) = \frac{{}^{13}C_1}{{}^{52}C_1} = \frac{13}{52} = \frac{1}{4}$$

When the card drawn is not replaced, the number of cards remaining (after $A$) $= 52 - 1 = 51$

Number of spades for the second draw $= 13 - 1 = 12$

Therefore, $P(B|A) = \frac{{}^{12}C_1}{{}^{51}C_1} = \frac{12}{51} = \frac{4}{7}$

Hence, required probability $P(A \cap B) = P(A)P(A|B) = \frac{1}{4}\cdot\frac{4}{7} = \frac{1}{7}$

**Example 5** If $P(A) = 0.4$, $P(A \cup B) = 0.7$ and $A$, $B$ are independent events, find $P(B)$.

**Solution** The events $A$ and $B$ are independent. Therefore, $P(A \cap B) = P(A)P(B)$

Now $P(A \cup B) = P(A) + P(B) - P(A \cap B)$

$$P(A \cup B) = P(A) + P(B) - P(A)P(B)$$

Therefore, $0.7 = P(A) + P(B)(1 - P(A)$

or $\quad 0.7 = 0.4 + P(B)(1 - 0.4)$

or $\quad 0.7 - 0.4 = P(B) \times 0.6$

i.e., $\quad 0.3 = 0.6P(B)$

Hence $P(B) = \frac{0.3}{0.6} = \frac{1}{2}$

**Example 6** $A$ can solve 90% of the problems given in a book and $B$ can solve 70% of the problems. What is the probability that at least one of them will solve a problem selected at random from the book?

**Solution** $P(A) = \frac{90}{100} = 0.9$, and $P(B) = \frac{70}{100} = 0.7$ (given)

where $A$ and $B$ denote the events in which $A$, $B$ can solve the problem.

$P(\overline{A}) = 1 - P(A) = 1 - 0.9 = 0.1$

$P(\overline{B}) = 1 - P(B) = 1 - 0.7 = 0.3$

$A$ and $B$ are independent events.

Probability that at least one of them can solve the problem $= 1 - P(\overline{A})P(\overline{B})$

$= 1 - (0.1)(0.3)$

$= 1 - 0.03 = 0.99$

**Example 7**: John and Jane took their driving tests today. The probability that John passes the test is $\frac{6}{10}$ and the probability that Jane passes the test is $\frac{7}{10}$. What is the probability that both pass their driving tests?

**Solution** Let $A$ denote the event in which John passes the driving test and $B$ denote the event in which Jane passes the driving test.

$A$ and $B$ are independent events.

We have $P(A) = \frac{6}{10}$, $P(B) = \frac{7}{10}$ (given)

Probability that both pass the test $= P(A \cap B) = P(A)P(B)$

$$= \frac{6}{10} \cdot \frac{7}{10} = \frac{42}{100} = \frac{21}{50}$$

**Example 8** $A$, $B$, $C$ are independent events and $P(A) = 0.3$, $P(B) = 0.2$, $P(C) = 0.1$. Find the probability that at least one of the events happen.

**Solution** $A$, $B$, $C$ are independent. Therefore, $\overline{A}$, $\overline{B}$, $\overline{C}$ are independent.

It is given that $P(A) = 0.3$, $P(B) = 0.2$, $P(C) = 0.1$.

Therefore, $P(\overline{A}) = 1 - P(A) = 1 - 0.3 = 0.7$

$P(\overline{B}) = 1 - P(B) = 1 - 0.2 = 0.8$

$P(\overline{C}) = 1 - P(C) = 1 - 0.1 = 0.9$

Hence, $P(\overline{A} \cap \overline{B} \cap \overline{C}) = P(\overline{A})P(\overline{B})P(\overline{C}) = 0.7 \times 0.8 \times 0.9$

Probability that at least one of the events happen $= P(A \cup B \cup C) = 1 - P(\overline{A})P(\overline{B})P(\overline{C})$

$= 1 - (0.7)(0.8)(0.9)$

$= 0.496$

**Example 9** Find the probability of drawing two aces in succession from a pack of cards if the first drawn card is replaced.

**Solution** Let $A$ denote the event in which the first card drawn is an ace and $B$ denote the event in which that card drawn after that event is also an ace.

The cards drawn are replaced; therefore, the events $A$ and $B$ are independent.

Number of cards in a pack = 52

We have 4 aces in a pack.

$$\text{Therefore, } P(A) = \frac{\text{Number of favourable cases}}{\text{Total number of cases}}$$

$$= \frac{{}^4C_1}{{}^{52}C_1} = \frac{4}{52} = \frac{1}{13}$$

Number of cards in the pack after the first drawn card is replaced = 52

Number of aces in the pack after the event $A$ = 4

$$\text{Therefore, } P(B|A) = P(B) = \frac{{}^4C_1}{{}^{52}C_1} = \frac{4}{52} = \frac{1}{13}$$

Probability of drawing two aces in succession with replacement $= P(A \cap B) = P(A)P(B)$

$$= \frac{1}{13} \cdot \frac{1}{13} = \frac{1}{169}$$

**Example 10** The odds against $A$ solving a problem are 8 to 6 and the odds in favour of $B$ solving the problem are 14 to 10. What is the probability that if both of them solve the problem, it will be solved?

**Solution** We have $P(\overline{A}) : P(A) = 8 : 6$

$$\text{Therefore, } P(\overline{A}) = \frac{8}{8+6} = \frac{8}{14} = \frac{4}{7}$$

$$P(A) = \frac{6}{8+6} = \frac{6}{14} = \frac{3}{7}$$

$$P(B) : P(\overline{B}) = 14 : 10$$

$$\text{Therefore, } P(B) = \frac{14}{14+10} = \frac{14}{24} = \frac{7}{12}$$

$$P(\overline{B}) = \frac{10}{14+10} = \frac{10}{24} = \frac{5}{12}$$

The probability that neither $A$ and $B$ can solve the problem

$$= P(\overline{A} \cap \overline{B}) = P(\overline{A})P(\overline{B})$$

$$= \frac{4}{7} \cdot \frac{5}{12} = \frac{5}{21}$$

The probability that the problem is solved when both $A$ and $B$ try

$$= P(A \cup B) = 1 - P(\overline{A} \cap \overline{B})$$

$$= 1 - \frac{5}{21} = \frac{16}{21}$$

**Example 11** A class has 10 boys and 5 girls. Three students are selected at random one after another. Find the probability that

(i) the first two are boys and the third is a girl.

(ii) the first and third are of the same gender and the second is of an opposite gender.

**Solution** Total number of students in the class $= 10B + 5G = 15$

(1) Let $E_1$, $E_2$ be the events in which boys are selected one after the another and $E_3$ be the event in which a girl is selected after selecting the boys.

We have $P(E_1) = \frac{10}{15}, P(E_2|E_1) = \frac{9}{14}, P(E_3|E_1E_2) = \frac{5}{13}$

Let $E$ denote the event of selecting 2 boys and a girl in succession.

$$\therefore P(E) = P(E_1)P(E_2|E_1)P(E_3|E_1E_2) = \frac{10}{15} \cdot \frac{9}{14} \cdot \frac{5}{13} = \frac{15}{91}$$

(2) Suppose $A$ is the event in which the first and the third are boys and the second is a girl. Let $E_1, E_2$ and $E_3$ denote events in which the first and the third are boys and the second is a girl. Then

$$P(A) = P(E_1)P(E_2|E_1)P(E_3|E_1E_2) = \frac{10}{15} \cdot \frac{9}{13} \cdot \frac{5}{14} = \frac{15}{91}$$

Similarly, if $B$ is the event in which the first and the third are girls and the second is a boy,

we get $P(B) = \frac{5}{15} \cdot \frac{10}{14} \cdot \frac{4}{13} = \frac{20}{273}$

Required probability $= P(A) + P(B) = \frac{15}{91} + \frac{20}{273} = \frac{65}{273}$

**Example 12** Determine (i) $P(B|A)$, (ii) $P(A|B^c)$ if $A$ and $B$ are events with

$$P(A) = \frac{1}{3},\ P(B) = \frac{1}{4},\ P(A \cup B) = \frac{1}{2}$$

**Solution** We have $P(A) = \frac{1}{3}$, $P(B) = \frac{1}{4}$, $P(A \cup B) = \frac{1}{2}$ (given)

$$P(A \cap B) = P(A) + P(B) - P(A \cup B)$$

$$= \frac{1}{3} + \frac{1}{4} - \frac{1}{2} = \frac{1}{12}$$

(i) $P(B|A) = \frac{P(A \cap B)}{P(A)} = \frac{(1/12)}{(1/3)} = \frac{1}{4}$

(ii) $P(B^c) = 1 - P(B) = 1 - \frac{1}{4} = \frac{3}{4}$

$$P(A \cap B^c) = P(A) - P(A \cap B) = \frac{1}{3} - \frac{1}{12} = \frac{1}{4}$$

$$P(A|B^c) = \frac{P(A \cap B^c)}{P(B^c)} = \frac{(1/3)}{(3/4)} = \frac{1}{3} \cdot \frac{4}{3} = \frac{4}{9}$$

**Example 13** Box *A* contains 5 red and 3 white marbles and box *B* contains 2 red and 6 white marbles. If a marble is drawn from each box, what is the probability that they are both of the same colour? (JNTU, Aug 2007)

**Solution** When both the marbles drawn are red:

Let $E_1$ be the event in which the marble drawn from box *A* is red and $E_2$ be the event in which the marble drawn from *B* is red.

We get $P(E_1) = \frac{1}{2} \cdot \frac{5}{8} = \frac{5}{16}$, $P(E_2) = \frac{1}{2} \cdot \frac{2}{8} = \frac{1}{8}$

$$\therefore P(E_1 \cap E_2) = P(E_1)P(E_2) = \frac{5}{16} \cdot \frac{1}{8} = \frac{5}{128}$$

When both the marbles drawn are white:

Let $E_3$ be the event in which the marble drawn from box *A* is white and $E_4$ be the event in which the marble drawn from *B* is white.

We get $P(E_3) = \frac{1}{2} \cdot \frac{3}{8} = \frac{3}{16}$, $P(E_4) = \frac{1}{2} \cdot \frac{6}{8} = \frac{3}{8}$

$$\therefore P(E_3 \cap E_4) = P(E_3)P(E_4) = \frac{3}{16} \cdot \frac{3}{8} = \frac{9}{128}$$

Let *E* be the event in which the marbles drawn from the boxes *A* and *B* are of the same colour.

Then we get $P(E) = P(E_1 \cap E_2) + P(E_3 \cap E_4) = \frac{5}{128} + \frac{9}{128} = \frac{7}{64}$

**Example 14** $A$ speaks truth 60% of the time and $B$ speaks truth 70% of the time. Find the probability that they will say the same thing while describing a single event.

**Solution** Let $E_1$ denote the event in which $A$ speaks the truth and $E_2$ denote the event in which $B$ speaks the truth. The events $E_1 \cap E_2$ and $(\overline{E_1} \cap \overline{E_2})$ are mutually exclusive.

$E_1 \cap E_2$ is the event in which both speak the truth.

$(\overline{E_1} \cap \overline{E_2})$ is the event in which both will not speak the truth.

$$P[(E_1 \cap E_2) \cup (\overline{E_1} \cap \overline{E_2})] = P(E_1 \cap E_2) + P(\overline{E_1} \cap \overline{E_2})$$

$$= P(E_1)P(E_2) + P(\overline{E_1})P(\overline{E_2})$$

$$= \frac{3}{5} \cdot \frac{7}{10} + \left(1 - \frac{3}{5}\right)\left(1 - \frac{7}{10}\right)$$

$$= \frac{21}{50} + \frac{2}{5} \cdot \frac{3}{10}$$

$$= \frac{21}{50} + \frac{6}{50} = \frac{27}{50}$$

**Example 15** $A$ can hit a target 3 times in 5 shots, $B$ hits the target 2 times in 5 shots and $C$ hits the target 3 times in 4 shots. Find the probability of the target being hit when all of them try.

**Solution** We have $P(A)$ = probability of the event in which $A$ hits a target $= \frac{3}{5}$

$P(B)$ = Probability of the event in which $B$ hits the target $= \frac{2}{5}$

$P(C)$ = Probability of the event in which $C$ hits the target $= \frac{3}{4}$

Therefore, $P(\overline{A}) = 1 - P(A) = 1 - \frac{3}{5} = \frac{2}{5}$

$$P(\overline{B}) = 1 - P(B) = 1 - \frac{2}{5} = \frac{3}{5}$$

$$P(\overline{C}) = 1 - P(C) = 1 - \frac{3}{4} = \frac{1}{4}$$

The probability of the event in which the target is not hit when $A$, $B$, $C$ all try

$$= P(\overline{A} \cap \overline{B} \cap \overline{C}) = P(\overline{A})P(\overline{B})P(\overline{C})$$

$$= \frac{2}{5} \cdot \frac{3}{5} \cdot \frac{1}{4} = \frac{3}{50}$$

The probability of the event in which target is hit is $P(A \cup B \cup C) = 1 - P(\overline{A} \cap \overline{B} \cap \overline{C})$

$$= 1 - \frac{3}{50} = \frac{47}{50}$$

**Example 16** $A$ and $B$ are two events such that $P(A) = \frac{1}{3}$, $P(B) = \frac{1}{4}$ and $P(A \cup B) = \frac{1}{2}$. Find $P(B|A)$.

**Solution** We have $P(A \cup B) = P(A) + P(B) - P(A \cap B)$

Substituting, we get $\frac{1}{2} = \frac{1}{3} + \frac{1}{4} - P(A \cap B)$

Therefore, we get $P(A \cap B) = \frac{1}{3} + \frac{1}{4} - \frac{1}{2} = \frac{1}{12}$

Hence, $P(B|A) = \frac{P(A \cap B)}{P(A)} = \frac{1/12}{1/3} = \frac{1}{4}$

**Example 17** If $A$ and $B$ are two events on $S$, with $P(A)$, $P(B) > 0$, prove the following

(i) $P(A|\overline{A}) = 0$

(ii) $P(A|B) = 0$, where $A$ and $B$ are mutually exclusive.

(iii) $P(A|B) = 1$, when $B \subset A$

**Solution** (i) $P(A|\overline{A}) = \frac{P(A \cap \overline{A})}{P(\overline{A})} = \frac{P(\phi)}{P(\overline{A})} = 0$ (since $P(\phi) = 0$)

(ii) $P(A|B) = \frac{P(A \cap B)}{P(B)} = \frac{P(\phi)}{P(B)} = 0$

(iii) $P(A|B) = \frac{P(A \cap B)}{P(B)} = \frac{P(B)}{P(B)} = 1$

**Example 18** If $A$ and $B$ are two events and $P(B) \neq 1$, prove that $P(A|\overline{B}) = \frac{P(A) - P(A \cap B)}{1 - P(B)}$.

and deduce that $P(A \cap B) \geq P(A) + P(B) - 1$.

**Solution** From the definition, we have $P(A|\overline{B}) = \frac{P(A \cap \overline{B})}{P(\overline{B})} = \frac{P(A) - P(A \cap B)}{1 - P(B)}$ (1)

Since $P(A|\overline{B}) = 1$, from (1) we get

$P(A) - P(A \cap B) = 1 - P(B)$

or $P(A) + P(B) - 1 = P(A \cap B)$

i.e., $P(A \cap B) = P(A) + P(B) - 1$

## Exercise 1.3

1. From a pack of cards, a card is drawn at random. After noting the card, it is replaced and the pack is well shuffled. Again if a card is drawn, find the probability of getting a card of clubs in the first draw and not a green card in the second draw.

2. $A$ and $B$ are independent events. The probability that $A$ and $B$ occur is $\frac{1}{6}$ and the probability that neither of them occur is $\frac{1}{3}$. Find the probability of occurrence of $A$.

3. Given $P(A) = \frac{1}{4}$, $P(B|A) = \frac{1}{2}$, $P(A|B) = \frac{1}{4}$. Find if

   (i) $A$ and $B$ are mutually exclusive.

   (ii) $A$ and $B$ are independent.

4. Find the probability of drawing 4 white and 2 black balls respectively in two drawings in succession without replacement from a bag containing 1 red, 4 black and 6 white balls.

5. From a pack of cards, two cards are drawn successively at random without replacement. Find the chance that the first is a king and the second is a queen.

6. $A$ speaks truth 75% of the time and $B$ speaks truth 80% of the time. In what percentage of the cases are they likely to contradict each other while stating the same fact?

7. A card is drawn from a pack. It is again replaced in the pack after noting it. After shuffling the pack, a card is taken again. What is the probability that the first card drawn may be hearts and the second may not be a queen?

8. There are 6 red and 5 balls in a bag. A ball is drawn from the bag at random and without replacing it, another ball is drawn. What is the probability that both the balls may be red?

9. Find the chance of drawing a king, a queen and a knave in that order from a pack in three consecutive draws, the cards drawn not being replaced.

10. A bag contains 4 red and 3 blue balls. Two drawings of two balls are made. Find the chance that the first drawings gives two red balls and the second drawing gives two blue balls

    (i) if the balls are returned to the bag after the first draw.

    (ii) if the balls are not returned.

11. A man and his wife appear for an interview for two posts. The probability of the husband's selection is $\frac{1}{7}$, and that of his wife's selection is $\frac{1}{5}$. What is the probability that only one of them is selected?

12. Two cards are drawn one by one without replacement from a well-shuffled pack. What is the probability that none of them is a heart.

13. What is the probability of drawing two aces in succession from a pack of 52 cards if the card drawn is not returned?

14. Find the probability of drawing two red balls in succession from a bag containing 4 red and 5 black balls when the ball that is drawn first (i) not replaced, (ii) replaced.

15. A problem in physics is given to 3 candidates $A$, $B$, $C$ whose chance of solving it are $\frac{1}{3}$, $\frac{1}{4}$ and $\frac{1}{5}$ respectively. Find the probability that the problem will be solved.

16. A university has to select an examiner from a list of 50 persons, 20 of whom are women and 30 are men. 10 of them know Hindi, and 40 not. 15 of them are teachers, and the remaining 35 not. What is the probability of the university selecting a woman teacher who knows Hindi?

17. The probabilities that students $A$, $B$, $C$, $D$ solve a problem are $\frac{1}{3}$, $\frac{2}{5}$, $\frac{1}{5}$ and $\frac{1}{4}$ respectively. If all of them try to solve the problem, what is the probability that the problem is solved?

18. A class has 10 boys and 5 girls. Three students are selected at random one after other. Find the probability that (i) the first two are boys and the third is a girl, (ii) first and third are of the same sex and the second is of the opposite sex.

19. Two marbles are drawn in succession from a box containing 10 red, 30 white, 20 blue and 15 orange marbles with replacement being made after each drawing. Find the probability that (i) both are white. (ii) the first is red and the second is white.

20. The students in a class are selected at random one after the other for an examination. Find the probability that the boys and girls sit alternatively if there are

    (i) 5 boys and 4 girls, (2) 4 boys and 4 girls.

21. Three boxes practically indistinguishable in appearance have two drawers each. Box 1 contains a gold coin in the first drawer and a silver coin in the other drawer. Box 2 contains a gold coin in each drawer. Box 3 contains a silver coin in each drawer. One box is chosen at random and one of its drawers is opened at random and a gold coin is found. What is the probability that the other drawer contains a coin of silver?

## Answers 1.3

(1) $\frac{3}{13}$ (2) $\frac{1}{2}$ or $\frac{1}{3}$ (3) (i) not mutually exclusive (ii) independent (4) $\frac{{}^6C_4}{{}^{11}C_4} \cdot \frac{{}^4C_2}{{}^7C_2}$

(5) $\frac{{}^{4}C_1}{{}^{52}C_1} \cdot \frac{{}^{4}C_1}{{}^{51}C_1}$ (6) 35% (7) 3/13 (8) 3/31

(9) $\frac{4}{52} \cdot \frac{4}{51} \cdot \frac{4}{50}$ (10) (i) $\frac{2}{49}$ (ii) $\frac{3}{35}$ (11) $\frac{2}{7}$ (12) $\frac{19}{34}$

(13) $\frac{1}{221}$ (14) (i) $\frac{4}{27}$ (ii) $\frac{16}{81}$ (15) 3/5 (17) $\frac{19}{25}$

(18) (i) $\frac{15}{91}$ (ii) $\frac{65}{27}$ (19) (i) $\frac{4}{25}$ (ii) $\frac{4}{75}$ (20) (i) $\frac{1}{126}$ (ii) $\frac{1}{35}$ (21) $\frac{1}{2}$

### Bayes' theorem

In this section, we consider Bayes' theorem named after the Thomas Bayes (1702–61). It is used to find posterior probabilities. Bayes' theorem makes it possible to revise probabilities regarding the result of experiment. Posterior probabilities are also known as the inverse probability theorem. By applying Bayes' theorem, a probability which is an unconditional probability becomes a posterior probability which is a conditional probability. Bayes' theorem is widely used in the theory of ranging fire (a branch of artillery science).

## 1.14 Bayes' Theorem

**Theorem 1.11** Let $E_1, E_2, E_3, \ldots, E_n$ be $n$ mutually and exhaustive events with non-zero probability of a random experiment. If $A$ is any arbitrary event of the sample space of the above experiment, then

$$P(E_i|A) = \frac{P(E_i)P(A|E_i)}{\sum_{j=1}^{n} P(E_j)P(A|E_j)}$$

*Proof* Let $S$ be the sample space of the random experiment. Then

$$S = E_1 \cup E_2 \cup E_3 \cup \ldots \cup E_n$$

Now, $A = S \cap A = (E_1 \cup E_2 \cup E_3 \cup \ldots \cup E_n) \cap A$

$$= (E_1 \cap A) \cup (E_2 \cap A) \cup \ldots \cup (E_n \cap A)$$

Therefore, we get $P(A) = P(E_1)P(A|E_1) + P(E_2)P(A|E_2) + \ldots + P(E_n)P(A|E_n)$

$$= \sum_{j=1}^{n} P(E_j)P(A|E_j) \qquad (1)$$

(By multiplication theorem)

Then for any event $E_i$

$$P(E_i|A) = \frac{P(A \cap E_i)}{P(A)}$$

$$= \frac{P(E_i \cap A)}{P(A)}$$

$$= \frac{P(E_i)P(A|E)}{\sum_{j=1}^{n} P(E_j)P(A|E_j)} \qquad \text{(using (1))}$$

**Remarks** The unconditional probability $P(E_1), P(E_2), \ldots, P(E_n)$ in the Bayes' theorem are called "a priori" or prior probabilities. The unconditional probability $P(E_1|A), P(E_2|A), \ldots, P(E_n|A)$ in the theorem are called "a posterior" or inverse probabilities and $P(A) = \sum_{j=1}^{n} P(E_j)P(A/E_j)$ is known as the total probability formula.

Bayes' theorem makes it possible to revise the probabilities of the hypothesis regarding the result of the experiment.

## Solved examples

**Example 1** Two urns identical in appearance, contain respectively 3 green and 2 black balls; and 2 green and 5 black balls. One urn is selected at random and a ball is drawn from it. What is the probability that it is black?

**Solution** Let $E_1$ be the event in which urn I is selected and $E_2$ be the event of selecting urn II. Then $P(E_1) = \frac{1}{2}$ and $P(E_2) = \frac{1}{2}$.

Let $E$ denote the event of selecting a black ball. We have $P(E|E_1) = \frac{2}{5}$, $P(E|E_2) = \frac{5}{7}$,

hence $P(E) = P(E_1)P(E|E_1) + P(E_2)P(E|E_2)$

$$= \frac{1}{2}\cdot\frac{2}{5} + \frac{1}{2}\cdot\frac{5}{7} = \frac{1}{5} + \frac{5}{14}$$

$$= 0.557$$

**Example 2** Assume that a factory has two machines. Past records show that machine I produces 20% of the items and machine II produces 80% of the items. Further, it shows that 6% of the items produced by machine I are defective and only 1% of the items produced by machine II were defective. If a defective item is drawn at random, what is the probability that it was produced by machine I?

**Solution** Let $E_1, E_2$ be the events of drawing an item produced by machines I and II respectively. Let $A$ denote the event of drawing a defective item.

We have $P(E_1) = \frac{20}{100} = \frac{1}{5}$, $P(E_2) = \frac{80}{100} = \frac{4}{5}$

Also, we have $P(A|E_1) = \frac{6}{100}$, $P(A|E_2) = \frac{1}{100}$

From Bayes' theorem, the required probability is

$$P(E_1|A) = \frac{P(E_1)P(A|E_1)}{P(E_1)P(A|E_1) + P(E_2)P(A|E_2)}$$

$$= \frac{\frac{1}{5}\cdot\frac{6}{100}}{\frac{1}{5}\cdot\frac{6}{100} + \frac{4}{5}\cdot\frac{1}{100}} = \frac{6}{6+4} = 0.6$$

Therefore, the probability that the defective item was produced by machine I is 0.6.

**Example 3** Suppose 5 men out of 100 and 25 out of 10,000 are colour blind. A person is chosen at random. What is the probability that the person is a male (assume that males and females are equal in number).

**Solution** Let $P(M)$ be the probability in which the person chosen is male and $P(W)$ be the probability in which the person chosen is a female. Then, we have

$$P(M) = \frac{1}{2} \text{ and } P(W) = \frac{1}{2}$$

Let $B$ denote a event in which the person chosen is a blind person.

Then $P(B|M) = \frac{5}{100} = 0.05$, $P(B|W) = \frac{25}{10000} = 0.0025$

The probability that the person chosen is a male

$$P(M|B) = \frac{P(M)P(B|M)}{P(M)P(B|M) + P(W)P(B|W)}$$

$$= \frac{0.05 \times \frac{1}{2}}{0.05 \times \frac{1}{2} + 0.0025 \times \frac{1}{2}}$$

$$= \frac{0.025}{0.025 + 0.00125} = 0.95$$

**Example 4** Three machines $A$, $B$, $C$ produce respectively 60%, 30%,10% of the total number of items of a factory. The percentage of respective outputs of these machines are respectively 2%, 3% and 4%. An item is selected at random and is found to be defective. Find the probability that the item was produced by machine $C$.

**Solution** Let $A$, $B$, $C$ denote the events in which the items are selected by machines $A$, $B$, $C$ respectively.

Then we have

$$P(A) = \frac{60}{100} = 0.6,\ P(B) = \frac{30}{100} = 0.3,\ P(C) = \frac{10}{100} = 0.1$$

Let $D$ denote the event in which the item selected at random is a defective item. Then

$$P(D|A) = \frac{2}{100} = 0.02,\ P(D|B) = 0.03,\ P(D|C) = 0.04$$

Applying Bayes' formula, we obtain

$$\begin{aligned} P(C|D) &= \frac{P(C)P(D|C)}{P(A)P(D|A) + P(B)P(D|B) + P(C)P(D|C)} \\ &= \frac{(0.1)(0.04)}{(0.6)(0.02) + (0.3)(0.03) + (0.1)(0.04)} \\ &= \frac{0.004}{0.012 + 0.009 + 0.004} = \frac{0.004}{0.025} \\ &= 0.16 \end{aligned}$$

**Example 5** The contents of urns I, II and III are as follows

Urn I : 1 white, 2 black and 3 red balls

Urn II : 2 white, 1 black and 1 red balls

Urn III : 4 white, 5 black and 3 red balls

One urn is chosen at random and 2 balls are drawn. They happened to be white and red. What is the probability that they come from urns I, II and III?

**Solution** Let $E_1$, $E_2$, $E_3$ be the events of choosing urn I, urn II, urn III respectively. Then $P(E_1) = \frac{1}{3},\ P(E_2) = \frac{1}{3},\ P(E_3) = \frac{1}{3}$.

Let $A$ be the event of drawing one white ball and one red ball simultaneously. Then, we have

$P(A|E_1) =$ Probability of drawing one white ball and one red ball from urn I

$$= \frac{{}^1C_1 \times {}^3C_1}{{}^6C_2} = \frac{3}{15} = \frac{1}{5}$$

$P(A|E_2)$ = Probability of drawing one white ball and one red ball from urn II

$$= \frac{{}^2C_1 \times {}^1C_1}{{}^4C_2} = \frac{1}{3}$$

$P(A|E_3)$ = Probability of drawing one white ball and one red ball from urn III

$$= \frac{{}^4C_1 \times {}^3C_1}{{}^{12}C_2} = \frac{2}{11}$$

## Exercise 1.4

1. State and prove Bayes' theorem.
2. A manufacturing company produces pipes in plants I and II with a daily production of 1500 and 2000 respectively. It was observed that 0.6% of the pipes produced by plant I and 0.8% of the pipes produced by machine II are defective. If a pipe is selected at random from a days production and is found to be defective, what is the probability that it has come from plant I.
3. The probability of $X$, $Y$ and $Z$ becoming managers are $\frac{4}{9}, \frac{2}{9}$ and $\frac{1}{3}$ respectively. The probabilities that the bonus scheme will be introduced if $X$, $Y$ and $Z$ become managers are $\frac{3}{10}$, $\frac{1}{2}$ and $\frac{4}{5}$ respectively.
   a. What is the probability that the bonus scheme will be introduced?
   b. If the bonus scheme has been introduced, what is the probability that the manager appointed was $X$?
4. A company manufactures scooters at two plants $A$ and $B$. Plant $A$ produces 80% and plant $B$ produces 20% of the total product. It was observed that 80% of the items produced at plant $A$ are standard in quality, while 65% of the scooters produced at plant $B$ are of standard quality. If a scooter manufactured by the company is selected at random and is found to be of standard quality, what is the probability that it was manufactured by plant $A$?
5. Two numbers are selected at random from the series 1, 2,..., $n$. What is the probability that the difference is no less than $m$ ($m > 0$).

   [Hint: $k$ denote the first selected number, $l$ denote the second selected number and $A$ denote the event consisting of the difference between $k$ and $l$ such that $k - l = m$.

   Let $E_k$ = the event that the first selected number is $k$), $(k = m+1, m+2, \ldots, n)$

   $$P(E_k) = \frac{1}{n},\ P(A|E_k) = (k-m)|n-1$$

   Hence $P(A) = \sum_{k=m+1}^{n} \frac{k-m}{n(n-1)} = \frac{(n-m)(n-m+1)}{2n(n-1)}$]

6. The first of three urns contains $a$ white and $b$ black balls, the second contains $c$ white and $d$ black balls, and the third contains $k$ white balls. An urn is chosen at random and a ball is drawn from it. If the ball drawn is found to be a white ball, find the probability that the ball came from (a) the first (b) the second or (c) the third urn.
7. In a certain college, 25% of boys and 10% of the girls are studying mathematics. The girls constitute 60% of the student body. (a) What is the probability that mathematics is being studied? (b) If a student is selected at random and is found to be studying mathematics, find the probability that the student is a girl, (c) a boy.
8. A company has two plants to manufacture scooters. Plant I manufactures 80% of the scooters and plant II manufactures 20%. At plant I, 85% of the scooters manufactured are rated to be of standard quality. What is the probability that a scooter selected at random came from plant I, if it is known that the scooter is of standard quality.
9. A car manufacturing factory has two plants. Plant A manufactures 70% of the cars and plant B manufactures 30%. At plant A, 80% of the cars are rated of standard quality and at plant B, 90% of the cars are rated of standard quality. A car is picked up at random and is found to be of standard quality. What is the probability that it has come from plant A?
10. In a bolt factory, machines $A$, $B$, $C$ manufacture respectively 25%, 35%, and 40% of the total. Of their output, 5%, 4% and 2% are defective bolts. A bolt is drawn at random from the produce and is found to be defective. What are the probabilities that it was manufactured by machines $A$, $B$, $C$.
11. Of three men, the chances that a politician, a business man or an academician will be appointed as a vice-chancellor of a university are 0.5, 0.3, 0.2 respectively. The probability that research is promoted by these persons if they are appointed as vice-chancellors are 0.3, 0.7, 0.8 respectively.
    (a) Determine the probability that research is promoted.
    (b) If research is promoted, what is that probability that the vice-chancellor is an academician.
12. In a factory machine, $A$ produces 40% of the output and machine $B$ produces 60%. On an average of 9 items in 1000 produced by A are defective and 1 item in 250 produced by $B$ is defective. An item drawn at random from a day's output is defective. What is the probability that it was produced by $A$ or $B$?
13. Police plan to enforce speed limits by using radar traps at 4 different locations within the city limits. The radar traps at each of the locations $L_1$, $L_2$, $L_3$ and $L_4$ are separated 40%, 30%, 20% and 10% of the time. A person speeding on his way to work has probabilities 0.2,0.1,0.5 and 0.2 respectively of passing through these locations. What is the probability that he will receive a speeding ticket?
14. A box contains 500 IC chips of which 100 are manufactured by company $X$ and the rest by company $Y$. It is estimated that 10% of the chips made by company $X$ and 5% made by company $Y$ are defective. If a randomly selected chip is found to be defective, find the probability that it came from company $X$.
15. A parent particle can be divided into 0, 1 or 2 particles with probabilities 1/4, 1/2, 1/4 respectively. This disappears after splitting. Beginning with one particle, let us denote $X_i$ as

the number of particles in the $i$th generation. Find (i) $P(X_2 > 0)$, (ii) the probability that $X_1 = 2$ given that $X_2 = 1$.

16. Urn I has 2 white and 3 black balls. Urn II has 4 white and 1 black and Urn III has 3 white and 4 black balls. An urn is selected at random and a ball drawn at random is found to be white. Find the probability that Urn I is selected.

17. A box contains 2000 components of which 5% are defective. A second box contains 500 components of which 40% are defective. Two other boxes contain 100 components. We select at random one of the above boxes and remove from it at random, a single component.

    (i) What is the probability that this component is defective?

    (ii) Finding that the selected component is defective, what is the probability that it was drawn from box 2.

## Answers 1.4

(2) 0.64

(3) 23/45, 6/23

(4) 68/81

(5) $\dfrac{(n-m)(n-m+1)}{2n(n-1)}$

(6) (a) $\left(\dfrac{a}{a+b}\right) \Big/ \left(\dfrac{a}{a+b} + \dfrac{c}{c+d} + 1\right)$

(b) $\left(\dfrac{c}{c+d}\right) \Big/ \left(\dfrac{a}{a+b} + \dfrac{c}{c+d} + 1\right)$

(c) $1 \Big/ \left(\dfrac{a}{a+b} + \dfrac{c}{c+d} + 1\right)$

(7) (a) $\dfrac{4}{25}$, (b) $\dfrac{3}{8}$, (c) $\dfrac{5}{8}$

(8) 0.84

(9) $\dfrac{50}{83}$

(10) 0.3623 , 0.405, 0.23

(11) (a) 1.8, (b) 4/13

(12) 0.6, 0.4

(13) 0.23

(14) 1/3

(15) 1/5

(16) 14/57

(17) (a) 0.1625, (b) 0.615

CHAPTER TWO

# Random Variables

## 2.1 Introduction

In this chapter we shall study variables defined on spaces associated with experiments in which the outcomes are uncertain. We know that a variable is symbol that can assume any of a prescribed set of values. The term "random variable" is more meaningful than the term "variable" and it is one of the basic concepts of the probability theory. The values of a random variable are real numbers associated with the outcome of an experiment and the concept of random variables is one of the most significant in the theory of probability.

**Definition 2.1** A random variable is a numerical valued variable defined on a sample space of an experiment. We shall denote a random variable by capital letters $X$, $Y$, $Z$ etc., and the corresponding small (lower case) letters are used to denote the numerical value taken by the random variable.

### Solved examples

**Example 1** Consider an experiment $A$ in which three coins are tossed simultaneously. Let $S$ be the sample space associated with the experiment. We have,

$$S = \{\text{HHH, HHT, HTH, THH, HTT, THT, TTH, TTT}\}$$

If we define a random variable $X$ as the number of heads, then the values of $X$ are 0, 1, 1, 1, 2, 2, 2 and 3, corresponding to the outcomes:

TTT (0 head), HTT (1 head) ,THT (1 head), TTH (1 head), HHT (2 heads), HTH (2 heads), THH (2 heads), HHH (3 heads).

We have the following table:

| Sample point | TTT | HTT | THT | TTH | HHT | HTH | THH | HHH |
|---|---|---|---|---|---|---|---|---|
| $X = x_i$ | 0 | 1 | 1 | 1 | 2 | 2 | 2 | 3 |

**Example 2** The number of calls from subscribers at the telephone exchange during a definite time period is a random variable.

**Example 3** Consider an experiment where 4 shots have to be fired at a target. The random variable is the number of hits.

In the above experiment, we have observed that each outcome is a simple event of the sample space $S$ and the corresponding value is a real number. Thus "a random variable is a rule that assigns one and only one numerical value to each simple event of an experiment" and we have the following definitions.

**Definition 2.2** Let $S$ be a sample space of a random experiment and $R$ denote the set of real numbers. Then, a real valued function $X$: $S \to R$ is called a random variable.

If $S$ is a sample space of an experiment, then we can define more than one random variable on $S$. The set $R_X = \{X(x) : x \in S\}$ is called the range of $X$.

If $X(x) \neq 0$ for all $X \in S$, then $\frac{1}{X(x)}$ and $|X(x)|$ are also random variables on $S$.

If $X$ and $Y$ are random variables on the sample space of an experiment, then $X+Y$, $X-Y$, and $XY$ are also random variables on $S$. For all real numbers $a$ and $b$, $aX+bY$ is a random variable on $S$.

Random variables are also known as stochastic variables. The term "variate" is also used for a random variable. Random variables are classified as follows

(1) Discrete random variable

(2) Continuous random variable

## 2.2 Discrete Random Variables

In statistics, we study variables such as heights of student, number of defective bulbs, number of children in a family, the number of printing mistakes in each page of a book and so on. Some of these quantities can vary only by finite increments, by observable jumps in value. Such values are called discrete random variables. The sample space $S$ of a discrete random variable contains either finite number of outcomes or those which are constantly infinite.

**Definition 2.3** A random variable which can take only finite number of values or countably infinite number of values is called a discrete random variable.

**Example 1** In an experiment of drawing four cards from a pack of cards, the random variable "the number of kings drawn" is a discrete random variable

**Example 2** Consider an experiment in which a coin is tossed four times. If $X$ denotes the number of heads obtained, then $X$ is a discrete random variable.

**Remarks** The sample space $S$ for a discrete random variable can be discrete, continuous or may contain both discrete and continuous points.

Discrete random variables are random variables whose range is finite or continuously infinite.

## 2.3 Probability Distribution for a Discrete Random Variable

The simplest distribution for a discrete random variable $X$ is an ordered series which is a table whose top row contains all the values of the random variables and whose bottom row contains the corresponding probabilities as shown in the following table:

| $X = x_i$ | $x_1$ | $x_2$ | ... | $x_i$ | ... |
|---|---|---|---|---|---|
| $P(X = x_i)$ | $p_1$ | $p_2$ | ... | $p_i$ | ... |

$x_1, x_2, \ldots, x_n$ are the values of the random variable $X$ and $p_i = P(X = X_i)$ are the corresponding probabilities where $\sum_1^n p_i = 1$. The above distribution is an ordered series. It can also be represented graphically as a frequency polygon or a histogram.

### 2.3.1 Probability mass function

**Definition 2.4** Let $X$ be a discrete random variable assuming the values $x = x_1, x = x_2, \ldots, x = x_n$ corresponding to the various outcomes of a random experiment. If the probability of occurrence of $X = x_i$, $(i = 1, 2, \ldots, n)$ is $P(X = x_i)$ such that

(1) $P(X = x_i) = p_i \geq 0$ for all $i$, $(1 \leq i \leq n)$

(2) $\sum_{i=1}^{n} P(X = x_i) = \sum_{i=1}^{n} p_i = 1$

then the function $P(X)$ is called the probability function of the random variable $X$ and the set $\{(x_1, P(X = x_1)), (x_2, P(X = x_2)), \ldots, (x_n, P(X = x_n))\}$ or simply $\{(x_1, p(x_1)), (x_2, P(x_2)), \ldots, (x_n, P(x_n))\}$ is called the probability distribution of the random variable $X$.

The probability function $P(X = x)$ is also denoted by $f(x)$ and is called the probability mass function of the discrete random variable $X$.

**Remark** Some authors refer to $f(x)$ as the frequency function or a probability function.

We have

(1) $f(x) = P(X = x)$

(2) $f(x) \geq 0$

(3) $\sum_x f(x) = 1$

**Definition 2.5 (Finite equiprobable space)** A finite probability distribution where each point $X = x_i$ has the same probability for all $i$ is called a finite equiprobable space or uniform space.

### 2.3.2 Distribution function

The probability $P(X \leq x)$ is the probability of the event $(X \leq x)$. It is a function of $x$. The function $P(X \leq x)$ is denoted by $F(x)$ and is called the cumulative probability distribution function of the random variable $x$.

Thus, $F(x) = P(X \leq x)$ , $(-\infty < x < \infty)$

$F(x)$ is often called the distribution function of $X$. $F(x)$ processes the following properties:

1. $F(-\infty) = 0$
2. $F(\infty) = 1$
3. $0 \leq F(x) \leq 1$
4. $F(x_1) \leq F(x_2)$ if $x_1 < x_2$
5. $F(x_1 \leq x \leq x_2) = F(x_2) - F(x_1)$
6. $F(x^+) = F(x)$

$F(X)$ can also be defined as follows.

**Definition 2.6** Let $X$ be a discrete random variable. Then the function

$$F(X) = P(X \leq x) = \sum_{t \leq x} f(t), \; -\infty < x < \infty$$

where $f(t)$ is the value of the probability distribution and is called the distribution function of $X$.

### 2.3.3 Additional properties of distribution function

**Property 1 (Interval property)** If $X$ is a random variable and $F(x)$ is the distribution function of $X$, then $P(a < X \leq b) = F(b) - F(a)$.

*Proof* The events $a < X \leq b$ and $X \leq a$ are disjoint and we have

$$(a < x \leq b) \cup (x \leq a) = P(x \leq b)$$

Hence, $P(a < x \leq b) + P(x \leq a) = P(x \leq b)$

or $$P(a < x \leq b) = P(x \leq b) - P(x \leq a)$$

$$= F(b) - F(a)$$

**Property 2** If $F(x)$ is the distribution function of a random variable $x$, then

$$P(a \le x \le b) = P(x = a) + F(b) - F(a).$$

*Proof* We have $P(a \le x \le b) = P(x = a) + P(a < x \le b)$

$$= P(x = a) + F(b) - F(a) \quad \text{(by Property (1))}$$

**Property 3** If $F(x)$ is the distribution function of a random variable $X$, then

$$P(a < x < b) = F(b) - F(a) - P(x = b).$$

*Proof* $P(a < x < b) = P(a < x \le b) - P(x = b)$

$$= F(b) - F(a) - P(x = b) \quad \text{(by Property (1))}$$

**Property 4** $P(a \le x < b) = F(b) - F(a) - P(x = b) + P(x = a)$

*Proof* We have $P(a \le x < b) = P(a < x < b) + P(x = a)$

$$= F(b) - F(a) - P(x = b) + P(x = a)$$

**Property 5 (Monotone increasing property)**

*Proof* If $F(x)$ is the distribution of a random variable $X$ and $a < b$, then $F(a) \le F(b)$.

Since $P(a < X \le b) \ge 0$, we have $F(b) - F(a) \ge 0$

or $F(b) \ge F(a)$

i.e., $F(a) \le F(b)$

**Example** Consider an experiment in which four coins are tossed. If we define a random variable as the number of heads obtained, then we have

$$P(X = 0) = \frac{1}{16};\ P(X = 1) = \frac{4}{16} = \frac{1}{4};\ P(X = 2) = \frac{6}{16} = \frac{3}{8}$$

$$P(X = 3) = \frac{4}{16} = \frac{1}{4};\ P(X = 4) = \frac{1}{16}$$

Therefore, we get

$$F(0) = P(X = 0) = \frac{1}{16}$$

$$F(1) = P(X = 0) + P(X = 1) = \frac{1}{16} + \frac{1}{4} = \frac{5}{16}$$

$$F(2) = P(X=0) + P(X=1) + P(X=2) = \frac{1}{16} + \frac{1}{4} + \frac{3}{8} = \frac{11}{16}$$

$$F(3) = P(X=0) + P(X=1) + P(X=2) + P(X=3) = \frac{1}{16} + \frac{1}{4} + \frac{3}{8} + \frac{1}{4} = \frac{15}{16}$$

$$F(4) = P(X=0) + P(X=1) + P(X=2) + P(X=3) + P(X=4) = 1$$

The distribution function $F(x)$ is given by

$$F(x) = \begin{cases} 0, & \text{for } X < 0 \\ \frac{1}{16}, & \text{for } 0 \le X < 1 \\ \frac{5}{16}, & \text{for } 1 \le X < 2 \\ \frac{11}{16}, & \text{for } 2 \le X < 3 \\ \frac{15}{16}, & \text{for } 3 \le X < 4 \\ 1, & \text{for } X \ge 4 \end{cases}$$

## 2.4 Mean and Variance of a Discrete Distribution

If $X$ is a discrete random variable, then the mean and variance of the discrete distribution can be defined as follows,

$$\text{Mean} = \mu = \sum_{i=1}^{n} x_i P(X = x_i) = \sum_{i=1}^{n} x_i p_i$$

$$\text{Variance} = Var[x] = \sum_{i=1}^{n} (x_i - \mu)^2 P(X = x_i)$$

or $\sigma^2 = \sum_{i=1}^{n} (x_i - \mu)^2 p_i$

where $\sigma$ is the standard deviation.

Since $\sum p_i = 1$, we get

$$\text{Var}[x] = \sum_{i=1}^{n} (x_i - \mu)^2 p_i$$

$$= \sum_{i=1}^{n} \left(x_i^2 + \mu^2 - 2x_i\mu\right) p_i$$

$$= \sum_{i=1}^{n} (x_i^2)\, p_i + \sum_{i=1}^{n} (\mu^2)\, p_i - \sum_{i=1}^{n} (2x_i\mu)\, p_i$$

$$= \sum_{i=1}^{n} (x_i^2)\, p_i + \mu^2 \sum_{i=1}^{n} p_i - 2\mu \sum_{i=1}^{n} x_i p_i$$

$$= \sum_{i=1}^{n} (x_i^2)\, p_i + \mu^2 - 2\mu.\mu$$

$$= \sum_{i=1}^{n} (x_i^2)\, p_i - \mu^2$$

Thus, $\sigma^2 = \sum_{i=1}^{n} (x_i^2)\, p_i - \mu^2$

## Solved examples

**Example 1** Two cards are drawn successively with replacement from a well-shuffled pack of cards. Find the probability distribution of the number of kings that can be drawn.

**Solution** Let $X$ denote the random variable "number of kings." Since, the number of cards drawn is 2, the random variable $X$ can take values 0, 1, 2.

Number of kings in a pack = 4

Number of cards in a pack = 52

When one card is drawn from the pack, the probability of getting a king is

$$= \frac{{}^4C_1}{{}^{52}C_1} = \frac{4}{52} = \frac{1}{13}$$

Probability of failure i.e., the probability of not getting a king is $1 - \frac{1}{13} = \frac{12}{13}$

Hence we get $P(x = 0) = P$ (no king) $= \frac{12}{13} \cdot \frac{12}{13} = \frac{144}{169}$

$P(x = 1) = P$ (one king) $= \frac{1}{13} \cdot \frac{12}{13} + \frac{12}{13} \cdot \frac{1}{13} = \frac{24}{169}$

$P(x = 2) = P$(2 kings)$= \frac{1}{13} \cdot \frac{1}{13} = \frac{1}{169}$

The required probability distribution is

| $x$ | 0 | 1 | 2 |
|---|---|---|---|
| $P(X=x)$ | $144/169$ | $24/169$ | $1/169$ |

**Example 2** Find the probability distribution of the number of white balls drawn when 3 balls are drawn without replacement from a bag containing 4 white and 6 red balls.

**Solution** Number of balls in the bag $= 4$ white $+$ 6 red $= 10$

Let $X$ be the random variable "number of white balls." Then $X$ can take values 0, 1, 2 and 3.

$$P(x=0) = P \text{ (no white ball)} = P(3 \text{ red balls }) = \frac{^6C_3}{^{10}C_3} = \frac{1}{6}$$

$$P(x=1) = P \text{ (1 white +2 red balls)} = \frac{^4C_2 \times {}^6C_2}{^{10}C_3} = \frac{1}{2}$$

$$P(x=2) = P(\text{2 white +1 red ball}) = \frac{^4C_1 \times {}^6C_1}{^{10}C_3} = \frac{31}{10}$$

$$P(x=3) = P(\text{3 white balls}) = \frac{^4C_3}{^{10}C_3} = \frac{1}{3}$$

The probability distribution is

| $x$ | 0 | 1 | 2 | 3 |
|---|---|---|---|---|
| $P(X=x)$ | 1/6 | 1/2 | 3/10 | 1/30 |

**Example 3** A random variable $X$ has the following probability distribution

| $X = x_i$ | 0 | 1 | 2 | 3 | 4 |
|---|---|---|---|---|---|
| $P(X = x_i)$ | $3k$ | $3k$ | $k$ | $2k$ | $6k$ |

Find (i) $k$, (ii) mean, (iii) $P(X > 2)$.

**Solution** (i) We have $\sum p_i = 1$

Therefore, $3k + 3k + 3 + 2k + 6k = 1$

or $15k = 1$

or $k = 1/15$

(ii) Mean $= \mu = \sum_i x_i p_i$

$(0)(3k) + (1)(3k) + (2)(k) + (3)(2k) + (4)(6k)$

$= 35k = 35(1/15) = 7/3$

(iii) $P(X > 2) = P(X = 3) + P(X = 4)$

$= 2k + 6k = 8k = 8(1/15) = 8/15$

**Example 4** A random variable $X$ has the following probability distribution

| $X = x_i$ | $-2$ | $-1$ | 0 | 1 | 2 | 3 |
|---|---|---|---|---|---|---|
| $P(X = x_i)$ | 0.1 | $k$ | 0.2 | $2k$ | 0.3 | $k$ |

Find (i) $k$, (ii) mean, (iii) variance of $X$.

**Solution** (i) We have $\sum p_i = 1$

i.e., $0.1 + k + 0.2 + 2k + 0.3 + k = 1$

or $4k + 0.6 = 1$

or $4k = 1 - 0.6 = 0.4$

We get $k = 0.1$.

(ii) Mean $= \mu = \sum_i x_i p_i$

$= (-2)(0.1) + (-1)(k) + (0)(0.2) + (1)(2k) + (2)(0.3) + (3)(k)$

$= 4k + 0.4 = 4(0.1) + 0.4 = 0.8$

(iii) Variance $= \sigma^2 = \sum_{i=1}^{n} (x_i^2)\, p_i - \mu^2$

$= (0.1)(-2)^2 + (k)(0.1)^2 + (0.2)(0)^2 + (2k)(1)^2 + (0.3)(2)^2 + (k)(3)^2 - (0.8)^2$

$= 0.4 + k + 0 + 2k + 1.2 + 9k - 0.64$

$= 12k + 1.6 - 0.64$

$= 12(0.1) + 1.6 - 0.64$

$= 2.16$

## Exercise 2.1

1. Define (a) random variable;
   (b) discrete random variable;

(c) density function;

(d) probability distribution.

2. Obtain the probability distribution of the total number of heads in three tosses of a coin.

3. A fair coin is tossed 4 times. Find the probability distribution of the number of heads.

4. A random variable $X$ has the following distribution

| $X$ | 0 | 1 | 2 | 3 | 4 |
|---|---|---|---|---|---|
| $P(X)$ | $2\lambda$ | $3\lambda$ | $\lambda$ | $2\lambda$ | $6\lambda$ |

Find (i) $P(0 < x < 2)$, (ii) $P(x > 2)$.

5. A bag contains two white, 3 red and 4 blue balls. Two balls are drawn at random from the bag. If the random variable $X$ denotes the "number of white balls" among the balls drawn, describe the probability distribution of $X$.

6. Three balls are drawn without replacement from a bag containing 5 white and 4 red balls. Find the probability distribution of the number of red balls drawn.

7. The probability distribution of a random variable $X$ is given below:

| $X$ | 0 | 1 | 2 |
|---|---|---|---|
| $P(X)$ | $3\lambda^2$ | $4\lambda - 10\lambda^2$ | $5\lambda - 1$ |

where $\lambda > 0$. Find (i) $\lambda$, (ii) $P(X \leq 1)$, (iii) $P(X > 0)$.

8. A random variable $X$ has the following probability distribution:

| $X$ | 1 | 2 | 3 | 4 |
|---|---|---|---|---|
| $P(X)$ | $k$ | $2k$ | $3k$ | $4k$ |

Find (i) $k$, (ii) $P(X < 3)$, (iii) $P(X \geq 3)$, (iv) mean.

9. A random variable $X$ has the following probability distribution :

| $X = x_i$ | $-2$ | $-1$ | 0 | 1 | 2 | 3 |
|---|---|---|---|---|---|---|
| $P(X = x_i)$ | 0.1 | $k$ | 0.2 | $2k$ | 0.3 | $k$ |

Find $k$, mean and variance.

10. A box contains 6 tickets. Two of the tickets carry a prize of Rs 5/- each, the other four a prize of Rs 1/- each. If one ticket is drawn, what is the mean value of the prize.

11. A die is tossed twice. Getting a "number greater than 4" is considered a success. For the probability distribution of the number successes, show that the mean is 2/3 and the variance is 4/9.

12. Obtain the probability distribution of the number of sixes in two tosses of a cubical die.

## Answers 2.1

2.

| $x$ | 0 | 1 | 2 | 3 |
|---|---|---|---|---|
| $P(x)$ | 1/8 | 3/8 | 3/8 | 1/8 |

3.

| $x$ | 0 | 1 | 2 | 3 | 4 |
|---|---|---|---|---|---|
| $P(x)$ | 1/16 | 1/4 | 3/8 | 1/4 | 1/16 |

4. (i) 3/14, (iii) 4/7

5.

| $X$ | 0 | 1 | 2 |
|---|---|---|---|
| $P(X=x)$ | 7/12 | 7/18 | 1/36 |

6.

| $X$ | 0 | 1 | 2 | 3 |
|---|---|---|---|---|
| $P(x)$ | 5/42 | 10/21 | 5/14 | 1/21 |

7. (i) $\lambda = \frac{1}{3}$, (ii) $P(X \leq 1) = \frac{1}{3}$, (iii) $P(X > 0) = \frac{8}{9}$

8. (i) $k = \frac{1}{10}$, (ii) $\frac{3}{10}$, (iii) $\frac{7}{10}$, (iv) $\mu = 3$

9. $K = 0$, $\mu = 0.8$, $\sigma^2 = 2.8$

10. $\frac{7}{3}$

12.

| $X$ | 0 | 1 | 2 |
|---|---|---|---|
| $P(X=x)$ | 25/36 | 10/36 | 1/36 |

## 2.5 Continuous Random Variable

Let $X$ be a random variable. If $X$ takes non-countable infinite number of values, then $X$ is called a continuous random variable. If $X$ is a continuous random variable then the range of $X$ is an interval on the real line.

**Example** The life of an electric bulb, the detection range of a radar etc., are examples of continuous random variables.

## 2.6 Probability Density Function

**Definition 2.7** Let $X$ be a continuous random variable. If for every $X$ in the range of $X$, we assign a real number $f(x)$ satisfying the conditions

1. $f(x) \geq 0$, for $-\infty < x < \infty$

2. $\int_{-\infty}^{\infty} f(x)dx = 1$

the function $f(x)$ is called a probability density function of $X$. It is also referred to as density function.

If $f(x)$ is a probability density function (p.d.f.) of $X$ then we have

$$P(a \leq x \leq b) = \int_{a}^{b} f(x)dx, \text{ for any real constants, } a \text{ and } b \text{ with } a \leq b$$

The set of values obtained from $\int_{a}^{b} f(x)dx$ for various possible intervals is called a continuous probability distribution for $X$.

## 2.7 Cumulative Distribution Function

**Definition 2.8** If $X$ is a continuous random variable having $f(x)$ as its probability density function, then the function given by

$$F(X) = P(X \leq x) = \int_{-\infty}^{x} f(x)dx, \text{ for } -\infty < x < \infty$$

is called the cumulative distribution function of $X$.

$F(x)$ is also referred to as the distribution function or the cumulative distribution of $X$.

**Remark** If the distribution function $f(x)$ of a random variable $X$ is continuous for every $x$ such that $F'(x)$ exists everywhere, except maybe, at individual particular points, then the random variable $X$ is said to be continuous.

The probability density function of a continuous random variable $X$ is the derivative of the distribution function i.e., $f(x) = F'(x)$

If $a$ and $b$ are real numbers with $a \leq b$, then we have $P(a \leq X \leq b) = F(b) - F(a)$

**Definition 2.9 (Mixed random variable)** Let $X$ be a random variable and $F(x)$ be the distribution function of $X$. If $F(x)$ continuously increases on certain intervals but has discontinuities at particular points, then the random variable is said to be mixed (Fig. 2.1).

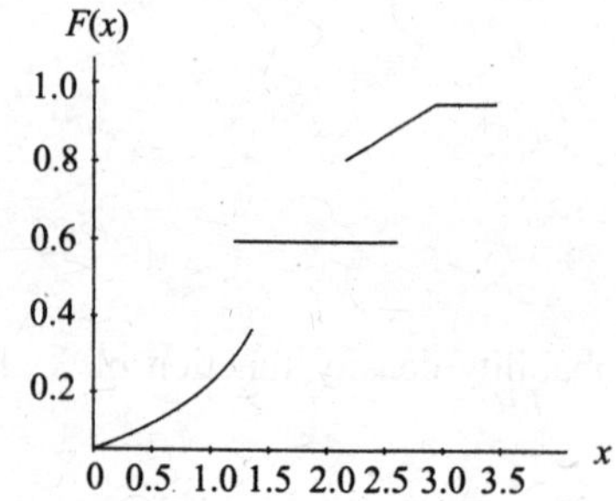


**Fig. 2.1** *Distribution function of a mixed random variable*

## 2.8 Mean and Variance of a Continuous Random Variable

Let $X$ be a continuous random variable. If $f(x)$ is the probability density function of $X$, then the mean and variance of $X$ are given as

$$\text{Arithmetic mean} = \mu = \int_{-\infty}^{\infty} xf(x)dx$$

$$\text{Variance} = \text{Var}\ [x] = v[x] = \sigma^2 = \int_{-\infty}^{\infty} x^2 f(x)dx - \mu^2$$

**Solved examples**

**Example 1** If $f(x) = \begin{cases} \dfrac{x}{6} + k, & 0 \le x \le 3 \\ 0, & \text{elsewhere} \end{cases}$

is a probability density function, find the value of $k$. Also find $P(1 \le x \le 2)$.

**Solution** $f(x)$ is a probability density function. Therefore, we have

$$\int_{-\infty}^{\infty} f(x)dx = 1$$

or $$\int_{-\infty}^{\infty} \left(\frac{x}{6} + k\right) dx = 1$$

or $\int_{-\infty}^{0}\left(\frac{x}{6}+k\right)dx+\int_{0}^{3}\left(\frac{x}{6}+k\right)dx+\int_{3}^{\infty}\left(\frac{x}{6}+k\right)dx=1$

or $0+\left[\frac{x^2}{12}+kx\right]_0^3+0=1$ or $\frac{9}{12}+3k=1$

or $3k=1-\frac{3}{4}=\frac{1}{4}$ or $k=\frac{1}{12}$

Now $P(1\le x\le 2)=\int_1^2 f(x)dx=\int_1^2\left[\frac{x}{6}+\frac{1}{12}\right]dx$

$$=\left[\frac{x^2}{12}+\frac{x}{12}\right]_1^2=\left(\frac{4}{12}+\frac{2}{12}\right)-\left(\frac{1}{12}+\frac{1}{12}\right)$$

$$=\frac{1}{3}+\frac{1}{6}-\frac{1}{6}=\frac{1}{3}$$

**Example 2** Given the cumulative distribution function

$$F(x)=\begin{cases}0, & \text{if } x>0\\ x^2, & \text{if } 0\le x\le 1\\ 1, & \text{if } x>1\end{cases}$$

(i) Find the probability density function.

(ii) Find $P(0.5< x\le 0.75)$.

**Solution:** (i) Since $f(x)=\frac{d}{dx}(F(x))$.

we get $f(x)=\begin{cases}0, & x>0\\ 2x, & 0\le x\le 1\\ 1, & x>1\end{cases}$

or simply $f(x)=\begin{cases}2x, & 0\le x\le 1\\ 0, & \text{otherwise}\end{cases}$

(ii) $P(0.5<x\le 0.75)=P(X\le 0.75)-P(X\le 0.5)$

$$=F(0.75)-F(0.5)$$

$$=(0.75)^2-(0.5)^2$$

$$=0.5625-0.25=0.3125$$

**Example 3** Find $k$ such that $f(x)$ is a probability density function of a continuous random variable $X$, where $f(x)$ is defined as follows.

$$f(x) = \begin{cases} kxe^{-x}, & 0 < x < 1 \\ 0, & \text{otherwise} \end{cases}$$

Also find the mean.

**Solution** Since $f(x)$ is a probability density function, we have $\int_{-\infty}^{\infty} f(x)dx = 1$

i.e., $$\int_0^1 kxe^{-x}dx = 1$$

or $$k\left[\frac{xe^{-x}}{(-1)} - \frac{e^{-x}}{(-1)^2}\right]_0^1 = 1$$

or $$k\left[(e^{-1} - 0) - (e^{-1} - e^0)\right] = 1$$

or $$k(1 - \frac{2}{e}) = 1$$

or $$k = \frac{e}{e-2}$$

(ii) Mean $= \mu = \displaystyle\int_{-\infty}^{\infty} xf(x)dx$

$$= \int_0^1 x\left[\frac{e}{e-2}xe^{-x}\right]dx = \frac{e}{e-2}\int_0^1 x^2e^{-x}dx$$

$$= \frac{e}{e-2}\left[\frac{x^2e^{-x}}{-1} - 2x\frac{e^{-x}}{(-1)^2} + 2\frac{e^{-x}}{(-1)^3}\right]_0^1$$

$$= \frac{e}{e-2}[-e^{-1} - 2e^{-1} - 2(e^{-1} - 1)]$$

$$= \frac{e}{e-2}\left[-\frac{1}{e} - \frac{2}{e} - \frac{2}{e} + 2\right]$$

$$= \frac{e}{e-2}\left[-\frac{5}{e} + 2\right] = \frac{2e-5}{e-2}$$

**Example 4** The distribution function of a random variable is given by

$$F(x) = \begin{cases} 1-(1+x)e^{-x}, & x \geq 0 \\ 0, & x < 0 \end{cases}$$

Find the corresponding density function of random variable $X$.

**Solution** We have $f(x) = \frac{d}{dx}(F(x)) = \frac{d}{dx}\left(1-(1+x)e^{-x}\right)$

$$= \frac{d}{dx}\left(1-e^{-x}-xe^{-x}\right) = 0-(-e^{-x})-(x(-e^{-x})+e^{-x})$$

$$= e^{-x}+xe^{-x}-e^{-x} = xe^{-x}$$

Hence, $f(x) = \begin{cases} xe^{-x}, & x \geq 0 \\ 0, & \text{otherwise} \end{cases}$

is the required density function.

**Example 5** Find the cumulative distribution function for the following probability function of a random variable $X$.

$$f(x) = \begin{cases} 6(x-x^2), & 0 \leq x \leq 1 \\ 0, & \text{otherwise} \end{cases}$$

**Solution** We have $F(x) = \int_{-\infty}^{x} f(x)dx = \int_{-\infty}^{0} f(x)dx + \int_{0}^{x} f(x)dx$

$$= 0 + 6\int_{0}^{x}(x-x^2)dx = 6\left|\frac{x^2}{2}-\frac{x^3}{3}\right|_0^x = 6\left(\frac{3x^2-2x^3}{6}\right) = 3x^2-2x^3$$

Hence, $F(x) = 3x^2-2x^3, 0 \leq x \leq 1$

**Example 6** For the following density function

$$f(x) = ae^{-|x|}, -\infty \leq x \leq \infty$$

of a random variable $X$. Find (i) $a$, (ii) mean, (iii) variance.

**Solution**

(i) $f(x)$ is a probability density function.

Therefore, $\int_{-\infty}^{\infty} f(x)dx = 1$

or $\int_{-\infty}^{\infty} a\epsilon^{-|x|}dx = 1$

or $2a\int_{0}^{\infty} e^{-|x|}dx = 1$ (since $e^{-|x|}$ is an even function.)

or $2a\int_{0}^{\infty} e^{-x}dx = 1$ or $2a\left|-e^{-x}\right|_{0}^{\infty} = 1$

or $2a\left|-(0-1)\right| = 1$

Hence $a = \dfrac{1}{2}$

(ii) Mean $= \int_{-\infty}^{\infty} f(x)dx$

i.e., $\mu = \int_{-\infty}^{\infty} x(\frac{1}{2}e^{-|x|})dx = \frac{1}{2}\int_{-\infty}^{\infty} xe^{-|x|}dx = 0$ (since $xe^{-|x|}$ is odd)

(iii) Variance $= \int_{-\infty}^{\infty} x^2 f(x)dx - \mu^2$

$$= \frac{1}{2}\int_{-\infty}^{\infty} x^2e^{-|x|}dx - 0^2$$

$$= \frac{1}{2}.2\int_{0}^{\infty} x^2e^{-|x|}dx \text{ (since } x^2e^{-|x|} \text{ is odd)}$$

$$= \int_{0}^{\infty} x^2e^{-|x|}dx$$

$$= \left[x^2\left(-e^{-x}\right) - 2xe^{-x} - 2e^{-x}\right]_0^{\infty}$$

$$= 0 - 0 - 2(0-1) = 2$$

## Exercise 2.2

1. A continuous random variable $X$ has a probability density function

   $f(x) = 3x^2,\ 0 < x \leq 1$

   $= 0$, otherwise

   Find $a$ and $b$, such that $P\ (x \leq b) = P(x > b)$ and $P(x > b) = 0.05$.

2. A random variable $X$ has the p.d.f

$$f(x) = 2x,\ 0 < x < 1$$
$$= 0, \text{ otherwise}$$

Find (i) $P\left(x < \frac{1}{2}\right)$, (ii) $P\left(\frac{1}{4} < x < \frac{1}{2}\right)$.

3. Find the cumulative distribution function for the following probability distribution function of a random variable $x$.

$$f(x) = \frac{x}{4}e^{-x/2},\ 0 < x < \infty$$
$$= 0, \text{ otherwise}$$

4. If the probability density function of a random variable is given by

$$f(x) = k(1 - x^2),\ 0 < x < 1$$
$$= 0, \text{ otherwise}$$

Find (i) $k$, (ii) the distribution function of the random variable.

5. If $X$ has the probability density function

$$f(x) = ke^{-3x},\ x > 0$$
$$= 0, \text{ otherwise}$$

Find $k$ and $P(0.5 \leq x \leq 1)$.

6. A continuous random variable $X$ that can assume any value between $x = 2$ and $x = 5$ has a density function given by $f(x) = k(1+x)$. Find $P(x < 4)$.

7. Find the probability density function for the random variable whose distribution function is given by

$$F(x) = 0, \text{ for } x \leq 0$$
$$= x, \text{ for } 0 < x < 1$$
$$= 1 \text{ for } x \geq 1$$

8. A continuous random variable $X$ has the following probability density function

$$f(x) = \frac{1}{2},\ -1 \leq x \leq 0$$
$$= \frac{1}{4}(2 - x),\ 0<x<2$$
$$= 0, \text{ elsewhere}$$

Obtain the distribution of $X$.

9. A random process gives measurements $x$ between 0 and 1 with a probability density function

$$f(x) = 12x^3 - 21x^2 + 10x,\ 0 \leq x \leq 1$$
$$= 0, \text{ elsewhere}$$

Find (i) $P\left(X \leq \frac{1}{2}\right)$, (ii) $P\left(X > \frac{1}{2}\right)$, (iii) a number $k$ such that $P(x \leq k) = \frac{1}{2}$.

10. Let $X$ be a continuous random variable with probability density function

$$\begin{aligned} f(x) &= ax,\ 0 \le x \le 1 \\ &= a,\ 1 \le x \le 2 \\ &= -ax + 3a,\ 2 \le x \le 3 \\ &= 0,\ \text{elsewhere} \end{aligned}$$

(i) Find the value of $a$, (ii) $P(X < 1.5)$.

## Answers 2.2

1. $a = \left(\frac{1}{2}\right)^{1/3}\ b = \left(\frac{19}{20}\right)^{1/3}$

2. $\frac{1}{4}, \frac{3}{16}$

3. $F(x) = 1 - e^{-x/2} - \frac{x}{2}e^{-x/2}, 0 < x < \infty$

4. $\frac{3}{2}, 1$

5. 1, 0.173

6. $\frac{2}{27}, \frac{16}{27}$

7. $f(x) = 1,\ 0 < x < 1$
   $= 0$, otherwise

8. $f(x) = \frac{x+1}{2},\ -1 \le x \le 0$
   $= \frac{1}{8}(4 + 4x - x^2),\ 0 < x < 2$
   $= 1$, if $x \ge 2$

9. $\frac{9}{16}, \frac{7}{16}, \frac{1}{2}$

10. $a = \frac{1}{2}, P(X < 1.5) = \frac{1}{2}$

**Note** If $X$ is continuous random variable and $f(x)$ is the probability density function of $X$ defined in the interval $(a, b)$ then

(i) The median $M$ of the distribution is obtained by solving $\int_a^M f(x)dx = \frac{1}{2}$ and $\int_M^\infty f(x)dx = \frac{1}{2}$, for $M$.

(ii) The mode of the distribution is the value of $x$ for which $f'(x) = 0$ and $f'(x) < 0$ $(a < x < b)$ hold i.e., $f(x)$ is maximum.

(iii) Mean deviation about the mean is given by

Mean deviation $= \int_a^b |x - \mu| f(x)dx$

(iv) Harmonic mean $= \int_a^b \frac{1}{x} f(x)dx$

(v) Geometric mean $= \log G = \int_a^b \log x f(x) dx$

where, $G$ is the geometric mean.

## 2.9 Joint Distributions

In the preceding sections, we have discussed one-dimensional random variables as real value functions defined over a sample space of an experiment. The distributions of one-dimensional random variables are known as univariate distributions. In this section, we shall be concerned with bivariate distributions. If $X$ and $Y$ are random variables defined over a sample space of an experiment, then $(X, Y)$ is a two-dimensional random variable or a bivariate random variable. A system of two random variables $X$ and $Y$ can be geometrically interpreted as a random point $(X, Y)$ on the $xy$-plane (Fig. 2.2).

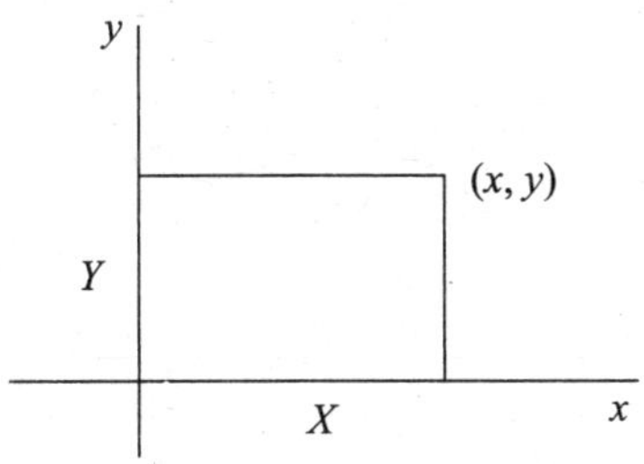


**Fig. 2.2** *Joint distribution*

If the number of possible values $(X, Y)$ is finite, then $(X, Y)$ is called a two-dimensional discrete random variable. If $R_X$ denotes the range space of $X$, and $R_Y$ is the range space of $Y$, then the Cartesian product of $R_X$ and $R_Y$ is the range space of $(X, Y)$. If $(X, Y)$ takes all the values on a region $R$, of the $xy$-plane, then $(X, Y)$ is called a two-dimensional continuous random variable.

If $X$ and $Y$ are discrete random variables, then $P(X = x, Y = y)$ is the probability of intersection of the events $X = x$ and $Y = y$. Similarly, we can define the probability of a continuous random variable. It is preferable to express the probability by the means of a function with the values $f(x, y) = P(X = x, Y = y)$ for any pair of values $(X, Y)$ within the range of the random variables $X$ and $Y$.

### 2.9.1 Joint probability function

**Definition 2.10 (Joint probability function)** If $X$ and $Y$ are discrete random variables, then the function $f(x_i, y_i) = P(X = x_i, Y = y_i) = P_{ij}$ is called the joint probability function for the discrete random variables $X$ and $Y$ if and only if $f(x_i, y_i)$ satisfies the following conditions.

(i) $f(x_i, y_i) = P_{ij} \geq 0$, for all $i,j$

(ii) $\sum_j \sum_i P_{ij} = 1$

The joint probability functions of the discrete random variables are also known as the joint probability mass functions of $X$ and $Y$.

### 2.9.2 Joint probability distribution of discrete random variables

**Definition 2.11 (Joint probability function)** If $X$ and $Y$ are discrete random variables, then the set of triples $\{x_i, y_j, p_{ij}\}$, $i = 1, 2, 3, \ldots, n$, $j = 1, 2, \ldots, m$ is called the joint probability distribution of $X$ and $Y$.

The joint probability distribution can be represented in the form of a table as shown below.

| X \ Y | $y_1$ | $y_2$ | . . | $y_j$ | . . | $y_m$ | $P(x_i)$ |
|---|---|---|---|---|---|---|---|
| $x_1$ | $p_{11}$ | $p_{12}$ | ... | $p_{1j}$ | ... | $p_{1m}$ | $p_{1\cdot}$ |
| $x_2$ | $p_{21}$ | $p_{22}$ | ... | $p_{2j}$ | ... | $p_{2m}$ | $p_2$ |
| | ..... | ..... | ..... | ..... | ..... | ..... | |
| | ..... | ..... | ..... | ..... | ..... | ..... | |
| $x_i$ | $p_{i1}$ | $p_{i2}$ | ... | $p_{ij}$ | ... | $p_{im}$ | $p_{i\cdot}$ |
| . | ..... | ..... | ..... | ..... | ..... | ..... | . |
| . | ..... | ..... | ..... | ..... | ..... | ..... | . |
| $x_n$ | $p_{n1}$ | $p_{n2}$ | ... | $p_{nj}$ | ... | $p_{nm}$ | $p_{n\cdot}$ |
| $P(y_j)$ | $p_{.1}$ | $p_{.2}$ | ... | $p_{.j}$ | ... | $P_{.m}$ | 1 |

### 2.9.3 Marginal probability function of a discrete random variable

**Definition 2.12 (Marginal probability function)** If $P(X = x_i,\ Y = y_i) = P_{ij}$ is the joint probability distribution of two discrete random variables $X$ and $Y$, the marginal probability function of $X$ is given by

$$P(X = x_i) = p_{i.} = p_{i1} + p_{i2} + \ldots + p_{ij} + \ldots$$

$$= \sum_j p_{ij} = p_{i.}$$

The marginal probability function of $X$ is also denoted by $f(x)$. The set $\{x_i,\ p_i\}$, is called the marginal distribution of $X$.

Similarly, the marginal probability function of $Y$ is given by

$$P(Y = y_i) = p_{.j} = \sum_i p_{ij} = p_{ij} + p_{2j} + \ldots + p_{nj}$$

The marginal probability function of $Y$ is also denoted by $f(y)$. The set $\{y_j,\ p_j\}$ is called the marginal distribution of $Y$.

### 2.9.4 Joint distributive function of discrete random variables

**Definition 2.13** If $X$ and $Y$ are discrete random variables, the function given by

$$F(x,y) = P(X \leq x, Y \leq y) = \sum_{s \leq x} \sum_{t \leq y} f(s,t) \text{ for } -\infty < x < \infty, -\infty < y < \infty$$

where $f(s, t)$ is the value of the joint probability distribution of $X$ and $Y$ at $(s, t)$ is called the joint distribution function.

$F(x, y)$ can be interpreted geometrically as the probability of a random point $(X, Y)$ falling in a quadrant whose vertex is $(x, y)$.

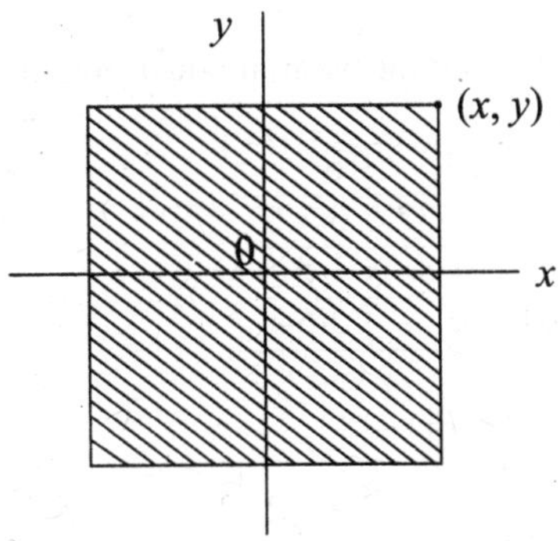


**Fig. 2.3** *Joint distributive function F(x, y)*

The joint distribution function $F(x, y)$ possesses the following properties

1. $F(\infty, y) = F(x, \infty) = 0$
2. $F(-\infty, -\infty) = 0$
3. $F(\infty, \infty) = 1$
4. $P(a_1 < X \leq b_1, a_2 < Y \leq b_2) = F(b_1, b_2) + F(a_1, a_2) - F(a_1, b_2) - F(b_1, a_2)$

## Solved examples

**Example 1** For the following joint distribution of $(X, Y)$, find (i) $F(1, 1)$, (ii) $F(1, 3)$.

| x\y | 1 | 2 | 3 | 4 |
|---|---|---|---|---|
| 0 | 1/24 | 1/12 | 1/12 | 1/24 |
| 1 | 1/12 | 1/6 | 1/6 | 1/12 |
| 2 | 1/24 | 1/12 | 1/12 | 1/24 |

**Solution** (i) We have $F(1,1) = P(X \leq 1, Y \leq 1)$

$$= P(0,1) + P(1,1)$$

$$= \frac{1}{24} + \frac{1}{12} = \frac{1+2}{24} = \frac{1}{8}$$

(ii) $F(1, 3) = P(0, 1) + P(0, 2) + P(0, 3) + P(1, 1) + P(1, 2) + P(1, 3)$

$$= \frac{1}{24} + \frac{1}{12} + \frac{1}{12} + \frac{1}{12} + \frac{1}{6} + \frac{1}{6} = \frac{15}{24} = \frac{5}{8}$$

**Example 2** From the following table for bivariate distribution, find (i) $P(x \leq 1)$, (ii) $P(y \leq 3)$.

| $x \backslash y$ | 1 | 2 | 3 | 4 | 5 | 6 |
|---|---|---|---|---|---|---|
| 0 | 0 | 0 | 1/32 | 2/32 | 2/32 | 3/32 |
| 1 | 1/16 | 1/16 | 1/8 | 1/8 | 1/8 | 1/8 |
| 2 | 1/32 | 1/32 | 1/64 | 1/64 | 0 | 2/64 |

**Solution** (i) $P(x \leq 1) = P(x = 0) + P(x = 1)$

$$= \left(0 + 0 + \frac{1}{32} + \frac{2}{32} + \frac{2}{32} + \frac{3}{32}\right) + \left(\frac{1}{16} + \frac{1}{16} + \frac{1}{8} + \frac{1}{8} + \frac{1}{8} + \frac{1}{8}\right)$$

$$= \frac{8}{32} + \frac{10}{16} = \frac{1}{4} + \frac{5}{8} = \frac{7}{8}$$

(ii) $P(y \leq 3) = P(y = 1) + P(y = 2) + P(y = 3)$

$$= \left[0 + \frac{1}{16} + \frac{1}{32}\right] + \left[0 + \frac{1}{16} + \frac{1}{32}\right] + \left[\frac{1}{32} + \frac{1}{8} + \frac{2}{64}\right]$$

$$= \frac{3}{32} + \frac{3}{32} + \frac{11}{64} = \frac{23}{64}$$

## 2.10 Conditional Probability Distribution

If $P(X = x_i,\ Y = y_j)$ is the probability function of discrete random variables $X$ and $Y$, then the conditional probability function of $X$ given $Y = y_j$ is defined as

$$P(X = x_i \,/\, Y = y)_j) = \frac{P(X = x_i, Y = y_j)}{P(Y = y_j)}$$

and the conditional probability function of $Y$, given $X = x_i$ is denoted by

$$P(Y = y_i \,/\, X{=}x_j) = \frac{P(X = x_i, Y = y_j)}{P(X = x_i)}$$

The conditional probability function of $X$ given $Y$, and the conditional probability function of $Y$ given $X$ are also denoted by $f(x/y)$ and $f(y/x)$ respectively.

The conditional probability distributions are univariate probability distributions.

## 2.11 Independent Random Variables

If $P(X = x, Y = y) = f(x, y)$ is the joint probability density function of discrete random variables $X$ and $Y$ and the marginal probability density functions $f_1(x)$ and $f_2(y)$ are such that $P(X = x, Y = y) = f_1(x)f_2(y)$, then $X$ and $Y$ are said to be independent random variables.

**Example** The two-dimensional random variable $(X,Y)$ has the joint density function

$$f(x, y) = \frac{x+2y}{27}, \; x = 0,1,2;\; y = 0,1,2$$

Find the conditional distribution of $Y$ for $X = x$. Also find the conditional distribution of $X$ given $Y = 1$.

**Solution** The joint probability distribution of $X$ and $Y$ in the tabular form is

| $x\backslash y$ | 1 | 2 | 3 |
|---|---|---|---|
| 0 | 0 | 1/27 | 2/27 |
| 1 | 2/27 | 3/27 | 4/27 |
| 2 | 4/27 | 5/27 | 6/27 |

We have $P(X = x) = f(x)$

The conditional probability distribution of $Y$ for $X = x$ is $f(y/x) = \dfrac{f(x,y)}{f(x)}$

where $f(x,y)$ is the joint probability distribution of $X$ and $Y$. Hence, we get

$$f(y=0/x=0) = \frac{f(X=x,y)}{f(x)} = \frac{f(0,0)}{f(x=0)} = 0$$

$$f(y=1/x=0) = \frac{f(0,1)}{f(x=0)} = \frac{\frac{1}{27}}{\frac{3}{27}} = \frac{1}{3} \qquad \left(\text{Since } f(x=0) = \frac{3}{27}\right)$$

$$f(y=1/x=0)=\frac{f(0,2)}{f(x=0)}=\frac{\frac{2}{27}}{\frac{3}{27}}=\frac{2}{3}$$

We have $f(x=1)=\frac{2}{27}+\frac{3}{27}+\frac{4}{27}=\frac{9}{27}$

Hence $f(y=0/x=1)=\frac{f(0,1)}{f(x=1)}=\frac{\frac{2}{27}}{\frac{9}{27}}=\frac{2}{9}$

$$f(y=1/x=1)=\frac{f(1,1)}{f(x=1)}=\frac{\frac{3}{27}}{\frac{9}{27}}=\frac{3}{9}$$

$$f(y=1/x=1)=\frac{f(2,1)}{f(x=1)}=\frac{\frac{4}{27}}{\frac{9}{27}}=\frac{4}{9}$$

Since $f(x=2)=\frac{4}{27}+\frac{5}{27}+\frac{6}{27}=\frac{15}{27}$

We get $f(y=0/x=2)=\frac{f(2,0)}{f(x=2)}=\frac{\frac{4}{27}}{\frac{15}{27}}=\frac{4}{15}$

$$f(y=1/x=2)=\frac{f(2,1)}{f(x=2)}=\frac{\frac{5}{27}}{\frac{15}{27}}=\frac{5}{15}$$

$$f(y=2/x=2)=\frac{f(2,2)}{f(x=2)}=\frac{\frac{6}{27}}{\frac{15}{27}}=\frac{6}{15}$$

Therefore, the conditional distribution of $Y$ for $X=x$ is

| $x\backslash y$ | 1 | 2 | 3 |
|---|---|---|---|
| 0 | 0 | 1/3 | 2/3 |
| 1 | 2/9 | 3/9 | 4/9 |
| 2 | 4/15 | 5/15 | 6/15 |

From the table, the conditional distribution of $X$ given $Y=1$ is

| $x/y$ | $f(x,y)/f(y=1)$ |
|---|---|
| 0 | 1/9 |
| 1 | 3/9 |
| 2 | 5/29 |

## 2.12 Joint Probability Function of Continuous Random Variables

**Definition 2.14** If $X$ and $Y$ are two continuous random variables, then the function $f(x,y)$ given by

$$P(a_1 \le x \le b_1, a_2 \le y \le b_2) = \int_{a_1}^{b_1}\int_{a_2}^{b_2} f(x,y)dydx$$

is called the joint probability function of $X$ and $Y$, if and only if it satisfies the following conditions.

(i) $f(x,y) \ge 0$, for $-\infty < x < \infty, -\infty < y < \infty$

(ii) $\int_{-\infty}^{\infty}\int_{-\infty}^{\infty} f(x,y)dydx = 1$

The joint probability function of $X$ and $Y$ is also called the joint probability density function of the continuous random variables $X$ and $Y$.

## 2.13 Joint Probability Distribution Function of Continuous Random Variables

**Definition 2.15** If $X$ and $Y$ are continuous random variables, the function given by

$$F(x,y) = P(X \le x, Y \le y) = \int_{-\infty}^{y}\int_{-\infty}^{x} f(s,t)dsdt \text{ for } -\infty < x < \infty,\ -\infty < y < \infty$$

where $f(s,t)$ is the value of the joint probability density function of $X$ and $Y$ at $(s, t)$ is called the joint probability distribution function or joint distribution function of $X$ and $Y$.

We have $f(x,y) = \dfrac{\partial^2}{\partial x \partial y} F(x,y)$

Hence the joint distribution function is obtained by integrating the joint probability function i.e.,

$$F(x,y) = \int_{-\infty}^{x}\int_{-\infty}^{y} f(t_1,t_2)dt_1dt_2$$

## 2.14 Marginal Distribution Function

Let $X$ and $Y$ be continuous random variables. If $F(x,y)$ is the joint distribution function of the random variables $X$ and $Y$, then the marginal distribution function of $X$ denoted by $F(x)$ is given by

$$F(x) = \int_{-\infty}^{x}\left(\int_{-\infty}^{\infty} f(x,y)dy\right)dx$$

and the marginal distribution function of $Y$ denoted by $F(Y)$ is given by

$$F(Y) = \int_{-\infty}^{y}\left(\int_{-\infty}^{\infty} f(x,y)dx\right)dy$$

where $f(x,y)$ is the joint probability function of $X$ and $Y$.

### 2.14.1 Marginal density functions

Let $X$ and $Y$ be continuous random variables and $f(x,y)$ be the value of their joint probability density at $(x,y)$. Then the function

$$f_1(x) = \int_{-\infty}^{\infty} f(x,y)dy \text{ for } -\infty < x < \infty$$

is called the marginal density function of $X$. Correspondingly, the function given by

$$f_2(y) = \int_{-\infty}^{\infty} f(x,y)dx \text{ for } -\infty < y < \infty$$

is called the marginal density function of $Y$.

**Example** The joint probability of continuous random variables $X$ and $Y$ is given by

$$f(x,y) = 8xy,\ 0 \le x \le 1, 0 \le y \le x$$

$$= 0, \text{ otherwise}$$

Find the marginal probability density functions $f(x)$ and $f(y)$.

**Solution** We have

$$f(x) = \int_{-\infty}^{\infty} f(x,y)dy \qquad \text{(By definition)}$$

$$= \int_{0}^{x} 8xydy \quad = 8x\left|\frac{y^2}{2}\right|_0^x \quad = 4x^3, \quad 0 \le x \le 1.$$

and $f(y) = \int_{-\infty}^{\infty} f(x,y)dx \quad = \int_{0}^{1} 8xydx$

$$= 8y\left|\frac{x^2}{2}\right|_0^1 \quad = 4y, \quad 0 \le y \le x$$

## 2.15 Conditional Probability Density Functions

**Definition 2.16** If $X$ and $Y$ are continuous random variables and $f(x,y)$ is the value of the joint density function of $X$ and $Y$ at $(x,y)$, then the conditional density of $Y$ given $X = x$ is defined as

$$f(y|x) = \frac{\partial}{\partial y}F(y/x) \quad = \frac{f(x,y)}{f(x)}, f(x) \neq 0$$

for $-\infty < x < \infty$

and the conditional density of $X$ given $Y = y$ is defined as

$$f(x|y) = \frac{f(x,y)}{f(y)}, f(y) \neq 0$$

**Example** The joint density function of continuous random variables $X$ and $Y$ is given by $f(x,y) = 2$, $0< x < y <1$. Find the marginal and conditional probability density functions.

**Solution** The marginal density function of $X$ given $y$ is

$$f(x) = \int_{-\infty}^{\infty} f(x,y)dy = \int_{x}^{1} 2dy = [2y]_x^1 = 2(1-x)$$

The marginal density function of $Y$ given $x$ is

$$f(y) = \int_{-\infty}^{\infty} f(x,y)dx = \int_{0}^{1} 2dx = [2x]_0^1 = 2$$

Conditional probability density function of $X$ given $y$ is

$$f(y|x) = \frac{f(x,y)}{f(x)} = \frac{2}{2(1-x)} = \frac{1}{1-x}, 0 < x < 1$$

Conditional probability density function of $Y$ given $x$ is

$$f(x|y) = \frac{f(x,y)}{f(y)} = \frac{2}{2} = 1$$

**Remarks** The conditional distribution of a random variable entering into a system is its distribution calculated under the condition that the other random variable has assumed a definite value.

The random variables $X$ and $Y$ are said to be mutually independent if the conditional probability distribution of one does not depend on the value of the other, i.e.,

$$f(x|y) = f(x) \text{ or } f(y|x) = f(y)$$

The above-mentioned properties of one-dimensional (univariate) and two-dimensional (bivariate) random variables can also be established in multi-dimensional random variables.

## Exercise 2.3

1. Find the constant $k$ so that the function

   $$f(x) = \frac{1}{k}, \ a \le x \le b$$
   $$= 0, \text{ otherwise}$$

   is a density function. Also find the cumulative distributive function (c. d. f) of the random variable $X$.

2. The probability distribution function of a random variable $X$ is

   $$f(x) = x, \ 0 < x < 1$$
   $$= 2 - x, 1 \le x \le 2$$
   $$= 0, \ x \ge 2$$

   Compute the cumulative distribution function of $X$.

3. A continuous random variable $X$ that can assume any value between $x = 2$ and $x = 5$ has a density function given by $f(x) = k(1 + x)$. Find $k$, $P(X < 4)$.

4. Suppose that the duration in minutes of a long distance telephone conversation follows an exponential density function

$$f(x) = \frac{1}{5}e^{-x/5}, \text{ for } x > 0$$

Find the probability that the duration of a conversation

(i) will exceed 5 minutes.

(ii) will be less than 3 minutes.

5. If $X$ is a continuous random variable with distribution

$$f(x) = kx, \ 0 \le x \le 5$$
$$= 0, \text{ otherwise}$$

Find (i) $k$, (ii) $P$ (1<x<3), (iii) $P(2 \le x \le 4)$, (iv) $P(x \le 3)$

6. Let $X$ be a continuous random variate with probability density function (p.d.f)

$$f(x) = ax, \ 0 \le x \le 1$$
$$= a, \ 1 \le x \le 2$$
$$= -ax + 3a, \ 2 \le x \le 3$$
$$= 0, \text{ otherwise}$$

(i) Determine the constant $a$.

(ii) Compute $P(X < 1.5)$.

7. The distribution function of a random variable $X$ is given by

$$F(x) = 1 - (1 + x)e^{-x}, \ x \ge 0$$
$$= 0, \text{ for } x < 0$$

Find the corresponding density function of $X$.

8. Show that $F(x) = 0, \ x < -a$

$$= \frac{1}{2}\left(\frac{x}{a} + 1\right), \ -a < x < a$$
$$= 0, \ x \ge a$$

is a distribution function.

9. Two random variables $X$ and $Y$ have the probability function

$$f(x, y) = Ae^{-(2x+y)}, \ x, y \ge 0$$
$$= 0, \text{ otherwise}$$

(i) Evaluate $A$. (ii) Find the marginal probability density functions.

10. If $f(x,y) = 2 - x - y,\ 0 \le x \le 1,\ 0 \le y \le 1$

$= 0$, elsewhere

Find (i) marginal probability function

(ii) conditional probability function

11. Let $X$ and $Y$ have joint density function

$f(x,y) = e^{-(x+y)},\ x > 0,\ y > \infty$

$= 0$, otherwise

Find $P(0 < x < 1/y = 2)$.

12. The joint density function of a bivariate distribution is given as

$f(x,y) = x + y,\ 0 \le x \le 1,\ 0 \le y \le 1$

$= 0$, elsewhere

Determine the marginal distributions of $x$ and $y$.

13. Let $X$ and $Y$ be joint distributed with probability density function

$f(x,y) = \frac{1}{4}(1 + xy),\ |x| < 1,\ |y| < 1$

$= 0$, elsewhere

Show that $X$ and $Y$ are independent.

14. The joint probability function of two random variables $X$ and $Y$ is

$f(x,y) = c(1 + xy),\ 0 \le x \le 6,\ 0 \le y \le 5$

$= 0$, elsewhere

Find (i) $c$, (ii) $F(0.1, 0.5)$, (iii) $f(x,3)$.

15. Let $X$ and $Y$ be two continuous random variables with joint probability density function

$f(x,y) = cx(x - y),\ 0 \le x \le 2,\ -x \le y \le x$

$= 0$, elsewhere

(i) Evaluate $c$, (ii) find $f(x)$ (iii) find $f(y/x)$.

16. Joint distribution of $X$ and $Y$ is given by

$F(x,y) = 4xy\ e^{-(x^2+y^2)},\ x \ge 0,\ y \ge 0$

Show that $X$ and $Y$ are independent. Find the conditional density of $X$ given $Y = y$.

17. The joint distribution of two random variables $X$ and $Y$ is given by the following table

| $X\backslash Y$ | 2 | 3 | 4 |
|---|---|---|---|
| 1 | 0.06 | 0.15 | 0.09 |
| 2 | 0.14 | 0.35 | 0.21 |

Determine the individual distributions of $X$ and $Y$. Also verify that $X$ and $Y$ are stochastically independent.

18. Determine the value of $k$ for which the function given by

$$f(x,y) = kxy, \text{ for } x = 1, 2, 3,\ y = 1, 2, 3$$

can serve as a joint probability distribution.

19. Given the joint probability density function

$$f(x,y) = \frac{3}{5}x(y+x),\ 0 \le x \le 1,\ 0 \le y \le 2$$
$$= 0, \text{ elsewhere}$$

of two random variables $X$ and $Y$. Find $P\{(X,Y) \in A\}$, where $A$ is the region

$$\{(x,y): 0 < X < \frac{1}{2},\ 1 < Y < 2\}$$

20. Find the joint probability density of the two random variables $X$ and $Y$ whose joint distribution is given by

$$f(x,y) = (1-e^{-x})(1-e^{-y}),\ x > 0,\ y > 0$$
$$= 0, \text{ otherwise}$$

Also use the joint probability density to determine $P(1 < X < 3,\ 1 < Y < 2)$

## Answers 2.3

1. $k = b - a,\ F(x) = \dfrac{x-a}{b-a}$

2. $F(x) = \dfrac{x^2}{2},\ 0 < x < 1$
$= 2x - \dfrac{x^2}{2} - 1,\ 1 \le x \le 2$
$= 1,\ x \ge 2$

3. $k = \dfrac{2}{27},\ P(X < 4) = \dfrac{16}{27}$

4. (i) $e^{-1}$, (ii) $1 - e^{-3/5}$

5. (i) $k = \dfrac{2}{25}$, (ii) $\dfrac{8}{25}$, (iii) $\dfrac{12}{25}$, (iv) $\dfrac{9}{25}$

6. (i) $a = \dfrac{1}{2}$, (ii) $\dfrac{1}{2}$

7. $f(x) = xe^{-x},\ x \ge 0$
$= 0,\ x < 0$

9. (i) $A = 2$, (ii) $f(x) = 2e^{-2x},\ x \ge 0,\ f(y) = e^{-y},\ y \ge 0$
$= 0$, otherwise $\quad = 0,\ y < 0$

10. (i) $f(x) = \frac{3}{2} - x$; $f(y) = \frac{3}{2} - y$

    (ii) $f(y|x) = \frac{2-x-y}{\frac{3}{2}-x}$, $f(x|y) = \frac{2-x-y}{\frac{3}{2}-y}$

11. $\frac{e-1}{e}$

12. $x+\frac{1}{2}$, $0 \le x \le 1$, $y+\frac{1}{2}$, $0 \le y \le 1$

14. (i) $\frac{1}{255}$, (ii) 0.0006102, (iii) $f(x,3) = \frac{1}{255}(1+3x)$, $0 \le x \le 6$
    $= 0$, otherwise

15. (i) $c = \frac{1}{8}$, (ii) $\frac{x^3}{4}$, $0 < x < 2$, (iii) $\frac{x-y}{2x^2}$, $0 < x < 2$, $-x < y < x$

16. $2xe^{-x^2}$

17.

| $X = x$ | 1 | 2 |
|---|---|---|
| $f(x)$ | 0.3 | 0.7 |

| $Y = y$ | 2 | 3 | 4 |
|---|---|---|---|
| $f(y)$ | 0.2 | 0.5 | 0.3 |

18. $k = \frac{1}{36}$

19. $\frac{11}{80}$

20. $f(x,y) = e^{-(x+y)}$, $x > 0$, $y > 0$
    $= 0$, elsewhere

## 2.16 Mathematical Expectation and Moments

In this section, we introduce special constants which serve to give a quantitative description of random variables. The constants of particular importance are mathematical expectations, variance and moments of various orders. We begin our study by defining mathematical expectation.

### Mathematical expectation

We shall first give a definition for discrete random variables.

**Definition 2.17** Let $X$ be a discrete random variable. Let $x_1, x_2, \ldots, x_n$ denote possible values of $X$ and let $p_1, p_2, \ldots, p_n$ denote the corresponding probabilities such that $\Sigma p_i = 1$. Then the mathematical expectation denoted by $E(X)$ is defined as

$$E(X) = \sum_{i=1}^{n} x_i p_i, \qquad i = 1, 2, 3, \ldots, n$$

If $X$ is a continuous random variable having $f(x)$ as its probability density function, then the mathematical expectation of $X$ is defined as

$$E(X) = \int_{-\infty}^{\infty} xf(x)dx$$

Mathematical expectation is also called the expectation or mean of the probability distribution.

**Remarks** It is possible to find mathematical expectation without computation.

Mathematical expectation does not exist for all random variables. Expectation is also denoted by $\mu$ (or by $\bar{x}$ ) and is defined by $\sum xP(X=x)$ (or by $E(X)=\sum xf(x)$, where $f(x)=P(X=x)$)

**Example** Consider an experiment in which a die is thrown. We have $S=\{1, 2, 3, 4, 5, 6\}$. Let $X$ be the number of points obtained in a single throw. Then the corresponding probabilities are $\frac{1}{6}, \frac{1}{6}, \frac{1}{6}, \frac{1}{6}, \frac{1}{6}$ and $\frac{1}{6}$

The expected value of $X$ is

$$E(X) = 1.\frac{1}{6} + 2.\frac{1}{6} + 3.\frac{1}{6} + 4.\frac{1}{6} + 5.\frac{1}{6} + 6.\frac{1}{6}$$

$$= \frac{21}{6} = \frac{7}{2}$$

## 2.16.1 Properties of mathematical expectation

**Theorem 2.1** If $X$ is a random variable and $k$ is a real number, then

(i) $E(k) = k$

(ii) $E(kx) = kE(x)$

(iii) $E(x+k) = E(x) + k$

*Proof* (i) If $X$ is a discrete random variable, then

$$E(k) = \sum_{i=1}^{n} kp_i = k\sum_{i=1}^{n} p_i = k.1 = k$$

If $X$ is a continuous random variable, then

$$E(k) = \int_{-\infty}^{\infty} kf(x)dx = k\int_{-\infty}^{\infty} f(x)dx = k.1 = k$$

Thus, $E(k) = k$

(ii) Let $X$ be a discrete random variable. Then

$$E(kx) = \sum_{i=1}^{n} kx_i p_i$$

$$= k\sum_{i=1}^{n} x_i p_i$$

$$= kE(x)$$

If $X$ is a continuous random variable, then

$$E(kx) = \int_{-\infty}^{\infty} kxf(x)dx \quad = k\int_{-\infty}^{\infty} xf(x)dx$$

$$= kE(x)$$

**Remark** In particular, we have $E(-x) = -E(x)$.

(iii) If $X$ is a discrete random variable then

$$E(X+k) = \sum_{i=1}^{n}(k+x_i)p_i \quad = \sum_{i=1}^{n} x_i p_i + \sum_{i=1}^{n} kp_i$$

$$= \sum_{i=1}^{n} x_i p_i + k\sum_{i=1}^{n} p_i$$

$$= E(X) + k.1 = E(X) + k$$

If $X$ is continuous random variable, then we have

$$E(x+k) = \int_{-\infty}^{\infty}(x+k)f(x)dx \quad = \int_{-\infty}^{\infty} xf(x)dx + \int_{-\infty}^{\infty} kf(x)dx$$

$$= \int_{-\infty}^{\infty} xf(x)dx \; + \; k\int_{-\infty}^{\infty} f(x)dx \quad = E(X) + k.1 = E(X) + k$$

Hence, proved.

**Remark** The mathematical expectation of $X$ need not be finite. For example, consider the probability of the discontinuous random variable $k$ defined by

$$p(k) = \frac{e^{-1}}{k!}\,, \; k = 0,\, 1,\, 2, \ldots$$

The expectation $E(k!)$ is

$$E(k!) = \sum_{k=0}^{\infty} k!\frac{e^{-1}}{k!} = \sum_{k=0}^{\infty} e^{-1}$$

$$= e^{-1} \sum_{k=0}^{\infty} 1 \text{ which is not finite.}$$

**Theorem 2.2** If $X$ and $Y$ are random variables, then $E(X+Y) = E(X) + E(Y)$

*Proof* Let $X$ and $Y$ be two discrete random variables. Let $x_1, x_2, \ldots, x_m$ be the values assumed by the random variable $X$. Let $p_1, p_2, \ldots, p_m$ denote the corresponding probabilities.

Let $y_1, y_2, \ldots, y_n$ be the values assumed by the random variable $Y$ and $p'_1, p'_2, \ldots, p'_n$ denote the corresponding probabilities. Also let $P(X = x_i, Y = y_j) = p_{ij}$.

By definition, we have

$$E(X+Y) = \sum_{i=1}^{m} \sum_{j=1}^{n} (x_i + y_j) p_{ij}$$

$$= \sum_{i=1}^{m} \sum_{j=1}^{n} x_i p_{ij} + \sum_{i=1}^{m} \sum_{j=1}^{n} y_j p_{ij}$$

$$= \sum_{i=1}^{m} x_i \left[ \sum_{j=1}^{n} p_{ij} \right] + \sum_{j=1}^{n} y_j \left[ \sum_{i=1}^{m} p_{ij} \right]$$

$$= \sum_{i=1}^{m} x_i p_i + \sum_{j=1}^{n} y_j p_j$$

$$= E(X) + E(Y) \tag{1}$$

Let $X$ and $Y$ be continuous random variables and let $f(x,y)$ be the probability density function (p.d.f) of $X$ and $Y$. Then

$$E(X+Y) = \int_{-\infty}^{\infty} \int_{-\infty}^{\infty} (x+y) f(x,y) dx dy$$

$$= \int_{-\infty}^{\infty} \int_{-\infty}^{\infty} x f(x,y) dx dy + \int_{-\infty}^{\infty} \int_{-\infty}^{\infty} y f(x,y) dx dy$$

$$= \int_{-\infty}^{\infty} x \left[ \int_{-\infty}^{\infty} f(x,y) dy \right] dx + \int_{-\infty}^{\infty} y \left[ \int_{-\infty}^{\infty} f(x,y) dx \right] dy$$

$$= \int_{-\infty}^{\infty} x f_1(x) dx + \int_{-\infty}^{\infty} y f_2(y) dy$$

($f_1(x), f_2(y)$ are marginal density functions of $X$ and $Y$ respectively)

$$= E(X) + E(Y) \qquad (2)$$

From (1) and (2), we conclude that

$$E(X+Y) = E(X) + E(Y)$$

The above theorem can be generalised and stated as follows

**Corollary 2.2.1** If $X$, $Y$, $Z$, ... are random variables, then

$$E(X+Y+Z+\ldots) = E(X) + E(Y) + E(Z) + \ldots$$

**Definition 2.18** Two jointly distributed random variables $X$ and $Y$ are statistically independent for each other if and only if the joint probability density function equals the product of the two marginal probability density functions,

i.e., $f(x, y) = f_1(x) f_2(y)$

where $f_1(x)$ and $f_2(y)$ are marginal probability density functions of $X$ and $Y$ respectively.

**Theorem 2.3** If $X$ and $Y$ are two independent random variables, then

$$E(XY) = E(X)E(Y)$$

*Proof* Let $X$ and $Y$ be two independent random variables and let $X$, $Y$ be discrete. Let $X$ assume the $m$ values $x_1, x_2, \ldots, x_m$. Let $p_1, p_2, \ldots, p_m$ denote the corresponding probabilities.

Let $Y$ assume the $n$ values $y_1, y_2, \ldots, x_m$ and corresponding probabilities be denoted by $p'_1, p'_2, \ldots, p'_n$.

$$\text{Then } E(XY) = \sum_{i=1}^{m} \sum_{j=1}^{n} p_i\, p'_j\, x_i x_j$$

$$= \sum_{i=1}^{m} x_i p_i \sum_{j=1}^{n} y_j p'_j$$

$$= E(X)E(Y)$$

When $X$ and $Y$ are continuous random variables which are independent, the theorem holds and is left as an exercise to the student.

**Generalisation** The above theorem can be extended to the case of several variables and stated as follows:

If $X$, $Y$, $Z$,... are independent random variables

$$E(XYZ\ldots) = E(X)E(Y)E(Z)\ldots$$

### 2.16.2 Variance

**Definition 2.19** Let $X$ be a discrete random variable having probability function $P(X = x)$, then the variance of $X$ is defined as the variance of $X = E\left[(X-\mu)^2\right]$, where $\mu$ is the mean of $X$. Since $\mu = E(X)$ the variance of $X$ can be written as

$$\text{Variance of } X = E\left[(X - E(X))^2\right]$$

Variance of $X$ is denoted by Var$[X]$ (or by $V[X]$) or $\sigma^2$.

If $x_1, x_2, \ldots, x_n$ are the values assumed by $X$, and $p_1, p_2, \ldots, p_n$ are the corresponding probabilities then

$$\sigma^2 = \text{Var}[X] = \sum_{i=1}^{n} (x_i - \mu)^2 p_i$$

If $X$ is continuous random variable, then the variance of $X$ is defined as follows.

**Definition 2.20** Let $X$ be a continuous random variable having probability density function (p.d.f) $f(x)$ then

$$\text{Var}[X] = \sigma^2 = \int_{-\infty}^{\infty} (x-\mu)^2 f(x)dx$$

### 2.16.3 Properties of variance

**Theorem 2.4** If $X$ is a random variable, then Var$[X] = E\left[x^2\right] - [E(X)]^2$.

*Proof* Let $X$ be discrete random variable.

By definition we have

$$\text{Var}[X] = V[X] = E\left[(X-\mu)^2\right] \quad = E\left[X^2 - 2X\mu + \mu^2\right]$$

$$= E\left[X^2\right] - 2E[X\mu] + E[\mu^2]$$

$$= E[X^2] - 2\mu E[X] + \mu^2 \qquad \text{(since } \mu^2 \text{ is a constant)}$$

$$= E\left[X^2\right] - 2E(X)E(Y) + [E(X)]^2 \qquad \text{(since } \mu = E(X))$$

Thus, Var$[X] = E\left[X^2\right] - [E(X)]^2$

The above property holds good for continuous variables also.

**Theorem 2.5** If $X$ is a random variable and $k$ is a real number, then

(i) $\text{Var}[kX] = k^2 v[X]$

(ii) $\text{Var}[X+k] = v[X]$

*Proof* (i) $\text{Var}[kX] = E\left[(kX)^2\right] - [E(kX)]^2 \quad = E\left[k^2X^2\right] - [kE(X)]^2$

$$= k^2E\left[X^2\right] - k^2\left[E(X)\right]^2 \quad = k^2\left[E\left[X^2\right] - [E(X)]^2\right]$$

$$\text{Var}[X+k] = k^2 v[X]$$

(ii) $\text{Var}[X+k] = E\left[(X+k)^2\right] - [E(X+k)]^2$

$$= E\left[X^2+2kX+k^2\right] - [E(X)+k]^2$$

$$= E\left[X^2\right] + 2kE[X] + E[k^2] - E(X)^2 - 2kE(X) - k^2$$

$$= E(X^2) + k^2 - [E(X)]^2 - k^2 \qquad \text{(since } E(k^2) = k^2\text{)}$$

$$= E(X^2) - [E(X)]^2$$

$$= \text{Var}[X]$$

Hence, proved.

**Remarks** The positive square root of variance is called the standard deviation of $X$.

1. Standard deviation $= \sigma = \sqrt{\text{Var}[X]}$

$$= \sqrt{E(X^2) - [E(X)]^2}$$

2. Since $E\left[(X-E(X))^2\right] \geq 0$, variance $X \geq 0$
3. $\text{Var}[X] = 0$ if and only if $X$ takes only one value with probability 1.
4. Variance of a constant is zero.
5. $\text{Var}[X]$ is invariant to the change of origin but variance is not invariant to the change of scale.
6. Let $X, Y$ be two random variables and $Y = \dfrac{X-k}{h}$, then $\sigma_Y^2 = \dfrac{1}{h^2}\sigma_x^2$, where $\sigma_X^2 = \text{Var}[X]$ and $\sigma_Y^2 = \text{Var}[Y]$.

## Solved examples

**Example 1** $X$ and $Y$ are two random variables such that $Y \leq X$. If $E(X)$ and $E(Y)$ exist, show that $E[Y] \leq E[X]$.

**Solution** We have $Y \leq X$ given i.e., $Y - X \leq 0$

Therefore, we get $E[Y - X] \leq 0$ (applying expectation on both sides)

i.e., $E[Y] - E[X] \leq 0$

or $E[Y] \leq E[X]$

**Example 2** If $X$ is a random variable and $E[X]$ exists, then show that $|E[X]| \leq E|X|$

**Solution** Since $X \leq |X|$, we have

$$E[X] \leq E|X| \tag{1}$$

Also, we have $-X \leq |X|$

Therefore, $E[-X] \leq E|X|$

$$-E[X] \leq E|X| \tag{2}$$

From (1) and (2), we get $|E[X]| \leq E|X|$

**Example 3** If $X$ and $Y$ are independent random variables with density functions

$f(x) = \dfrac{8}{x^3}$ for $x > 2$ and

$f(y) = 2y$ for $0 < y < 1$

Find $E[XY]$.

**Solution** Since $X$ and $Y$ are independent random variables

$$f(x, y) = f(x)f(y)$$

Therefore, $E[XY] = \int_2^{\infty}\int_0^1 xyf(x, y)dxdy$

$$= \int_2^{\infty}\int_0^1 xy\left(\frac{8}{x^3}\right)(2y)\,dxdy = 16\int_2^{\infty}\left(\frac{1}{x^2}\int_0^1 y^2dy\right)dx$$

$$= 16\int_2^{\infty}\frac{1}{x^2}\left[\frac{y^3}{3}\right]_0^1 dx \quad = \frac{16}{3}\left[\frac{x^{-1}}{-1}\right]_2^{\infty}$$

$$= \frac{16}{3}\left[\frac{1}{x}\right]_2^{\infty} = \frac{16}{3}\left[0-\frac{1}{2}\right] = \frac{8}{3}$$

**Example 4** If $X$ takes the values $x_n = (-1)^n 2^n n^{-1}$ for $n = 1, 2, \ldots$ with probabilities $p_n = 2^{-n}$, then show that $E[X] = -\log 2$.

**Solution** Using the definition $E[X] = \sum x_i p_i$, we get

$$E[x] = \sum_{i=1}^{\infty} 2^{-n}(-1)^n 2^n n^{-1} \quad = \sum_{i=1}^{\infty}\frac{(-1)^n}{n}$$

$$= -1 + \frac{1}{2} - \frac{1}{3} + \frac{1}{4} - \ldots \quad = -\left[1 - \frac{1}{2} + \frac{1}{3} - \frac{1}{4} + \ldots\right]$$

$$= -\log 2$$

**Example 5** Find variance for the following probability distribution

| $X$: | 8 | 12 | 16 | 20 | 24 |
|---|---|---|---|---|---|
| $P(X)$ : | $\frac{1}{8}$ | $\frac{1}{6}$ | $\frac{3}{8}$ | $\frac{1}{4}$ | $\frac{1}{12}$ |

**Solution** We have $E[X] = \sum_i x_i p_i$

$$= 8.\frac{1}{8} + 12.\frac{1}{6} + 16.\frac{3}{8} + 20.\frac{1}{4} + 24.\frac{1}{12}$$

$$= 1 + 2 + 6 + 5 + 2 = 16$$

Mean $= E[X] = 16$

Variance $= E(X^2) - [E(X)]^2$

$$= \sum_i X_i^2 p_i - [E(X)]^2$$

$$= 8^2.\frac{1}{8} + 12^2.\frac{1}{6} + 16^2.\frac{3}{8} + 20^2.\frac{1}{4} + 24^2.\frac{1}{12} - 16^2$$

$$= 8 + 24 + 96 + 100 + 48 - 256$$

$$= 276 - 256 = 20$$

Therefore, $\text{Var}[X] = 20$

**Example 6** Let the variable $X$ have the distribution $P(X=0)=P(X=2)=p, P(X=1)=1-2p$, for $0 \leq p \leq \frac{1}{2}$. For what values of $p$ is the Var$[X]$ maximum?

**Solution** The probability distribution of $X$ is

| $X:$ | 0 | 1 | 2 |
|---|---|---|---|
| $P(X=x):$ | $p$ | $1-2p$ | $p$ |

Therefore, expectation $= E(X) = 0.p + 1.(1-2p) + 2.p$

$$= 0 + 1 - 2p + 2p = 1$$

Variance = Var$[X] = E(X^2) - [E(X)]^2$

$$= 0^2.p + 1^2.(1-2p) + 2^2.p - 1^2$$

$$= 0 + 1 - 2p + 4p - 1$$

$$= 2p$$

Clearly Var$[X]$ is maximum when $p = \dfrac{1}{2}$

(since, the maximum value Var$[X]$ can take (1) in $0 \leq p \leq \dfrac{1}{2}$ is at $p = \dfrac{1}{2}$)

**Example 7** A coin is tossed until a head appears. What is the expectation of the number of tosses required?

**Solution** A head can appear in the first toss, or in the second toss or in the third toss, and so on.

The favourable cases are

H, TH, TTH, TTTH,...

The probability in which head appears in the first toss $= \dfrac{1}{2}$

The probability in which head appears in the second toss $= \left(\dfrac{1}{2}\right)^2$

The probability of getting head in the $n$th toss $= \left(\dfrac{1}{2}\right)^n$

Let $X$ denote the number of tosses required to get the first head. The probability distribution of $X$ is

| $X$ : | 1 | 2 | 3 | ... | n | ... |
|---|---|---|---|---|---|---|
| $P(X=x)$ : | $\frac{1}{2}$ | $\left(\frac{1}{2}\right)^2$ | $\left(\frac{1}{2}\right)^3$ | ... | $\left(\frac{1}{2}\right)^n$ | ... |

The expectation of number of tosses required is $E(X) = \sum_i x_i p_i$

i.e., $E(X) = 1.\frac{1}{2} + 2.\left(\frac{1}{2}\right)^2 + 3.\left(\frac{1}{2}\right)^3 + \ldots + n.\left(\frac{1}{2}\right)^n + \ldots$

$$= \frac{1}{2}\left[1 + 2 + 3\left(\frac{1}{2}\right)^2 + 4\left(\frac{1}{2}\right)^3 + \ldots + n\left(\frac{1}{2}\right)^{n-1} + \ldots\right]$$

$$= \frac{1}{2}\left[1 - \frac{1}{2}\right]^{-2}$$

$$= \frac{1}{2}\left(\frac{1}{2}\right)^{-2} = \frac{2^2}{2} = 2$$

**Example 8** (i) What is the expected value of the number of points obtained in a single throw with an ordinary die. Also find the variance.

(ii) What is the mathematical expectation of the sum of points obtained on $n$ dice.

**Solution** (i) Let $X$ be the random variable "the number of points" obtained.

$X$ assumes the values 1, 2, 3, 4, 5 and 6, with probabilities $\frac{1}{6}$ in each case.

Hence the expectation of $X$ is

$$E(X) = 1.\frac{1}{6} + 2.\frac{1}{6} + 3.\frac{1}{6} + 4.\frac{1}{6} + 5.\frac{1}{6} + 6.\frac{1}{6}$$

$$= \frac{1}{6}(1+2+3+4+5+6) = \frac{21}{6} = \frac{7}{2}$$

(ii) Let $x_i$ denote the number of points obtained in the $i$th die. Then, $E(x_i) = \frac{7}{2}$, $i = 1, 2, 3, \ldots, n, \ldots$

The sum of points on $n$ dice is $x_1 + x_2 + \ldots + x_n$

Therefore, $E(x_1 + x_2 + \ldots\ldots + x_n) = E(x_1) + E(x_2) + \ldots + E(x_n)$

$$= \frac{7}{2} + \frac{7}{2} + \ldots + n \text{ terms}$$

$$= n\left(\frac{7}{2}\right) = \frac{7}{2}n$$

Hence the expectation of sum of points on $n$ dice $= \frac{7}{2}n$

**Example 9** If $X$ and $Y$ are random variables having joint density function

$$f(x, y) = 4xy, \quad 0 \le x \le 1, 0 \le y \le 1$$

$$= 0, \text{ otherwise}$$

Verify that (i) $E(X + Y) = E(X) + E(Y)$

(ii) $E(XY) = E(X)E(Y)$

**Solution** We have

$$E(X) = \int_{-\infty}^{\infty}\int_{-\infty}^{\infty} xf(x,y)dxdy = \int_{-\infty}^{\infty}\int_{-\infty}^{\infty} x(4xy)dxdy = 4\int_{-\infty}^{\infty}\int_{-\infty}^{\infty} x^2ydxdy$$

$$= 4\int_{x=0}^{1}\int_{y=0}^{1} x^2ydxdy = 4\int_{y=0}^{1} y\left(\int_{x=0}^{1} x^2dx\right)dy = 4\int_{y=0}^{1} y\left[\frac{x^3}{3}\right]_0^1 dy$$

$$= \frac{4}{3}\int_{y=0}^{1} ydy = \frac{4}{3}\left[\frac{y^2}{2}\right]_0^1$$

$$= \frac{4}{6} = \frac{2}{3}$$

Similarly, $E(Y) = \int_{-\infty}^{\infty}\int_{-\infty}^{\infty} yf(x,y)dxdy = \int_{-\infty}^{\infty}\int_{-\infty}^{\infty} y(4xy)dxdy = 4\int_{-\infty}^{\infty}\int_{-\infty}^{\infty} xy^2dxdy$

$$= 4\int_{x=0}^{1}\int_{y=0}^{1} xy^2dxdy = \frac{2}{3}$$

Now $E(X + Y) = \int_{-\infty}^{\infty}\int_{-\infty}^{\infty} (x + y)f(x,y)dxdy$

$$= \int_{-\infty}^{\infty}\int_{-\infty}^{\infty} (x + y)(4xy)dxdy = 4\int_{0}^{1}\int_{0}^{1} (x^2y + xy^2)dxdy$$

$$= 4\int_{y=0}^{1}\left(\int_{x=0}^{1}(x^2y+xy^2)dx\right)dy = 4\int_{y=0}^{1} y\left[\frac{x^3}{3}y+\frac{x^2}{2}y^2\right]_0^1 dy$$

$$= 4\int_{y=0}^{1}\left[\frac{y}{3}+\frac{y^2}{2}\right]dy = 4\left[\frac{y^2}{6}+\frac{y^3}{6}\right]_0^1 = 4\left(\frac{1}{6}+\frac{1}{6}\right)$$

$$= 4\left(\frac{2}{6}\right) = \frac{4}{3}$$

Therefore, $E(X+Y) = \frac{4}{3} = \frac{2}{3}+\frac{2}{3} = E(X)+E(Y)$

Also, $E(XY) = \int_{-\infty}^{\infty}\int_{-\infty}^{\infty}(xy)f(x,y)dxdy$

$$= \int_{-\infty}^{\infty}\int_{-\infty}^{\infty}(xy)(4xy)dxdy = 4\int_{0}^{1}\int_{0}^{1}x^2y^2dxdy$$

$$= 4\int_{y=0}^{1} y^2\left[\frac{x^3}{3}\right]_0^1 dy = \frac{4}{3}\int_{y=0}^{1} y^2dy = \frac{4}{3}\left[\frac{y^3}{3}\right]_0^1$$

$$= \frac{4}{3}\cdot\frac{1}{3} = \frac{4}{9}$$

Thus, $E(XY) = \frac{4}{9} = \frac{2}{3}\cdot\frac{2}{3} = E(X)E(Y)$

Hence, verified.

**Example 10** Find the expectation of the number of failures preceding the first success in an infinite series of independent trials with constant probability $p$ of success in a trial.

**Solution** Let $X$ denote the number of failures preceding the first success. Then, $X$ takes the values 0, 1, 2, 3, 4,. . .

The corresponding probabilities are

$$p,\ qp,\ q^2p,\ q^3p,\ q^4p,\ldots$$

where $q = 1-p$.

We have the probability density function $f(X) = q^n p$, where $n =$ 0, 1, 2, 3. . .

Hence, the expectation is

$$E(x) = 0.p + 1.qp + 2.q^2p + 3q^3p + 4q^4p + \dots$$
$$= pq(1 + 2q + 3q^2 + 4q^3 + \dots)$$
$$= pq(1-q)^{-2}$$
$$= pq.p^{-2}$$
$$= \frac{pq}{p^2} = \frac{q}{p} = \frac{1-p}{p} = \frac{1}{p} - 1$$

**Theorem 2.6 (Cauchy–Schwartz inequality)**

If $X$, $Y$ are random variables taking real values, then

$$[E[XY]]^2 \le E[X^2]E[Y^2]$$

*Proof* Let $t$ be a real variable.

Consider the expression $(X + tY)^2$.

$(X + tY)^2$ is a non-negative for all values of $X$ and $Y$.

Hence, $E(X + tY)^2 \ge 0$ for all $t$.

i.e., $E(X^2 + 2tXY + t^2Y^2) \ge 0$ for all $t$

i.e., $E(X^2) + 2tE(XY) + t^2E(Y^2) \ge 0$ for all $t$

$t^2E(Y^2) + 2tE(XY) + E(X^2)$ is of the form, $at^2 + bt + c$, a quadratic in $t$.

We have $a = E(Y^2)$, $b = 2E(XY)$, $c = E(X^2)$

$at^2 + bt + c \ge 0$, implies $b^2 - 4ac \le 0$

Therefore, $4[E[XY]]^2 - 4E[X^2]E[Y^2] \le 0$

or $[E[XY]]^2 - E[X^2]E[Y^2] \le 0$

or $[E[XY]]^2 \le E[X^2]E[Y^2]$

Hence, proved.

**Theorem 2.7 (Chebyshev's inequality)**

If $X$ is a random variable with mean $\mu$ and variance $\sigma^2$ then

$$P(|x-\mu| \geq k) \leq \frac{\sigma^2}{k^2}$$

or $P(|x-\mu| \leq k) \geq 1 - \dfrac{\sigma^2}{k^2}$

where $k$ is a real number.

*Proof* **Case (i)** Let $X$ be a discrete random variable and $f(x)$ denote the probability function of $X$.

Then, $\text{Var}[X] = \sigma^2 = E\left[(x-\mu)^2\right]$

$$= \Sigma (x-\mu)^2 f(x) \tag{1}$$

Right-hand side of the sum given by (1) is non- negative.

Hence, $\sigma^2 \geq \sum_{|x-\mu|\geq k} (x-\mu)^2 f(x) \geq \sum_{|x-\mu|\geq k} k^2 f(x)$

i.e., $\sigma^2 \geq k^2 \sum_{|x-\mu|\geq k} f(x)$ (2)

But $\sum_{|x-\mu|\geq k} f(x) = P(|x-\mu| \geq k)$

Therefore, from (2) we get

$$\sigma^2 \geq k^2 P(|x-\mu| \geq k)$$

i.e., $\dfrac{\sigma^2}{k^2} \geq P(|x-\mu| \geq k)$

Thus, we get $P(|x-\mu| \geq k) \leq \dfrac{\sigma^2}{k^2}$

**Case (ii)** Let $X$ be a continuous random variable and $f(x)$ be the probability density function of $X$. Then $\sigma^2 = \int_{-\infty}^{\infty} (x-\mu)^2 f(x)dx.$

Clearly the above integrand is positive. Therefore, when the range of integration is reduced, the value of the integral decreases.

Thus, we have $\sigma^2 \geq \int\limits_{|x-\mu|\geq k} (x-\mu)^2 f(x)dx \geq \int\limits_{|x-\mu|\geq k} k^2 f(x)dx$

or $\sigma^2 \geq k^2 \int\limits_{|x-\mu|\geq k} f(x)dx$

But $\int\limits_{|x-\mu|\geq k} f(x)dx = P(|x-\mu| \geq k)$

Hence, we get $\sigma^2 \geq k^2 P(|x-\mu| \geq k)$

or $P(|x-\mu| \geq k) \leq \dfrac{\sigma^2}{k^2}$ (3)

Also, we have $P(|x-\mu| \leq k) + P(|x-\mu| \geq k) = 1$

Therefore, we get

$$P(|x-\mu| \leq k) = 1 - P(|x-\mu| \geq k)$$

From (3), we get

$$P(|x-\mu| \leq k) \geq 1 - \frac{\sigma^2}{k^2}$$

Hence, proved.

**Remark**

1. $P(|x-\mu| \leq k) \geq 1 - \dfrac{\sigma^2}{k^2}$ can be written as $P(\mu - k < x < \mu + k) \geq 1 - \dfrac{\sigma^2}{k^2}$
2. $\dfrac{\sigma^2}{k^2}$ is called an upper bound of $P(|x-\mu| \geq k)$

## Solved examples

**Example 1** If the Chebyshev's inequality for the random variable $X$ is given by $P(-2 < X < 8) \geq \dfrac{21}{25}$, find $E(X)$ and $\text{Var}[X]$.

**Solution** Comparing $P(-2 < X < 8) \geq \dfrac{21}{25}$

with $P(\mu - k < x < \mu + k) \geq 1 - \dfrac{\sigma^2}{k^2}$, we get

$$\mu - k = -2 \tag{1}$$

$$\mu + k = 8 \tag{2}$$

and $1 - \dfrac{\sigma^2}{k^2} = \dfrac{21}{25}$ (3)

Adding (1) and (2), we get

$$2\mu = 6 \text{ or } \mu = 3$$

Now $\mu + k = 8$ gives $3 + k = 8$ or $k = 5$

Substituting in (3), we get $1 - \dfrac{\sigma^2}{k^2} = \dfrac{21}{25}$

i.e., $1 - \dfrac{\sigma^2}{5^2} = \dfrac{21}{25}$

or $1 - \dfrac{21}{25} = \dfrac{\sigma^2}{5^2}$

or $\dfrac{4}{25} = \dfrac{\sigma^2}{5^2}$

or $\sigma^2 = 4$

Hence, $E(X) = 3$, $\text{Var}[X] = 4$

**Example 2** The number of components manufactured in a factory during a one month period is a random variable with mean 600 and variance 100. What is the probability that the production will be between 500 and 700 over a month

**Solution** We have $\mu = 600, \sigma^2 = 100$

$P(500 < X < 700) = P(600 - 100 < X < 600 + 100)$

Comparing with $P(\mu - k < x < \mu + k)$, we get $k = 100$

Using Chebyshev's inequality, we get

$$P(|x - 600| < k) \geq 1 - \frac{\sigma^2}{k^2} = 1 - \frac{100}{100^2} = 1 - \frac{1}{100}$$

or $P(|x - 600| < 100) \geq 1 - 0.01 = 0.99$

i.e, $P(|x - 600| < 100) \geq 0.99$

Hence, the probability that the production over a month is between 500 and 700 is 0.99.

**Definition 2.21** Let $X$ and $Y$ be two discrete random variables and $f(x,y)$ be the value of the joint probability distribution of $X$ and $Y$ at $(x,y)$. If $(x,y)$ is a real valued function of $(X,Y)$, then the expectation of $g(X,Y)$ is given by

$$E(g(X,Y)) = \sum_x \sum_y g(x,y) f(x,y) dxdy$$

If $X$ and $Y$ are continuous random variables and $f(x,y)$ is the value of their joint probability density of $X$ and $Y$ at $(x,y)$, then

$$E(g(X,Y)) = \int_{-\infty}^{\infty} \int_{-\infty}^{\infty} g(x,y) f(x,y) dxdy$$

For example, consider the joint density of $X$ and $Y$ given by

$$F(x,y) = \frac{2}{5}(x+2y), \quad 0 < x < 1, \quad 1 < y < 2$$

$$= 0, \quad \text{elsewhere}$$

The expected value of $g(x,y) = \frac{x}{y^3}$ is

$$E\left[\frac{x}{y^3}\right] = \frac{2}{5} \int_1^2 \int_0^1 \frac{x}{y^3}(x+2y) dxdy$$

$$= \frac{2}{5} \int_1^2 \left[ \int_0^1 \left( \frac{x^2}{y^3} + \frac{2xy}{y^3} \right) dx \right] dy$$

$$= \frac{2}{5} \int_1^2 \left[ \frac{x^3}{3y^3} + \frac{x^2}{y^2} \right]_0^1 dy$$

$$= \frac{2}{5} \int_1^2 \left[ \frac{1}{3y^3} + \frac{1}{y^2} \right] dy$$

$$= \frac{2}{5} \left[ \frac{1}{3} \left( \frac{y^{-2}}{-2} \right) + \frac{y^{-1}}{(-1)} \right]_1^2$$

$$= \frac{2}{5} \left[ \frac{-1}{6y^2} - \frac{1}{y} \right]_1^2$$

$$= \frac{1}{4}$$

### 2.16.4 Covariance

**Definition 2.21** Let (*X*, *Y*) be a two-dimensional (bivariate) random variable. Then, covariance between *X* and *Y* denoted by Cov(*X*, *Y*) is defined as follows

$$\text{Cov}(X,Y) = E[(X - E(X))(Y - E(Y))]$$

where $E(X)$ is the expectation of $X$ and $E(Y)$ is the expectation of $Y$.

From the definition of covariance, we have

$$\begin{aligned}\text{Cov}(X,Y) &= E[(X - E(X))(Y - E(Y))] \\ &= E[XY - XE(Y) - YE(X) + E(X)E(Y)] \\ &= E(XY) - E(X)E(Y) - E(Y)E(X) + E(X)E(Y) \\ &= E(XY) - E(X)E(Y)\end{aligned}$$

**Remarks** (i) If *X* and *Y* are independent,

$$\begin{aligned}\text{Cov}(X,Y) &= E(XY) - E(X)E(Y) \\ &= E(X)E(Y) - E(X)E(Y) = 0\end{aligned}$$

But the $\text{Cov}(X,Y) = 0$ does not imply that $X$ and $Y$ are independent.

(ii) If *a*, *b* are real constants, then

$$\text{Cov}(aX, bY) = ab\ \text{Cov}(X, Y)$$

and

$$\text{Cov}(X + a, Y + b) = \text{Cov}(X, Y)$$

**Example** For the following joint probability distribution, find Cov(*X*, *Y*).

| $x\backslash y$ | 1 | 2 | 3 | $p_i$ |
|---|---|---|---|---|
| 0 | 2/12 | 1/12 | 1/12 | 4/12 |
| 1 | 1/12 | 1/12 | 2/12 | 4/12 |
| 2 | 3/12 | 1/12 | 0 | 4/12 |
| $p_{.j}$ | 6/12 | 3/12 | 3/12 | 1 |

**Solution** From the given table, we get

$$E(X) = 0.\frac{4}{12} + 1.\frac{4}{12} + 2.\frac{4}{12} = \frac{12}{12} = 1$$

$$E(Y) = 1.\frac{6}{12} + 2.\frac{3}{12} + 3.\frac{3}{12}$$

$$= \frac{6}{12} + \frac{6}{12} + \frac{9}{12} = \frac{21}{12} = \frac{7}{4}$$

$$E(XY) = 0.1.\frac{1}{12} + 0.2.\frac{1}{12} + 0.3.\frac{1}{12} + 1.1\frac{1}{12} + 1.2.\frac{1}{12} + 1.3.\frac{2}{12} + 2.1.\frac{3}{12} + 2.2.\frac{1}{12} + 2.3.0$$

$$= 0 + 0 + 0 + \frac{1}{12} + \frac{2}{12} + \frac{6}{12} + \frac{6}{12} + \frac{4}{12} + 0 = \frac{19}{12}$$

Hence, $\text{Cov}(X,Y) = E(XY) - E(X)E(Y)$

$$= \frac{19}{12} - 1.\frac{7}{4}$$

$$= \frac{19}{12} - \frac{7}{4} = \frac{-2}{12} = \frac{-1}{6}$$

## Exercise 2.4

1. Let $X$ be a random variable with the following probability distribution. Find $E(k)$

| $X$: | 3 | 6 | 9 |
|---|---|---|---|
| $P(X = x)$: | 1/6 | 1/2 | 1/3 |

2. If $X$ is a random variable and the density function of $X$ is

$f(x) = e^{-x}, x \geq 0$

$= 0$, otherwise

Find (i) $E(X)$, (ii) $E(X^2)$, (iii) $E(X-1)^2$, (iv) $\text{Var}[X]$

3. Define expectation of a random variable.

4. A random variable $X$ is given by

| $X$ | −2 | 3 | 1 |
|---|---|---|---|
| $P(X = x)$ | 1/3 | 1/2 | 1/6 |

Find (i) $E(X)$ (ii) $\text{Var}[X]$

5. Find the mean and variance for the distribution

| $X$ | −1 | 0 | 1 | otherwise |
|---|---|---|---|---|
| $P(X = x)$ | 1/8 | 3/8 | 1/2 | 0 |

6. The distribution of the number of raisins in a cookie is given by the following table. Find the mean and variance of raisins in a cookie.

| No. of raisins ($x$) | 0 | 1 | 2 | 3 | 4 | 5 |
|---|---|---|---|---|---|---|
| Probability $P(x)$ | 0.05 | 0.1 | 0.2 | 0.4 | 0.15 | 0.1 |

7. If $X$ is a random variable defined by the density function

$$f(x) = 3x^2,\ 0 \le x \le 1$$
$$= 0, \quad \text{otherwise}$$

Find (i) $E(X)$, (ii) $E(3X-2)$, (iii) $E(X^2)$, (iv) $\text{Var}[X]$

8. If $X$ and $Y$ are random variables each having the density functions

$$f(x) = e^{-x},\ x \ge 0$$
$$= 0, \quad \text{otherwise}$$

Find (i) $E(X+Y)$, (ii) $E(X^2+Y^2)$, (iii) $E(XY)$, (iv) $\text{Var}[X]$ and $\text{Var}[Y]$.

9. Two unbiased dice are thrown. Find the expected value of the sum of the numbers obtained on the top faces of them.

10. Find the expectation of the number of failures preceding the first successes in an infinite series of independent trials with constant probability $p$ of successes in each trial.

11. The density function of a continuous random variable $X$ is given by

$$f(x) = \frac{1}{2}x,\ 0 < x < 2$$
$$= 0, \quad \text{otherwise}$$

Find $E(X)$, $\text{Var}[X]$ and $E(3x^2 - 2x)$.

12. For the density function

$$f(x) = \frac{1}{4} \text{ for } x = -1$$
$$= \frac{1}{4} \text{ for } x = 0$$
$$= \frac{1}{2} \text{ for } x = 2$$

Find $E(X)$ and $\text{Var}[X]$

13. If $a > 0$ and $n \ge 0$, find $E(X)$ and $\text{Var}[X]$ when

$$f(x) = \left[\frac{a^{n+1}.\, x^n}{n!}\right] e^{-ax}$$

14. From the following table, find (i) $k$, (ii) $E(2X+3)$, (iii) $\text{Var}(x)$

| X | −3 | −2 | −1 | 0 | 1 | 2 | 3 |
|---|---|---|---|---|---|---|---|
| $P(X=x)$ | 0.05 | 0.10 | $2k$ | 0 | 0.30 | $k$ | 0.10 |

15. The function $f(x) = \frac{1}{\pi}.\frac{1}{1+x^2}$, $-\infty < x < \infty$ defines the probability density function of $X$. Find $E(X)$.

16. If $X$ is a random variable and $Y = \dfrac{(x-\mu)}{\sigma}$ is also a random variable, show that

(i) $E(Y) = 0$, (ii) $\text{Var}(Y) = 1$.

17. For a random variable with $f(x) = \dfrac{1}{2}e^{-|x|}$, find its standard deviation.

18. A random variable $X$ with unknown probability distribution has an average of 14 and variance 9. Find the probability that $X$ will be between 7 and 21.

19. A random variable $X$ has the density function $e^{-x}$ for $x \geq 0$. Show that

Chebyshev's inequality gives $P(|x-1| > 2) < \dfrac{1}{4}$.

20. If a random variable $X$ has mean of 9 and variance of 4, find $P(|x-9| < 3)$.

### Answers 2.4

| | | | |
|---|---|---|---|
| 1. | $\dfrac{13}{2}$ | 2. | (i) 1 (ii) 2 (iii) 1 (iv) 1 |
| 4. | (i) 1 (ii) 5 | 5. | (i) $\dfrac{3}{8}$ (ii) $\dfrac{5}{8}$ |
| 6. | 2.8 , 1.56 | 7. | (i) $\dfrac{3}{9}$ (ii) $\dfrac{1}{4}$ (iii) $\dfrac{3}{5}$ (iv) $\dfrac{3}{80}$ |
| 8. | (i) $\dfrac{1}{2}$ (ii) 1 (iii) $\dfrac{1}{4}$ (iv) $\dfrac{1}{4}, \dfrac{1}{4}$ | 9. | 7 |
| 10. | $\dfrac{q}{p}$ | 11. | $\dfrac{4}{3}, \dfrac{2}{9}, \dfrac{10}{3}$ |
| 12. | $\dfrac{3}{4}, \dfrac{27}{16}$ | 13. | $\dfrac{n+1}{a}, \dfrac{n+1}{a^2}$ |
| 14. | (i) 0.15 (ii) 3.5 (iii) 2.887 | 15. | $E(X)$ does not exist |
| 17. | $\sigma = \sqrt{2}$ | 18. | $\dfrac{40}{49}$ |
| 19. | $\dfrac{5}{9}$ | | |

## 2.17 Moments

In this section, we introduce moments of a random variable which serve to describe the shape of the distribution of a random variable.

## Moments about arithmetic mean (central moments)

**Definition 2.22** If $X$ is a discrete random variable, the $r$th moment of $X$ about the mean $\mu$, denoted by $\mu_r$ is defined by

$$\mu_r = E(x-\mu)^r = \sum_{i=1}^{n} (x_i-\mu)^r p_i, \; r = 0, 1, 2, \ldots$$

If $X$ is a continuous random variable, the $r$th moment about the mean is defined by

$$\mu_r = E(x-\mu)^r = \int_{-\infty}^{\infty} (x-\mu)^r f(x)dx$$

**Remarks** (i) Since $\mu = E(X)$, we can write

$$\mu_r = E[X - E[x]]^r$$

(ii) If $x_1, x_2, \ldots, x_n$ are in dual series, i.e., values of a random variable, with mean $\mu$, then the $r$th moment about $\mu$ is

$$\mu_r = \frac{1}{n}\sum_{i=1}^{n} (x_i-\mu)^r, \; r = 0, 1, 2, \ldots$$

From the definition of rth moment we have,

$$\mu_r = \sum_{i=1}^{n} (x_i-\mu)^r p_i$$

Therefore, we get

$$\mu_0 = \sum_{i=1}^{n} (x_i-\mu)^0 p_i = \sum p_i = 1$$

$$\mu_1 = \sum_{i=1}^{n} (x_i-\mu) p_i = \sum_{i=1}^{n} x_i p_i - \sum_{i=1}^{n} \mu p_i$$

$$= E[X] - \mu \sum p_i = E[X] - \mu.1$$

$$= E[X] - E[X] = 0 \text{ (since } \mu = E(X))$$

Therefore, $\mu_0 = 1$, $\mu_1 = 0$ for any variable $X$.

$$\mu_2 = \sum_{i=1}^{n} (x_i-\mu)^2 p_i$$

$$= \sum_{i=1}^{n} (x_i^2 - 2\mu x_i + \mu^2)\, p_i$$

$$= \sum_{i=1}^{n} x_i^2 p_i - 2\mu \sum_{i=1}^{n} x_i p_i + \mu^2 \sum_{i=1}^{n} p_i$$

$$= E(X^2) - \mu^2$$

or $\mu_2 = E(X^2) - [E(x)]^2$

$\mu_2$ is called the variance of the distribution of $X$. It is denoted by $\sigma^2$. The positive square root of $\sigma^2$, (i.e., variance) is called the standard deviation. The moments about mean are also called the central moments or simply the moments of the distribution. $\mu_1$, $\mu_2$ , $\mu_3$ , $\mu_4$ are called the first four central moments of the distribution.

**Example** Find the first four moments of the following series about mean: 3, 6, 8, 10, 18.

**Solution** We have $x_1 = 3$, $x_2 = 6$, $x_3 = 8$, $x_4 = 10$, $x_5 = 18$ (given)

$$\text{Mean} = \mu = \frac{3+6+8+10+18}{5} = \frac{45}{5} = 9$$

$$\mu_1 = \frac{1}{5}\sum_{i=1}^{5} (x_i - \mu) = \frac{1}{5}[(3-9)+(6-9)+(8-9)+(10-9)+(18-9)] = 0$$

$$\mu_2 = \frac{1}{5}\sum_{i=1}^{5} (x_i - \mu)^2 = \frac{1}{5}\left[(3-9)^2+(6-9)^2+(8-9)^2+(10-9)^2+(18-9)^2\right]$$

$$= \frac{1}{5}[36+9+1+1+81] \quad = \frac{128}{5} = 25.6$$

$$\mu_3 = \frac{1}{5}\sum_{i=1}^{5} (x_i - \mu)^3 = \frac{1}{5}\left[(3-9)^3+(6-9)^3+(8-9)^3+(10-9)^3+(18-9)^3\right]$$

$$= \frac{1}{5}[-216+(-27)+(-1)+1+729] \quad = \frac{486}{5} = 97.2$$

$$\mu_4 = \frac{1}{5}\sum_{i=1}^{5} (x_i - \mu)^4 = \frac{1}{5}\left[(3-9)^4+(6-9)^4+(8-9)^4+(10-9)^4+(18-9)^4\right]$$

$$= \frac{1}{5}[1296+81+1+1+6561]$$

$$= \frac{7940}{5} = 1588$$

The first four moments about the mean are 0, 97.2, 25.6 and 1588.

### 2.17.1 Moments about an arbitrary number

**Definition 2.23** Let $X$ be a discrete random variable and $A$ be an arbitrary number. The $r$th moment about $A$ of $X$ denoted by $\mu'_r$ is defined as

$$\mu'_r = E\left[(x-A)^r\right] = \sum_{i=1}^{n} (x_i - A)^r p_i, r = 0,1,2,\ldots$$

If $X$ is a continuous random variable, then $\mu'_r$ is defined as

$$\mu'_r = \int_{-\infty}^{\infty} (x-A)^r f(x)dx, r = 0,1,2,\ldots$$

The moments about an arbitrary number are also called raw moments.

**Relation between $\mu_r$ and $\mu'_r$**

If $\mu_r$ denotes the $r$th moment of a distribution about the mean and $\mu'_r$ denotes the $r$th moment about an arbitrary number, then

$$\mu_r = \mu'_r - {}^rC_1\mu'_{r-1}\mu'_1 + {}^rC_2\mu'_{r-2}(\mu'_2)^2 + \ldots + (-1)^r(\mu'_1)^r$$

Substituting $r = 2,3,4,\ldots$ we get

$$\mu_2 = \mu'_2 - {}^2C_1\mu'_1\mu'_1 + {}^2C_2\left(\mu'_1\right)^2 = \mu'_2 - \left(\mu'_1\right)^2$$

$$\mu_3 = \mu'_3 - {}^3C_1\mu'_2\mu'_1 + {}^3C_2\mu'_1\left(\mu'_1\right)^2 - {}^3C_3\left(\mu'_1\right)^3$$

$$= \mu'_3 - 3\mu'_2\mu'_1 + 2\left(\mu'_1\right)^2$$

$$\mu_4 = \mu'_4 - {}^4C_1\mu'_3\mu'_1 + {}^4C_2\mu'_2\left(\mu'_1\right)^2 - {}^4C_3\mu'_1\left(\mu'_1\right)^3 + {}^4C_4\left(\mu'_1\right)^4$$

$$= \mu'_4 - 4\mu'_3\mu'_1 + 6\mu'_2\left(\mu'_1\right)^2 - 3\left(\mu'_1\right)^4$$

**Remarks** (i) $\mu'_0 = \sum_{i=1}^{n} (x_i - A)^0 p_i$

$$= \sum p_i = 1$$

(ii) $\mu'_1 = \sum_{i=1}^{n} (x_i - A)\, p_i$

$$= \sum_{i=1}^{n} x_i p_i - A \sum_{i=1}^{n} p_i$$

$$= E(X) - A$$

$$\therefore \mu = \mu_1' + A$$

**Example** The first four moments of a distribution about $x = 2$ are 1, 2.5, 5.5 and 26 respectively. Find the values of the first four central moments.

**Solution** We have $\mu_1' = 1, \mu_2' = 2.5, \mu_3' = 5.5, \mu_4' = 26$ and $A = 2$.

Thus, $\mu_1 = 0, \mu_2 = \mu_2' - (\mu_1')^2 = (2.5) - 1^2 = 1.5$

$$\mu_3 = \mu_3' - 3\mu_2'\mu_1' + 2(\mu_1')^2$$

$$= 5.5 - 3(2.5)(1) + 2(1)^3$$

$$= 5.5 - 7.5 + 2 = 0$$

$$\mu_4 = \mu_4' - 4\mu_3'\mu_1' + 6\mu_2'(\mu_1')^2 - 3(\mu_1')^4$$

$$= 26 - 4(5.5)(1) + 6(2.5)(1) + 3(1)^4$$

$$= 26 - 22 + 15 - 3 = 16$$

$$\mu_1 = 0, \mu_2 = 1.5, \mu_3 = 0, \mu_4 = 16$$

## 2.17.2 Moments about origin

**Definition 2.24** The $r$th moment about origin of a random variable $X$, denoted by $V_r$ is the expected value of $X^r$, symbolically

$$V_r = E[X^r] = \sum_{i=1}^{n} x_i^r p_i$$

for $r = 0, 1, 2, \ldots$, when $X$ is discrete, and

$$V_r = E[X^r] = \int_{-\infty}^{\infty} x^r f(x)dx$$

when $X$ is continuous.

**Relation between $\mu_r$ and $V_r$**

If $\mu$ is the arithmetic mean of $X$, then the distribution

$$V_r = \mu_r + {}^rC_1\mu_{r-1}\mu + {}^rC_2\mu_{r-2}\mu^2 + \ldots + \mu^r$$

Putting $r = 1, 2, 3, \ldots$, we get

$$V_1 = \mu, V_2 = \mu_2 + \mu^2, V_3 = \mu_3 + 3\mu_2\mu + \mu^3$$

**Remark** From the definition of moment about arbitrary number $A$, we have

$$\mu'_r = E[(x-A)^r] = \sum_{i=1}^{n} (x_i - A)^r p_i, r = 0, 1, 2, \ldots$$

Taking $A = 0$, we obtain

$$\mu'_r = E[X^r] = \sum_{i=1}^{n} (x_i)^r p_i$$

i.e., $\mu'_r = V_r$

## 2.18 Skewness

If the curve of a distribution is bell-shaped, then the distribution is said to be a symmetrical distribution. In a symmetrical distribution, the value of mean, median and mode coincide (Fig. 2.4).

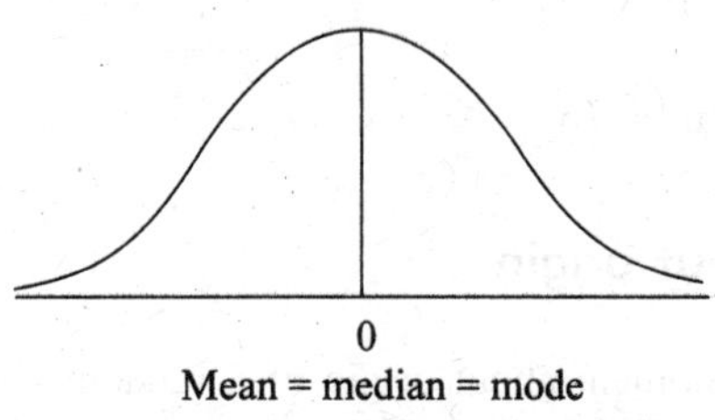


**Fig. 2.4** *Symmetrical distribution*

"Skewness" is a word used for lack of symmetry". Skewness of a distribution is defined as the tendency of a distribution to depart from symmetry.

Most distributions are not symmetrical. Curve of an asymmetric distribution may have a longer "tail" to the right than to the left or, it may have a longer "tail" to the left than to the right.

Relative measure of skewness is called coefficient of skewness. A measure of skewness is defined by

$$Sk_m = \frac{E\left[(x-\mu)^3\right]}{\sigma^3} = \frac{\mu_3}{\sqrt{\mu_2^3}}$$

If we define $\beta_1 = \frac{\mu_3^2}{\mu_2^3}$, the moment coefficient of skewness $= \frac{\mu_3}{\sqrt{\mu_2^3}} = \pm\beta_1$, where the sign of $\sqrt{\beta_1}$ is taken as that of $\mu_3$.

The moment coefficient of skewness is also denoted by $\gamma_1$. Thus

$$\gamma_1 = \frac{\mu_3}{\sqrt{\mu_3^3}} = \pm\sqrt{\beta_1}$$

If measure of skewness (Fig. 2.5) is positive, we say that the distribution is positively skewed or right tailed. And if the measure of skewness is negative, we say that the distribution is negatively skewed or left tailed.

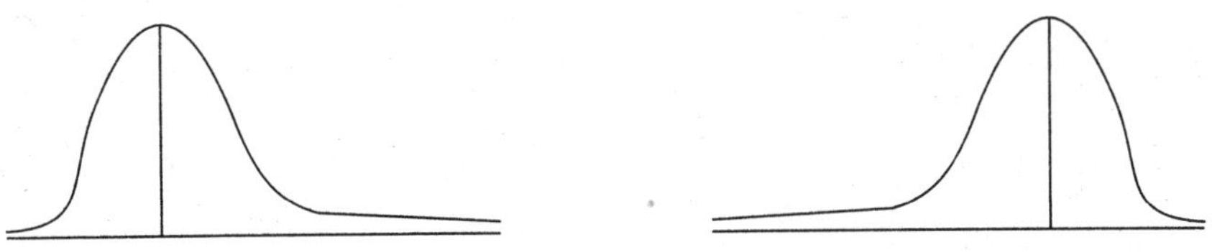

(a) Positively skewed distribution (b) Negatively skewed distribution

**Fig. 2.5** *Skewed distribution*

If arithmetic mean of the distribution is greater than the mode, the distribution is positively skewed (i.e., right tailed), and if the arithmetic mean is less than the value of the mode, the distribution is negatively skewed (i.e., left tailed). In a symmetrical distribution, the quartiles are equidistant from the median.

## 2.19 Other Methods of Measuring Skewness

Skewness by the method of moments was discussed in the preceding section. We now introduce other methods of finding skewness.

### Karl Pearson's method

In a symmetrical distribution, the value of arithmetic mean is equal to the value of mode. Karl Pearson's method is based on this fact and the coefficient of skewness is defined as follows

$$Sk_p = \frac{\text{A.M} - \text{Mode}}{\text{S.D}}$$

where A.M = Arithmetic mean

S.D = Standard deviation

$Sk_p$ = Karl Pearson's coefficient of skewness

Since mode = 3 median − 2 A.M., the Karl Pearson's coefficient of skewness is

$$Sk_p = \frac{3(\text{A.M} - \text{Median})}{\text{S.D}}$$

**Bowley's method**

Bowley's coefficient of skewness is based on the fact that in a symmetric distribution the quartiles are equidistant from the median.

The coefficient of skewness is defined as follows

$$Sk_B = \frac{Q_3 + Q_1 - 2(\text{Median})}{Q_3 - Q_1}$$

where $Q_1$ = lower quartile

$Q_3$ = upper quartile

$Sk_B$ = Bowley's coefficient of skewness

Bowley's coefficient of skewness lies between −1 and +1

For a symmetrical distribution $Sk_B = 0$

**Kelly's method**

Kelly's coefficient of skewness is defined by

$$S_k = \frac{P_{90} + P_{10} - 2(\text{Median})}{P_{90} - P_{10}}$$

where $P_{10}$ = 10th percentile

$P_{90}$ = 90th percentile

The value of Kelly's coefficient of skewness lies between −1 and 1.

## 2.20 Kurtosis

The word "kurtosis" is used to describe the peakedness of a curve. The measure of peakedness and flatness of the curve of a distribution, relative to the curve of a normal distribution, is called the measure of kurtosis. The curve of a normal distribution is neither flat nor sharply peaked. The measure of kurtosis tells us the extent to which a distribution is more peaked or more flat topped than the normal curve. If the curve of a frequency distribution is more peaked than the normal curve, it is said to be "leptokurtic". The curve of a frequency distribution is called "platykurtic" if it is more flat-topped than the normal curve. If the curve of a frequency is neither flat nor sharply peaked, then the curve of the distribution is called "mesokurtic". The measure of kurtosis of a frequency distribution is based upon the fourth moment about the mean of the distribution. The measure of kurtosis is denoted by $\beta_2$ and is defined as

$$\beta_2 = \frac{\mu_4}{\mu_2^2}$$

where $\mu_2$ and $\mu_4$ are respectively the second and fourth central moments about the mean. Kurtosis is also known as convexity of a curve (or bulginess).

If $\beta_2 < 3$, the distribution is said to be "platykurtic".

If $\beta_2 = 3$, the distribution is said to be "mesokurtic".

If $\beta_2 > 3$, the distribution is said to be "leptokurtic".

The kurtosis of distribution (Fig. 2.6) is also measured by considering the value of $\beta_2 - 3$ which is denoted by $\gamma_2$.

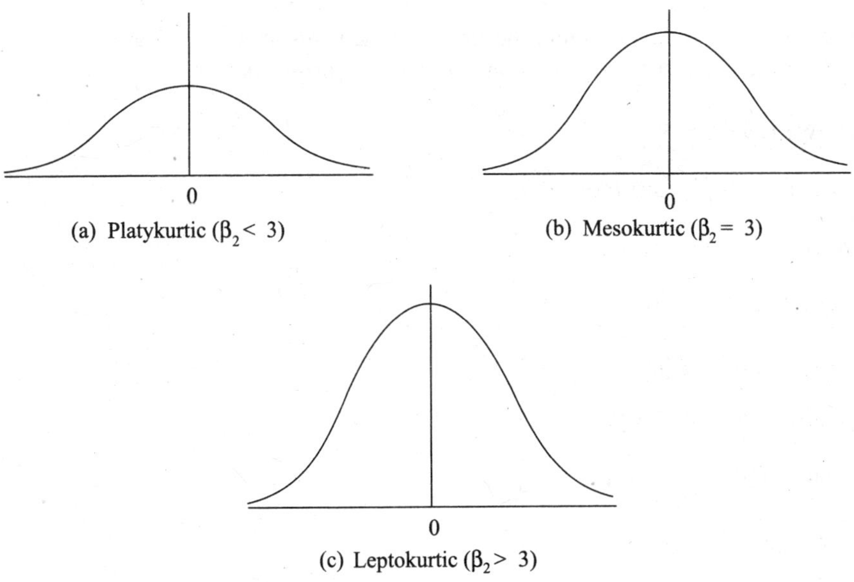


**Fig. 2.6** *Kurtosis of the distribution*

Thus, $\gamma_2 = \beta_2 - 3$.
If $\gamma_2 = 0$, the distribution is mesokurtic.
If $\gamma_2 < 0$, the distribution is platykurtic.
If $\gamma_2 > 0$, the distribution is leptokurtic.

## Solved examples

**Example 1** The first four moments of a distribution about $x = 4$ are 1, 4, 10 and 45. Comment upon the nature of the distribution.

**Solution** We have $\mu_1' = 1$, $\mu_2' = 4$, $\mu_3' = 10, \mu_4' = 45$ and $A = 4$.

The mean of the distribution is $\mu = \mu_1' + A = 1 + 4 = 5$

$$\text{Moment of skewness} = \frac{\mu_4}{\mu_2^2} = \frac{0}{\sqrt{3^3}} = 0$$

i.e., $\gamma_1 = 0$

Therefore, the distribution is symmetrical kurtosis,

$$\beta_2 = \frac{\mu_4}{\mu_2^2} = \frac{26}{3^2} = 2.89$$

Since $\beta_2 < 3$, the distribution is platykurtic.

**Example 2** In a certain distribution, the first four moments about $x = 5$ are 2, 20, 40 and 50. Calculate and state whether the distribution is leptokurtic or platykurtic.

**Solution** We have $A = 5$.

$$\mu_1' = 2, \mu_2' = 20, \mu_3' = 40, \mu_4' = 50$$

Hence, we get $\mu_1 = 0$ (always),

$$\mu_2 = \mu_2' - (\mu_1')^2 = (20) - 2^2 = 16$$

$$\mu_3 = \mu_3' - 3\mu_2'\mu_1' + 2(\mu_1')^2$$

$$= 40 - 3(2)(20) + 2(2)^3$$

$$= 40 - 120 + 16 = -64$$

$$\mu_4 = \mu_4' - 4\mu_3'\mu_1' + 6\mu_2'(\mu_1')^2 - 3(\mu_1')^4$$

$$= 50 - 4(40)(2) + 6(20)(2^2) + 3(2)^4$$

$$= 56 - 320 + 480 - 480$$

$$= 162$$

$$\beta_1 = \frac{\mu_3^2}{\mu_2^3} = \frac{(-64)^2}{16^3} = 1$$

$$\beta_2 = \frac{\mu_4}{\mu_2^2} = \frac{162}{16^2} = 0.63$$

Since $\beta_2 < 3$, the curve of the distribution is platykurtic.

**Example 3** The density of a random variable $X$ is given by $f(x) = kx(2-x), 0 \le x \le 2$. Find (i) $k$, (ii) mean, (iii) variance, (iv) $\beta_1$ and $\beta_2$, (v) $r$th moment.

**Solution** (i) Since $\int_{-\infty}^{\infty} f(x)dx = 1$ i.e., $\int_0^2 f(x)dx = 1$, we have $\int_0^2 kx(2-x)dx = 1$

$$\text{or } k\left[x^2 - \frac{x^3}{3}\right]_0^2 = 1$$

$$\text{or } k\left[4 - \frac{8}{3}\right] = 1 \quad \text{or } \frac{4k}{3} = 1$$

$$\therefore k = \frac{3}{4}$$

(ii) Mean $= E(X) = \int_{-\infty}^{\infty} xf(x)dx \quad = \frac{3}{4}\int_0^2 (2x^2 - x^3)dx$

$$= \frac{3}{4}\left[\frac{2}{3}x^3 - \frac{x^4}{4}\right]_0^2 \quad = \frac{3}{4}\left[\frac{16}{3} - \frac{16}{4}\right]$$

$$= \frac{3}{4}.16\left[\frac{1}{3} - \frac{1}{4}\right] = 3.4.\frac{(4-3)}{3.4}$$

$$= 1$$

(iii) $\mu_r' = \int_0^2 x^r f(x)dx \quad = \frac{3}{4}\int_0^2 x^r(2x - x^2)dx$

$$= \frac{3}{4}\int_0^2 (2x^{r+1} - x^{r+2})dx \quad = \frac{3}{4}\left[\frac{2}{r+2}x^{r+2} - \frac{x^{r+3}}{r+3}\right]_0^2$$

$$= \frac{3}{4}\left[\frac{2}{r+2}.2^{r+2} - \frac{2^{r+3}}{r+3}\right]$$

$$= \frac{3}{4}\left[\frac{2^{r+3}}{(r+2)(r+3)}\right]$$

$$= \left[\frac{3.2^{r+1}}{(r+2)(r+3)}\right]$$

$$\therefore E(X) = \mu_1' = \frac{3.2^2}{3.4} = 1 \text{ i.e., } \mu = 1$$

(iv) We have $\mu_2' = \frac{3.8}{4.5} = \frac{6}{5}$

$$\mu_3' = \frac{3.2^4}{5.6} = \frac{8}{5}$$

$$\mu_4' = \frac{3.2^5}{6.7} = \frac{16}{7}$$

Hence $\mu_1$=1,

$$\mu_2 = \mu_2' - \left(\mu_1'\right)^2 = \frac{6}{5} - 1^2 = \frac{1}{5}$$

$$\mu_3 = \mu_3' - 3\mu_2'\mu_1' + 2\left(\mu_1'\right)^2$$

$$= \frac{8}{5} - \frac{18}{5} + 2 = \frac{18-18}{5} = 0$$

$$\mu_4 = \mu_4' - 4\mu_3'\mu_1' + 6\mu_2'\left(\mu_1'\right)^2 - 3\left(\mu_1'\right)^4$$

$$= \frac{16}{7} - \frac{32}{5} + \frac{36}{5} - 3 = \frac{3}{35}$$

Therefore, $\beta_1 = \frac{\mu_3^2}{\mu_2^3} = 0$

and $\beta_2 = \frac{\mu_4}{\mu_2^2} = \frac{15}{7}$

## Exercise 2.5

1. The first four central moments of a distribution are 0, 2.5, 0.7 and 18.75. Test the skewness and kurtosis of the distribution.

2. Find the first moments about the mean for the series 4, 5, 6, 1, 4.

3. The first four moments of a distribution, about the value 35 are $-1.8$, 240, $-1020$ and 14400. Find the values of $\mu_1, \mu_2, \mu_3$ and $\mu_4$.

4. The first four moments of a distribution about $x = 4$ are the $-1.5$, 17, $-30$ and 108. Find the moments $\mu_1, \mu_2, \mu_3$ and $\mu_4$ about the mean.

5. The first three critical moments of a distribution are 0, 2.5 and 0.7. Find the value of the moment of coefficient of skewness.

6. The first four moments of a distribution about the value 5 of the variable are 2, 20, 40 and 50. Calculate the moment of skewness.

7. The first four central moments of a distribution are 0, 2.5, 0.7 and 18.75. Test the kurtosis of the distribution.

8. The first three central moments of a distribution are 0, 15, $-31$. Find the moment of coefficient of skewness.

9. Calculate the first four moments about the mean for the series 4, 7, 10, 13, 16, 19, 22.

10. The first four moments of a distribution about $x = 4$ are $-1.5$, 17, 30 and 108. Find the moments about the mean.

11. $X$ is a random variable whose density function is

$$f(X) = Ae^{x}, \quad 0 < x < \infty$$
$$= 0, \quad \text{otherwise}$$

Find the value of (i) $A$, (ii) mean of $X$, (iii) variance of $X$, (iv) third moment about the mean, (v) kurtosis, (vi) $r$th moment about the origin.

12. The first three moments of a distribution about the value 2 are 1, 16, $-40$. Find the mean and the variance.

13. The first three moments of a distribution about the value 3 are 2, 10, $-30$. Show that the moments about $x = 0$ are 5, 31, 141. Find the mean and the variance.

14. Given the p.d.f $f(x) = \begin{cases} cx(2-x), & 0 \le x \le 1 \\ 0, & \text{elsewhere} \end{cases}$

find $c$ and obtain (i) $k$th moment about the origin.

(ii) first three moments about the mean.

(iii) the mean and the variance.

15. Show that for the exponential distribution $P = y_0 e^{-x/a} dx, 0 \le x < \infty, \sigma > 0$, $y_0$ being a constant, the mean and the standard deviation are equal to $\sigma$ and that the interquartile range is $\sigma \log_e 3$. Also find $\mu_r^1$ and show that $\beta_1 = 4$ and $\beta_2 = 9$.

[Hint: $y_0 \int_0^\infty e^{\frac{-x}{\sigma}} dx = 1 \Rightarrow y_0 = \frac{1}{\sigma}$

$$\mu_r' = \frac{1}{\sigma} \int_0^\infty x^r e^{-x/\sigma} dx = r! \sigma^r$$

We get $\mu_1' = \sigma, \mu_2' = 2\sigma^2, \mu_3' = 6\sigma^2, \mu_4' = 9\sigma^4$

Hence $\beta_1 = \frac{\mu_3^2}{\mu_2^3} = 4, \beta_2 = \frac{\mu_4}{\mu_2^2} = 9$

$$\frac{1}{\sigma} \int_0^{Q_1} e^{-x/\sigma} dx = \frac{1}{4} \text{ gives } e^{-Q_1/\sigma} = \frac{3}{4}$$

$$\frac{1}{\sigma} \int_0^{Q_3} e^{-x/\sigma} dx = \frac{3}{4} \text{ gives } e^{-Q_3/\sigma} = \frac{1}{4}$$

Hence $e^{\frac{Q_3 - Q_1}{\sigma}} = 3$ or $Q_3 - Q_1 = \sigma \log_e 3$]

16. For the triangular distribution $dp = \frac{1}{a}\left[1 - \frac{|x-b|}{a}\right] dx, |x-b| < a$. Show that the mean is $b$ and the variance is $\frac{a^2}{6}$.
17. For the rectangular distribution $dp = dx, 1 \le x \le 2$, show that A.M > G.M > H.M.
18. For the triangular distribution, with the density function $f(x) = 1 - |1-x|, 0 < x < 2$, show that the mean is 1 and variance is 1/6

## Answers 2.5

1. Positively skewed: $\beta_1 = 0.031$, mesokurtic, ($\beta_2 = 3$)
2. $\mu_1 = 0, \mu_2 = 2.8, \quad \mu_3 = -3.6, \mu_4 = 19.6$
3. $\mu_1 = 0, \mu_2 = 236.76, \quad \mu_3 = 264.36, \quad \mu_4 = 141290.11$
4. 0, 14.75, 39.75, 142.31
5. 0.177
6. −1

7. Mesokurtic ($\beta_2 = 3$)

8. $-0.53$

9. 0, 36, 0, 2268

10. 0, 14.75, 39.75, 142.31

11. (i) $A = 1$, (ii) 1, (iii) 2, (iv) $\mu_3 = 2$, (v) $a$ (leptokurtic), (vi) $r!$

12. 3, 15

13. $\mu = 5, \sigma^2 = 6$

14. $c = \frac{3}{4}$ (i) $\frac{3}{4}\left[\frac{2^{k+3}}{(k+2)(k+3)}\right]$,

 (ii) $\mu_1 = 0$, $\mu_2 = \frac{1}{5}$, $\mu_3 = 0$,

 (iii) Mean $= 1$, Variance $= \frac{1}{5}$

## 2.21 Moment Generating Function

**Definition 2.25** The moment generating function of a random variable $X$, denoted by $M_x(t)$ is defined by

$$M_x(t) = E[e^{tx}] = \sum_x e^{tx} P(X = x)$$

when $X$ is a discrete random variable, and

$$M_x(t) = E[e^{tx}] = \int_{-\infty}^{\infty} e^{tx} f(x)dx \qquad (t \text{ is an independent variable})$$

when $X$ is a continuous random variable.

The moment generating function of $X$ is also denoted by $M(t)$. The moment generating function may exist but the moments may not exist.

## 2.22 Properties of Moment Generating Function

**Theorem 2.8**
If $a, c$ are constants and $X$ is a random variable, then

(i) $M_{cx}[t] = M_x[ct]$

(ii) $M_{\frac{x-a}{c}}[t] = e^{-at/c} M_x\left[\frac{t}{c}\right]$

*Proof* (i) By definition, we have

$$M_{cx}[t] = E[e^{t(cx)}]$$

$$= E[e^{x(ct)}] = M_x[ct]$$

(ii) $M_{\frac{x-a}{c}}[t] = E\left[e^{t\left(\frac{x-a}{c}\right)}\right] = E\left[e^{\frac{tx}{c}}.e^{-\frac{at}{c}}\right] = e^{\frac{-at}{c}} E\left[e^{x\left(\frac{t}{c}\right)}\right]$

$$= e^{\frac{-at}{c}} M_X\left(\frac{t}{c}\right)$$

**Theorem 2.9** If $M_X[t]$ and $M_y[t]$ are the moment generating functions of the independent random variables $X$ and $Y$, then

$$M_{X+Y}[t] = M_X[t] M_Y[t]$$

i.e., the moment generating function of the sum of two independent random variables is equal to the product of their respective moment generating functions.

*Proof* $M_{X+Y}[t] = E\left[e^{t(X+Y)}\right]$

$$= E\left[e^{tX+tY}\right] = E\left[e^{tX} e^{tY}\right]$$

$$= E\left[e^{tX}\right] E\left[e^{tY}\right] = M_X[t] M_y[t]$$

Hence, proved.

**Theorem 2.10** If $X$ is a random variable, then

$$M_x[t] = 1 + \mu_1' t + \mu_2' \frac{t^2}{2!} + \mu_3' \frac{t^3}{3!} + \ldots + \mu_r' \frac{t^r}{r!} + \ldots$$

where $\mu_r'$, $(r = 1, 2, 3, \ldots)$ are the moments about the origin.

*Proof* $M_x[t] = E[e^{tx}]$

$$= E\left[1 + tx + \frac{t^2x^2}{2!} + \frac{t^3x^3}{3!} + \ldots + \frac{t^rx^r}{r!} + \ldots\right]$$

$$= E(1) + E(tx) + E\left(\frac{t^2x^2}{2!}\right) + E\left(\frac{t^3x^3}{3!}\right) + \ldots + E\left(\frac{t^rx^r}{r!}\right) + \ldots$$

$$= 1 + tE(x) + \frac{t^2}{2!}E(x^2) + \frac{t^3}{3!}E(x^3) + \ldots + \frac{t^r}{r!}E(x^r) + \ldots \quad (1)$$

From the definition, we have

$$E[(x-A)^r] = \mu'_r, \text{ where } A \text{ is an arbitrary number.}$$

Putting $A = 0$, we get $E[x^r] = \mu'_r$

$$E(x) = \mu'_1, E(x^2) = \mu'_2, E(x^3) = \mu'_3, \ldots$$

Substituting in (1), we get

$$M_x[t] = 1 + \mu'_1 t + \mu'_2 \frac{t^2}{2!} + \mu'_3 \frac{t^3}{3!} + \ldots + \mu'_r \frac{t^r}{r!} + \ldots$$

## Solved examples

**Example 1** The probability density function of the random variable $X$ has the following probability law

$$P(x) = \frac{1}{2\theta} e^{-\left|\frac{x-\theta}{\theta}\right|}, -\infty < x < \infty$$

Find the m.g.f of $X$. Hence find $E(X)$ and $\text{Var}[X]$

**Solution** The moment generating function of $X$ is

$$M_x[t] = E[e^{tx}] = \int_{-\infty}^{\infty} e^{tx} \frac{1}{2\theta} e^{-\left|\frac{x-\theta}{\theta}\right|} dx$$

$$= \frac{1}{2\theta} \int_{-\infty}^{\theta} e^{tx} e^{-\left|\frac{x-\theta}{\theta}\right|} dx + \frac{1}{2\theta} \int_{\theta}^{\infty} e^{tx} e^{-\left|\frac{x-\theta}{\theta}\right|} dx$$

$$= \frac{1}{2\theta} \int_{-\infty}^{\theta} e^{tx} e^{\left(\frac{x-\theta}{\theta}\right)} dx = + \frac{1}{2\theta} \int_{\theta}^{\infty} e^{tx} e^{-\left(\frac{x-\theta}{\theta}\right)} dx$$

$$= \frac{e^{-1}}{2\theta} \int_{-\infty}^{\theta} e^{\left(t+\frac{1}{\theta}\right)x} dx + \frac{e}{2\theta} \int_{\theta}^{\infty} e^{\left(t-\frac{1}{\theta}\right)x} dx$$

$$= \frac{e^{-1}}{2\theta} \left| \frac{e^{\left(t+\frac{1}{\theta}\right)x}}{t+\frac{1}{\theta}} \right|_{-\infty}^{\theta} - \frac{e}{2\theta} \left| \frac{e^{\left(t-\frac{1}{\theta}\right)x}}{t-\frac{1}{\theta}} \right|_{\theta}^{\infty}$$

$$= \frac{1}{2\theta} \cdot \frac{e^{t\theta}}{\left(\frac{1+t\theta}{\theta}\right)} + \frac{1}{2\theta} \cdot \frac{e^{t\theta}}{\left(\frac{1-t\theta}{\theta}\right)}$$

$$= \frac{e^{t\theta}}{2}\left[\frac{1}{1+t\theta} + \frac{1}{1-t\theta}\right] = \frac{e^{t\theta}}{1-(t\theta)^2}$$

$$= e^{t\theta}\left[1 - t^2\theta^2\right]^{-1}$$

$$= e^{t\theta}\left[1 + t^2\theta^2 + \ldots\right]$$

$$= \left(1 + t\theta + \frac{\theta^2 t^2}{2!} + \ldots\right)\left(1 + \theta^2 t^2 + \ldots\right)$$

$$= \left(1 + t\theta + \frac{3\theta^2 t^2}{2!} + \ldots\right) \qquad (1)$$

$$\mu_1' = \text{Coefficient of } k \text{ in (1)} = \theta$$

$$\mu_2' = \text{Coefficient of } \frac{t^2}{2!} \text{ in (1)} = 3\theta^2$$

Hence, $\mu_2 = \mu_2' - (\mu_1')^2 = 3\theta^2 - \theta^2 = 2\theta^2$

$E(X) = \text{mean} = \theta$, $\text{Var}[X] = 2\theta^2$

**Example 2** Find the moment generating function of a random variable $X$ having the density function

$$f(x) = \frac{x}{2}, \quad 0 \le x \le 2$$
$$= 0, \quad \text{otherwise}$$

and use it to find the first four moments about its origin.

**Solution** We have $M_x[t] = E[e^{tx}] = \int_{-\infty}^{\infty} e^{tx} f(x)dx$

$$= \int_0^2 e^{tx}\frac{x}{2}dx = \frac{1}{2}\int_0^2 e^{tx}xdx$$

$$= \frac{1}{2}\left[x\frac{e^{tx}}{t} - 1.\frac{e^{tx}}{t^2}\right]_0^2$$

$$= \frac{1}{2}\left[\frac{2\,(e^{2t})}{t} - \frac{(e^{2t}-1)}{t^2}\right]$$

$$= \frac{1}{2}\left[\frac{2\,(e^{2t})}{t} + 1 - \frac{e^{2t}}{t^2}\right]$$

$$= \frac{1}{2t^2}\left[1 + 2te^{2t} - e^{2t}\right]$$

$$\therefore M_x(t) = \frac{1}{2t^2}\left[1 + 2te^{2t} - e^{2t}\right]$$

Consider $M_x(t) = \frac{1}{2t^2}\left[1 + 2te^{2t} - e^{2t}\right]$

$$= \frac{1}{2t^2} + \frac{1}{t}e^{2t} - \frac{1}{2t^2}e^{2t}$$

$$= \frac{1}{2t^2} + \frac{1}{t}\left[1 + \frac{2t}{1!} + \frac{2^2t^2}{2!} + \ldots\right] - \frac{1}{2t^2}\left[1 + \frac{2t}{1!} + \frac{2^2t^2}{2!} + \ldots\right]$$

$$= \frac{1}{2t^2} + \left[\frac{1}{t} + 2 + \frac{2^2t}{2!} + \frac{8t^2}{3!} + \frac{16t^3}{4!} + \ldots\right] + \left[\frac{-1}{2t^2} - \frac{1}{t} + \frac{2}{2!} - \frac{2^2t}{3!} - 2^3\frac{t^2}{4!}\ldots\right]$$

$$= 1 + \left(2 - \frac{2}{3}\right)t + \left(\frac{4}{3} - \frac{1}{3}\right)t^2 + \left(\frac{2}{3} - \frac{2}{15}\right)t^3 + \ldots$$

$$= 1 + \frac{4}{3}t + t^2 + \frac{8}{15}t^3 + \frac{2}{9}t^4 + \ldots$$

Therefore,

$M_x(t) = 1 + \frac{4}{3}t + 2\frac{t^2}{2!} + \frac{16}{5}\frac{t^3}{3!} + \frac{16}{3}\frac{t^4}{4!} + \ldots$ is the required moment generating function.

Since $M_x(t) = 1 + \mu_1' t + \mu_2'\frac{t^2}{2!} + \mu_3'\frac{t^3}{3!} + \mu_4'\frac{t^4}{4!} + \ldots$

Comparing, we get $\mu_1' = \frac{4}{3}, \mu_2' = 2, \mu_3' = \frac{16}{5}, \mu_4' = \frac{16}{3}$

## Exercise 2.6

1. Define moment generating function.

2. Find the m.g.f of a random variable $X$ having the density function

$$f(x) = \frac{x}{2}, 0 \leq x \leq 2$$
$$= 0, \quad \text{otherwise}$$

and find the first four moments about the origin.

3. A random variable $X$ assumes values $\frac{1}{2}$ and $\frac{1}{2}$ with probability $\frac{1}{2}$ each. Find the moment generating function and four moments about the origin

4. A random variable has the density function

$$f(x) = e^{-x}, x \geq 0$$
$$= 0, \quad \text{otherwise}$$

Determine the m.g.f about the origin and also about the mean.

5. A random variable $X$ has the probability distribution $f(x) = \frac{1}{8}{}^3C_x$ for $x = 0, 1, 2$ and 3. Find the moment generating function of this random variable and use it to determine $\mu_1'$ and $\mu_2'$.

6. If $a$ and $b$ are constants, prove the following

   (i) $M_{x+a}[t] = e^{at} M_x(t)$

   (ii) $M_{\frac{x+a}{b}}[t] = e^{at/b} M_x\left[\frac{t}{b}\right]$

## Answers 2.6

2. $\frac{4}{3}, 2, \frac{16}{5}, \frac{16}{3}$

3. $M_x(t) = 1 + \frac{t^2}{2!2^2} + \frac{t^4}{4!2^4} + \ldots$

$$\mu_1' = \frac{4}{3}, \mu_2' = 2, \mu_3' = \frac{16}{5}, \mu_4' = \frac{16}{3}$$

4. $M_x(t) = 1 + t + 2\frac{t^2}{2!} + 6\frac{t^3}{3!} + \ldots$

   $\mu_1' = 1, \mu_2' = 2, \mu_3' = 6, \mu_4' = 24$

   $\mu_1 = 0, \mu_2 = 1, \mu_3 = 2, \mu_4 = 9$

5. $E\left(X^k\right) = \frac{b^{k+1} - a^{k+1}}{(k+1)(b-a)}$

   $\mu_1' = \frac{b+a}{2}, \mu_2' = \frac{b^2 + ab + a^2}{3}, \mu_3' = \frac{(b+a)(b^2 + a^2)}{4}$

6. $\frac{3}{2}, 3, M_x(t) = \frac{1}{8}\left(1 + e^t\right)^3$

## CHAPTER THREE

# Probability Distributions

## 3.1 Introduction

In this chapter we shall study some discrete and continuous probability distributions, which are derived using the theory of probability for the outcomes of a conceptual experiment. We first discuss the following discrete distributions.

(i) Binomial distribution

(ii) Poisson distribution

(iii) Geometric distribution

(iv) Uniform distribution

We begin our study by defining Bernoulli's distribution.

A random experiment with only two possible outcomes, success or failure is called a binomial trial and a random variable $X$ which takes either zero or 1 as its value is called a Bernoulli's variable. The corresponding distribution is called the Bernoulli's distribution.

It was discovered in 1713 by James Bernoulli and is defined as follows.

Let $X$ be a Bernoulli random variable. If the probability of success is denoted by $p$, and the probability of failure is denoted by $q = 1 - p$, and the probability mass function is defined by $P(X = x) = p^x q^{1-x}$, $x = 0, 1$

$= 0$, otherwise

then the probability distribution of $X$ is called Bernoulli's distribution (Table 3.1). Thus we have

**Table 3.1** *Bernoulli's distribution*

| $X = x_i$ | 0 | 1 |
|---|---|---|
| $P(X = x_i)$ | $1 - p$ | $p$ |

The mean of Bernoulli's distribution is $\mu = E(X) = \sum_i x_i p_i$

$$= 0(1-p) + 1.p = p$$

The variance of Bernoulli's distribution is $\sigma^2 = \text{Var}(X) = \sum_i (x_i - \mu)^2 p_i$

$$= (0-p)^2(1-p) + (1-p)^2 p$$

$$= (1-p)\ (p^2 + p(1-p))$$

$$= (1-p)(p^2 + p - p^2)$$

$$= p(1-p)$$

$$= pq$$

Standard distribution is $\sigma = \sqrt{pq}$

If $n$ is the number of Bernoulli's trials, then Bernoulli's theorem states that the probability of $x$ successes is ${}^nC_x p^x q^{n-x}$.

## 3.2 Binomial Distribution

Binomial distribution is a discrete distribution. It is a commonly used probability distribution that has been developed to represent various discrete phenomenon which occur in business, social sciences, natural sciences and medical research.

Binomial distribution is widely used due to its relation with binomial distribution. The following should be satisfied for the application of binomial distribution.

1. The experiment consists of $n$ identical trials, where $n$ is finite.
2. There are only two possible outcomes in each trial i.e., each trial is a Bernoulli's trial. We denote one outcome by $S$ (for success) and the other by $F$ (for failure).
3. The probability of $S$ remains the same from trial to trail. The probability of $S$ (success) is denoted by $p$ and the probability of failure by $q$ (where $p+q=1$)
4. All the trials are independent.
5. The binomial random variable $x$ is the number of succeses in $n$ trials. If $X$ denotes the number of successes in $n$ trials under the conditions stated above, then $x$ is said to follow binomial distribution with parameters $n$ and $p$.

**Definition 3.1 (Binomial distribution)** A discrete random variable taking the values 0, 1, 2,. . ., $n$ is said to follow binomial distribution with parameters $n$ and $p$ if its probability mass function is given by

$$P(X=x)=P(x)^nC_xp^xq^{n-x}, \quad x=0,1,2,\ldots,n$$

$$0<p<1,\ q=1-p$$

$$=0, \quad \text{otherwise.}$$

If $x$ follows a binomial distribution with parameters $n$ and $p$ symbolically, we express $X \sim B(n,p)$.

**Remark** We have $\sum_{x=0}^{n} P(X=x)=\sum_{x=0}^{n} {}^nC_xp^xq^{n-x}$

$$=(q+p)^n=1$$

The probabilities are the terms in the binomial expansion of $(q+p)^n$ (i.e., $(p+q)^n$), hence the name binomial distribution given to it.

### 3.2.1 Expected frequencies and fitting of a binomial distribution

If we take large $N$ sets of $n$ (Bernoulli) trials each, then the expected or theoretical frequencies of getting $x$ successes is given by

$$N.P(X=x)=N^nC_xp^xq^{n-x}$$

$$=N.^nC_xq^xp^{n-x} \quad x=0,1,2,\ldots n$$

The theoretical frequencies of getting 0 success, 1 success, …, $n$ successes are respectively the 1st, 2nd,…, $(n+1)$th terms in the expansion of $N(p+q)^n$ (i.e., $N(q+p)^n$).

The expected or theoretical frequencies of the binomial distribution are as shown in Table 3.2.

**Table 3.2** *Theoretical or expected frequencies*

| Number of success | Expected frequency |
|---|---|
| 0 | $N.q^n$ |
| 1 | $N.^nC_1pq^{n-1}$ |
| 2 | $N.^nC_2p^2q^{n-2}$ |
| .... | |
| $x$ | $N.^nC_xp^xq^{n-x}$ |
| .... | |
| $n$ | $N.p^n$ |

### 3.2.2 Recurrence relation

Let $X$ be a binomial variable. Since $X$ follows a binomial distribution, we have

$$P(x) = P(X = x) = {}^nC_x p^x q^{n-x} \text{ for } x = 0, 1, 2, \ldots, n$$

Replacing $x$ by $x+1$ we get

$$P(x+1) = P(X = x+1) = {}^nC_{x+1} p^{x+1} q^{n-x-1} \quad \text{for } x = 0, 1, 2, \ldots\ldots, n-1$$

Dividing $P(x+1)$ by $P(x)$ we get

$$\frac{P(x+1)}{P(x)} = \frac{{}^nC_{x+1} q^{n-x-1} p^{x+1}}{{}^nC_x q^{n-x} p^x} = \frac{n!}{(x+1)(n-x-1)!} \cdot \frac{x!(n-x)!}{n!}$$

$$= \frac{(n-x)(n-x-1)!x!}{(x+1)x!(n-x-1)} \cdot \frac{p}{q}$$

$$\text{or } P(x+1) = \frac{(n-x)}{(x+1)} \cdot \frac{p}{q} . P(x) \text{ for } x = 0, 1, 2, \ldots, n-1$$

Hence the required recurrence relation is

$$P(x+1) = \frac{(n-x)}{(x+1)} \cdot \frac{p}{q} . P(x) \text{ for } x = 0, 1, 2, \ldots, n-1$$

Using the above recurrence relation, we can write

$$P(1) = n . \frac{p}{q} . P(0)$$

$$P(2) = \frac{n-1}{2} \cdot \frac{p}{q} . P(1)$$

$$P(3) = \frac{n-2}{3} \cdot \frac{p}{q} . P(2)$$

The recurrence relation is used to find the expected or theoretical frequencies i.e., for fitting the binomial distribution of a given data.

### 3.2.3 Moments, skewness and kurtosis of the binomial distribution

Taking an arbitrary origin at 0 successes, we get

$$\mu_1' = \sum_{x=0}^{n} xP(x)$$

$$= \sum_{x=0}^{n} x . {}^nC_x p^x q^{n-x}$$

$$= \sum_{x=0}^{n} x.\frac{n!}{x!(n-x)!}p^x q^{n-x}$$

$$= np\sum_{x=1}^{n} \frac{(n-1)!}{(x-1)!n-1-(x-1))!}p^{x-1}q^{n-x}$$

$$= np\sum_{x=1}^{n} {}^{n-1}C_{x-1}p^{x-1}q^{n-1-(x-1)}$$

$$= np(p+q)^{n-1}$$

$$= np.1 = np$$

$\therefore$ Mean $= \mu' = E(X) = np$

$$\mu_2' = E(X^2) = E(X(X-1)) + E(X)$$

$$= \sum_{x=0}^{n} x^{2n}C_x p^x q^{n-1}$$

$$= \sum_{x=0}^{n} (x + x(x-1))^n C_x p^x q^{n-x}$$

$$= \sum_{x=0}^{n} x^n C_x p^x q^{n-x} + \sum_{x=0}^{n} x(x-1)^n C_x p^x q^{n-x}$$

$$= np + n(n-1)p^2 \sum_{x=2}^{n} {}^{n-2}C_{x-2}p^{x-2}q^{n-x}$$

$$= np + (n^2p^2 - np^2)\sum_{x=2}^{n} {}^{n-2}C_{x-2}p^{x-2}q^{(n-2)-(x-2)}$$

$$= np + (n^2p^2 - np^2)(p+q)^{n-2}$$

$$= n^2p^2 - np^2 + np$$

Variance $= v[x] = \mu_2' - {\mu_1'}^2$

i.e., $\mu_2 = E(X^2) - [E(X)]^2$

$$= n^2p^2 - np^2 + np - (np)^2$$

$$= np - np^2$$

$$= np(1-p)$$

$= npq$

Again, we have $\mu_3' = \sum_{x=0}^{n} x^3 p(x)$

$$= \sum_{x=0}^{n} (x(x-1)(x-2) + 3x(x-1) + x)p + x$$

$$\text{(since } x^3 = x(x-1)(x-2) + 3x(x-1) + x)$$

$$= \sum_{x=0}^{n} x(x-1)(x-2) + 3\sum_{x=0}^{n} x(x-1) + p(x) + \sum_{x=0}^{n} xP(x)$$

$$= n(n-1)(n-2)p^3(q+p)^{n-3} + 3n(n-1)p^2(q+p)^{n-2} + np(q+p)^{n-1}$$

$$= n(n-1)(n-2)p^3 + 3n(n-1)p^2 + np$$

Since $\mu_3 = \mu_3' - 3\mu_2'\mu_1' + 2\mu_1'^3$, we get $\mu_3 = n(n-1)(n-2)p^3 + 3n(n-1)p^2 + np - 3(n^2p^2 - np^2 + np)(np) + 2n^3p^3$

$$= npq(q-p)$$

Similarly, we get $\mu_4' = n(n-1)(n-2)(n-3)p^4 + 6n(n-1)(n-2)p^3 + 7n(n-1)p^2 + np$

Since $\mu_4 = \mu_4' - 4\mu_3'\mu_1' + 6\mu_2\mu'^2 - 3\mu_1'^4$, we get $\mu_4 = npq(1 + 3(n-2)pq)$

Hence we get $\beta_1 = \dfrac{\mu_3^2}{\mu_2^3} = (npq)^2 \dfrac{(q-p)^2}{(npq)^3} = \dfrac{(1-2p)^2}{npq}$

$$r_1 = \frac{1-2p}{\sqrt{npq}}$$

$$\beta_2 = \frac{\mu_4}{\mu_2^2} = 3 + \frac{1-6pq}{npq} \text{ and } r_2 = \frac{1-6pq}{npq}$$

When $p < \frac{1}{2}$, the distribution is positively skewed and when $p > \frac{1}{2}$, the distribution is negatively skewed. When $p = \frac{1}{2}$, the distribution is symmetric.

Since coefficient of kurtosis is given by $r_2 = \beta_2 - 3 = \dfrac{1-6pq}{npq}$

If $6pq < 1$ i.e., $pq < \frac{1}{6}$, $r_2 > 0$, the distribution is leptokurtic.

If $6pq = 1$ i.e., $pq = \frac{1}{6}$, $r_2 = 0$, the distribution is mesokurtic.

If $6pq > 1$ i.e., $pq > \frac{1}{6}$, $r_2 < 0$, the distribution is platykurtic.

### 3.2.4 Moment generating function of a binomial distribution

From the definition, we have $P(X = x) = {}^nC_x p^x q^{n-x}$

Therefore, $M(t) = E(e^{tx})$

$$= \sum_{1}^{n} e^{tx}\, {}^nC_x p^x q^{n-x}$$

$$= \sum_{1}^{n} {}^nC_x (pe^t)^x q^{n-x}$$

$$= (q + pe^t)^n \qquad (1)$$

or $\quad M(t) = q + p\left(1 + t + \frac{t^2}{2!} + \frac{t^3}{3!} + \ldots\right)^n$

$$= \left(q + p + pt + \frac{pt^2}{2!} + \frac{pt^3}{3!} + \ldots\right)^n$$

$$= \left(1 + pt + \frac{pt^2}{2!} + \frac{pt^3}{3!} + \ldots\right)^n \qquad (2)$$

is the m.g.f for the binomial distribution.

### 3.2.5 Characteristics of a binomial distribution

Binomial distribution is a discrete probability distribution. It is applied when the number of repeated trials of any experiment is finite and fixed. Each trial is a Bernoulli trial. The trials are all independent. The probability of success of any trial is the same. It is denoted by $p$. If $n$ denotes the number of trials, $n$ and $p$ are called the parameters of the binomial distribution. The mean of a Binomial distribution is $np$ and the variance is $npq$. If $p = 0.5$, the binomial distribution is symmetric. If $p$ is very small and less than 0.5, then the Binomial distribution is skewed to the right. If $p > 0.5$, the distribution is skewed to the left.

## Solved examples

**Example 1** A die is thrown 4 times. Getting a number greater than 2 is a success. Find the probability of getting (i) exactly one success, (ii) less than 3 successes.

**Solution** We have $n = 4$

$P =$ probability of getting a number greater than $2 = \frac{4}{6} = \frac{2}{3}$

$\therefore \quad q = 1 - p = 1 - \frac{2}{3} = \frac{1}{3}$

i. Probability of getting one success $= P(X = 1)$

$$= {}^4C_1 \left(\frac{2}{3}\right)^1 \left(\frac{1}{3}\right)^{4-1}$$

$$= 4.\frac{2}{3}.\frac{1}{27} = \frac{8}{81}$$

ii. Probability of getting less than 3 successes $= P(X < 3) = P(X = 0) + P(X = 1) + P(X = 2)$

$$= {}^4C_0 \left(\frac{2}{3}\right)^0 \left(\frac{1}{3}\right)^4 + {}^4C_1 \left(\frac{2}{3}\right)^1 \left(\frac{1}{3}\right)^{4-1} + {}^4C_2 \left(\frac{2}{3}\right)^2 \left(\frac{1}{3}\right)^{4-2}$$

$$= \frac{1}{81} + 4.\frac{2}{3}.\frac{1}{27} + 6.\frac{4}{9}.\frac{1}{9}$$

$$= \frac{1+8+24}{81} = \frac{33}{81}$$

**Example 2** If the chance that any one of 5 telephone lines is busy at any instant is 0.01, what is the probability that all the lines are busy? What is the probability that more than 3 lines are busy?

**Solution** It is given that $n = 5,\ p = 0.01$

$q = 1 - p = 1 - 0.01 = 0.99$

Therefore, the probability that all the lines are busy $= P(X = 5)$

$$= {}^5C_5\,(0.01)^5\,(0.99)^0$$

$$= (0.01)^5$$

Probability that more than 3 lines are busy $= P(X > 3) = P(X = 4) + P(X = 5)$

$$= {}^5C_4\,(0.01)^4\,(0.99)^1 + {}^5C_5\,(0.01)^5\,(0.99)^0$$

$$= 5(0.01)^4(0.99) + (0.01)^5$$

$$= (0.01)^4(5 \times 0.99 + 0.01)$$

$$= (0.01)^4(4.95 + 0.01)$$

$$= (0.01)^4(4.96)$$

**Example 3** A box contains 100 tickets each bearing one of the numbers from 1 to 100. If 5 tickets are drawn successively with replacement from the box, find the probability that all the tickets bear a number divisible by 10.

**Solution** The numbers which are divisible by 10 are 10, 20, 30,....,100.

There are 10 numbers (between 1 and 100 (inclusive)) which are divisible by 10.

We have, $n = 10, p = \frac{10}{100} = \frac{1}{10}, q = 1 - p = 1 - \frac{1}{10} = \frac{9}{10}$

Probability that all the tickets drawn bear the number divisible by 10 $= P(X = 5)$

$$= {}^5C_5 \left(\frac{1}{10}\right)^5 \left(\frac{9}{10}\right)^0 = \frac{1}{10^5}$$

**Example 4** A die is thrown three times. Getting a "3" or a "6" is considered to be success. Find the probability of getting at least two successes.

**Solution** Probability of getting a 3 or 6 $= p = \frac{2}{6} = \frac{1}{3}$

Probability of getting at least two successes $= P(X \geq 2) = P(X = 2) + P(X = 3)$

$$= {}^3C_2 \left(\frac{1}{3}\right)^2 \left(\frac{2}{3}\right)^1 + {}^3C_3 \left(\frac{1}{3}\right)^3 \left(\frac{2}{3}\right)^0$$

$$= 3.\frac{1}{9}.\frac{2}{3} + \frac{1}{27} = \frac{7}{27}$$

**Example 5** If 20% of the bolts produced by a machine are defective, determine the probability that out of 4 bolts chosen at random (i) 1, (ii) 0, will be defective.

**Solution** Probability that the bolt is defective $= p = \frac{20}{100} = 0.2$

$$q = 1 - p = 1 - 0.2 = 0.8$$

We have $n = 4$

(i) $P(X = 1) = {}^4C_1 \,(0.2)^1 \,(0.8)^3$

$$= 4.(0.2)(0.8)^3 = 0.4096$$

(ii) $P(X = 0) = {}^4C_0 \,(0.2)^0 \,(0.8)^4 = 0.4096$

**Example 6** Out of 1000 families of 3 children each, how many families would you expect to have two boys and one girl, assuming that boys and girls are equally likely.

**Solution** Probability of having a boy $= p = \frac{1}{2}$

We have $q = 1 - p = 1 - \frac{1}{2} = \frac{1}{2}, n = 3, N = 1000$

Expected number of families having two boys and one girl $= 1000.{}^3C_2\left(\frac{1}{2}\right)^2\left(\frac{1}{2}\right)^1$

$= 1000.3.\frac{1}{4}.\frac{1}{2} = 375$

**Example 7** The average percentage of failures in a certain examination is 40. What is the probability that out of a group of 6 candidates, at least 4 pass in the examination?

**Solution** Let $p$ be the probability that a candidate passes the examination.

$\therefore \quad p = \frac{60}{100}$ (given)

We have $p = 0.6, q = 0.4, n = 6$.

$P(X \geq 4) =$ Probability that at least 4 candidates pass the examination

$$= P(X = 4) + P(X = 5) + P(X = 6)$$

$$= {}^6C_4(0.6)^4(0.4)^2 + {}^6C_5(0.6)^5(0.4) + {}^6C_6(0.6)^6(0.4)^0$$

$$= (15)(0.6)^4(0.4)^2 + (6)(0.6)^5(0.4) + (0.6)^6$$

$$= 0.5443$$

**Example 8** $X$ follows a binomial distribution such that $4P(x = 4) = P(x = 2)$. If $n = 6$, find $p$ the probability of success.

**Solution** We have $4P(x = 4) = P(x = 2)$ (given)

or $\quad 4.{}^6C_4.p^4q^2 = {}^6C_2.p^2q^4$

or $\quad 4p^2 = q^2$

or $\quad 4p^2 = (1 - p)^2 = 1 + p^2 - 2p$

or $\quad 3p^2 + 2p - 1 = 0$

or $\quad p=\dfrac{-2\pm\sqrt{4+12}}{(2)(3)}=\dfrac{-2\pm 4}{6}=\dfrac{-6}{6}$ or $\dfrac{2}{6}$

$$=-1 \text{ or } \frac{1}{3}$$

$p$ cannot be negative, therefore we consider $p=\dfrac{1}{3}$

Thus, $p=\dfrac{1}{3}$ (since $p=-1$ is inadmissible)

**Example 9** Find the maximum $n$ such that the probability of getting no head in tossing a coin $n$ times is greater than 0.1.

**Solution** Let $p$ denote the probability of getting a head. Then

$$p=\frac{1}{2},\ q=1-p=1-\frac{1}{2}=\frac{1}{2}$$

Let $n$ be the number of trials, such that $P(X=0)>0.1$

i.e., $\quad {}^{n}C_{0}p^{0}q^{n-0}>0.1$

or $\quad \left(\dfrac{1}{2}\right)^{n}>0.1$

When $n=1$, we get $\dfrac{1}{2}>0.5$

$n=2$, we get $\left(\dfrac{1}{2}\right)^{2}=0.25>0.1$

$n=3$, we get $\left(\dfrac{1}{2}\right)^{3}=0.125>0.1$

$n=4$, we get $\left(\dfrac{1}{2}\right)^{4}=0.0625<0.1$

and the maximum value such that $P(X=0)>0.1$ holds is 3. Hence the maximum $n$ such that $P(X=0)>0.1$ is $n=3$.

**Example 10** The probability of a man hitting a target is $\dfrac{1}{2}$. How many times must he fire so that the probability of hitting the target at least once is more than 90%?

**Solution** Let $p$ denote the probability of hitting the target and $n$ denote the number of times he fired.

We have $p = \frac{1}{2}$ (given)

$$q = 1 - p = 1 - \frac{1}{2} = \frac{1}{2}$$

For the probability of hitting the target at least once to be greater than 90%.

i.e., $P(X \geq 1) > \frac{90}{100}$

or $1 - P(X < 1) > \frac{9}{10}$ or $1 - P(X = 0) > \frac{9}{10}$

or $1 - {}^{n}C_{0}p^{0}q^{n-0} > \frac{9}{10}$

or $1 - \left(\frac{1}{2}\right)^{n} > \frac{9}{10}$

or $1 - \frac{9}{10} > \left(\frac{1}{2}\right)^{n}$

or $\frac{1}{10} > \frac{1}{2^{n}}$

i.e., $\frac{1}{2^{n}} < \frac{1}{10}$

or $2^{n} > 10$

for $n = 1,\ 2^{1} = 2 < 10$

for $n = 2,\ 2^{2} = 4 < 10$

for $n = 3,\ 2^{3} = 8 < 10$

for $n = 4,\ 2^{4} = 16 > 10$

Hence the least value of $n$ such that $P(X \geq 1) > \frac{9}{10}$ is 4.

**Example 11** If the sum of the mean and variance of a binomial distribution of 5 trials is $\frac{9}{5}$, find the binomial distribution.

**Solution** Let $n$ be the number of trials.

Let $p$ be the probability of success and $q$ be the probability of failure.

Then, we have $np+npq=\frac{9}{5}$ (given), $n=5$

or $5p+5p(1-p)=\frac{9}{5}$

or $25p+25p-25p^2=9$

or $25p^2-50p+9=0$

or $p=\frac{50\pm\sqrt{2500-900}}{(2)(25)}=\frac{50\pm 40}{50}=\frac{9}{5}$ or $\frac{1}{5}$

$p=\frac{9}{5}>1$ is inadmissible. Therefore, $p=\frac{1}{5}$.

The required binomial distribution is $P(X=x)={}^5C_x\left(\frac{1}{5}\right)^x\left(\frac{4}{5}\right)^{5-x}, 0\le x\le 5$

**Example 12** If the probability of a defective bolt is $\frac{1}{10}$, (i) find the mean, (ii) variance, (ii) moment of coefficient skewness, (iv) kurtosis for the distribution of defective bolts in a total of 400.

**Solution** $p=$ Probability of a defective bolt $=\frac{1}{10}$ (given)

We have $n=400$, $p=\frac{1}{10}$, $q=\frac{9}{10}$

Therefore,

(i) Mean $=E(X)=np=400.\frac{1}{10}=40$

(ii) Variance $=npq=400.\frac{1}{10}.\frac{9}{10}=36$

(iii) Moment of coefficient skewness $r_1=\frac{q-p}{\sqrt{npq}}=\frac{\frac{9}{10}-\frac{1}{10}}{\sqrt{36}}=\frac{8/10}{6}$

$$=\frac{8}{60}=\frac{2}{15}=0.1333$$

(iv) Coefficient of kurtosis $r_2=\frac{1-6pq}{npq}=\frac{1-6.\frac{1}{10}.\frac{9}{10}}{36}=\frac{1-0.54}{36}$

$$=\frac{0.46}{36}=0.01277$$

**Example 13** Using the recurrence formula of the binomial distribution, compute $P(x$ successes) for $x = 1, 2, 3, 4, 5$ given $n = 5$ and $p = \frac{1}{6}$.

**Solution** We have $p = \frac{1}{6}, q = 1 - p = 1 - \frac{1}{6} = \frac{5}{6}, n = 5$

Therefore,

$$P(X = x) = {}^nC_x p^x q^{n-x}, 0 \le x \le n$$

$$= {}^5C_x \left(\frac{1}{6}\right)^x \left(\frac{5}{6}\right)^{5-x} \quad , 0 \le r \le 5$$

The recurrence formula of binomial distribution is $P(x+1) = \frac{n-x}{x+1} \cdot \frac{p}{q} . P(x), 0 \le x \le n-1$

We have $P(X = 0) = P(0) = {}^5C_0 \left(\frac{1}{6}\right)^0 \left(\frac{5}{6}\right)^{5-0} = \frac{3125}{7776}$

Using the recurrence formula we get,

$$P(1) = \frac{5-0}{0+1} \cdot \frac{(1/6)}{(5/6)} P(0) = 5 . \frac{1}{5} P(0) = \frac{3125}{7776}$$

$$P(2) = \frac{5-1}{1+1} \cdot \frac{(1)}{(5)} P(1) = \frac{2}{5} \cdot \frac{3125}{7776} = \frac{1250}{7776}$$

$$P(3) = \frac{5-2}{2+1} \cdot \frac{(1)}{(5)} P(2) = \frac{3}{3} \cdot \frac{1}{5} \cdot \frac{1250}{7776} = \frac{250}{7776}$$

$$P(4) = \frac{5-3}{3+1} \cdot \frac{(1)}{(5)} P(3) = \frac{2}{4} \cdot \frac{1}{5} \cdot \frac{250}{7776} = \frac{25}{7776}$$

$$P(5) = \frac{5-4}{4+1} \cdot \frac{(1)}{(5)} P(4) = \frac{1}{5} \cdot \frac{1}{5} \cdot \frac{25}{7776} = \frac{1}{7776}$$

The required probabilities are $\frac{3125}{7776}, \frac{1250}{7776}, \frac{250}{7776}, \frac{25}{7776}$ and $\frac{1}{7776}$.

**Example 14** In 256 sets of twelve tosses of a coin, in how many cases may one expect eight heads and four tails?

**Solution** Probability of getting a head in a single toss $= p = \frac{1}{2}$

We have $n = 12$, $q = \frac{1}{2}$, $x =$ number of heads.

Probability of getting 8 heads and 4 tails $= P(X=8) = {}^{12}C_8 p^8 q^4$

$$= {}^{12}C_8 \left(\frac{1}{2}\right)^8 \left(\frac{1}{2}\right)^4$$

Expected value of getting 8 heads and 4 tails in 256 tosses is $N.P(X=8) = 256\ {}^{12}C_8 \left(\frac{1}{2}\right)^{12}$

$$= 31$$

**Example 15** Fit a binomial distribution and calculate the expected frequencies for the following data.

| $x$ | 0 | 1 | 2 | 3 | 4 | 5 |
|---|---|---|---|---|---|---|
| $f$ | 10 | 20 | 30 | 15 | 15 | 10 |

**Solution** Mean of the given distribution is

$$\mu = E(X) = \frac{(0)(10)+(1)(20)+(3)(15)+(4)(15)+(5)(10)}{10+20+30+15+15+10} = \frac{235}{100} = 2.35$$

We have $n = 5$.

Since $\mu = np = 2.35$, we get $5p = 2.35$

or $p = \frac{2.35}{5} = 0.47$

Hence $q = 1 - p = 1 - 0.47 = 0.53$

$$P(X=0) = {}^nC_0 p^0 q^{n-0}$$

$$= {}^5C_0 (0.47)^0 (0.53)^{5-0}$$

$$= (0.53)^5$$

$$= 0.0418$$

$$N = \sum_{i=1}^{5} f_i = 10+20+30+15+15+10 = 100$$

Therefore, expected frequencies are

$$N.P(X=0) = 100 \times (0.0418) = 4.18 \cong 4$$

$$N.P(X=1) = 100 \times {}^5C_1 (0.47)(0.53)^4$$

$$= 100 \times 5(0.47)(0.53)^4 = 18.53 \cong 19$$

$$N.P(X=2) = 100 \times {}^5C_2(0.47)^2(0.53)^3$$
$$= 32.87 \cong 33$$
$$N.P(X=3) = 100 \times {}^5C_3(0.47)^3(0.53)^2$$
$$= 29.15 \cong 29$$
$$N.P(X=4) = 100 \times {}^5C_4(0.47)^4(0.53)$$
$$= 12.93 \cong 13$$
$$N.P(X=5) = 100 \times {}^5C_5(0.47)^5(0.53)$$
$$= 2.29 \cong 2$$

The expected frequencies are 4, 19, 33, 29, 13, 2.

The required binomial distribution is

| $x$ | 0 | 1 | 2 | 3 | 4 | 5 |
|---|---|---|---|---|---|---|
| $N.P(X=x)$ | 4 | 19 | 33 | 29 | 13 | 2 |

That is, the required binomial distribution of the fit is $N(p+q)^n = 100(0.53+0.47)^5$.

**Example 16** Five dice were thrown 96 times and the number of times an odd number actually turned one in the experiment is given below. Calculate the expected frequencies.

| Number of dice showing 1 or 3 or 5 | 0 | 1 | 2 | 3 | 4 | 5 |
|---|---|---|---|---|---|---|
| Observed frequency | 1 | 10 | 24 | 35 | 18 | 8 |

**Solution** Probability of getting an odd number when a dice is thrown $= \frac{3}{6} = \frac{1}{2}$

$$\therefore q = \frac{1}{2}$$

We have $N = 1+10+24+35+18+8 = 96$

$$n = 5$$

The expected (theoretical) frequencies are

$$N.P(X=0) = 96 \times {}^5C_0\left(\frac{1}{2}\right)^0\left(\frac{1}{2}\right)^{5-0} = 3$$

$$N.P(X=1) = 96 \times {}^5C_1\left(\frac{1}{2}\right)^1\left(\frac{1}{2}\right)^{5-1} = 15$$

$$N.P(X=2) = 96 \times {}^5C_2 \left(\frac{1}{2}\right)^2 \left(\frac{1}{2}\right)^{5-2} = 30$$

$$N.P(X=3) = 96 \times {}^5C_3 \left(\frac{1}{2}\right)^3 \left(\frac{1}{2}\right)^{5-3} = 30$$

$$N.P(X=4) = 96 \times {}^5C_4 \left(\frac{1}{2}\right)^4 \left(\frac{1}{2}\right)^{5-4} = 15$$

$$N.P(X=5) = 96 \times {}^5C_5 \left(\frac{1}{2}\right)^5 \left(\frac{1}{2}\right)^{0} = 3$$

The binomial distribution of fit is $N(q+p)^n = 96\left(\frac{1}{2}+\frac{1}{2}\right)^5$

**Example 17** Show that for the binomial distribution $\mu_{r+1} = pq\left(n\mu_{r-1} + \frac{d\mu_r}{dp}\right)$, where $\mu_r$ is the $r$th moment about the mean. Hence obtain $\mu_2$, $\mu_3$ and $\mu_4$.

**Solution** We know that the mean of the binomial distribution is $\mu = np$

Therefore, $\mu_r = \sum_{x=0}^{n} (x-\mu)^r P(x)$

$$= \sum_{x=0}^{n} (x-np)^r P(x)$$

$$= \sum_{x=0}^{n} (x-np)^r . {}^nC_x P^x q^{n-x}$$

Differentiating with respect to $p$ we get

$$\frac{d\mu_r}{dp} = \sum_{x=0}^{n} -rn(x-np)^{r-1} . {}^nC_x P^x q^{n-x} + \sum_{x=0}^{n} \left[(x-np)^r \left[{}^nC_x p^{x-1} q^{n-x} - {}^nC_x p^x (n-x) q^{n-x-1}\right]\right]$$

$$= -nr \sum_{x=0}^{n} (x-np)^{r-1} . {}^nC_x p^x q^{n-x} + \frac{1}{pq}\left[\sum_{x=1}^{n} {}^nC_x p^x q^{n-x} (x-np)^r (xq - np + xp)\right]$$

$$= -nr\mu_{r-1} + \frac{1}{pq}\left[\sum_{x=0}^{n} {}^nC_x p^x q^{n-x} (x-np)^{r+1}\right]$$

or $$\frac{d\mu_r}{dp} + nr\mu_{r-1} = \frac{1}{pq}\left[\sum_{x=0}^{n} {}^nC_x p^x q^{n-x} (x-np)^{r+1}\right]$$

or $\quad pq\dfrac{d\mu_r}{dp} + nr\mu_{r-1} = \mu_{r+1} \qquad (1)$

Hence proved.

Putting $r = 1, 2, 3$ successively, we get

$$\mu_2 = pq\left(\frac{d\mu_1}{dp} + n\mu_0\right) = pq(0+n) = npq$$

$$\mu_3 = pq\left(\frac{d\mu_2}{dp} + n\mu_1\right) = p(nq - np + 0) = npq(q - p)$$

and $\mu_4 = pq\left(\dfrac{d\mu_3}{dp} + 3\mu_2\right) = pq(n(6p^2 - 6p + 1) + 3n(np)(1 - p)$

$$= npq(1 - 6p(1 - p) + 3npq)$$

$$= npq(1 - 6pq) + 3n^2p^2q^2$$

## Exercise 3.1

1. Eight coins are tossed simultaneously, find the probability of getting at least 6 heads.
2. Assuming that it is true that 2 out of 10 industrial accidents are due to fatigue, find the probability that exactly 2 of 8 industrial accidents will be due to fatigue.
3. The average percentage of failures in a certain examination is 40. What is the probability that out of a group of 6 candidates, at least 4 pass in the examination.
4. $X$ is a binomial random variable. If $4P(X = 4) = P(X = 2)$, find $p$.
5. In a binomial distribution, the mean and standard deviation are 12 and 2 respectively, find $n$ and $p$.
6. An oil exploration firm finds that 55 of the wells it drills yield a deposit of natural gas. If it drills 6 wells, find the probability that at least one well will yield gas.
7. A fair coin is tossed ten times. Find the probability of getting at least seven heads.
8. Ten coins are thrown simultaneously. Find the probability of getting at least seven heads.
9. In four throws with a pair of dice, what is the probability of throwing a doublet at least twice?
10. In a throw of 4 dice, find the probability that at least one die shows up 4.

11. The probability of a man hitting a target is $\frac{1}{4}$. He fires 7 times. What is the probability of his hitting at least twice the target?

12. The items produced by a firm are supposed to be 5% defective. What is the probability that 8 items will contain less than 2 defective items?

13. Assuming that half the population is vegetarian so that the chance of an individual being vegetarian is $\frac{1}{2}$ and assuming that 100 investigators can take samples of 10 individuals to see whether they are vegetarian, how many investigators would you expect to report that three people or less were vegetarian?

14. A box contains 100 tickets each bearing one of the numbers from 1 to 100. If 5 tickets are drawn successively with replacement from the box, find the probability that all the tickets bear a number divisible by 10.

15. If the probability of success is $\frac{1}{20}$, how many trials are necessary in order that the probability of at least one success is just greater than $\frac{1}{2}$?

16. If the probability that a man aged 60 will live to be 70 is 0.65, what is the probability that out of 10 men now 60, at least 7 will live to be 70?

17. The mean and variance of a binomial variable $X$ with parameters $n$ and $p$ are 16 and 8. Find $P(X \geq 1)$ and $P(X > 2)$.

18. In 256 sets of 12 tosses of a coin, how many sets can one expect to have 8 heads and 4 tails?

19. If 3 of 20 tyres are defective and 4 of them are randomly chosen for inspection, what is the probability that only one of the defective tyre will be included?

20. The mean and variance of a binomial distribution are 4 and $\frac{4}{3}$ respectively. Find $P(X \geq 1)$

21. Assume that 50% of all engineering students are good in mathematics, determine the probability that among 18 engineering students (i) exactly 10, (ii) at least 10, (iii) at most 8, (iv) at least 2 and at most 9, are good in mathematics.

22. Six dice are thrown 729 times. How many times would you expect at least three dice to show a 5 or 6?

23. Out of 800 families with 5 children each, how many would you expect to have (i) either 2 or 3 boys? Assume equal probabilities for boys and girls.

24. The mean of a binomial distribution is 3 and variance is $\frac{9}{4}$. Find (i) the value of $n$, (ii) $P(X \geq 7)$, (iii) $P(1 \leq X \leq 6)$.

25. A pair of dice is thrown. Find the probability of getting a sum of 11 (i) once, (ii) twice.

26. Fit a binomial distribution for the data.

| $X$ | 0 | 1 | 2 | 3 | 4 | 5 |
|---|---|---|---|---|---|---|
| $f(X)$ | 38 | 144 | 342 | 287 | 164 | 25 |

27. Seven coins are tossed at a time 128 times. The number of heads observed at each throw is recorded and the results are given below.

| No. of heads | 0 | 1 | 2 | 3 | 4 | 5 | 6 | 7 |
|---|---|---|---|---|---|---|---|---|
| Frequency | 7 | 6 | 19 | 35 | 30 | 23 | 7 | 1 |

Fit a binomial distribution to the data assuming that the coins are (i) biased, (ii) unbiased.

28. Fit a binomial distribution to the following frequency distribution

| $x$ | 0 | 1 | 2 | 3 | 4 | 5 | 6 |
|---|---|---|---|---|---|---|---|
| $f$ | 13 | 25 | 52 | 38 | 32 | 16 | 4 |

29. A newly generated virus attacking a computer system will corrupt 4 files out of 20 files which are opened in an hour. If 20 files are opened in an hour, find the probability that

    (i) at least 10 files are corrupted.

    (ii) exactly 3 files are corrupted.

    (iii) all the files are corrupted.

    (iv) all the files are safe.

30. Fit a binomial distribution for the following data

| $x$ | 0 | 1 | 2 | 3 | 4 | 5 |
|---|---|---|---|---|---|---|
| $f(x)$ | 2 | 14 | 20 | 34 | 22 | 8 |

31. Fit a binomial distribution whose mean is 9 and standard deviation is $\frac{3}{2}$.

32. The sum of mean and variance of a binomial distribution is 15 and the sum of their squares is 117. Find the distribution.

33. If on an average 1 vessel in every 10 is wrecked. Find the probability that out of 5 vessels expected to arrive, at least 4 will arrive safely.

34. The probability that a bulb produced by a factory will fuse after 160 days of use is 0.06. Find the probability that out of 5 such bulbs, at most one bulb will fuse after 160 days in use.

35. Find the expectation of the number of heads in 15 tosses of a coin.

36. The incidence of an occupational disease in an industry is such that the workmen have a 20% chance of suffering from it. What is the probability that out of 6 workmen, 4 or more will contact the disease?

37. The probability of a defective bolt is 0.2. Find (i) mean, (ii) standard deviation of the distribution, (iii) coefficient of skewness $\beta_1$ and, (iv) coefficient of kurtosis $\beta_2$.

38. Show that the mean of a binomial distribution is always greater than its variance.

39. Determine the binomial distribution whose mean is 5 and standard deviation is $\sqrt{2.5}$.

40. The mean of a binomial distribution is 20 and standard deviation is 4. Find out $n$,$p$ and $q$.

## Answers 3.1

(1) $\frac{37}{256}$ (2) 0.29 (3) 0.5443 (4) $\frac{1}{3}$

(5) $n = 15, p = \frac{2}{3}$ (6) $1-(0.95)^6$ (7) 0.1719 (8) 0.1719

(9) $\frac{171}{1296}$ (10) $\frac{671}{1296}$ (11) $\frac{4547}{8192}$ (12) $\left(\frac{19}{20}\right)^7\left(\frac{27}{20}\right)$

(13) 17 (14) $\frac{1}{10^5}$ (15) 14 (16) 0.5139

(17) $1-\frac{1}{2^{32}}$ and 0.999 (18) 31 (19) 0.3685 (20) 0.998

(21) (i) 0.617

(ii) $\left(\frac{1}{2}\right)^{18}\left({}^{18}C_{10}+{}^{18}C_{11}+\ldots+{}^{18}C_{18}\right)$

(iii) $\left(\frac{1}{2}\right)^{18}\left({}^{18}C_{0}+{}^{18}C_{1}+\ldots+{}^{18}C_{8}\right)$

(iv) $\left(\frac{1}{2}\right)^{18}\left({}^{18}C_{2}+{}^{18}C_{3}+\ldots+{}^{18}C_{9}\right)$

(22) 233 (23) (i) 250 (ii) 25 (iii) 500

(24) (i) $n$ =12, (ii) 0.1446, (iii) 0.82 (25) (i) 0.105, (ii) 0.003

(26) 33, 162, 316, 309, 151, 29

(27) (i) $128(0.517+0.483)^7$ (ii) 1, 7, 21, 35, 35, 21, 7, 1

(28) 6, 28, 56, 60, 36, 12, 2

(29) (i) 0.0000045 (ii) 0.001073 (iii) $\frac{1}{512}$ (iv) 0.9999

(30) 2, 10, 26, 34, 22, 6 (31) ${}^{12}C_x\left(\frac{3}{4}\right)^x\left(\frac{1}{4}\right)^{12-x}$ $(0 \le x \le 12)$

(32) ${}^{27}C_x\left(\frac{1}{3}\right)^x\left(\frac{2}{3}\right)^{27-x}$ $(0 \le x \le 27)$ (33) 0.9185

(34) 0.9681 (35) 7.5 (36) 0.016

(37) (i) 200, (ii) 12.6, (iii) $\beta_1 = 0.225$, (iv) $\beta_2 = 3.0025$

(39) ${}^{10}C_x\left(\frac{1}{2}\right)^x\left(\frac{1}{2}\right)^{10-x}$ $(0 \le x \le 10)$ (40) $p = 0.2, q = 0.8, n = 100$

## 3.3 Poisson Distribution

### 3.3.1 Introduction

A type of probability distribution useful in describing the number of events that will occur in a specific period of time or in a specific area or volume is the Poisson distribution. It is a probability distribution discovered by French mathematician Simon Denis Poisson (1781–1840) in the year 1837. The following are examples of random variables for which the Poisson probability distribution provides a good model.

1. The number of car accidents in a year, on a road.
2. The number of earthquakes in a year.
3. The number of breakdowns of an electronic computer.
4. The number of printing mistakes at each page of a book
5. The number of defective screws per box of 10 screws.

### 3.3.2 Conditions under which Poisson distribution is used

The conditions for the applicability of the Poisson distribution are the same as those for the applicability of the Binomial distribution. The additional requirement is that the probability of success is very small. The conditions are:

1. The random variable $X$ is discrete.
2. The number of trials is indefinitely very large i.e., $n \to \infty$.
3. The probability of success in a trail is very small i.e., $p$ is close to zero.
4. The product of $n$ and $p$ is constant.

A random variable which counts the number of successes in an experiment is called a Poisson variable.

### 3.3.3 Poisson probability function

In this section, we show that the Poisson distribution is the limiting form of a binomial distribution and define Poisson distribution.

Let an experiment satisfying the conditions for Poisson distribution be formed. Let $n$ denote the number of trials, and $p$ denote the probability of success. We assume that $n$ is indefinitely very large and $p$ is very small.

Using the definition of binomial distribution, we have,

$P(X = x) =$ Probability of $x$ successes

$$= {}^nC_x p^x q^{n-x} \quad , 0 \le x \le n \text{ , and } q = 1 - p$$

Let $np = \lambda$, then $p = \frac{\lambda}{n}$ and $q = 1 - p = 1 - \frac{\lambda}{n}$

Therefore, we get

$$P(X = x) = {}^nC_x p^x q^{n-x}$$

$$= \frac{n!}{x!(n-x)!}\left(\frac{\lambda}{n}\right)^x\left(1-\frac{\lambda}{n}\right)^{n-x}$$

$$= \frac{n(n-1)\ldots(n-x+1)(n-x)!}{x!(n-x)!}\left(\frac{\lambda}{n}\right)^x\left(1-\frac{\lambda}{n}\right)^{n-x}$$

$$= \frac{\left(\frac{n}{n}\right)\left(\frac{n-1}{n}\right)\ldots\left(\frac{n-(x-1)}{n}\right)}{x!}(\lambda)^x\left(1-\frac{\lambda}{n}\right)^n\left(1-\frac{\lambda}{n}\right)^{-x}$$

$$= \frac{\lambda^x}{x!}\left(1-\frac{1}{n}\right)\left(1-\frac{2}{n}\right)\ldots\left(1-\frac{x-1}{n}\right)\left(1-\frac{\lambda}{n}\right)^n\left(1-\frac{\lambda}{n}\right)^{-x}$$

$$\therefore \operatorname*{lt}_{n\to\infty} P(X = x) = \operatorname*{lt}_{n\to\infty} \left[\frac{\lambda^x}{x!}\left(1-\frac{1}{n}\right)\left(1-\frac{2}{n}\right)\ldots\left(1-\frac{x-1}{n}\right)\left[\left(1-\frac{\lambda}{n}\right)^{\frac{-n}{\lambda}}\right]^{-\lambda}\left(1-\frac{\lambda}{n}\right)^{-x}\right.$$

$$= \operatorname*{lt}_{n\to\infty}\frac{\lambda^x}{x!}\operatorname*{lt}_{n\to\infty}\left(1-\frac{1}{n}\right)\left(1-\frac{2}{n}\right)\ldots\left(1-\frac{x-1}{n}\right)\operatorname*{lt}_{\frac{-n}{\lambda}\to\infty}\left(\left(1-\frac{\lambda}{n}\right)^{\frac{-n}{\lambda}}\right)^{-\lambda}\operatorname*{lt}_{n\to\infty}\left(1-\frac{\lambda}{n}\right)^{-x}$$

$$= \frac{\lambda^x}{x!}(1-0)(1-0)\ldots\ldots..(1-0).e^{-\lambda}.(1-0)$$

$$= \frac{\lambda^x e^{-\lambda}}{x!}$$

Therefore, when $n$ is indefinitely very large

$$P(X = x) = \frac{\lambda^x e^{-\lambda}}{x!}, x = 0, 1, 2, 3, \ldots$$

which is called the Poisson probability function. The corresponding distribution is called the Poisson distribution. $\lambda$ is called the parameter of the distribution.

**Definition 3.2 (Poisson distribution)** A discrete random variable $X$ taking the values 0, 1, 2,... is said to follow Poisson distribution with parameter $\lambda$, if its probability mass function is given by

$$P(X=x)=\frac{\lambda^x e^{-\lambda}}{x!},\ x=0,1,2,3,\ldots$$
$$=0, \quad \text{otherwise}$$

In the above definition $P(X=x)$ is called the Poisson probability function and the corresponding Poisson distribution is

| $X$ | 0 | 1 | 2 | ... | $x$ | ... |
|---|---|---|---|---|---|---|
| $P(X=x)$ | $\frac{\lambda^0 e^{-\lambda}}{0!}$ | $\frac{\lambda^1 e^{-\lambda}}{1!}$ | $\frac{\lambda^2 e^{-\lambda}}{2!}$ | ... | $\frac{\lambda^x e^{-\lambda}}{x!}$ | ... |

### 3.3.4 Poisson frequency distribution

If an experiment satisfying the requirements of Poisson distribution is repeated $N$ times, the expected frequency distribution of getting $x$ successes is given by

$$N.P(X=x)=N.\frac{\lambda^x e^{-\lambda}}{x!},\quad x=0,1,2,3,\ldots..$$

**Remarks**

1. Since $e^{-\lambda}>0$, $P(X=0)\geq 0$ for all $x=0,1,2,3,\ldots$

2. $\sum\limits_{x=0} P(X=x)=\sum\limits_{x=0}^{\infty}\frac{e^{-\lambda}\lambda^x}{x!}$

$$=e^{-\lambda}\sum_{x=0}^{\infty}\frac{\lambda^x}{x!}$$

$$=e^{-\lambda}\left[1+\frac{\lambda}{1!}+\frac{\lambda^2}{2!}+\ldots\right]$$

$$=e^{-\lambda}e^{\lambda}=e^0=1$$

From (1) and (2), $P(X=x)=\frac{\lambda^x e^{-\lambda}}{x!}$, $x=0,1,2,3,\ldots$ is a probability mass function.

### 3.3.5 Moments of a Poisson distribution

Taking an arbitrary origin at 0 successes, we have

$$\mu_1^1 = \sum_{x=0}^{\infty} x.\frac{e^{-\lambda}\lambda^x}{x!} = \sum_{x=1}^{\infty} \frac{e^{-\lambda}\lambda^x}{(x-1)!}$$

$$= e^{-\lambda}\left[\lambda + \frac{\lambda^2}{1!} + \frac{\lambda^3}{2!} + \ldots\right]$$

$$= \lambda e^{-\lambda}\left[1 + \frac{\lambda}{1!} + \frac{\lambda^2}{2!} + \ldots\right]$$

i.e., $E(X) = \lambda e^{-\lambda} e^{\lambda} = \lambda$

Thus, mean $= \mu_1' = E(X) = \lambda$

$$\mu_2' = \sum_{x=0}^{\infty} x^2.\frac{e^{-\lambda}\lambda^x}{x!} = \sum_{x=0}^{\infty} [x(x-1)+x]\frac{e^{-\lambda}\lambda^x}{x!}$$

$$= \sum_{x=0}^{\infty} [x(x-1)]\frac{e^{-\lambda}\lambda^x}{x!} + \sum_{x=0}^{\infty} [x]\frac{e^{-\lambda}\lambda^x}{x!}$$

$$= e^{-\lambda}\lambda^2 \sum_{x=2}^{\infty} \frac{\lambda^{x-2}}{(x-2)!} + E(X)$$

$$= e^{-\lambda}\lambda^2 e^{\lambda} + \lambda = \lambda^2 + \lambda$$

Hence, $\mu_2 = \text{Var}[x] = \mu_2^1 - \mu'^2 = \lambda^2 + \lambda - \lambda^2$

$$= \lambda$$

$$\mu_3' = \sum x^3 \frac{e^{-\lambda}\lambda^x}{x!}$$

$$= e^{-\lambda}\sum \lambda^x \frac{[x(x-1)(x-2) + 3x(x-1) + x]}{x!}$$

$$= e^{-\lambda}\lambda^3 \sum_{x=3}^{\infty} \frac{\lambda^{x-3}}{(x-3)!} + e^{-\lambda}3\lambda^2 \sum_{x=2}^{\infty} \frac{\lambda^{x-2}}{(x-2)!} + e^{-\lambda}\lambda \sum_{x=1}^{\infty} \frac{\lambda^{x-1}}{(x-1)!}$$

$$= e^{-\lambda}\lambda^3 e^{\lambda} + e^{-\lambda}(3\lambda^2)e^{\lambda} + e^{-\lambda}\lambda e^{\lambda}$$

$$= \lambda^3 + 3\lambda^2 + \lambda$$

$$\mu_4' = \sum x^4 \frac{e^{-\lambda}\lambda^x}{x!}$$

$$= e^{-\lambda}\sum \frac{\lambda^x}{x!}[x(x-1)(x-2)(x-3) + 6x(x-1)(x-2) + 7x(x-1) + x]$$

$$= e^{-\lambda}\lambda^4 \sum_{x=4}^{\infty} \frac{\lambda^{x-4}}{(x-4)!} + e^{-\lambda}6\lambda^3 \sum_{x=3}^{\infty} \frac{\lambda^{x-3}}{(x-3)!} + e^{-\lambda}7\lambda^2 \sum_{x=2}^{\infty} \frac{\lambda^{x-2}}{(x-2)!} + e^{-\lambda}\lambda \sum_{x=1}^{\infty} \frac{\lambda^{n-1}}{(x-1)!}$$

$$= e^{-\lambda}\lambda^4 e^{\lambda} + e^{-\lambda}(6\lambda^3)e^{\lambda} + e^{-\lambda}(7\lambda^2)e^{\lambda} + e^{-\lambda}\lambda e^{\lambda}$$

$$= \lambda^4 + 6\lambda^3 + 7\lambda^2 + \lambda$$

Using the values of $\mu'_1, \mu'_2, \mu'_3, \mu'_4$ we get,

$$\mu_3 = \mu'_3 - 3\mu'_2\mu'_1 + 2{\mu'_1}^3$$

$$= \lambda^3 + 3\lambda^2 + \lambda - 3(\lambda^2 + \lambda)\lambda + 2\lambda^3$$

$$= \lambda$$

and $\mu_4 = \mu'_4 - 4\mu'_3\mu'_1 + 6\mu'_2{\mu'_1}^2 - 3{\mu'_1}^4$

$$= \lambda^4 + 6\lambda^3 + 7\lambda^2 + \lambda - 4(\lambda^3 + 3\lambda^2 + \lambda)\lambda + 6(\lambda^2 + \lambda)\lambda^2 - 3\lambda^4$$

$$= 3\lambda^2 + \lambda$$

We have $\beta_1 = \dfrac{\mu_3^2}{\mu_2^3} = \dfrac{\lambda^2}{\lambda^3} = \dfrac{1}{\lambda}$, $\gamma_1 = \sqrt{\beta_2} = \dfrac{1}{\sqrt{\lambda}}$

and $\beta_2 = \dfrac{\mu_4{}^2}{\mu_2} = \dfrac{3\lambda^2 + \lambda}{\lambda^2} = 3 + \dfrac{1}{\lambda}$

$$\gamma_2 = \beta_2 - 3 = \frac{1}{\lambda}$$

The values of $\mu_1$, $\mu_2$, $\mu_3$, $\mu_4$ and $\beta_1$, $\beta_2$, $\gamma_1$, $\gamma_2$ are also called the constants of Poisson distribution. We observe that the variance and mean of Poisson distribution are equal in magnitude.

### 3.3.6 Recurrence relation

If $X$ is Poisson variate with $\lambda$ as the parameter, then we have

$$P(X = x) = \frac{\lambda^x e^{-\lambda}}{x!}, \; x = 0, 1, 2, 3, \ldots$$

and $P(X = x+1) = \dfrac{\lambda^{x+1}e^{-\lambda}}{(x+1)!} = \dfrac{e^x e^{-\lambda}.e}{(x+1)x!}$

Dividing, we get $\dfrac{P(X = x+1)}{P(X = x)} = \dfrac{\lambda}{x+1}$

Therefore, we obtain $P(X = x+1) = \dfrac{\lambda}{x+1}P(X = x)$ or simply $P(x+1) = \dfrac{\lambda}{x+1}P(x)$

which is called the relation between $P(x)$ and $P(x+1)$. It is also known as the recurrence relation between probabilities of Poisson distribution.

### 3.3.7 Characteristics of a Poisson distribution

1. A Poisson distribution is a discrete distribution.
2. If $n$, the number of trials, is large and $p$ is small, the distribution gives a close approximation to the binomial distribution.
3. The distribution depends mainly on the value of the mean (i.e., $\lambda$).
4. A Poisson distribution has only one parameter viz. $\lambda$.
5. The experiment consists of counting the number of times a particular event occurs during a given unit of time or given area or volume.
6. The probability that an event occurs in a given unit of time, area or volume is the same for all units.

### 3.3.8 Moment generating function of the Poisson distribution

The moment generating function of a random variable $X$ is $M(t)=\sum_{x=0}^{\infty} e^{tx}P(X=x)$

If $X$ follows a Poisson distribution, then

$$M(t)=\sum_{x=0}^{\infty} e^{tx}\frac{\lambda^x e^{-\lambda}}{x!}$$

$$=\sum_{x=0}^{\infty}\frac{e^{-\lambda}(\lambda e^t)^x}{x!}$$

$$=e^{-\lambda}\sum_{x=0}^{\infty}\frac{(\lambda e^t)^x}{x!}$$

$$=e^{-\lambda}\left[1+\lambda e^t+\frac{\lambda^2(e^t)^2}{2!}+\ldots\right]$$

$$=e^{-\lambda}e^{\lambda e^t}$$

$$=e^{\lambda(e^t-1)}$$

Hence, the moment generating function of the Poisson distribution is $M(t)=e^{\lambda(e^t-1)}$.

### 3.3.9 Reproductive property of the Poisson distribution

**Theorem 3.1** If two independent random variables $X$ and $Y$ have Poisson distributions with means $\lambda_1$ and $\lambda_2$ respectively, their sum $X+Y$ is a Poisson variable with mean $\lambda_1+\lambda_2$.

*Proof* Let $M_X(t)$ and $M_Y(t)$ be the moment generating functions of $X$ and $Y$ and let $M(t)$ be the moment generating function of $X+Y$, then

$$M_X(t) = e^{\lambda_1(e^t-1)} \text{ and } M_Y(t) = e^{\lambda_2(e^t-1)}$$

The moment generating function of the sum $X+Y$ is given by

$$M(t) = M_X(t)M_Y(t) = e^{\lambda_1(e^t-1)}\, e^{\lambda_2(e^t-1)}$$

$$= e^{(\lambda_1+\lambda_2)(e^t-1)} \tag{1}$$

Equation (1) is the moment generating function of a Poisson distribution with mean $\lambda_1+\lambda_2$. Hence, the theorem.

**Solved examples**

**Example 1** There are 50 telephone lines in an exchange. The probability of them being busy is 0.1. What is the probability that all the lines are busy?

**Solution** Let $X$ be the Poisson variable "number of busy lines in the exchange".

By Poisson distribution,

$$P(X=x) = \frac{\lambda^x e^{-\lambda}}{x!},\ x = 0, 1, 2, 3, \ldots$$

We have $n = 50$, $p = 0.1$

$\lambda = np = (50)(0.1) = 5$

Hence, $P$ (all lines are busy) $= P(X = 50) = \dfrac{5^{50}e^{-5}}{50!}$

**Example 2** The probability that a bomb dropped from an envelope will strike a certain target is $\frac{1}{5}$. If 6 bombs are dropped, find the probability that

(i) exactly 2 will strike the target.

(ii) at least 2 will strike the target.

Given $e^{-1.2} = 0.3012$

**Solution** We have $n = 6$, $p = \frac{1}{5}$

$$\lambda = np = (6)(\frac{1}{5}) = 1.2$$

By Poisson distribution $P(X = x) = \frac{\lambda^x e^{-\lambda}}{x!}, x = 0, 1, 2, 3, \ldots$

(i) $P(X = 2) = P(\text{exactly 2 will strike the target})$

$$= \frac{(1.2)^2 e^{-1.2}}{2!} = \frac{0.3012 \times 1.414}{2}$$

$$= 0.2169$$

(ii) $P(\text{at least 2 will strike the target}) = P(X \geq 2) = 1 - P(X < 2)$

$$= 1 - [P(X = 0) + P(X = 1)$$

$$= 1 - \frac{(1.2)^0 e^{-1.2}}{0!} - \frac{(1.2)^1 e^{-1.2}}{1!}$$

$$= 1 - e^{-1.2}(1 + 1.2)$$

$$= 1 - (0.3012)(2.2) = 0.3374$$

**Example 3** If the probability that an individual suffers a bad reaction from an injection of a given serum is 0.001, determine the probability that out of 2000 individuals, exactly 3 individuals suffer a bad reaction.

**Solution** We have $n = 2000$, $p = 0.001$ (given)

$$\lambda = np = (2000)(0.001) = 2$$

By Poisson distribution,

$$P(X = x) = \frac{\lambda^x e^{-\lambda}}{x!}, \; x = 0, 1, 2, 3, \ldots$$

$P(X = 3) =$ Probability that 3 individuals suffer bad reaction $= \frac{2^3 e^{-2}}{3!} = 0.1805$

**Example 4** For a Poisson distribution, show that $\mu_{r+1} = r\lambda\mu_{r-1} + \lambda\frac{d\mu_r}{d\lambda}$, where $\lambda$ is the mean.

**Solution** The $r$th moment of the distribution is

$$\mu_r = \sum_{x=0}^{\infty} (x-\lambda)^r P(X=x)$$

$$= \sum_{x=0}^{\infty} (x-\lambda)^r \frac{\lambda^x e^{-\lambda}}{x!}$$

Differentiating with respect to $\lambda$, we get

$$\frac{d\mu_r}{d\lambda} = -r\sum_{x=0}^{\infty}(x-\lambda)^{r-1}\frac{e^{-\lambda}\lambda^x}{x!} + \sum_{x=0}^{\infty}(x-\lambda)^r\left[\frac{x\lambda^{x-1}e^{-\lambda} - \lambda^x e^{-\lambda}}{x!}\right]$$

$$= -r\mu_{r-1} + \sum_{x=0}^{\infty}(x-\lambda)^r\left[\frac{\lambda^{x-1}e^{-\lambda}(x-\lambda)}{x!}\right]$$

$$\therefore \lambda\frac{d\mu_r}{d\lambda} = -r\lambda\mu_{r-1} + \sum_{x=0}^{\infty}\left[\frac{\lambda^x e^{-\lambda}(x-\lambda)^{r+1}}{x!}\right]$$

$$= -r\lambda\mu_{r-1} + \mu_{r+1}$$

Hence, $\lambda\dfrac{d\mu_r}{d\lambda} + r\lambda\mu_{r-1} = \mu_{r+1}$

i.e., $\mu_{r+1} = r\lambda\mu_{r-1} + \lambda\dfrac{d\mu_r}{d\lambda}$

**Example 5** Suppose that $P(X=2) = \dfrac{2}{3}P(X=1)$, find $P(X=0)$.

**Solution** $P(X=2) = \dfrac{2}{3}P(X=1)$ (given)

$$\frac{\lambda^2 e^{-\lambda}}{2!} = \frac{2}{3}\frac{\lambda^1 e^{-\lambda}}{1!}$$

or $\dfrac{\lambda^2}{2!} = \dfrac{2}{3}\lambda$

or $3\lambda^2 - 4\lambda = 0$ or $\lambda(3\lambda - 4) = 0$

$$\lambda = 0 \text{ or } \lambda = \frac{4}{3}$$

We consider $\lambda = \dfrac{4}{3}$, since $\lambda = 0$ is impossible

$$P(X=0) = \frac{\lambda^0 e^{-\lambda}}{0!} = e^{-4/3}$$

**Example 6** A car-hire firm has two cars, which it hires out day by day. The number of demands for a car on each day is distributed as a Poisson distribution with mean 1.5. Calculate the proportion of days on which neither car is used and the proportion of days on which some demand is refused. $(e^{-1.5} = 0.2231)$.

**Solution** We have $\lambda = 1.5$ (given)

The proportion of days when no car will be required is

$$P(X = 0) = \frac{\lambda^0 e^{-\lambda}}{0!} = e^{-\lambda} = e^{-1.5} = 0.2231$$

The probability that no car, one car , two cars will be required

$$= P(X = 0) + P(X = 1) + P(X = 2)$$

$$= e^{-\lambda} + e^{-\lambda}\frac{\lambda}{1!} + e^{-\lambda}\frac{\lambda^2}{2!}$$

$$= e^{-\lambda}(1 + \frac{\lambda}{1!} + \frac{\lambda^2}{2!})$$

$$= e^{-1.5}(1 + 1.5 + 1.125)$$

$$= (0.2231)(3.625)$$

$$= 0.80873$$

Probability that some demand is refused

$$= 1 - P(x \leq 2)$$

$$= 1 - 0.80873$$

$$= 0.19127$$

**Example 7** In a Poisson distribution $P(X = x)$ for $x = 0$ is 10%. Find the mean, given that $\log_e 10 = 2.3026$.

**Solution** Let $\lambda$ be the mean

$$P(X = x) = P(X = 0) = \frac{\lambda^0 e^{-\lambda}}{0!} = e^{-\lambda} = \frac{10}{100}$$

$$\therefore e^{-\lambda} = 0.1 = \frac{1}{10} \text{ or } \therefore e^{\lambda} = 10$$

Hence $\lambda = \log_e 10 = 2.3026$

**Example 8** At a busy traffic intersection, the probability $p$ of an individual car having an accident is very small, say, $p = 0.0001$. However during a certain part of the day, a large number of cars, say 1000 pass through the intersection. Under these conditions, what is the probability of two or more accidents occurring during that period?

**Solution** Here, we have $n = 1000$, $p = 0.0001$

$$\therefore \lambda = np = 1000 \times 0.0001 = 0.1$$

$$P(X \geq 2) = 1 - P(X < 2) = 1 - [P(X = 0) + P(X = 1)]$$

$$= 1 - \left[\frac{e^{-0.1}(0.1)^0}{0!} + \frac{e^{-0.1}(0.1)^1}{1!}\right]$$

$$= 1 - e^{-0.1}(1 + 0.1)$$

$$= 1 - (0.9048)(1.1)$$

$$= 0.00472$$

**Example 9** Using Poisson distribution, find the probability that the ace of spades will-be drawn from a pack of well-shuffled cards at least once in 104 consecutive trials.

**Solution** Probability that the card drawn is an ace of spades $= \dfrac{1}{52}$

Since $n = 104$,

Mean of the distribution $= \lambda = np = 104 \times \dfrac{1}{52} = 2$

Probability of getting ace of spades at least once $= P(X \geq 1) = 1 - P(X = 0)$

$$= 1 - \frac{\lambda^0 e^{-\lambda}}{0!} = 1 - e^{-2}$$

$$= 1 - 0.136$$

$$= 0.864$$

**Example 10** Probability of getting no misprint in a page of a book is $e^{-4}$. What is the probability that a page contains more than two misprints?

**Solution** It is given that $P(X = 0)$ i.e., probability of getting no misprint $= e^{-4}$

i.e., $\therefore \dfrac{\lambda^0 e^{-\lambda}}{0!} = e^{-4}$

or $e^{-\lambda} = e^{-4}$

i.e., $\lambda = 4$

Probability that a page contains more than 2 misprints $= P(X > 2) = 1 - P(X \leq 2)$

$$= 1 - [P(X = 0) + P(X = 1) + P(X = 2)]$$

$$= 1 - \left[e^{-4} + e^{-4}.4 + e^{-4}\frac{4^2}{2!}\right]$$

$$= 1 - e^{-4}[1 + 4 + 8]$$

$$= 0.7618$$

**Example 11** Six coins are tossed 6400 times using Poisson distribution. What is the approximate probability of getting six heads 10 times?

**Solution** Probability of getting a head when a coin is tossed $= \dfrac{1}{2}$

Probability of getting 6 heads when 6 coins are tossed $= \left(\dfrac{1}{2}\right)^6 = \dfrac{1}{64}$

$$\text{Mean} = \lambda = np = 6400 \times \frac{1}{64} = 100$$

Probability of getting six heads 10 times $= P(X = 10) = \dfrac{\lambda^{10} e^{-\lambda}}{10!} = \dfrac{e^{-100}(100)^{10}}{100!}$

**Example 12** Suppose 2% of the people on an average are left-handed. Find (i) the probability of finding 3 or more left-handed people, (ii) the probability of finding none or one left-handed person.

**Solution** Let $x$ be the random variable of the number of left handed people.

$$\text{The mean} = \lambda = 2\% = \frac{2}{100} = 0.02$$

Since $P(X = x) = \dfrac{e^{-\lambda}\lambda^x}{x!}$, $x = 1, 2, 3, \ldots$

(i) $P(X \geq 3) = 1 - [P(X = 0) + P(X = 1) + P(X = 2)]$

$$= 1 - e^{-0.02}[1 + 0.02 + \frac{0.02^2}{2}]$$

$$= 1 - e^{-0.02}[1.0303]$$

$$= 1.307 \times 10^{-6}$$

(ii) $P(X \leq 1) = P(X = 0) + P(X = 1)$

$$= e^{-0.02} + e^{-0.02}[0.02]$$

**Example 13** If $X$ is a Poisson variate such that $3P(X = 4) = \frac{1}{2}P(X = 2) + P(X = 0)$, find (i) the mean of $X$, (ii) $P(X \leq 2)$.

**Solution**

(i) Since $X$ is a Poisson variate, we have

$$P(X = x) = \frac{e^{-\lambda}\lambda^x}{x!}, \ x = 1, 2, 3, \ldots$$

Consider $3P(X = 4) = \frac{1}{2}P(X = 2) + P(X = 0)$ (given)

or $\dfrac{e^{-\lambda}\lambda^4}{4!} = \dfrac{1}{2}\dfrac{e^{-\lambda}\lambda^2}{2!} + \dfrac{e^{-\lambda}\lambda^0}{0!}$

or $\dfrac{\lambda^4}{8} = \dfrac{\lambda^2}{4} + 1$

or $\lambda^4 - 2\lambda^2 - 8 = 0$

or $(\lambda^4 - 4)(\lambda^2 + 2) = 0$

Since $\lambda > 0$, we get $\lambda = 2$

Mean of the Poisson variate $\lambda = 2$

i.e., $\lambda = 2$

(ii) $P(X \leq 2) = P(X = 0) + P(X = 1) + P(X = 2)$

$$= \frac{e^{-2}2^0}{0!} + \frac{e^{-2}2^1}{1!} + \frac{e^{-2}2^2}{2!}$$

$$= 0.675$$

**Example 14** Average number of accidents on any day on a national highway is 1.8. Determine the probability that the number of accidents (i) is at least one, (ii) at most one.

**Solution** Average number of accidents = 1.8

i.e., $\lambda =$ mean $= 1.8$

i. $P(X \geq 1) =$ Probability that the number of accidents is at least one

$$= 1 - P(X = 0)$$

$$= 1 - e^{-1.8}$$

$$= 1 - 0.1653 = 0.8347$$

ii. Probability that the number of accidents is at most one $= P(X \leq 1)$

$$= P(X = 0) + P(X = 1)$$

$$= \frac{e^{-1.8}(1.8)^0}{0!} + \frac{e^{-1.8}(1.8)^1}{1!}$$

$$= e^{-1.8}(1 + 1.8)$$

$$= 0.4628$$

**Example 15** Using the recurrence formula, find the probabilities when $x =$ 0, 1, 2, 3, 4 and 5, if the mean of the Poisson distribution is 3.

**Solution** We have $\lambda = 3$ (given)

$$P(X = 0) = P(0) = e^{-\lambda} = e^{-3}$$

The recurrence formula is $P(x+1) = \dfrac{\lambda}{x+1} P(x)$

Substituting $x = 0, 1, 2, 3, 4$ and 5, we get

$$P(0+1) = P(1) = \frac{\lambda}{0+1} P(0) = 3(e^{-3}) = 0.147$$

$$P(1+1) = P(2) = \frac{\lambda}{1+1} P(1) = \frac{3}{2}(0.147) = 0.2205$$

$$P(2+1) = P(3) = \frac{\lambda}{2+1} P(2) = \frac{3}{3}(0.2205) = 0.2205$$

$$P(3+1) = P(4) = \frac{\lambda}{3+1} P(3) = \frac{3}{4}(0.2205) = 0.1653$$

$$P(4+1) = P(5) = \frac{\lambda}{4+1} P(4) = \frac{3}{5}(0.1653) = 0.099$$

$$P(5+1) = P(6) = \frac{\lambda}{5+1} P(5) = \frac{3}{6}(0.099) = 0.0495$$

$P(1) = 0.147, \quad P(2) = 0.2205, \quad P(3) = 0.2205$

$P(4) = 0.1653, \quad P(5) = 0.099, \quad P(6) = 0.0495$

**Example 16** Wireless sets are manufactured with 25 soldered joints each. On the average, 1 joint in 500 is defective. How many sets can be expected to be free from defective joints in a consignment of 10,000 sets?

**Solution** Probability that a joint is defective $= p = \dfrac{1}{500}$

We have $n = 25$, $p = \dfrac{1}{500}$

Mean $= \lambda = np = 25 \times \dfrac{1}{500} = 0.05$

By Poisson distribution $P(X = x) = \dfrac{e^{-\lambda}\lambda^x}{x!}$, $x = 1, 2, 3, \ldots$

Probability that no joint is defective $= P(0) = \dfrac{e^{-\lambda}\lambda^0}{0!} = e^{-0.05}$

Thus, the expected number of sets free from defects in 10,000 sets

$= 10,000 \times P(0) = 10,000 \times e^{-0.05}$

$= 10,000(0.9512)$

$= 9512.0$

9512 sets are expected to be free from defects.

**Example 17** Fit a Poisson distribution to the following data and calculate the theoretical (expected) frequencies.

| $x$ | 0 | 1 | 2 | 3 | 4 |
|---|---|---|---|---|---|
| $f(x)$ | 122 | 60 | 15 | 2 | 1 |

**Solution** Mean of the distribution $= \dfrac{0.122 + 1.60 + 2.15 + 3.2 + 4.1}{122 + 60 + 15 + 2 + 1}$

$$\bar{x} = \frac{60 + 30 + 6 + 4}{200} = \frac{100}{200}$$

$\therefore \lambda = \bar{x} = 0.5$

We have $N = 122 + 60 + 15 + 2 + 1 = 200$

$N = 200$, $\lambda = 0.5$

$$P(0) = P(X=0) = \frac{e^{-\lambda}\lambda^0}{0!} = e^{-0.5}$$

Their theoretical frequencies are obtained as follows

$$N.P(0) = 200.e^{-0.5} = 121$$

$$N.P(1) = 200.\frac{e^{-0.5}(0.5)^1}{1!} = 61$$

$$N.P(2) = 200.\frac{e^{-0.5}(0.5)^2}{2!} = 15$$

$$N.P(3) = 200.\frac{e^{-0.5}(0.5)^3}{3!} = 3$$

$$N.P(4) = 200.\frac{e^{-0.5}(0.5)^4}{4!} = 0$$

Hence, the expected (theoretical frequencies ) are 121, 61, 15, 3, 0.

**Example 18** The variance of the Poisson distribution is 2. Find the distribution for $x = 0, 1, 2, 3, 4$ and 5, from the recurrence relation of the distribution ($e^{-2} = 0.1353$).

**Solution** The variance of the Poisson distribution $\lambda = 2$ (given)

We have $P(0) = e^{-2} = 0.1353$ (given)

The recurrence relation is $P(x+1) = \dfrac{\lambda}{x+1}P(x)$

Hence we get

$$P(1) = \frac{\lambda}{0+1}P(0) = \frac{2}{1}e^{-2} = 2 \times 0.1353 = 0.2706$$

$$P(2) = \frac{\lambda}{1+1}P(1) = \frac{2}{2}(0.2706) = 0.2706$$

$$P(3) = \frac{\lambda}{2+1}P(2) = \frac{2}{3}(0.2706) = 0.1804$$

$$P(4) = \frac{\lambda}{3+1}P(3) = \frac{2}{4}(0.1804) = 0.0902$$

$$P(5) = \frac{\lambda}{4+1}P(4) = \frac{2}{5}(0.0902) = 0.361$$

**Example 19** Assuming that the typing mistakes per page committed by a typist follows a Poisson distribution, find the theoretical frequencies for the following distribution of typing mistakes.

| No. of mistakes per page | 0 | 1 | 2 | 3 | 4 | 5 |
|---|---|---|---|---|---|---|
| No. of pages | 40 | 30 | 20 | 15 | 10 | 5 |

**Solution** We have $N = 40 + 30 + 20 + 15 + 10 + 5 = 120$

$n = 5$

$$\text{Mean of the distribution} = \lambda = \frac{0.40 + 1.30 + 2.20 + 3.15 + 4.10 + 5.5}{120}$$

$$= \frac{30 + 40 + 45 + 40 + 25}{120} = \frac{180}{120} = 1.5$$

Using Poisson distribution we get

$$P(0) = \frac{e^{-\lambda}\lambda^0}{0!} = e^{-\lambda} = e^{-1.5} = 0.22313$$

$$P(1) = \frac{e^{-\lambda}\lambda^1}{1!} = e^{-\lambda}\lambda = e^{-1.5}(1.5) = 0.334695$$

$$P(2) = \frac{e^{-\lambda}\lambda^2}{2!} = \frac{e^{-1.5}(1.5)^2}{2} = 0.25$$

$$P(3) = \frac{e^{-\lambda}\lambda^3}{3!} = \frac{e^{-1.5}(1.5)^3}{6} = 0.13$$

$$P(4) = \frac{e^{-\lambda}\lambda^4}{4!} = \frac{e^{-1.5}(1.5)^4}{24} = 0.05$$

$$P(5) = \frac{e^{-\lambda}\lambda^5}{5!} = \frac{e^{-1.5}(1.5)^5}{120} = 0.01$$

The theoretical (expected) frequencies are

$NP(0) = (120)(0.22313) = 27,$

$NP(1) = (120)(0.334695) = 40,$

$NP(2) = (120)(0.25) = 30,$

$NP(3) = (120)(0.13) = 16,$

$NP(4) = (120)(0.05) = 6,$

$NP(5) = (120)(0.01) = 1$

i.e., 27, 40, 30, 16, 6 and 1

**Example 20** If $X$ and $Y$ are two independent Poisson random variables such that $\text{Var}[X]+\text{Var}[Y] = 3$. Find $P(X+Y<2)$.

**Solution** Let $X$ and $Y$ follow a Poisson distribution.

Let $\lambda_1$ be the mean of $X$ and $\lambda_2$ be the mean of $Y$.

Then by additive property $X+Y$ follows Poisson distribution with mean $\lambda_1+\lambda_2$.

We have $\text{Var}[X]+\text{Var}[Y] = \lambda_1+\lambda_2 = 3$ (given)

$\therefore \lambda = \lambda_1+\lambda_2 = 3$ (say)

Now $P(X+Y<2) = P(X+Y=0)+P(X+Y=1) = e^{-\lambda}+\dfrac{e^{-\lambda}\lambda}{1!}$

$$= e^{-3}+e^{-3}.3 = e^{-3}(1+3)$$

$$= 4e^{-3} = 0.199$$

**Example 21** In a company, a board receives 15 calls per minute. If the board is capable of handling at the most 25 calls per minute, what is the probability that the board will saturate in one minute.

**Solution** Number of calls received = 15 per minute

Therefore, $\lambda = 15$

The board reaches the saturation point if the number of calls exceeds 25.

Hence, the probability of saturation $= 1-\sum\limits_{x=0}^{25}\dfrac{e^{-15}(15)^x}{x!}$

$$= 1-\left[e^{-15}+\frac{e^{-15}.15}{1!}+\ldots+\frac{e^{-15}(15)^{25}}{25!}\right]$$

$$= 1-0.994 = 0.006$$

## Exercise 3.2

1 If a random variable $X$ follows a Poisson distribution such that $P(X=1) = P(X=2)$. Find (i) mean, (ii) $P(X=0)$.

2 A hospital receives patients at the rate of 3 patients per minute on an average. What is the probability of receiving no patients in a one minute interval?

3 Suppose a book of 285 pages contains typographical errors. If these errors are randomly

distributed throughout the book what is the probability that 10 pages, selected at random will be free from errors. ($e^{-0.735} = 0.4975$)

4 In a Poisson distribution if $3P(X = 2) = P(X = 4)$, find $P(X = 3)$.

5 Between the hours 2 p.m and 4 p.m, the average number of phone calls per minute coming into the switch board of a company is 2.35. Find the probability that during one particular minute, there will be at most 2 phone calls.

6 Out of 1000 balls, 50 are red and the rest are white. If 60 balls are picked at random, what is the probability of picking up (i) 3 red balls, (ii) not more than 3 red balls in the sample? Assume Poisson distribution for the number of red balls picked up in the sample where $e^{-3} = 0.0498$.

7 A manufacturer, who produces medicine bottles, finds that 0.1% of the bottles are defective. The bottles are packed in boxes containing 500 boxes. How many boxes will contain (i) no defectives, (ii) at least 2 defectives. (given $e^{-0.5} = 0.60650$)

8 The probability that a man aged 50 years will die within a year is 0.01125. What is the probability that out of 12 such men, at least 11 will reach their fifty-first birthday ($e^{-0.135} = 0.87371$).

9 A pair of dice is thrown 6 times. If getting a total of 7 is considered to be a success, what is the probability of at least 4 successes?

10 There are 20% chances for a worker of an industry to suffer from an occupational disease. 50 workers were selected at random and examined for the occupational disease. Find the probability that (i) only one worker is found suffering from the disease, (ii) more than 3 are suffering from the disease, (iii) none suffer from the disease.

11 If a random variable X follow a Poisson distribution such that $P(X = 2) = 9P(X = 4) + 90P(X = 6)$, find the mean and variance of X.

12 A random variable has a Poisson distribution such that $P(1) = P(2)$. Find the

(i) mean of the distribution

(ii) $P(4)$

(iii) $P(X \geq 1)$

(iv) $P(1 < X < 4)$

13 Show that three successive values of a Poisson variate cannot have equal probability of success.

14 Fit a Poisson distribution for the following data and calculate the expected frequencies.

| $x$ | 0 | 1 | 2 | 3 | 4 |
|---|---|---|---|---|---|
| $f(x)$ | 109 | 65 | 22 | 3 | 1 |

15 A distributor of bean seeds determines from extensive tests that 5% of a large batch of seeds will not germinate. He sells the seeds in packets of 200 and generates 90% germination. Determine the probability that a particular packet will violate the guarantee.

16 In a year, the number of accidents to the taxi drivers in a city follows a Poisson distribution with mean 3. Out of 1000 taxi drivers, find approximately the number of drivers with (i) no accidents in a year, (ii) more than 3 accidents in a year.

17 If the mean of a Poisson distribution is 4, find (i) standard deviation, (ii) $\beta_1$, (iii) $\beta_2$, (iv) $\mu_3$, (v) $\mu_4$.

18 100 car radios are inspected as they come off the production line and the number of defects per set is recorded below.

| No. of defects | 0 | 1 | 2 | 3 | 4 |
|---|---|---|---|---|---|
| No. of sets | 79 | 18 | 2 | 1 | 0 |

Find the Poisson distribution to the above data and calculate the expected frequencies of 0, 1, 2, 3 and 4 defects.

19 A book contains 100 misprints distributed randomly throughout its 100 pages. What is the probability that a page observed at random contains at least two misprints. Assume Poisson distribution.

20 If $X$ and $Y$ are independent Poisson random variables with means 2 and 4 respectively, find (i) $P\left[\frac{X+Y}{2}<1\right]$, (ii) $P[3(X+Y)\geq 9]$.

21 For a Poisson distribution, $P(X=1)=0.03$ and $P(X=2)=0.2$, find (i) $P(X=0)$ and (ii) $P(X=3)$.

22 Fit a Poisson distribution to the following data which gives the number of yeast cells per square for 400 squares.

| No of cells per square | 0 | 1 | 2 | 3 | 4 | 5 | 6 | 7 | 8 | 9 | 10 |
|---|---|---|---|---|---|---|---|---|---|---|---|
| No. of squares | 103 | 143 | 98 | 42 | 8 | 4 | 2 | 0 | 0 | 0 | 0 |

23 The number of items collected at a centre follows a Poisson distribution, where the probability of no items collected is 0.4. Find the probability that 3 or less items are collected.

24 A shop keeper receives 5 bad currencies every day. What are the probabilities that he will receive (i) 3 bad currencies on a particular day, (ii) 8 bad currencies on two consecutive days?

## Answers 3.2

(1) $\lambda = 2$, $P(X = 0) = 0.1353$

(2) 0.04979

(3) 0.4795

(4) $36e^{-6}$

(5) 0.58285

(6) (i) 0.2241 (ii) 0.6474

(7) (i) 61 (ii) 9

(8) 0.9917

(9) $\dfrac{406}{46650}$

(10) (i) $10\left(\dfrac{4}{5}\right)^{49}$
(ii) $1-\left(\dfrac{4^{47}}{550}\right)(25364)$
(iii) $\left(\dfrac{4}{5}\right)^{50}$

(11) mean = 21, variance = 1

(12) (i) 2 (ii) 0.0902 (iii) 0.8647 (iv) 0.4511

(14) 108, 67, 66, 20, 4, 11, 1

(15) 0.0016

(16) (i) 50 (ii)350

(17) (i) 2 (ii) 0.25 (iii) 3.25 (iv) 4 (v) 52

(18) 78, 20, 2, 0, 0

(19) 0.264

(20) (i) 0.017352 (ii) 0.93803

(21) (i) 0.00225 (ii) 0.888899

(22) 107,14,93,41,14,4,0,0,0,0

(23) 0.9856

(24) 0.1113

## 3.4 Discrete Uniform Distribution

If a random variable $X$ can take on $k$ different values with equal probability, we say that $X$ has a discrete uniform distribution, which is defined as follows.

**Definition 3.3** Let $X$ be a discrete random variable taking the values $1,2,3,\ldots,n$.

$X$ is said to follow a uniform distribution if its probability mass function is given by

$$P(X = x) = \frac{1}{k}, \; x = 1, 2, 3, \ldots k$$
$$= 0, \quad \text{otherwise}$$

$K$ is called the parameter of the distribution. Discrete uniform distribution is applied whenever all the values of the random variable $X$ are equally likely. The mean of the discrete uniform distribution is $\frac{k+1}{2}$ and the variance is $\frac{k^2-1}{12}$. Coefficient of skewness of the uniform distribution is zero. Hence uniform distribution is symmetric.

**Example** Let $X$ denote the number on the face of an unbiased die when it is rolled.

We have $P(X = x) = \frac{1}{6}, \; x = 1, 2, 3, \ldots 6$

$$= 0, \quad \text{otherwise}$$

Clearly, $X$ follows uniform distribution.

## 3.5 Negative Binomial and Geometric Distributions

Consider an experiment consisting of repeated Bernoulli (independent) trials. The number of the trial on which the $k$th success occurs is a random variable having a negative binomial distribution which is defined as follows.

**Definition 3.4** A random variable $X$ has a negative binomial distribution if and only if

$$P(X = x) = {}^{x-1}C_{k-1} p^k q^{x-k}, \text{ for } x = k, k+1, k+2, \ldots$$

where $p$ is the probability of success, $q$ in the probability of failure given by $q = 1 - p$

In this case, the random variable $X$ is called the negative binomial random variable. The negative binomial distribution is also called a Pascal or waiting time distribution.

The mean of a negative binomial distribution is $\frac{kq}{p}$ and the variance of a negative binomial distribution is $\frac{k(1-p)}{p^2}$.

The moment generating function of a negative binomial distribution is $p^k(1 - qe^t)^{-k}$. For a negative binomial distribution Var[X]> $E(X)$. If $k = 1$, the distribution is called a geometrical distribution.

### 3.5.1 Geometric distribution

**Definition 3.5** A random variable $X$ is said to follow geometric distribution, if it assumes only non-negative values and its probability mass function is

$$P(X = x) = pq^{x-1},\ x = 1, 2, 3, \ldots \text{ where } q = 1 - p$$

The mean of a geometric distribution is $\frac{1}{p}$ and the variance is $\frac{2q}{p^2}$. The moment generating function of a geometric distribution is $\frac{pe^t}{1-(1-p)e^t}$.

**Example** A typist types 3 letters erroneously for every 100 letters. What is the probability that the tenth letter typed is the first erroneous letter?

**Solution** We have

$$p = \frac{3}{100} = 0.03$$

$$q = 1 - p = 0.97$$

$$x = 10$$

$$\begin{aligned} P(X = x) &= pq^{x-1} \\ &= (0.03)(0.97)^{10-1} \\ &= (0.03)(0.97)^9 \end{aligned}$$

## Exercise 3.3

1 If the probability is 0.75 that an applicant for a driver's license will pass the road test at any given try, what is the probability that an applicant will finally pass the test on the fifth try?

2 A typist types 2 letters erroneously for every 100 letters. What is the probability that the tenth letter typed is the first erroneously typed letter?

3 If the probability that a child exposed to a certain contagious disease will catch it is 0.40, what is the probability that the tenth child exposed to the disease will be the third to catch it?

4 Find the probability that in tossing four coins, one will get either all heads or tails for the third time on the seventh toss.

5 The probability of a student passing a subject is 0.6. What is the probability that he will pass in the subject on his third attempt?

6 In a company, a group of particular items is accepted if more items are tested before the first defective is found. If $m = 5$, and the group consists of 15% defective items, what is the probability that it will be accepted.

**Answers 3.3**

(1) $(0.75)(0.25)^4$ (2) $(0.02)(0.98)^9$

(3) ${}^9C_2(0.40)^3(0.60)^7$ (4) ${}^6C_2\left(\frac{1}{8}\right)^3\left(\frac{7}{8}\right)^3$

(5) $(0.6)(0.4)^2$ (6) 0.4437

## 3.6 Continuous Probability Distributions

In this section, we discuss the method of finding the probability distributions associated with continuous random variables and also how to use the mean and standard deviation to describe these distributions.

### 3.6.1 Uniform distribution

**Definition 3.6** Let $X$ be a continuous random variable. $X$ is said to have uniform distribution, if its probability density function is given by

$$f(x) = k, \; a < x < b$$

$$= 0, \quad \text{otherwise}$$

where $k$ is a constant given by $\int_a^b f(x)dx = 1$

$$\text{i.e.,} \quad \int_a^b k dx = 1$$

$$\text{or} \quad k[x]_a^b = 1 \quad \text{or} \quad k(b-a) = 1$$

$$\text{or} \quad k = \frac{1}{b-a}$$

Hence, $f(x) = \dfrac{1}{b-a}, \quad a < x < b$

$= 0, \quad$ otherwise

Uniform distribution is also known as rectangular distribution.

The distribution function $f(x)$ of $X$ is given by

$$F(x) = \begin{cases} 0, -\infty < x < a \\ \dfrac{x-a}{b-a}, a \le x \le b \\ 1, b < x < \infty \end{cases}$$

Clearly $F(x)$ is not continuous at the two end points, $x = a$ and $x = b$. Hence $F(x)$ is not differentiable at $x = a$ and $x = b$.

The graphs of $f(x)$ and $F(x)$ are shown in Fig. 3.1 given below.

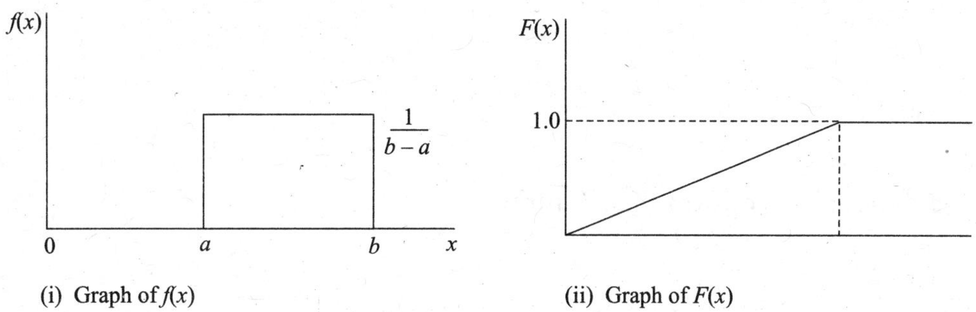


**Fig. 3.1** *Uniform probability density function $f(x)$ and its distribution function $F(x)$*

The probability density function of a uniform variate $X\ (-a, a)$ is given by

$$f(x) = \begin{cases} \dfrac{1}{2a}, -a < x < a \\ 0, \text{ otherwise} \end{cases}$$

### 3.6.2 Moments of uniform distribution

We have $\mu'_r = r$th moment about the origin

$$= \int_a^b x^r f(x)dx$$

$$= \int_a^b x^r \frac{1}{b-a} dx = \frac{1}{b-a}\left[\frac{x^{r+1}}{r+1}\right]_a^b$$

$$= \frac{b^{r+1} - a^{r+1}}{(r+1)(b-a)}$$

Putting $r = 1, 2, 3, 4$, we obtain,

$$\mu_1' = \frac{b^2 - a^2}{2(b-a)} = \frac{b+a}{2}$$

$$\mu_2' = \frac{b^3 - a^3}{3(b-a)} = \frac{b^2 + ba + a^2}{2}$$

$$\mu_3' = \frac{b^4 - a^4}{4(b-a)} = \frac{(b^2 + a^2)(b+a)}{4}$$

$$\mu_4' = \frac{b^5 - a^5}{5(b-a)} = \frac{b^4 + b^3a + b^2a^2 + ba^3 + a^4}{5}$$

### 3.6.3 Mean of uniform distribution

$$\mu_1' = E(X) = \frac{b+a}{2}$$

### 3.6.4 Variance of uniform distribution

$$\mu_2 = \mu_2' - \mu_1'^2 = \frac{b^2 + ab + a^2}{3} - \frac{(b+a)^2}{4}$$

$$= \frac{1}{12}(4b^2 + 4ab + 4a^2 - 3b^2 - 3a^2 - 6ab)$$

$$= \frac{1}{12}(b^2 + a^2 - 2ab)$$

$$= \frac{(b-a)^2}{12}$$

### 3.6.5 Moment generating function of the uniform distribution

We have $M(t) = E[e^{tx}]$

$$= \int_a^b e^{tx} \frac{1}{b-a} dx = \frac{e^{bt} - e^{at}}{(b-a)t}$$

$$= \frac{\left(1 + \frac{bt}{1!} + \frac{b^2t^2}{2!} + \ldots\right) - \left(1 + \frac{at}{1!} + \frac{a^2t^2}{2!} + \ldots\right)}{(b-a)t}$$

$$= 1 + \frac{b+a}{2}.t + \frac{b^3 - a^3}{3(b-a)} \cdot \frac{t^2}{2} + \frac{b^4 - a^4}{4(b-a)} \cdot \frac{t^3}{3} + \ldots$$

## Solved examples

**Example 1** If $X$ is uniformly distributed over $(-a, a)$, find $a$ so that $P(X>1) = \frac{1}{3}$.

**Solution** Therefore, the probability density function of $X$ is

$$f(x) = \begin{cases} \frac{1}{2a}, -a < x < a \\ 0, \text{ otherwise} \end{cases}$$

We have $P(X > 1) = \frac{1}{3}$ (given)

i.e., $\int_1^a f(x)dx = \frac{1}{3}$ or $\int_1^a \frac{1}{2a} dx = \frac{1}{3}$

or $\frac{1}{2a} [x]_1^a = \frac{1}{3}$

or $\frac{a-1}{2a} = \frac{1}{3}$

i.e, $3a - 3 = 2a$

Hence $a = 3$

**Example 2** Subway trains run on a certain line every half an hour between midnight and 6 in the morning. What is the probability that a man entering the station at a random time during this period will have to wait at least 20 minutes?

**Solution** Let the man's waiting time for the next train be denoted by $X$.

Clearly $X$ is a random variable which is uniformly distributed in (0, 30) when the man arrived at the station. The p.d.f of $X$ is given by

$$f(x) = \begin{cases} \frac{1}{30}, 0 < x < 30 \\ 0, \text{ otherwise} \end{cases}$$

Probability that a man entering the station at random waits at least 20 minutes

$$= P(X = 20)$$

$$= \int_{20}^{30} f(x)dx = \int_{20}^{30} \frac{1}{30}dx$$

$$= \frac{1}{30}[x]_{20}^{30} = \frac{30-20}{30}$$

$$= \frac{10}{30} = \frac{1}{3}$$

## 3.7 Exponential and Negative Exponential Distribution

**Definition 3.7** Let $X$ be a continuous random variable. $X$ is said to have an exponential distribution if the p.d.f of $x$ is given by

$$f(x) = \begin{cases} e^{-x}, x \geq 0 \\ 0, \text{ otherwise} \end{cases}$$

$X$ is said to have a negative exponential distribution with parameter, $\lambda > 0$ if the p.d.f is given by

$$f(x) = \begin{cases} \lambda e^{-\lambda x}, x > 0 \\ 0, x \leq 0 \end{cases}$$

Mean of negative exponential distribution is $\frac{1}{\lambda}$ and the variance is $\frac{1}{\lambda^2}$. The moment generating function of the negative exponential distribution is $\frac{\lambda}{\lambda - t}, (\lambda > t)$. The $r$th moment about the origin is $\frac{r!}{\lambda^r}, r = 1, 2, 3, \ldots$

## 3.8 Normal Distribution

Normal distribution is a continuous distribution. It was first discovered by Abraham De Moivre (1667–1754). He derived normal probability function as the limiting form of the binomial distribution. Simultaneously, Karl Friedrich Gauss (1777–1853) also derived normal distribution as a law of errors. Hence the normal probability distribution is also called as the Gaussian distribution.

**Definition 3.8** A continuous random variable $X$ is said to have a normal distribution if its probability function is given by $f(x) = \frac{1}{\sigma\sqrt{2\pi}} e^{-\frac{1}{2}\left[\frac{x-\mu}{\sigma}\right]^2}, (-\infty < x < \infty)$,

where $\mu$ = mean of the normal random variable $X$

$\sigma$ = standard deviation of the normal variable $X$

and are the parameters of the distribution. If $X$ follows normal distribution, then it is denoted by $X \sim N(\mu, \sigma)$.

### 3.8.1 Standard normal variable

Computing the area over intervals under the normal probability distribution is a difficult task. Consequently, we will use a table of areas as a function of Z-score. Z-score gives the distance between a measurement and the mean in units equal to the standard deviation. $Z$ is always a normal random variable with mean 0 and standard deviation 1. For this reason, $Z$ is often referred to as the standard normal random variable. $Z$ is a continuous random variable.

**Definition 3.9** The standard normal random variable $Z$ is defined by

$$Z = \frac{x-\mu}{\sigma}$$

where $\mu$ = mean of the normal random variable $X$

$\sigma$ = Standard deviation of the normal variable $X$

$X$ = normal random variable

### 3.8.2 Standard normal distribution

**Definition 3.10** A random variable $Z = \frac{x-\mu}{\sigma}$ is said to have a standard normal distribution if its probability function is defined by

$$f(z) = \frac{1}{\sqrt{2\pi}} e^{-\frac{1}{2}Z^2}, \ (-\infty < Z < \infty)$$

A standard normal variate is denoted by $N$ (0,1). A standard normal distribution is also known as a $Z$-distribution or a unit normal distribution. Standardisation of a normal distribution helps us to make use of the tables of the area of the standard curve.

$$\frac{1}{\sqrt{2\pi}}e^{-\frac{1}{2}z^2}$$

for various points along the $x$-axis.

The area between the points, say $Z_1$ and $Z_2$ under the standard normal curve will represent the probability that $Z$ will lie between $Z_1$ and $Z_2$. It is denoted by

$$P(Z_1 \leq Z \leq Z_2)$$

The general equation of the normal curve is

$$y = f(x) = \frac{1}{\sigma\sqrt{2\pi}}e^{-\frac{1}{2}\left[\frac{x-\mu}{\sigma}\right]^2}, (-\infty < x < \infty)$$

If the corresponding total frequency is $N$, then

$$\frac{N}{\sigma\sqrt{2\pi}}e^{-\frac{1}{2}\left[\frac{x-\mu}{\sigma}\right]^2}$$

is the equation of the best fitting normal probability curve.

### 3.8.3 Distribution function (Z) of a standard normal variate

The distribution function $\phi(Z)$ of a standard normal variate, $Z$ is defined by

$$\phi(Z) = P(Z \leq z) = \int_{-\infty}^{Z} f(t)dt$$

$$= \frac{1}{\sqrt{2\pi}}\int_{-\infty}^{Z} e^{-\frac{1}{2}Z^2}$$

where $Z = \dfrac{x-\mu}{\sigma}$

**Properties of $\phi(Z)$**

1. $\phi(-Z) = 1 - \phi(Z)$
2. $P(a \leq X \leq b) = \phi\left(\dfrac{b-\mu}{\sigma}\right) - \phi\left(\dfrac{a-\mu}{\sigma}\right)$
3. $P(Z \leq a) = P(Z \geq -a)$

### 3.8.4 Area under normal curve

The curve of a normal distribution is unimodal and bell shaped, with the highest point over the mean $\mu$. It is symmetrical about a vertical line through $\mu$.

The normal curve (Fig. 3.2) has the following features.

(a) The curve is symmetrical about the coordinate at its mean which locates the peak of the bell.

(b) The values of mean, median and mode are equal.

(c) The curve extends from $-\infty$ to $\infty$.

(d) The area covered between $\mu-\sigma$ and $\mu+\sigma$ is 0.6826 (i.e., 68.26% of the total area)

(e) The area covered between the limits $\mu-2\sigma$ and $\mu+2\sigma$ is 0.9544 (i.e., 95.44% of area).

(f) The area covered between $\mu-3\sigma$ and $\mu+3\sigma$ is 0.9974 (i.e., 99.74% of area).

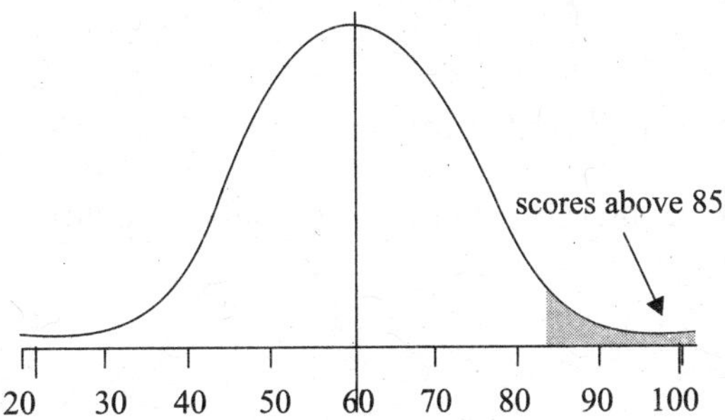


**Fig. 3.2** *Normal curve*

### 3.8.5 Area under standard normal curve

Let $X$ be a normal random variable and $Z$ is the corresponding standard normal variable. Then $Z$ and $X$ have identical curves. The curve of $Z$ is called the standard normal curve. The area bounded by a standard normal variable $Z$, the $Z$ axis and the ordinates at $Z=\infty$ and any positive value of $Z$ is provided by a standard table.

The standard normal curve covers 68.26% area between $Z=-1$ and $Z=1$. Therefore, $P(-1\leq Z\leq 1)=0.6826$.

The curve covers 95.44% of the total area between $Z=-2$ and $Z=2$. Therefore, $P(-2\leq Z\leq 2)=0.9544$.

The area covered between $Z=-3$ and $Z=3$ is 99.74% of the total area. Hence, $P(-3\leq Z\leq 3)=0.9474$.

### 3.8.6 Properties of normal curve

1. The mean $\mu$ and variance $\sigma^2$ of a normal distribution are called the parameters of the distribution.
2. The mean, median and mode of a normal distribution coincide with each other.
3. In a normal distribution, mean deviation about mean is approximately equal to $\frac{4}{5}$ times its standard deviation.
4. In a normal distribution, the quartiles $Q_1$ and $Q_3$ are equidistant from the median.
5. The normal curve is bell shaped and is symmetrical about $X = \mu$
6. The area bound by the normal curve and the $X$-axis equal to 1.
7. The tails of the curve of a normal distribution extend identically in both sides of $x = \mu$ and never touch the $X$-axis.
8. The probability that $x$ lies between $a$ and $b$ is equal to the area bounded by the curve $X$-axis and ordinates $X = a$ and $X = b$.

### 3.8.7 Mean of normal distribution

Consider the normal distribution with $\mu$ and $\sigma$ as the parameters, then

$$f(x) = \frac{1}{\sigma\sqrt{2\pi}} e^{-\frac{1}{2}\left[\frac{x-\mu}{\sigma}\right]^2}, (-\infty < x < \infty)$$

Mean of $X = E[x]$

Therefore, $\mu = E[X] = \int\limits_{-\infty}^{\infty} xf(x)dx$

$$= \frac{1}{\sigma\sqrt{2\pi}} \int\limits_{-\infty}^{\infty} x.e^{-\frac{1}{2}\left[\frac{x-\mu}{\sigma}\right]^2} dx$$

Putting $z = \dfrac{x-\mu}{\sigma}$, we have

$$dz = \frac{dx}{\sigma} \text{ i.e., } dx = \sigma dz$$

and $x = \mu + \sigma z$

Hence we get $\mu = \dfrac{1}{\sqrt{2\pi}} \int\limits_{-\infty}^{\infty} (\sigma z + \mu).e^{-\frac{1}{2}z^2} dz$

$$= \frac{\sigma}{\sqrt{2\pi}} \int_{-\infty}^{\infty} z e^{-\frac{1}{2}z^2} dz + \frac{\mu}{\sqrt{2\pi}} \int_{-\infty}^{\infty} e^{-\frac{1}{2}z^2} dz$$

$$= 0 + \frac{\mu}{\sqrt{2\pi}} \int_{-\infty}^{\infty} z e^{-\frac{1}{2}z^2} dz = \frac{2\mu}{\sqrt{2\pi}} \int_0^{\infty} e^{-\frac{1}{2}z^2} dz$$

$$= \frac{\sqrt{2}\mu}{\sqrt{\pi}} \cdot \frac{\sqrt{\pi}}{\sqrt{2}} = \mu$$

### 3.8.8 Variance of normal distribution

Variance $= E(X-\mu)^2$

i.e., $\text{Var}[X] = \int_{-\infty}^{\infty} (x-\mu)^2 f(x) dx$

$$= \frac{1}{\sigma\sqrt{2\pi}} \int_{-\infty}^{\infty} (x-\mu)^2 e^{-\frac{1}{2}\left[\frac{x-\mu}{\sigma}\right]^2} dx$$

Putting $z = \frac{x-\mu}{\sigma}$, we get

$$\text{Var}[X] = \frac{1}{\sigma\sqrt{2\pi}} \int_{-\infty}^{\infty} \sigma^2 z^2 e^{-\frac{1}{2}z^2} (\sigma dz)$$

Putting $\frac{z^2}{2} = t$, we obtain

$$\text{Var}[X] = \frac{2\sigma^2}{\sqrt{2\pi}} \int_0^{\infty} 2t e^{-t} \frac{dt}{\sqrt{2t}} = \frac{2\sigma^2}{\sqrt{\pi}} \int_0^{\infty} e^{-t} \sqrt{t} dt$$

$$= \frac{2\sigma^2}{\sqrt{\pi}} \int_0^{\infty} e^{-t} t^{\frac{3}{2}-1} dt$$

$$= \frac{2\sigma^2}{\sqrt{\pi}} \Gamma\left(\frac{3}{2}\right) = \frac{2\sigma^2}{\sqrt{\pi}} \frac{3}{2} \Gamma\left(\frac{1}{2}\right)$$

$$= \frac{\sigma^2}{\sqrt{\pi}} \cdot \sqrt{\pi}$$

Hence, variance $= \sigma^2$

### 3.8.9 Mode of normal distribution

We have $f(x) = \frac{1}{\sigma\sqrt{2\pi}} e^{-\frac{1}{2}\left[\frac{x-\mu}{\sigma}\right]^2}$

Applying logarithms on both sides, we get

$$\log f(x) = -\log \sigma\sqrt{2\pi} - \frac{1}{2\sigma^2}.(x-\mu)^2$$

Differentiating with respect to $x$ we get

$$\frac{f'(x)}{f(x)} = -\frac{1}{\sigma^2}(x-\mu)$$

or $f'(x) = -\frac{1}{\sigma^2}(x-\mu)f(x)$ (1)

Again differentiating, we get

$$f''(x) = -\frac{1}{\sigma^2}[1.f(x) + (x-\mu)f'(x)]$$

$$= -\frac{f(x)}{\sigma^2}\left[1 - \left(\frac{x-\mu}{\sigma}\right)^2\right] \quad \text{(using (1))}$$

$f'(x) = 0$ gives $x = \mu$ (since $f(x) \neq 0$)

At $x = \mu$, we have $f''(x) = -\frac{1}{\sigma^2} f(\mu)$

$$= -\frac{1}{\sigma^2}\frac{1}{\sigma\sqrt{2\pi}}$$

$$= -\frac{1}{\sigma^3\sqrt{2\pi}} < 0$$

Hence, $x = \mu$, is the mode of the distribution.

### 3.8.10 Median of the normal distribution

Let $M$ denote the median of normal distribution

$$\int_{-\infty}^{M} f(x)dx = \frac{1}{2}$$

$$\text{i.e., } \frac{1}{\sigma\sqrt{2\pi}}\int_{-\infty}^{\mu} e^{-\frac{1}{2}\left[\frac{x-\mu}{\sigma}\right]^2} dx + \frac{1}{\sigma\sqrt{2\pi}}\int_{\mu}^{M} .e^{-\frac{1}{2}\left[\frac{x-\mu}{\sigma}\right]^2} dx = \frac{1}{2} \tag{1}$$

Consider $\frac{1}{\sigma\sqrt{2\pi}}\int_{-\infty}^{\mu} e^{-\frac{1}{2}\left[\frac{x-\mu}{\sigma}\right]^2} dx$

Putting $\frac{x-\mu}{\sigma} = z$, we get $dx = \sigma dz$

$$\text{Therefore, } \frac{1}{\sigma\sqrt{2\pi}}\int_{-\infty}^{\mu} e^{-\frac{1}{2}\left[\frac{x-\mu}{\sigma}\right]^2} dx = \frac{1}{\sqrt{2\pi}}\int_{-\infty}^{0} e^{-\frac{1}{2}z^2} dz$$

$$= \frac{1}{\sqrt{2\pi}}\int_{0}^{\infty} e^{-\frac{1}{2}z^2} dz \text{ (by symmetry)}$$

$$= \frac{1}{\sqrt{2\pi}}\frac{\sqrt{\pi}}{\sqrt{2}} = \frac{1}{2} \tag{2}$$

From (1) and (2), we obtain

$$\frac{1}{2} + \frac{1}{\sigma\sqrt{2\pi}}\int_{\mu}^{M} e^{-\frac{1}{2}\left[\frac{x-\mu}{\sigma}\right]^2} dx = \frac{1}{2}$$

$$\text{Thus } \frac{1}{\sigma\sqrt{2\pi}}\int_{\mu}^{M} e^{-\frac{1}{2}\left[\frac{x-\mu}{\sigma}\right]^2} dx = 0$$

$$\text{or } \int_{\mu}^{M} e^{-\frac{1}{2}\left[\frac{x-\mu}{\sigma}\right]^2} dx = 0$$

which is possible when $\mu = M$

Hence $\mu = M$

i.e., Median of normal distribution $= \mu$

### 3.8.11 Moment generating function of normal distribution with respect to origin

$$\text{We have } M_x(t) = E(e^{tx}) = \int_{-\infty}^{\infty} e^{tx} f(x)dx = \frac{1}{\sigma\sqrt{2\pi}}\int_{-\infty}^{\infty} e^{tx} e^{-\frac{1}{2}\left[\frac{x-\mu}{\sigma}\right]^2} dx$$

Putting $\frac{x-\mu}{\sigma} = z$, we get

$$M_x(t) = \frac{e^{\mu t}}{\sqrt{2\pi}} \int_{-\infty}^{\infty} e^{-\frac{1}{2}z^2+(t\sigma z)} dz$$

$$= \frac{e^{\mu t + \frac{1}{2}t^2\sigma^2}}{\sqrt{2\pi}} \int_{-\infty}^{\infty} e^{-\frac{1}{2}(z^2 - t\sigma)^2} dz$$

$$= \frac{e^{\mu t + \frac{1}{2}t^2\sigma^2}}{\sqrt{2\pi}} .2 \frac{\sqrt{\pi}}{\sqrt{2}}$$

$$= e^{\mu t + \frac{1}{2}t^2\sigma^2}$$

## 3.8.12 Mean deviation of normal distribution

We have mean deviation $= \int_{-\infty}^{\infty} |x-\mu| f(x) dx$

$$= \frac{1}{\sigma\sqrt{2\pi}} \int_{-\infty}^{\infty} e^{-\frac{1}{2}\left[\frac{x-\mu}{\sigma}\right]^2} |x-\mu| dx$$

$$= \frac{\sigma}{\sqrt{2\pi}} \int_{-\infty}^{\infty} e^{-\frac{1}{2}z^2} |z| dz, \text{ where } z = \frac{x-\mu}{\sigma},$$

$$= \frac{2\sigma}{\sqrt{2\pi}} \int_{0}^{\infty} e^{-\frac{1}{2}z^2} |z| dz \quad \text{(since the integral is even )}$$

$$= \sigma\sqrt{\frac{2}{\pi}} = \frac{4}{5}\sigma \text{ (approximately)}$$

## Solved examples

**Example 1** A normal distribution has a mean 20 and a standard deviation of 4. Find out the probability that a value of $X$ lies between 20 and 24.

**Solution** We have $\mu = 20$, and $\sigma = 4$ (given)

$$Z = \frac{x-\mu}{\sigma} = \frac{x-20}{4}$$

When $X = 20$, $Z = \dfrac{20 - 20}{4} = 0$

and when $X = 24$, $Z = \dfrac{24 - 20}{4} = 1$

Therefore, when $X$ lies between 20 and 24, the value of $Z$ lies between 0 and 1.

Hence $P(0 \leq Z \leq 1) = 0.3413$

**Example 2** The mean and standard deviation of a normal variable $x$ are 50 and 4 respectively. Find the values of the corresponding standard normal variable when $x$ is equal to 42, 54 and 84.

**Solution** We have $\mu = 50$, and $\sigma = 4$ (given)

$$Z = \frac{x - \mu}{\sigma} = \frac{x - 50}{4}$$

When $X = 42$, $Z = \dfrac{42 - 50}{4} = -2$

When $X = 54$, $Z = \dfrac{54 - 50}{4} = 1$

When $X = 84$, $Z = \dfrac{84 - 50}{4} = 8.5$

**Example 3** Find the area under the standard normal curve which lies

(i) to the left of $Z = 1.73$.

(ii) to the right of $Z = -0.66$.

(iii) between $Z = 1.25$ and $Z = 1.67$.

(iv) between $Z = -1.45$ and $Z = 1.45$.

**Solution**

(i) Area to the left of $Z = 1.73$

$= 0.5 +$ area between $Z = 0$ and $Z = 1.73$

$= 0.5 + 0.4582$

$= 0.9582$ (using the table)

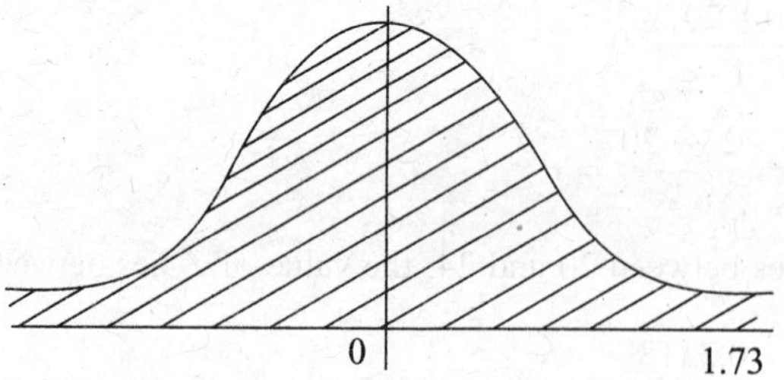


(ii) Area to the right of $Z = -0.66$

$= \text{Area from } Z = -0.66 \text{ to } Z = 0 + 0.5$

$= 0.2454 + 0.5$

$= 0.7454$ (using the table)

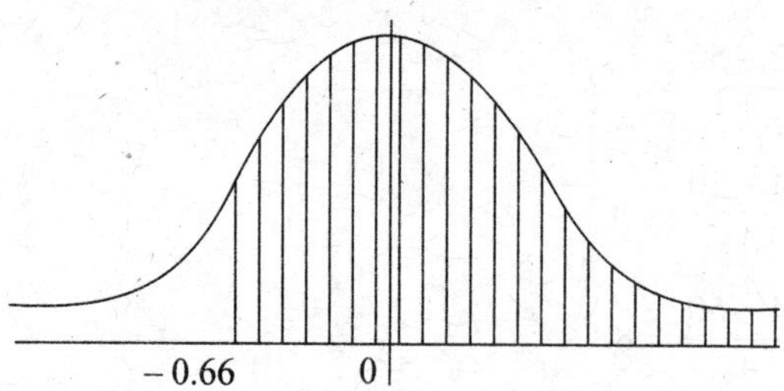


(iii) Area between $Z = 1.25$ and $Z = 1.67$

$= \text{Area between } Z = 0 \text{ and } Z = 1.67$

$-(\text{Area between } Z = 0 \text{ and } Z = 1.25)$

$= 0.4525 - 0.3944$

$= 0.0581$ (using the table)

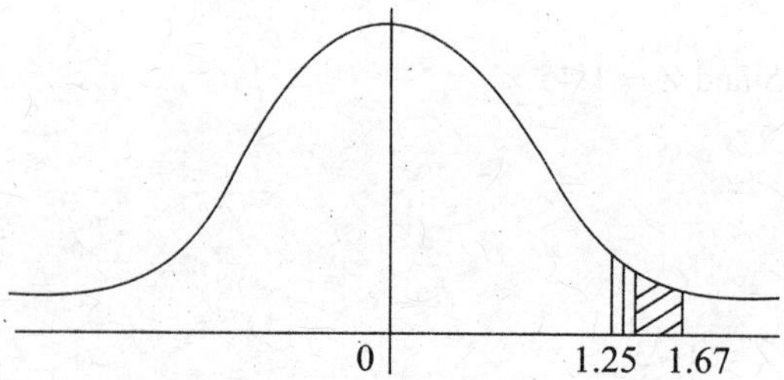


(iv) Area between $Z = -1.45$ and $Z = 1.45$

$= \text{Area between } Z = 0 \text{ and } Z = 1.45$

$+ (\text{Area between } Z = 0 \text{ and } Z = 1.45)$(by symmetry)

$= 2$ (Area between $Z = 0$ and $Z = 1.45$)

$= 2(0.4265)$

$= 0.8530$ (using the table)

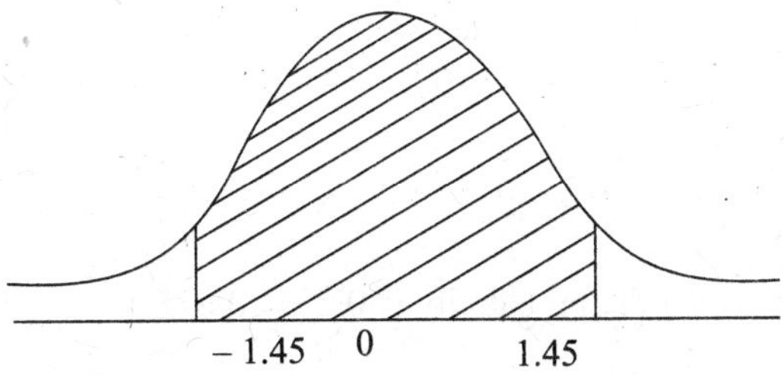

**Example 4** The ABC company uses a machine to fill boxes with soap powder. Assume that the net weight of the boxes of soap is normally distributed with mean 15 ounces and standard deviation 0.8 ounces.

(i) What proportion of boxes will have net weight of more than 14 ounces?

(ii) 25% of the boxes will be heavier than a certain net weight $\omega$ and 75% of the boxes will be lighter than this net weight. Find $\omega$.

**Solution** We have $\mu = 15$ ounce, $\sigma = 0.8$ ounce

Let Z be the standard normal variable

$$Z = \frac{x-\mu}{\sigma} = \frac{x-15}{0.8}$$

(i) When $x = 14$, $Z = \dfrac{14-15}{0.8} = \dfrac{-1}{0.8} = -1.25$

When $x = 14$, the probability that $x \geq 14 = P(x \geq 14)$

$= P(Z > -1.25)$

$= P(-1.25 \leq Z \leq 0) + P(Z \geq 0)$

$= P(0 \leq Z \leq 1.25) + 0.5$

$= 0.3944 + 0.5$

$= 0.8944$

89.44% of the boxes will have their net weight greater than 14.

(ii) Taking $x = \omega$, we have $Z = \dfrac{x-\mu}{\sigma} = \dfrac{\omega - 15}{0.8}$

Probability of Z being greater than $\dfrac{\omega - 15}{0.8}$ is 0.25.

Therefore, $0.25 = 0.5 - P\left(0 \le Z \le \dfrac{\omega - 15}{0.8}\right)$

i.e., $P\left(0 \le Z \le \dfrac{\omega - 15}{0.8}\right) = 0.5 - 0.25 = 0.25$

The value of Z for which the probability is $0.25 = 0.675$

Therefore, $\dfrac{\omega - 15}{0.8} = 0.675$

or $\omega = 0.675 \times 0.8 + 15$

$= 15.54$ ounces

**Example 5** Let $X$ denote the number of successes in a test. If $X$ is normally distributed with mean 100 and standard deviation 15, find the probability that $x$ does not exceed 130.

**Solution** We have $\mu = 100$ and $\sigma = 15$

Let $Z = \dfrac{x-\mu}{\sigma} = \dfrac{x-100}{15}$

Probability that $x$ does not exceed $130 = P(X \le 130)$

$$= P\left(\frac{x-100}{15} \le \frac{130-100}{15}\right)$$

$$= P(Z \le 2)$$

$$= P(-\infty < Z \le 0) + P(0 \le Z \le 2)$$

$$= 0.5 + 0.4772$$

$= 0.9772$ (from the area table)

**Example 6** Assume the mean height of soldiers to be 68.22 inches with a variance of 10.8 inches. How many soldiers in a regiment of 1000 would you expect to be over 6 feet tall?

**Solution** We have $\mu = 68.22$, $\sigma^2 = 10.8$, $\sigma = \sqrt{10.8} = 3.286$

$X = 6$ feet $= 6 \times 12 = 72$ inches

When $X = 72$, $Z = \dfrac{x-\mu}{\sigma} = \dfrac{72-68.22}{3.286} = 1.1503$

$$P(X > 72) = P(Z > 1.1503)$$
$$= 0.5 - P(0 < Z < 1.1503)$$
$$= 0.5 - 0.3749$$
$$= 0.1251$$

Expected number of soldiers over 6 feet tall in a regiment of $1000 = 1000 \times 0.1251$

$$= 125.1 = 125 \text{ soldiers}$$

**Example 7** What is the probability that a standard normal variate Z will be lying between 1.25 and 2.75?

**Solution** $P(1.25 < Z < 2.75) = P(0 < Z < 2.75) - P(0 < Z < 1.25)$

= Area between Z = 0 and Z = 2.75 – (Area between Z = 0 and Z = 1.25)

$= 0.4970 - 0.3944$

$= 0.1026$

**Example 8** In a normal distribution exactly 7 per cent of the items are under 35 and 89% of the items are under 63. What are the values of the mean and standard deviation of the distribution?

**Solution** Let $\mu$ be the mean and $\sigma$ be the standard deviation of the distribution.

The area lying to the left of the ordinate at $x = 35$ is 7% i.e., 0.07. The corresponding values of $Z$ is negative.

The area lying to the right of the ordinate at $x = 35$ up to the mean $\mu$ is

$$= 0.5 - 0.07 = 0.43$$

Hence, the values of Z corresponding in the area 0.43 is 1.48

i.e., $\dfrac{35-\mu}{\sigma} = -1.48$ or $\mu - 1.48\,\sigma = 35$ (1)

Similarly, the area lying to the left of the ordinate at $x = 63$ up to the mean is 0.39 (i.e., 39%).

The value of $Z$ corresponding to the area 0.39 is 1.23.

i.e., $\dfrac{63-\mu}{\sigma} = 1.23$ or $\mu + 1.23\sigma = 63$ (2)

From (1) and (2), we get

$$\mu = 50.3;\ \sigma = 1.33$$

**Example 9** The income of a group of 10,000 persons was found to be normally distributed with mean equal to Rs.750, and standard deviation equal to Rs.50. What was the lowest income among the richest 250?

**Solution** We have $\mu = 750$, $\sigma = 50$ (given)

Let $x'$ be the lowest income among the richest 250 people.

Then $P(x \geq x') = \dfrac{250}{10,000}$ (given)

i.e., $P\left(\dfrac{x-\mu}{\sigma} \geq \dfrac{x'-750}{50}\right) = 0.025$

or $P\left(Z \geq \dfrac{x'-750}{50}\right) = 0.025$

We obtain $\left(\dfrac{x'-750}{50}\right) = 1.96$

or $x' - 750 = 50 \times 1.96$

or $x' = 750 + 98 = 848$

Hence, Rs. 848 is the lowest income among the richest 250 people.

**Example 10** If $\log_e x$ is normally distributed with mean 1 and variance 4, find $P(\frac{1}{2} < X < 2)$ given that $\log_e 2 = 0.693$.

**Solution** Here we have $\mu = 1$, variance $= \sigma^2 = 4$ (given)

$\therefore \sigma = \sqrt{4} = 2$

Let $X = \log_e x$

$$Z = \frac{X-\mu}{\sigma} = \frac{X-1}{2} = \frac{\log_e X - 1}{2}$$

$$\text{When } x = \frac{1}{2},\ Z_1 = \frac{\log_e \frac{1}{2} - 1}{2} = \frac{\log_e 1 - \log_e 2 - 1}{2}$$

$$= \frac{-\log_e 2 - 1}{2} = \frac{0.693 - 1}{2}$$

$$= -0.8465$$

$$\text{When } x = 2,\ Z_2 = \frac{\log_e 2 - 1}{2} = \frac{0.693 - 1}{2}$$

$$= -0.1535$$

$$P(\frac{1}{2} < X < 2) = P(Z_1 < Z < Z_2)$$

$$= P(-0.8465 < Z < -0.1535)$$

$$= P(0.1535 < Z < 0.8465)$$

$$= \text{Area between } Z = 0 \text{ and } Z = 0.8465$$

$$-(\text{Area between } Z = 0 \text{ and } Z = 0.1535)$$

$$= 0.2996 - 0.0596$$

$$= 0.24$$

**Example 11** In a normal distribution, 31% of the items are under 45 and 8% of the items are over 64. Find the mean and standard deviation of the distribution.

**Solution** Let $X$ be the normal variable. Let $\mu$ denote the mean and let $\sigma$ be the standard deviation of $X$.

$$\text{Let } Z = \frac{x - \mu}{\sigma}$$

$$\text{It is given that } P(X < 45) = \frac{31}{100} = 0.31$$

$$\text{and } P(X > 64) = \frac{8}{100} = 0.08$$

Since $P(X < 45) < 0.5$, $X$ lies to the left of $X = \mu$. Therefore the value of $Z$ is negative, say $Z_1$.

$$\text{i.e., when } x = 45, \text{ let } Z = \frac{45 - \mu}{\sigma} = -Z_1,\ (Z_1 > 0) \text{ (say)}$$

Similarly $Z$ is positive when $X = 64$

Let $Z = \dfrac{64-\mu}{\sigma} = Z_2 (Z_2 > 0)$

We have $P(X < 45) = 0.31$

i.e., $P(Z < -Z_1) = 0.31$

or $P(Z > Z_1) = 0.31$

or $0.5 - P(0 \le Z \le Z_1) = 0.31$

i.e., $P(0 \le Z \le Z_1) = 0.19$

$Z_1 = 0.5$ (from the area table)

Therefore, $\dfrac{45-\mu}{\sigma} = -0.5$

or $45 - \mu = -0.5\sigma$ (1)

Also, we have $P(X > 64) = 0.08$

i.e., $P(Z > Z_2) = 0.08$

or $0.5 - P(0 \le Z \le Z_2) = 0.08$

$P(0 \le Z \le Z_2) = 0.42$

Hence, $Z_2 = 1.4$ (from the area table)

Therefore, $\dfrac{64-\mu}{\sigma} = 1.4$

or $64 - \mu = 1.4\sigma$ (2)

Solving (1) and (2), we get

$\mu = 50$ and $\sigma = 10$

**Example 12** The mean inside diameter of a sample of 200 washers produced by a machine is 0.502 cm and the standard deviation is 0.005 cm. The purpose for which these washers are intended allows a maximum tolerance in the diameter of 0.496 to 0.508 cm. Otherwise, the washers are considered to be defective washers. Assuming that the diameters are normally distributed, find the percentage of defective washers.

**Solution** We have $\mu = 0.502$ cm and $\sigma = 0.005$ cm

Let $Z$ be the standard normal variable.

$$Z = \frac{x-\mu}{\sigma} = \frac{x-0.502}{0.005}$$

Probability that a washer is not defective

$$= P(0.496 \leq X \leq 0.508)$$

$$= P\left(\frac{0.496-0.502}{0.005} \leq \frac{X-0.502}{0.005} \leq \frac{0.508-0.502}{0.005}\right)$$

$$= P(-1.2 \leq X \leq 1.2)$$

$$= 2.P(0 \leq X \leq 1.2) \text{ (By symmetry)}$$

$$= 2.(0.3849) = 0.7698$$

Probability that a washer is defective

$$= 1 - \text{ Probability that a washer is not defective}$$

$$= 1 - 0.7698$$

$$= 0.2302$$

Therefore, the percentage of defective washers $= 23.02$

**Example 13** The mean of a normal distribution is 60 and 6% of the values are greater than 70. Find the standard deviation of the distribution.

**Solution** Let $X$ denote the normal random variable.

We have $P(X > 70) = \frac{6}{100} = 0.06$ (given)

Therefore, $P\left(\frac{x-\mu}{\sigma} > \frac{70-\mu}{\sigma}\right) = 0.06$

i.e., $P\left(Z > \frac{70-60}{\sigma}\right) = 0.06$

or $P\left(Z > \frac{10}{\sigma}\right) = 0.06$

Let $Z_1 = \frac{10}{\sigma}$, then $\frac{10}{\sigma} = 1.56$ (from the table of standard normal probabilities)

We get $\sigma = \frac{10}{1.56} = \frac{1000}{156} = 6.4.$

Hence, the standard deviation is 6.4.

**Example 14** Fit a normal curve to the following data

| $x$ | 6 | 7 | 8 | 9 | 10 | 11 | 12 |
|---|---|---|---|---|---|---|---|
| $f(x)$ | 3 | 6 | 9 | 13 | 8 | 5 | 4 |

**Solution** Mean of the distribution is $\mu = \dfrac{\sum x_i f_i}{\sum f_i}$

$$\text{i.e., } \mu = \frac{(6)(3)+(7)(6)+(8)(9)+(9)(13)+(10)(8)+(11)(5)+(12)(4)}{3+6+9+13+8+5+4}$$

$$= \frac{18+42+72+117+80+55+48}{48} = \frac{432}{48} = 9$$

$$\sigma = \text{Standard deviation} = \sqrt{\frac{\sum f_i x_i^2}{\sum f_i} - \mu^2}$$

$$= \sqrt{\frac{(3)(36)+(6)(49)+(9)(64)+(13)(81)+(8)(100)+(5)(121)+(4)(144)}{48} - 9^2}$$

$$= \sqrt{2.5833} = 1.607$$

The required best fitting normal curve is

$$Y = \frac{N}{\sigma\sqrt{2\pi}} e^{-\frac{1}{2}\left[\frac{x-\mu}{\sigma}\right]^2}$$

$$= \frac{48}{(1.607)\sqrt{2\pi}} e^{-\left[\frac{(x-9)^2}{5.1666}\right]}$$

$$= 11.916.e^{-\left[\frac{(x-9)^2}{5.1666}\right]}$$

**Fitting a normal distribution**

If $N$ denotes the sum of all frequencies of a normal distribution, the normal curve is given by $y = \dfrac{N}{\sigma\sqrt{2\pi}} e^{-\frac{1}{2}\left[\frac{x-\mu}{\sigma}\right]^2}$.

The area under the normal curve is equal to $N$.

To fit a normal curve to a distribution, there are two methods namely

1. Ordinate method

2. Area method

These methods are explained with the help of a few examples.

## Solved examples

**Example 1** Fit a normal curve to the distribution

| $x$: | 1 | 3 | 5 | 7 | 9 |
|---|---|---|---|---|---|
| $f$: | 1 | 2 | 3 | 2 | 1 |

**Solution** Let $\mu$ denote the mean and $\sigma$ denote the standard deviation of the distribution

We have, $N = \sum f_i = 1+2+3+2+1 = 9$

Mean of the distribution is $\bar{x} = \dfrac{\sum x_i f_i}{\sum f_i} = \dfrac{1.1+3.2+5.3+7.2+9.1}{9}$

$$= \frac{45}{9} = 5$$

$$\sigma^2 = \text{variance} = \frac{\sum x_i^2 f_i}{\sum f_i} - \mu^2$$

$$= \frac{1^2.1+3^2.2+5^2.3+7^2.2+9^2.1}{9} - 5^2$$

$$= \frac{1+18+75+98+81}{9} - 25$$

$$= \frac{273}{9} - 25$$

$$= \frac{273-225}{9} = \frac{48}{9}$$

$$\therefore \quad \sigma = \sqrt{\frac{48}{9}} = \frac{4\sqrt{3}}{3}$$

The equation of best fitting curve is

$$y = \frac{N}{\sigma\sqrt{2\pi}} e^{-\frac{1}{2}\left[\frac{x-\mu}{\sigma}\right]^2}$$

$$= \frac{9}{\frac{4\sqrt{3}}{3}\sqrt{2\pi}} e^{-\frac{1}{2}\left[\frac{x-5}{\frac{48}{9}}\right]^2}$$

$$= \frac{9\sqrt{3}}{4\sqrt{2\pi}} e^{-\frac{9}{96}[x-5]^2}$$

$$= \frac{9\sqrt{3}}{4\sqrt{2\pi}} e^{-\frac{3}{32}[x-5]^2}$$

**Example 2** Fit a normal distribution to the following data by the method of ordinates.

| Class interval | 59.5−62.5 | 62.5−65.5 | 65.5−68.5 | 68.5−72.5 | 71.5−74.5 |
|---|---|---|---|---|---|
| Frequency ($f$) | 5 | 18 | 42 | 27 | 8 |

**Solution**

| Class interval | $f$ | $x_i$ (midpoint) | $f_i x_i$ | $f_i x_i^2$ |
|---|---|---|---|---|
| 59.5−62.5 | 5 | 61 | 305 | 18605 |
| 62.5−65.5 | 18 | 64 | 1152 | 73728 |
| 65.5−68.5 | 42 | 67 | 2814 | 188538 |
| 68.5−71.5 | 27 | 70 | 1890 | 132300 |
| 71.5−74.5 | 8 | 73 | 584 | 42632 |
| | 100 | | 6745 | 455803 |

We have $\sum f_i = N = 100$

$\sum f_i x_i = 6745$

$\sum f_i x_1^2 = 455803$

Mean of the distribution is $\bar{x} = \dfrac{\sum x_i f_i}{\sum f_i} = \dfrac{6745}{100} = 67.45$

$$\sigma = \text{Standard deviation} = \sqrt{\frac{\sum f_i x_i^2}{\sum f_i} - \mu^2}$$

$$= \sqrt{\frac{455803}{100} - (67.45)^2} = 2.92$$

Hence $\sigma = 2.92$

The ordinates are obtained at various distances from the mean. We use the formula

$$f(x) = \frac{1}{\sqrt{2\pi}} e^{-\frac{z^2}{2}}$$

where $Z = \dfrac{x - \mu}{\sigma}$

We have the following table.

| CI | Midpoint | $Z = \dfrac{x-\mu}{\sigma}$ | $f(x) = \dfrac{1}{\sqrt{2\pi}} e^{-\frac{z^2}{2}}$ | $y = \dfrac{N}{\sigma} f(Z)$ | $\dfrac{N}{\sigma} f(Z) \times C$ |
|---|---|---|---|---|---|
| 59.5–62.5 | 61 | −2.21 | 0.0347 | 1.188 | 3.565 |
| 62.5–65.5 | 64 | −1.18 | 0.1984 | 6.794 | 20.382 |
| 65.5–68.5 | 67 | −0.15 | 0.3943 | 13.50 | 40.50 |
| 68.5–71.5 | 70 | 0.87 | 0.2725 | 9.333 | 27.999 |
| 71.5–74.5 | 73 | 1.90 | 0.0656 | 2.247 | 6.741 |

The expected (i.e., theoretical) frequencies are 3.565, 20.382, 40.50, 27.999, 6.741 i.e., 4, 20, 41, 28 and 7.

**Example 3** Fit a normal distribution to the following data by the area method.

| CI | 60–65 | 65–70 | 70–75 | 75–80 | 80–85 | 85–90 | 90–95 | 95–100 |
|---|---|---|---|---|---|---|---|---|
| Frequency ($f$) | 3 | 21 | 150 | 335 | 326 | 135 | 26 | 4 |

**Solution** We find the values of mean and standard deviation as follows.

| Class interval | $f$ | $x_i$ (midpoint) | $f_i x_i$ | $f_i x_i^2$ |
|---|---|---|---|---|
| 60–65 | 3 | 62.5 | 187.5 | 11718.75 |
| 65–70 | 21 | 67.5 | 1417.5 | 95681.25 |
| 70–75 | 150 | 72.5 | 10875.0 | 788437.5 |
| 75–80 | 335 | 77.5 | 25962.5 | 2012093.75 |
| 80–85 | 326 | 82.5 | 26895.0 | 2218837.5 |
| 85–90 | 135 | 87.5 | 11812.5 | 1033593.75 |
| 90–95 | 26 | 92.5 | 2405.0 | 22462.5 |
| 95–100 | 4 | 97.5 | 390.0 | 38025.0 |
| | $N = 1000$ | | 79945.0 | 64200850.0 |

We have $\sum f_i = N = 1000$

$$\sum f_i x_i = 79945$$

$$\sum f_i x_1^2 = 64200850.0$$

Mean of the distribution is $\mu = \dfrac{\sum x_i f_i}{\sum f_i} = \dfrac{79945}{1000} = 79.945$

$$\sigma^2 = \text{variance} = \frac{\Sigma x_i^2 f_i}{\Sigma f_i} - \mu^2$$

$$= \frac{64200850}{1000} - (79.945)^2$$

$$= 6420.85 - 639120$$

$$= 29.65$$

$$\therefore \sigma = \sqrt{29.65} = 5.4451$$

Since, the normal distribution takes values from $-\infty$ to $\infty$, the given data can be modified as follows.

| CI | $f$ | Lower limit | $Z = \frac{x-\mu}{\sigma}$ | Area $\phi(Z)$ | $\Delta\phi(Z) = \phi(Z+1) - \phi(Z)$ | $N.\Delta\phi(Z)$ |
|---|---|---|---|---|---|---|
| $-\infty$ to 60 | 0 | $-\infty$ | $-\infty$ | 0 | 0.0001 | 0.1 |
| 60–65 | 3 | 60 | −3.6629 | 0.0001 | 0.003 | 3 |
| 65–70 | 21 | 65 | −2.7447 | 0.0031 | 0.0313 | 31.3 |
| 70–75 | 150 | 70 | −1.8264 | 0.0344 | 0.1497 | 149.7 |
| 75–80 | 335 | 75 | −0.9081 | 0.1841 | 0.3199 | 319.9 |
| 80–85 | 326 | 80 | 0.0101 | 0.5040 | 0.3172 | 317.2 |
| 85–90 | 135 | 85 | 0.9283 | 0.8212 | 0.1459 | 145.9 |
| 90–95 | 26 | 90 | 1.8466 | 0.9671 | 0.03 | 30 |
| 95–100 | 4 | 95 | 2.7649 | 0.9671 | 0.0028 | 2.8 |
| $100-\infty$ | $\infty$ | 100 | 3.6831 | 0.9999 | | |

The expected frequencies are 0.1, 3, 31.3, 149.7, 319.9, 317.2, 145.9, 30, 2.8 i.e., 0, 3, 31, 150, 320, 317, 146, 30, 3.

**Example 4** Fit a normal distribution for the following data by the ordinates' method

| Class interval | 150–160 | 160–170 | 170–180 | 180–190 | 190–200 | 200–210 | 210–220 | 220–230 | 230–240 |
|---|---|---|---|---|---|---|---|---|---|
| Frequency | 9 | 24 | 51 | 66 | 72 | 48 | 21 | 6 | 3 |

**Solution** In order to fit a normal distribution, we compute the values of mean ($\sigma$) and standard deviation ($\sigma$) of the distribution. Using the values (of $\mu$ and $\sigma$) obtained, we compute.

$$Z = \frac{x-\mu}{\sigma}$$

For the given values of $x$, after finding the values of $Z$ and using the table of areas, the values of $\phi(Z)$ are obtained. The ordinate $Z$ for any value of $\phi(Z)$ is obtained by the relation

$$Z = \frac{N}{\sigma}\phi(Z)$$

The theoretical frequencies can be calculated by using the above relation.

| Class interval | $f$ | $x_i$ (midpoint) | $f_ix_i$ | $f_ix_i^2$ |
|---|---|---|---|---|
| 150–160 | 9 | 155 | 1345 | 216225 |
| 160–170 | 24 | 165 | 3960 | 653400 |
| 170–180 | 51 | 175 | 8925 | 156187 |
| 180–190 | 66 | 185 | 12210 | 2258850 |
| 190–200 | 72 | 195 | 14040 | 2737800 |
| 200–210 | 48 | 205 | 9840 | 2017200 |
| 210–220 | 21 | 215 | 4515 | 970725 |
| 220–230 | 6 | 225 | 1350 | 303750 |
| 230–240 | 3 | 235 | 705 | 165675 |
| | 300 | | 56940 | 10885500 |

Mean of the distribution is $\mu = \frac{\Sigma x_i f_i}{\Sigma f_i} = \frac{56940}{300} = 189.8$

$$\sigma = \text{Standard deviation} = \sqrt{\frac{\Sigma f_i x_i^2}{\Sigma f_i} - \mu^2}$$

$$= \sqrt{\frac{10885500}{300} - 189.8^2} = 16.15$$

Hence $\sigma = 16.15$

Using the values of $\mu$ and $\sigma$, we fit the normal distribution for the given data as follows (where $C$ denotes the length of the class interval)

| CI | Midpoint | $Z = \frac{x-\mu}{\sigma}$ | Value of the ordinate | $\frac{N}{\sigma}\phi(Z)$ $\phi(Z)$ | $\left[\frac{N}{\sigma}\phi(Z)\right] \times C$ |
|---|---|---|---|---|---|
| 150–160 | 155 | 2.15 | 0.0396 | 0.7356 | 7.35~7 |
| 160–170 | 165 | 1.5356 | 0.1238 | 2.299 | 22.99~23 |
| 170–180 | 175 | 0.9164 | 0.2637 | 4.902 | 49 |
| 180–190 | 185 | 0.297 | 0.3825 | 7.1113 | 71 |
| 190–200 | 195 | 0.322 | 0.379 | 7.05 | 71 |
| 200–210 | 205 | 0.942 | 0.2565 | 4.77 | 48 |
| 210–220 | 215 | 1.562 | 0.1182 | 2.198 | 22 |
| 220–230 | 225 | 2.181 | 0.0371 | 0.689 | 7 |
| 230–240 | 235 | 2.80 | 0.0079 | 0.15 | 2 |

Hence the expected frequencies are 7, 23, 49, 71, 71, 48, 22, 7 and 2.

**Example 5** Fit a normal distribution to the following data and calculate the theoretical frequencies by the area method.

| Class interval | 59.5−62.5 | 62.5−65.5 | 65.5−68.5 | 68.5−72.5 | 71.5−74.5 |
|---|---|---|---|---|---|
| No. of students | 5 | 18 | 42 | 27 | 8 |

**Solution**

| Class interval | $f$ | $x_i$ (midpoint) | $f_i x_i$ | $f_i x_i^2$ |
|---|---|---|---|---|
| 59.5−62.5 | 5 | 61 | 305 | 18605 |
| 62.5−65.5 | 18 | 64 | 1152 | 73728 |
| 65.5−68.5 | 42 | 67 | 2814 | 188538 |
| 68.5−71.5 | 27 | 70 | 1890 | 132300 |
| 71.5−74.5 | 8 | 73 | 584 | 42632 |
| | 100 | | 6745 | 455803 |

We have $\sum f_i = N = 100$

$$\sum f_i x_i = 6745$$

$$\sum f_i x_1^2 = 455803$$

Mean of the distribution is $\bar{x} = \dfrac{\sum x_i f_i}{\sum f_i} = \dfrac{6745}{100} = 67.45$

$$\sigma = \text{Standard deviation} = \sqrt{\frac{\sum f_i x_i^2}{\sum f_i} - \mu^2}$$

$$= \sqrt{\frac{455803}{100} - (67.45)^2} = 2.92$$

Hence $\sigma = 2.92$

The expected frequencies can be computed as shown in the table given below.

| Lower limits | $Z = \dfrac{x-\mu}{\sigma}$ | Area under the normal curve $\phi(Z)$ | Area included in each class $\Delta\phi(Z)$ | Expected frequency $N.\Delta\phi(Z)$ |
|---|---|---|---|---|
| 59.5 | −2.72 | 0.4967 | | |
| 62.5 | −1.70 | 0.4554 | 0.0413 | 4.13∼4 |
| 65.5 | −0.67 | 0.2486 | 0.2068 | 20.68∼21 |
| 68.5 | 0.36 | 0.1406 | 0.3892 | 38.92∼39 |
| 71.5 | 1.39 | 0.4177 | 0.2771 | 27.71∼28 |
| 74.5 | 2.4 | 0.4920 | 0.0743 | 7.43∼7 |

Hence the theoretical frequencies are 4, 21, 39, 28 and 7.

**Example 6** Find the points of inflection of a normal curve. Find the asymptote of the curve.

**Solution** The equation of the normal curve is

$$y = f(x) = \frac{1}{\sigma\sqrt{2\pi}} e^{-\frac{1}{2}\left[\frac{x-\mu}{\sigma}\right]^2} \tag{1}$$

Differentiating (1) with respect to $x$, we get

$$f'(x) = \frac{dy}{dx} = \frac{1}{\sigma\sqrt{2\pi}} e^{-\frac{1}{2}\left[\frac{x-\mu}{\sigma}\right]^2}\left[-\frac{(x-\mu)}{\sigma^2}\right]$$

$$= -\frac{1}{\sigma\sqrt{2\pi}\sigma^2}(x-\mu).e^{-\frac{1}{2}\left[\frac{x-\mu}{\sigma}\right]^2}$$

Again differentiating, we get

$$f''(x) = \frac{d^2y}{dx^2} = \frac{1}{\sigma\sqrt{2\pi}}\left[e^{-\frac{1}{2}\left[\frac{x-\mu}{\sigma}\right]^2}\left[-\frac{(x-\mu)}{\sigma^2}\right]^2 - \frac{1}{\sigma^2}e^{-\frac{1}{2}\left(\frac{x-\mu}{\sigma}\right)^2}\right] = 0$$

$$\frac{d^2y}{dx^2} = 0 \quad \text{when} \quad \left(\frac{x-\mu}{\sigma}\right)^2 - 1 = 0$$

i..e, $(x-\mu)^2 = \sigma^2$

or $(x-\mu) = \pm\sigma$

or $x = \mu \pm \sigma$

Differentiating $\frac{d^2y}{dx^2}$ with respect to $x$, we get

$$\frac{d^3y}{dx^3} = \frac{1}{\sigma\sqrt{2\pi}}\left[e^{-\frac{1}{2}\left[\frac{x-\mu}{\sigma}\right]^2}\left[-\frac{(x-\mu)}{\sigma^2}\right]^3 + \frac{3(x-\mu)}{\sigma^2}e^{-\frac{1}{2}\left(\frac{x-\mu}{\sigma}\right)^2}\right]$$

Hence $\frac{d^3y}{dx^3} = \pm\frac{2}{\sigma^4\sqrt{2\pi}}e^{-\frac{1}{2}}$ when $\mu = x \pm \sigma$

i.e., $\frac{d^3y}{dx^3} \neq 0$ when $\mu = x \pm \sigma$

The values of $y = f(x)$ at $x = \mu \pm \sigma$ is $\frac{1}{\sigma\sqrt{2\pi}}e^{-\frac{1}{2}}$

Therefore, the points of inflection are $\left(\mu \pm \sigma, \frac{1}{\sigma\sqrt{2\pi}}e^{-\frac{1}{2}}\right)$

The line $y = 0$, meets the curve at $x = \pm\infty$.

Hence the curve has $x$-axis for its asymptote.

## Exercise 3.4

1 If $X$ is a normal variable with mean 25 and standard deviation 5. Find the probability that (i) $15 \le X \le 30$ and (ii) $|x-30| \ge 10$.

2 Find the probability that the standard normal variate lies between 0 to 1.5.

3 $X$ is normally distributed. The mean of $X$ is 12 and the S.D is 4. Find the probability of the following, (i) $x \ge 20$, (ii) $0 \le x \le 12$.

4 In a normal distribution, 31% of the items are under 45 and 8% are over 64. Find the mean and the standard deviation.

5 A large number of measurements are normally distributed with mean 65.5" and S.D 6.2". Find the percentage of measurement that fall between 54.8" and 68.8".

6 In a distribution exactly normal, 7% of the items are under 35 and 89% are under 63. What are the values of mean and standard deviation of the distribution?

7 The weekly wages of 1000 workers are normally distributed around a mean of Rs.70 with a standard deviation of Rs.5. Estimate the number of workers whose weekly wages will be

(a) between Rs. 70 and Rs. 72

(b) between Rs. 68 and Rs. 72

(c) more than Rs. 75

(d) less than Rs. 63

(e) more than Rs. 80

8 The mean heights of students in a class is 159 cms. With a standard deviation of 20 cms, find how many students' heights lie between 150 cms and 170 cms, if there are 100 students in the class.

9 A manufacturer knows from experience that the resistance of the resistors he produces is normal with mean 100 ohms and standard deviation 20 hours. What percentage of resistors will have resistance between 96 ohms and 104 ohms?

10 $X$ is a normal variate with mean 30 and standard deviation 5. Find the probability of (i) $26 \le x \le 40$, (ii) $x \ge 45$.

11 In a sample of 1000 students, the mean of the common score is 14 and the S.D is 2.5. Assuming the distribution to be normal, find (i) how many students score between 12 and 15, (ii) how many score above 18, (iii) how many score below 8.

12 Students of a class were given a mechanical application test. Their marks were found to be normally distributed with mean 60 and standard deviation 5. What per cent of students scored (i) more than 60 marks, (ii) less than 56 marks, (iii) between 45 and 65 marks?

13 If the diameters of ball bearings are normally distributed with mean as 0.614 and standard deviation 0.0025 cms, determine the percentage of ball bearings with diameter (i) between 0.610 cms and 0.618 cms, (ii) greater than 0.617 cms, (iii) less than 0.608 cms, (iv) equal to 0.615 cms.

14 The life of electronic tubes of certain types may be assumed to be normally distributed with mean 155 hours and standard deviation 19 hours. What is the probability that the life of a randomly chosen tube is (i) less than 117 hours, (ii) between 136 hours and 194 hours, (iii) more than 395 hours?

15 Obtain the equation of the normal probability curve that may be fitted to the following data

| $x$: | 4 | 6 | 8 | 10 | 12 | 14 | 16 | 18 | 20 | 22 | 24 |
|---|---|---|---|---|---|---|---|---|---|---|---|
| $f$: | 1 | 7 | 15 | 22 | 25 | 43 | 38 | 20 | 13 | 15 | 1 |

16 Fit a normal curve to the following data and obtain the expected frequencies.

| $x$: | 8.60 | 8.59 | 8.58 | 8.57 | 8.56 | 8.55 | 8.54 | 8.53 | 8.52 |
|---|---|---|---|---|---|---|---|---|---|
| $f$: | 2 | 3 | 4 | 9 | 10 | 8 | 4 | 1 | 1 |

17 Fit a normal curve to the following data.

| $x$: | 1 | 2 | 3 | 4 | 5 | 6 | 7 | 8 | 9 | 10 | 11 | 12 |
|---|---|---|---|---|---|---|---|---|---|---|---|---|
| $f$: | 2 | 10 | 19 | 25 | 40 | 44 | 41 | 28 | 25 | 15 | 5 | 1 |

18 Prove that for the normal distribution the quartile deviation, mean deviation and the standard deviation are approximately in the ratio 10 : 12: 15.

19 Find the equation of the best fitting normal curve that may be fitted to the following distribution

| $x$ | 1 | 3 | 5 | 7 | 9 |
|---|---|---|---|---|---|
| $f$ | 1 | 2 | 3 | 2 | 1 |

20 Fit a normal curve to the following data.

| $x$: | 1 | 2 | 3 | 4 | 5 | 6 | 7 | 8 | 9 | 10 | 11 | 12 |
|---|---|---|---|---|---|---|---|---|---|---|---|---|
| $f$: | 2 | 10 | 19 | 25 | 40 | 44 | 41 | 28 | 25 | 15 | 5 | 1 |

21 Fit a normal distribution to the following data by the ordinates' method.

| CI: | 150–160 | 160–170 | 170–180 | 180–190 | 190–200 | 200–210 | 210–220 | 220–230 | 230–240 |
|---|---|---|---|---|---|---|---|---|---|
| $f$: | 9 | 24 | 51 | 66 | 72 | 48 | 21 | 6 | 3 |

22 Find the theoretical frequencies for the distribution

| $x$ | 2 | 4 | 6 | 8 | 10 |
|---|---|---|---|---|---|
| $f$ | 1 | 4 | 6 | 4 | 1 |

24 Obtain the equation of the normal probability curve that may be fitted to the following data and calculate the theoretical frequencies.

| $x$: | 4 | 6 | 8 | 10 | 12 | 14 | 16 | 18 | 20 | 22 | 24 |
|---|---|---|---|---|---|---|---|---|---|---|---|
| $f$: | 1 | 7 | 15 | 22 | 25 | 43 | 38 | 20 | 13 | 15 | 1 |

## Answers 3.4

1. (i) 0.8185 (ii) 0.16
2. 0.4327
3. (i) 0.0228 (ii) 0.9972 (iii) 0.4987
4. $\mu = 50, \sigma = 10$
5. 6690
6. 50.3, 10.33
7. 155, 235, 159, 81, 23
8. 38
9. 95
10. 0.7653, 0.0014
11. 0.4435, 0.0547, 0.0082
12. 50, 21.19, 84
13. 89.04, 11.51% , 0.82%
14. 0.0288, 0.8211, 0
15. $e^{-\left(\frac{x-13.85}{29.34}\right)^2}$
16. $(9.8)e^{-0.163}e^{(x-8.563)^2}$, 0.99, 2.24, 5.63, 7.5, 9.5, 7.5, 5.63, 2.24, 0.79
17. $e^{-0.99(x-6.259)^2}$
19. $\dfrac{9\sqrt{3}}{4\sqrt{2\pi}}e^{-3\left(\frac{x-5}{32}\right)^2}$
20. $\dfrac{113}{\sqrt{2\pi}}e^{-0.099(x-6.259)^2}$
21. 7, 23, 49, 71, 71, 48, 22, 7, 2
22. 0.79, 2.24, 5.63, 7.5, 9.5, 7.5, 5.63, 2.24, 0.79

23. 0.864, 3.872, 6.3824, 3.872, 0.864

24. $y = \dfrac{1}{3.83\sqrt{2\pi}} e^{-\left(\frac{(x-13.85)^2}{29.355}\right)}$

## Linear combination of independent normal variables

If $\lambda_1, \lambda_2, \ldots, \lambda_n$ are independent normal variables, and $a_1, a_2, \ldots, a_n$ are real constants, then $a_1x_1 + a_2x_2 + \ldots + a_nx_n$ is also a normal variate.

*Proof* Since $M_x(t) = e^{t\mu + \frac{1}{2}t^2\sigma^2}$,

We have $M(t)_{a_ix_i} = M_X(a_it) = e^{at\mu_i + \frac{1}{2}a_i^2t^2\sigma_i^2}$

$$= e^{t(a_i\mu_i) + \frac{1}{2}t^2(a_i\sigma_i)^2}$$

$$M(t)_{\sum a_ix_i} = M(t)_{a_1x_1 + a_2x_2 + \ldots + a_nx_n}$$

$$= M_{a_1x_1}(t).M_{a_2x_2}(t)\ldots M_{a_nx_n}(t)$$

$$= e^{t(a_1\mu_1) + \frac{1}{2}t^2(a_1\sigma_1)^2}.e^{t(a_2\mu_2) + \frac{1}{2}t^2(a_2\sigma_2)^2}\ldots e^{t(a_n\mu_n) + \frac{1}{2}t^2(a_n\sigma_n)^2}$$

$$= e^{t\sum_1^n(a_i\mu_i) + \frac{1}{2}t^2\sum_1^n(a_i\sigma_i)^2}$$

is the moment generating function of a normal variate with mean $\sum a_i\mu_i$ and variance, $\sum a_i^2\sigma_i^2$

Hence, the theorem.

## Normal approximation to binomial distribution

The normal distribution can be used to approximate the binomial distribution. Consider a random variable $X$. If $X$ follows a binomial distribution, then we have

$$P(X = x) = {}^nC_xp^xq^{n-x}$$

which is very large, then

$$P(X_1 < X < X_2) = \sum_{x=x_1}^{n=x_2} {}^nC_xp^xq^{n-x}$$

(i) When $p = q = \dfrac{1}{2}$, $n$ may or may not be large

Let $Z = \dfrac{x-\mu}{\sigma}$

where $\mu$ mean $= np$ and S.D $= \sigma = \sqrt{npq}$

$$\text{Then, } P(X_1 < X < X_2) = P\left(\frac{X_1-\mu}{\sigma} < \frac{X-\mu}{\sigma} < \frac{X_2-\mu}{\sigma}\right)$$

$$= P(Z_1 < Z < Z_2) = \int_{z_1}^{z_2} \phi(z)dz$$

which can be obtained from the area table.

(ii) When $p \neq q$, and $n$ is large

The real class interval is $(x - \frac{1}{2}, x + \frac{1}{2})$

For $X$ to lie in the interval, we consider the interval $(x - \frac{1}{2}, x + \frac{1}{2})$

The corresponding values of $Z$ are $\dfrac{x_1 - \frac{1}{2} - \mu}{2}$ and $\dfrac{x_2 - \frac{1}{2} - \mu}{2}$, where $\mu = np$ and $\sigma = \sqrt{npq}$.

The required probability is $\int_{z_1}^{z_2} \phi(z)dz$.

## Solved examples

**Example 1** Find the probability of getting 1 or 3 or 4 or 5 in throwing a die 5 to 7 times among 9 trials.

**Solution** We have $n = 9$

$$p = 4.\frac{1}{6} = \frac{2}{3}$$

$$q = 1 - p = 1 - \frac{2}{3} = \frac{1}{3}$$

$$\text{Mean } \mu = np = 9.\frac{2}{3} = 6$$

$$\sigma^2 = \text{variance} = npq = 9.\frac{2}{3}.\frac{1}{3} = 2$$

$$\therefore \sigma = \sqrt{2} = 1.414$$

$$x_1 = 5,\ x_2 = 7$$

$$Z_1 = \frac{x_1 - \frac{1}{2} - \mu}{\sigma} = \frac{5 - \frac{1}{2} - 6}{1.414} = \frac{4.5 - 6}{1.414}$$

$$= \frac{-1.5}{1.414} = -1.0608$$

$$Z_2 = \frac{x_2 - \frac{1}{2} - \mu}{\sigma} = \frac{7 - \frac{1}{2} - 6}{1.414} = \frac{7.5 - 6}{1.414}$$

$$= \frac{1.5}{1.414} = 1.0608$$

$$P(X_1 < X < X_2) = P(X_1 - \frac{1}{2} < X < X_2 + \frac{1}{2})$$

$$\therefore P(4.5 < X < 7.5) = P(-1.0608 < Z < 1.0608)$$

$$= P(-1.0608 < Z < 0) + P(0 < Z < 1.0608)$$

$$= 2[P(0 < Z < 1.0608)] = 2(0.3554)$$

$$= 0.7108$$

**Example 2** 8 coins are tossed together. Find the probability of getting 1 to 5 heads in a single toss.

**Solution** We have $n = 8$

$$p = \frac{1}{2} \text{ and } q = \frac{1}{2}$$

$$\text{Mean } \mu = np = 8.\ \frac{1}{2} = 4$$

$$\sigma^2 = \text{variance} = npq = 8.\ \frac{1}{2}.\frac{1}{2} = 2$$

$$\therefore \sigma = \sqrt{2} = 1.414$$

$$x_1 = 1,\ x_2 = 5$$

$$Z_1 = \frac{x_1 - \frac{1}{2} - \mu}{\sigma} = \frac{1 - \frac{1}{2} - 4}{1.414}$$

$$= \frac{-3.5}{1.414} = -2.4752$$

$$Z_2 = \frac{x_2 - \frac{1}{2} - \mu}{\sigma} = \frac{5 - \frac{1}{2} - 4}{1.414}$$

$$= \frac{3.5}{1.414} = 2.4752$$

$$\therefore P(1 \le X \le 5) = P(-2.475 \le Z \le 2.475)$$

$$= 2[P(0 \le Z \le 2.475)] = 2(0.4932)$$

$$= 0.9864$$

**Example 3** Random variable $X$ is normally distributed with $\mu = 12$ and standard deviation 2. Find $P(9.6 < x < 13.8)$, given that for $\frac{x}{\sigma} = 0.9$, $A = 0.3159$ and for $\frac{x-\mu}{\sigma} = 1.2$, $A = 0.3849$.

**Solution** We have $P(9.6 < x < 13.8) = \int_{9.6}^{13.8} \frac{1}{\sigma\sqrt{2\pi}} e^{-\frac{1}{2}\left[\frac{x-\mu}{\sigma}\right]^2} dx$ (1)

$$\mu = 12, \sigma = 2\ , Z = \frac{x-\mu}{\sigma} = \frac{x-12}{2}$$

When $X = 9.6$ , $Z = Z_1 = \frac{9.6-12}{2} = -1.2$

And $X = 13.8$, $Z = Z_2 = \frac{13.8-12}{2} = 0.9$

Substituting in (1) we get

$$P(9.6 < x < 13.8) = \int_{9.6}^{13.8} \frac{1}{2\sqrt{2\pi}} e^{-\frac{1}{2}\left[\frac{x-12}{2.4}\right]^2} dx$$

$$= \frac{1}{\sqrt{2\pi}} \int_{-1.2}^{0.9} e^{-\frac{1}{2}z^2} dx$$

$$= \frac{1}{\sqrt{2\pi}} \int_{-1.2}^{0} e^{-\frac{1}{2}z^2} dx + \frac{1}{\sqrt{2\pi}} \int_{0}^{0.9} e^{-\frac{1}{2}z^2} dx$$

$$= 0.3849 + 0.3159 = 0.7008$$

### Characteristic function

Let $x$ be a continuous random variable. The characteristic function of $x$ for the continuous probability distribution with density function $f(x)$ is defined as

$$E[e^{itx}] = \int_{-\infty}^{\infty} e^{itx} f(x)dx$$

$$= \int_{-\infty}^{\infty} e^{itx} dF(x), (i = \sqrt{-1})$$

where $t$ is a real parameter. The characteristic function of $x$ is denoted by $\phi_x(t)$ or $\phi(t)$.

(1) $\phi(t)$ is continuous in $t$

(2) $\phi(0) = 1$

(3) $\phi(t)$ is defined in every finite $t$ interval

(4) $|\phi(t)| \leq 1$

(5) $\phi(t)$ and $\phi(-t)$ are conjugate

## Solved examples

**Example 1** Find the characteristic function of the exponental random variable

$$f(x) = ae^{-ax}, \quad x \geq 0, a > 0$$

$$= 0, \quad \text{otherwise}$$

**Solution** We have $\phi(t) = E(e^{itx})$

$$\text{i.e., } \phi(t) = \int_0^{\infty} e^{itx}(ae^{-ax})dx$$

$$= a\int_0^{\infty} e^{-(a-it)x} dx$$

$$= \frac{a}{a-it}$$

**Example 2** Find the characteristic function of a binomial distribution.

**Solution** The characteristic function of a binomial distribution is given by

$$\phi(t) = \sum_{n=0}^{\infty} e^{itx}\, {}^nC_x p^x q^{n-x}$$

$$= \sum_{x=0}^{n} {}^nC_x$$

**Example 3** A random variable $x$ has the rectangular distribution

$$f(x) = \frac{1}{2k} \text{ for } a-k \le x \le a+k$$

$$= 0 \text{ otherwise}$$

Find the characteristic function of the random variable.

**Solution** The characteristic function is

$$\phi(t) = \frac{1}{2k} \int_{a-k}^{a+k} e^{itx} dx$$

$$= \frac{e^{it(a+k)} - e^{it(a-k)}}{2itk} = \frac{e^{ita}\left[e^{itk} - e^{-itk}\right]}{2itk}$$

$$= \frac{1}{kt} e^{ita} . \sin tk$$

**Example 4** Show that $\phi(-t)$ is in complex conjugate of $\phi(t)$.

**Solution** We have $\phi(t) = E\left(e^{itx}\right)$

$$= E(\cos tx + i \sin tx) = E(\cos tx) + iE \sin tx \qquad (1)$$

$$\phi(-t) = E\left(e^{-itx}\right)$$

$$= E[\cos tx - i\sin tx]$$

$$= E[\cos tx] - iE(\sin tx) \qquad (2)$$

From (1) and (2),

$\phi(-t)$ is in complex conjugate of $\phi(t)$

## 3.9 Gamma distribution

Let $x$ be a continuous random variable. $x$ is said to follow a Gamma distribution with parameter $\lambda$, if its probability function is given by

$$P(X = x) = P(x) = \frac{e^{-x}x^{\lambda-1}}{\Gamma\lambda},\ \lambda>0,\ 0< x < \infty$$

$$= 0, \text{ otherwise}$$

The above distribution is known as Gamma distribution of first kind.

### 3.9.1 Moment generating function of Gamma distribution

$$M_X(t) = E[e^{tx}] = \int_0^\infty e^{tx} f(x)dx$$

$$= \int_0^\infty e^{tx}\frac{e^{-x}x^{\lambda-1}}{\Gamma\lambda}dx$$

$$= \frac{1}{\Gamma\lambda}\int_0^\infty e^{-(1-t)x}x^{\lambda-1}dx \qquad (1)$$

If we put $(1-t)x = \theta$, we get $(1-t)dx = d\theta$

When $x = 0$, $\theta = 0$ and when $x = \infty$, $\theta = \infty$.

From (1) we get,

$$M_X(t) = \frac{1}{(1-t)^\lambda}\frac{1}{\Gamma\lambda}\int_0^\infty e^{-\theta}\theta^{\lambda-1}d\theta$$

$$= \frac{1}{(1-t)^\lambda}\frac{1}{\Gamma\lambda}\Gamma\lambda$$

$$= \frac{1}{(1-t)^\lambda} = (1-t)^{-\lambda}, \text{ when } |t| < 1$$

### 3.9.2 Mean of Gamma distribution

Since $M_X(t) = (1-t)^{-\lambda}$

We get $M'_X(t) = -\lambda(1-t)^{-\lambda-1}(-1)$

$$= \lambda(1-t)^{-\lambda-1}$$

Hence $\mu'_1 = M'_X(0) = \lambda$

Since $M'_X(t) = \lambda(1-t)^{-\lambda-1}$, we get

$$M''_X(t) = (-\lambda-1)\lambda(1-t)^{-\lambda-2}(-1)$$

$$= (\lambda+1)\lambda(1-t)^{-\lambda-2}$$

$$\therefore \quad \mu'_2 = M''_X(0) = \lambda(\lambda+1)$$

$$\text{Variance} = \mu'_2 - \mu'^2$$

$$= \lambda(\lambda+1) - \lambda^2$$

$$= \lambda^2 + \lambda - \lambda^2 = \lambda$$

Hence mean of Gamma distribution

= variance of Gamma distribution = $\lambda$

The first four central moments of a Gamma distribution of the first kind are

$$\mu_1 = 0,\ \mu_2 = \lambda,\ \mu_3 = 2\lambda \text{ and } \mu_4 = 3\lambda(\lambda+2)$$

Skewness of distribution is given by

$$\beta_1 = \frac{\mu_3^2}{\mu_2^3} = \frac{|2\lambda|^2}{\lambda^3} = \frac{4}{\lambda}$$

Kurtosis of the distribution is given by

$$\beta_2 = \frac{\mu_4}{\mu_2^2} = \frac{3\lambda^2+6\lambda}{\lambda^2} = 3+\frac{6}{\lambda}$$

### 3.9.3 Gamma distribution of the second kind

A random variable $x$ is said to follow a Gamma distribution of the second kind if it has the following density function.

$$f(x) = P(X=x) = \frac{a^\lambda}{\Gamma\lambda} e^{-ax} x^{\lambda-1}, a>0, \lambda>0, 0<x<\infty$$

$$= 0, \text{ otherwise}$$

## 3.10 Beta Distribution

Let $x$ be a continuous random variable. $x$ is said to follow a Beta distribution of the first kind if its probability density function is given by

$$f(x) = \frac{1}{\beta(m,n)} . x^{m-1}(1-x)^{n-1}, m,n>0, 0<x<1$$

where $m$, $n$ are parameters of the distribution.

We have $\beta(m,n) = \dfrac{\Gamma m \Gamma n}{\Gamma(m+n)}$

### 3.10.1 Moments of Beta distribution of the first kind about the origin

The $r$th moment about origin

$$= \mu_r' = E(X^r) = \frac{1}{\beta(m,n)} \int_0^1 x^{m+r-1}(1-x)^{n-1} dx$$

$$= \frac{\beta(m+r,n)}{\beta(m,n)}$$

The moment generating function is given by

$$M_X(t) = E(e^{tx}) = \sum_{r=0}^{\infty} \frac{(m+r-1)(m+r-2)\ldots m}{(m+n+r-1)(m+n+r-2)\ldots(m+n)} \cdot \frac{t^r}{r!}$$

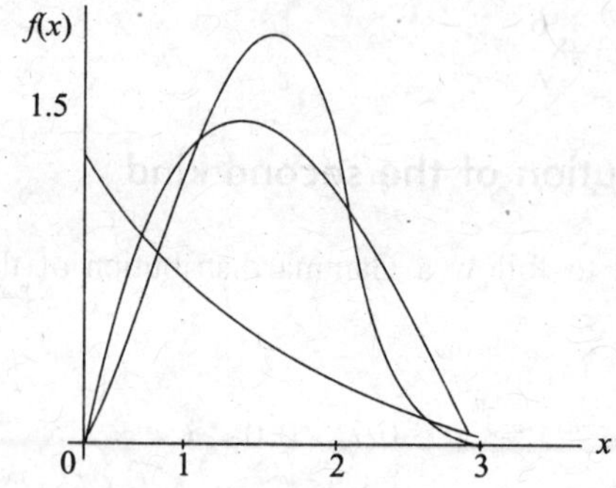


**Fig. 3.3** *Density function of a Beta distribution of the first kind for various values of m and n*

### 3.10.2 Beta distribution of the second kind

A continuous random variable $x$ is said to follow Beta distribution of the second kind if its probability density function is given by

$$f(x) = \frac{1}{\beta(m,n)} \cdot \frac{x^{m-1}}{(1+x)^{m+n}}, m,n > 0, 0 < x < \infty$$

$$= 0, \text{ otherwise}$$

The $r$th moment of Beta distribution of the second kind is given by

$$\mu_r' = E(X^r) = \frac{1}{\beta(m,n)} \int_0^\infty \frac{x^r x^{m-1}}{(1+x)^{m+n}} dx$$

$$= \frac{\Gamma(m+r)\Gamma(n-r)}{\Gamma m \Gamma n}$$

## 3.11 Weibull Distribution

The random variable $x$ has a Weibull distribution if its probability density function has the form

$$f(x) = \frac{\beta}{\alpha}\left(\frac{x-\nu}{\alpha}\right)^{\beta-1} e^{-\left(\frac{x-\nu}{\alpha}\right)^{\beta}}, x = \nu$$

$$= 0, \text{ otherwise}$$

The three parameters of Weibull's distribution are $\nu\,(-\infty < \nu < \infty)$, $\alpha$ and $\beta$, where $\alpha > 0$ and $\beta > 0$.

$\nu$ is called the location parameter.

$\alpha$ is the called scale parameter.

$\beta$ is the called the shape parameter

When $\nu = 0$, the Weibull's p.d.f becomes

$$f(x) = \frac{\beta}{\alpha}\left(\frac{x}{\alpha}\right)^{\beta-1} e^{-\left(\frac{x}{\alpha}\right)^{\beta}},\ x \geq 0$$

$$= \frac{\beta}{\alpha^{\beta-1}} x^{\beta-1} e^{-\left(\frac{x}{\alpha}\right)}$$

$$= 0, \text{ otherwise}$$

Figure 3.4 shows several Weibull densities for $\nu = 0$, and $\alpha = \frac{1}{2}$.

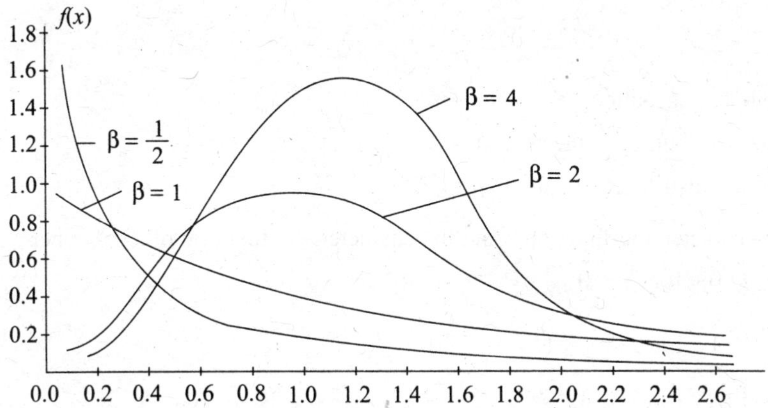


**Fig. 3.4** *Weilbull probability density function for* $\nu = 0,\ \alpha = \frac{2}{2}$

When $\nu = 0$, and $\beta = 1$, the Weibull distribution reduces to

$$f(x) = \frac{1}{\alpha} e^{-x/\alpha},\ x \geq 0$$

$$= 0, \text{ otherwise}$$

which is an exponential distribution with parameter $\lambda = \frac{1}{\alpha}$.

The mean and variance of the Weibull distribution are given by the following expressions

$$E(X) = \nu + \alpha\Gamma\left(\frac{1}{\beta} + 1\right)$$

$$V[X] = \alpha^2\left[\Gamma\left(\frac{2}{\beta} + 1\right) - \left[\Gamma\left(\frac{1}{\beta} + 1\right)\right]^2\right]$$

Thus, the location parameters have no effect on the variance, however the mean is increased or decreased by $\nu$.

The cumulative distribution function of the Weibull distribution is given by

$$F(x) = 0,\ x < \nu$$

$$= 1 - e^{-\left(\frac{x-\nu}{\alpha}\right)^{\beta}}\ x \geq \nu$$

**Example** The time it takes for an aircraft to clear the runway at a major international airport has a Weibull distribution with $\nu = 1.34$ minutes. $\beta = 0.5$ and $\alpha = 0.04$ minutes. Determine the probability that an incoming airplane will take more than 1.5 minutes to land and clear the runway.

**Solution** We have $P(X = 1.5) = F(1.5) = 1 - e^{-\left(\frac{1.5-1.34}{0.04}\right)^{0.5}}$

$$= 1 - e^{-2} = 1 - 0.135 = 0.865$$

$P(X > 1.5)$ = Probability that an incoming airplane will take more than 1.5 minutes $= 1 - 0.865 = 0.135$

## Exercise 3.5

1 Define

(a) Weibull distribution

(b) Beta distribution of the first kind

(c) Gamma distribution of the first kind

2 Define characteristic function. Find the characteristic function of $f(x)$, where

$$f(x) = \begin{cases} 0, & \text{for } x < 0 \\ 1 & \text{for } 0 \leq x \leq 1 \\ 0 & \text{for } x > 2 \end{cases}$$

3 Find the characteristic function for $f(n) = \dfrac{1}{\sqrt{2\pi}} e^{\frac{x^2}{2}}$

4 Show that the distribution for which the characteristic function $e^{|t|}$ has the density function $f(x) = \dfrac{1}{\pi(1+x^2)}, \; -\infty < x < \infty.$

## Answers 3.5

2. $\dfrac{e^{it} - 1}{e^{it}}$ 3. $e^{-\frac{t^2}{2}}$

CHAPTER FOUR

# Sampling Distributions

## 4.1 Introduction

In this chapter, we introduce sampling fundamentals. Sampling is the process of obtaining information about an entire population by examining a part of it. A sample study is usually less expensive and produces results at a relatively faster speed. Sampling remains the only way when a population contains infinitely many members. It usually enables the estimation of the sampling errors and thus assists in obtaining information about some characteristics of the population. Sampling theory is the study of relationships existing between a population and samples drawn from the population. The theory is applicable only to random variables. Sampling theory deals with two types of problems viz. estimation and testing of hypothesis. Before discussing the theory in detail, we should be familiar with some fundamental definitions concerning sampling.

## 4.2 Population

**Definition 4.1** A population is defined as a large collection of individuals or attributes or numerical data. It is also known as a universe.

A population may be finite or infinite. One containing a finite number of individuals or members is said to be a finite population. A population or universe with infinite number of members is called an infinite population. The size of the population is usually designated as $N$.

*A universe* (population) may be real or hypothetical. The universe of concrete objects is called a real universe i.e., if the universe is known and is in existence at the time of sampling then it is called a *real universe.*

A *hypothetical universe* may be defined as the collection of all possible ways in which specified events can happen. It is built up by repeating the event any number of trials. A hypothetical universe population may be imagined around any event. When a coin is tossed any number of times, the population of heads and tails obtained is a hypothetical universe. Another example of a hypothetical population is the population of all values which the bank rate can have in ten years time. The population of concrete subjects is known as the existent population.

## 4.3 Sample

**Definition 4.2** A sample is defined as a finite subset of a population. The size of a population is denoted by $n$. If the size of a sample is less than or equal to 30, then the sample is said to be a small sample. If the size of the sample is greater than 30, the sample is said to be a large sample.

**Definition 4.3** The process of selecting a sample from a population is called sampling.

A descriptive measure computed from the data of a sample is called a statistic. The statistics of a population are known as parameters.

If there are $n$ observations say $x_1, x_2, \ldots, x_n$ in a sample, then the mean of the sample is given by $\bar{x} = \frac{\sum_{i=1}^{n} x_i}{n}$ and the variance of the sample is given by $\sigma^2 = \frac{\Sigma (x_i - \bar{x})^2}{n-1}$.

Sampling where a member can be selected more than once is called sampling with replacement. If $N$ is the size of a population, then the number of samples of size $n$ with replacement is $N^n$.

Suppose we have a population of size $N = 5$, it is possible to draw samples of size $n = 2$. The number of possible samples is equal to $N^n = 5^2 = 25$.

If a member of the population cannot be chosen more than once, the sampling is known as sampling without replacement. When drawing samples of size $n$ from an infinite population of size $N$ without replacement and ignoring the order in which the sample is drawn, the number of possible samples is given by ${}^NC_n = \frac{N!}{n!\,(N-n)!}$.

## 4.4 Methods of Sampling

There are different ways of selecting a sample. The following are some of the important methods employed in sampling.

(1) Purposive or deliberate sampling
(2) Random sampling
(3) Simple sampling
(4) Stratified sampling
(5) Area sampling
(6) Cluster sampling
(7) Sequential sampling

We now briefly explain a few methods.

### 4.4.1 Purposive sampling

Sampling in which the sampling units are selected with a purpose of benefitting certain elements in view is called purposive sampling. In this method, the sample obtained is based on certain pre-determined criteria. The fixation of criteria and deliberate choice of the sampling units which fit into the criteria bring in the personal elements and introduce bias in the system. Purposive method gives us a sample which is typical or representative of the population.

### 4.4.2 Random sampling

In this method of sampling, the individuals are selected in such a way that each individual has the same chance of being selected. The random sample may vary from the average, whereas the purposive sample does not. As the sample becomes larger, the random sample becomes more and more like the parent, whereas, using a bias in the purposive sample, in general, does not lead to the population. The selection of random sampling is very much facilitated by the use of a table of random numbers. Random numbers are distributed uniformly and independently in the interval (0,1). Random digits are uniformly distributed on the set $\{0,1,2,\ldots,9\}$. They can be used to form random numbers by selecting a proper number of digits for each random number and placing decimal points to the left of the value selected.

### 4.4.3 Simple sampling

A random sampling in which each event has the same chance of success and in which the chances of success of each event are independent of the success or failure of events in preceding trials is said to be a simple sampling. This simple sampling is a particular form of a random sampling. In the case of tossing of a coin, the chance of getting a head is not effected by the success or failure to obtain it on previous trials and remains constant irrespective of the number of trials. Hence, the tossing of a coin is a case of a simple sampling.

### 4.4.4 Stratified sampling

The sample which is the aggregate of the sampled units of each of the stratum is called the stratified sample. When the population is heterogeneous, it would be desirable to classify the whole universe into a number of relatively more homogenous stratums for the purpose of sampling. Each stratum is then sampled and would properly represent the whole strata. This eliminates the differences between the strata and thereby reduces sampling error. Such a method adds to the precision of the sample estimate when strata variability is the least.

### 4.4.5 Mean and standard deviation of the simple sampling of attributes

Sampling of attributes may be regarded as the drawing of samples from a population containing attributes. Consider a population of size $N$. Let a sample of size $n$ be drawn from the population. This is identical with that of a series of $n$ independent trials with constant probability of success. Let $p$ denote the probability of success and $q$ denote the probability of failure. Then, we have $p+q=1$. The probability of 0, 1, 2,......$n$ successes are the terms in the binomial expansion of $(q+p)^n$. The distribution so obtained is called the sampling distribution of the number of successes in the sample.

The mean number of successes is therefore $np$ and the variance of the number of successes is $npq$. The standard deviation is $\sqrt{npq}$.

The mean of the distribution is also known as the expected value. The mean proportion of successes is $\frac{np}{n} = p$ and the standard deviation of the proportion of successes is $\sqrt{\frac{pq}{n}}$. The value $np$ is called the expected value of the number of successes in the sample. The divergence of the number of successes from the expected value in any random sampling is known as fluctuation of sampling.

### 4.4.6 Test of significance for large samples

Suppose a large number of samples is classified according to the frequencies of an attribute, it gives rise to a binomial distribution. When the size of the sample $n$ is large, the binomial distribution tends to a normal distribution. Majority of the numbers of the sample lie within a range of $\pm 3\sigma$ on each side of the mean.

If the number of successes in the large sample differs from the expected value $np$ by more than $3\sqrt{npq}$, we call the difference highly significant, in which case the truth of the hypothesis is improbable.

## 4.5 Standard Error

**Definition 4.4** The standard deviation of a sampling distribution of a statistic is called a standard error. It is denoted by S.E.

Standard error is used to measure the variability of the values of a statistic computed from the samples of the same size drawn from the population.

Standard deviation and standard error have the same meaning. When the average amount of variability of the observations of a population is computed, it is called the standard deviation. But when the average amount of the variability of the observations of a sampling distribution is computed, it is called the standard error.

The standard error of the sampling mean $(\bar{x})$ is given by $\frac{\sigma}{\sqrt{n}}$, where $\sigma$ is the standard deviation of the population and $n$ is the sample size.

The standard error of the sample mean is denoted by $\frac{\sigma}{\sqrt{n}}$.

$$\therefore \text{S.E.}\,(\bar{x}) = \sigma_{\bar{x}} = \frac{\sigma}{\sqrt{n}}, \text{ when } \sigma \text{ is known}$$

When $\sigma$ is not known, the standard error of mean is $\text{S.E.}(\bar{x}) = \sigma_{\bar{x}} = \frac{s}{\sqrt{n}}$, where $s$ is the standard deviation of the sample.

Samples of size $n$ can be drawn from a population of size $N$. For each sample we may compute the mean and variance etc. Each of these is called a statistic. The value of a statistic is different for different samples and is treated as a random variable. The distribution of the statistic is defined as follows:

"The distribution of all possible values that can be assumed by some statistic calculated from samples of the same size randomly drawn from the same population is called the sampling distribution of the statistic".

The mean and variance of a sampling distribution are known as important characteristics of a sampling distribution. These characteristics of a sampling distribution help us to test a hypothesis and draw a statistical inference from it.

They are also used to set up the confidence limits of the parameters within which the parameter values are expected to lie.

When the standard deviation is known, the standard error of difference of the sample mean $\overline{x_1}$ and $\overline{x_2}$ drawn from the same population is given by

$$\text{S.E.}(\overline{x_1} - \overline{x_2}) = \sqrt{\sigma^2\left(\frac{1}{n_1} + \frac{1}{n_2}\right)}$$

where $n_1$ and $n_2$ are the sizes of the two samples drawn from the same population.

The standard error of difference of the two sample means $\overline{x_1}$ and $\overline{x_2}$ drawn from different populations with standard deviations $\overline{\sigma_1}$ and $\overline{\sigma_2}$ respectively is defined as

$$\text{S.E.}(\overline{x_1} - \overline{x_2}) = \sqrt{\frac{\sigma_1^2}{n_1} + \frac{\sigma_2^2}{n_2}}$$

where $n_1$ and $n_2$ are the sizes of the two samples.

In general, the standard error of mean is used for large samples. It is used to measure the variability of the values of a statistic from its mean. Standard error is used to set up confidence limits and is also used to test the hypothesis and draw conclusions from it.

## 4.6 Probable Error

The probable error of a distribution is 0.67449 times the standard error.

## 4.7 Central Limit Theorem

**Theorem 4.1** Let $x_1, x_2, \ldots, x_n$ be $n$ independent random variables all of which have the same distribution. Let the common expectation and variance be $\mu$ and $\sigma^2$ respectively. Let $\bar{x} = \sum_{i=1}^{n} \frac{x_i}{n}$. Then the distribution of $\bar{x}$ approaches the normal distribution with mean $\mu$ and variance $\frac{\sigma^2}{n}$ as $n \to \infty$ i.e., variate

$$Z = \frac{\bar{x} - \mu}{\sigma|\sqrt{n}}$$

has standard normal distribution.

*Proof* The moment generating function of $Z$ about the origin is given by

$$M_z(t) = E(e^{tz}) = E\left[e^{t\left[\frac{\bar{x}-\mu}{\sigma/\sqrt{n}}\right]}\right]$$

$$= e^{-\mu t\sqrt{n}/\sigma} E(e^{t\bar{x}\sqrt{n}|\sigma})$$

$$= e^{-\mu t\sqrt{n}/\sigma} E\left(e^{t\sqrt{n}/\sigma\left(\frac{x_1+x_2+\ldots+x_n}{n}\right)}\right)$$

$$= e^{-\mu t\sqrt{n}/\sigma} E\left(e^{t1/\sigma\sqrt{n}(x_1+x_2+\ldots+x_n)}\right)$$

$$= e^{-\mu t\sqrt{n}/\sigma} M(x_1+x_2+\ldots+x_n)\frac{t}{\sigma\sqrt{n}}$$

$$= e^{-\mu t\sqrt{n}/\sigma}\left[M_z\left(\frac{t}{\sigma\sqrt{n}}\right)\right]^n$$

Since the random variables are independent and have the same moment generating function, applying logarithms, we get

$$\log M_z(t) = -\mu t\frac{\sqrt{n}}{\sigma} + n\log M_z\left(\frac{t}{\sigma\sqrt{n}}\right)$$

$$= -\mu t\frac{\sqrt{n}}{\sigma} + n\log\left[1 + \frac{\mu_1' t}{\sigma\sqrt{n}} + \frac{\mu_2'}{2!}\cdot\left(\frac{t}{\sigma\sqrt{n}}\right)^2 + \ldots\right]$$

$$= -\mu t\frac{\sqrt{n}}{\sigma} + n\left[\frac{\mu_1' t}{\sigma\sqrt{n}} + \frac{\mu_2'}{2!}\cdot\left(\frac{t^2}{\sigma^2 n}\right) + \ldots - \frac{1}{2}\left(\frac{\mu_1' t}{\sigma\sqrt{n}} + \ldots\right)^2 + \ldots\right]$$

$$= -\mu t\frac{\sqrt{n}}{\sigma} + \left[\frac{\mu_1' t\sqrt{n}}{\sigma} + \frac{\mu_2'}{2!}\cdot\left(\frac{t^2}{\sigma^2}\right) + \ldots - \frac{1}{2}\left(\frac{\mu_1' t}{\sigma} + \ldots\right)^2 + \ldots\right]$$

$$= \frac{t^2}{2} + O\left(n^{-1/2}\right) \text{ (since } \mu_2' - \mu_1' = \sigma^2 \text{ and } \mu_1' = \mu$$

As $n \to \infty$, we have $\log M_z(t) \to \dfrac{t^2}{2}$

$$M_z(t) = e^{\frac{t^2}{2}}$$

which is the moment generating function of a standard normal variable.

Hence, the random variable $z$ converges to $N$. Therefore, it follows that the limiting distribution of it is normal with mean $\bar{x}$ and variance $\dfrac{\sigma^2}{n}$.

## 4.8 Finite Population Correction

When the sampling is without replacement from a finite population, the sampling distribution of $\bar{x}$, will have mean $\mu$ and variance

$$\frac{\sigma^2}{n} \cdot \frac{N-n}{N-1}$$

The factor $\dfrac{N-n}{N-1}$ is called the finite population correction factor or multiplier. It can be ignored when the sample size is small in comparison with the population size. It is also ignored when $n/N \leq 0.05$.

If the sample size is large, the central limit theorem applies and the sampling distribution of $\bar{x}$ will be approximately normally distributed.

## 4.9 Sampling Fluctuation

Any parameter expected from a population is always fixed and is a constant. Therefore, there cannot be a function in the value of the parameter computed from the population. But this is true in the case of a statistical measure. The statistic computed from different samples of the sample size drawn from the same parent population may be different. Any statistic computed from each of such a sample will also vary from sample to sample, which is known as sampling fluctuation.

## 4.10 Sampling Distribution of the Mean ($\sigma$ Unknown)

When the size of a sample is large, and the standard deviation of the population is unknown, we can replace the standard deviation of the population by the standard deviation $s$ of the sample. $\sigma$ is given by $\sigma = \sqrt{s^2} = \sqrt{\dfrac{\sum_{i=1}^{n} (x_i - \bar{x})^2}{n-1}}$.

If the size of the sample is small, we assume that the sample is taken from a normal population and substitute $s$ in place of $\sigma$, where $\sigma$ is unknown.

### Solved examples

**Example 1** Suppose we have a population of size $N = 5$. If all possible samples of size 2 are drawn from this population, find the finite population correction factor.

**Solution** We have $N = 5$, $n = 2$

The finite population correction factor $= \dfrac{N-n}{N-1}$

$$= \frac{5-2}{5-1} = \frac{3}{4} = 0.75$$

**Example 2** Find the correction factor if $n = 5$ and $N = 100$

**Solution** The correction factor is $\dfrac{N-n}{N-1} = \dfrac{100-5}{100-1} = \dfrac{95}{99} = 0.959$

**Example 3** A population consists of five numbers 6, 8, 10, 12 and 14. If all samples of size 2 are drawn from this population with replacement, find

(a) the total number of samples with replacement

(b) the mean of the population

(c) the standard deviation of the population

(d) the mean of the sampling distribution of means and

(e) the standard deviation of the sampling distribution of means i.e., standard error of means.

**Solution**

(a) We have $N = 5$, $n = 2$

The total number of samples that can be drawn from the population $= N^n = 5^2 = 25$

(b) Mean of the population $= \dfrac{\Sigma x_i}{N}$

i.e., $\mu = \dfrac{6+8+10+12+14}{5} = \dfrac{50}{5} = 10$

(c) The standard deviation of the population

$$\sigma = \sqrt{\frac{\Sigma(x-\mu)^2}{N}}$$

$$\sigma = \sqrt{\frac{(6-10)^2+(8-10)^2+(10-10)^2+(12-10)^2+(14-10)^2}{5}}$$

$$\sigma = \sqrt{\frac{16+4+0+4+16}{5}} = \sqrt{\frac{40}{5}} = \sqrt{8} = 2.82$$

(d) All the possible samples of size 2, with replacement from the given population are follows:

| | | Second draw | | | | |
|---|---|---|---|---|---|---|
| | | 6 | 8 | 10 | 12 | 14 |
| First draw | 6 | (6, 6) | (6, 8) | (6, 10) | (6, 12) | (6, 14) |
| | 8 | (8, 6) | (8, 8) | (8, 10) | (8, 12) | (8, 14) |
| | 10 | (10, 6) | (10, 8) | (10, 10) | (10, 12) | (10, 14) |
| | 12 | (12, 6) | (12, 8) | (12, 10) | (12, 12) | (12, 14) |
| | 14 | (14, 6) | (14, 8) | (14, 10) | (14, 12) | (14, 14) |

Number of samples = 25

Sampling distribution of means

| | | | | |
|---|---|---|---|---|
| 6 | 7 | 8 | 9 | 10 |
| 7 | 8 | 9 | 10 | 11 |
| 8 | 9 | 10 | 11 | 12 |
| 9 | 10 | 11 | 12 | 13 |
| 10 | 11 | 12 | 13 | 14 |

Mean of the sampling distribution of means $\mu_{\bar{x}} = \dfrac{6+7+8+9+10+\ldots+14}{25}$

(i.e., sum of all the means of the sample/$N^n$)

$$= \frac{250}{25} = 10$$

(e) The standard deviation of the sampling distribution of means

$$\sigma_{\bar{x}} = \sqrt{\frac{(6-10)^2+(8-10)^2+(10-10)^2+(12-10)^2+(14-10)^2}{25}}$$

$$= \sqrt{\frac{100}{25}} = \sqrt{4} = 2$$

Or $\sigma_{\bar{x}} = \sigma/\sqrt{n} = \sqrt{4} = 2$

**Example 4** A population consists of five numbers 2, 3, 6, 8, 11. If all samples of size 2 are drawn from this population without replacement, find

(a) the mean of the population

(b) the standard deviation of the population

(c) the mean of the sampling distribution of means and

(d) the standard deviation of the sampling distribution of means.

**Solution**

(a) Mean of the population $= \dfrac{\sum x_i}{N}$

i.e., $\mu = \dfrac{2+3+6+8+11}{5} = \dfrac{30}{5} = 6$

(b) The standard deviation of the population

$$\sigma = \sqrt{\frac{\sum (x-\mu)^2}{N}}$$

$$\sigma = \sqrt{\frac{(2-6)^2+(3-6)^2+(6-6)^2+(8-6)^2+(11-6)^2}{5}}$$

$$\sigma = \sqrt{\frac{16+9+0+4+25}{5}} = \sqrt{\frac{54}{5}} = \sqrt{10.8} = 3.286$$

(c) All the possible samples of size 2, without replacement from the given population:

(2,3) (2,6) (2,8) (2,4)
(3,6) (3,8) (3,11)
(6,8) (6,11)
(8,11)

Number of samples $= {}^{N}C_n = {}^{5}C_2 = 10$

(d) Sampling means are

2.5 4 5 6.5
4.5 5 7
7 8.5
9.5

Mean of the sampling distribution of means

$$\mu_{\bar{x}} = \frac{2.5+4+5+6.5+4.5+5+7+7+8.5+9.5}{10}$$

$$= \frac{60}{10} = 6$$

The standard deviation of the sampling distribution of means

$$\sigma_{\bar{x}} = \sqrt{\frac{(2.5-6)^2+(4-6)^2+(5-6)^2+(6.5-6)^2+\ldots+(9.5-6)^2}{10}}$$

$$= \sqrt{\frac{40.5}{10}} = \sqrt{4.05} = 2.012$$

**Example 5** A finite population consists of the number 3, 6, 9, 15, 27.

(1) List all possible samples of size 3 without replacement.

(2) Find the mean of the population.

(3) Find the standard deviation of the population.

(4) Calculate the mean of the sampling distribution of means.

(5) Find the standard deviation of sampling distributions of means.

**Solution**

(1) The possible samples of size 3 without replacement are

(3, 6, 9) (3, 6, 15) (3, 6, 27) (3, 9, 15) (3, 9, 27) (3, 15, 27)

(6, 9, 15) (6, 9, 27) (6, 15, 27)

(9, 15, 27)

Number of samples $= {}^{N}C_n = {}^{5}C_3 = 10$

(2) Mean of the population $= \dfrac{\sum x_i}{N}$

i.e., $\mu = \dfrac{3+6+9+15+27}{5} = \dfrac{60}{5} = 12$

(3) The standard deviation of the population

$$\sigma = \sqrt{\frac{\Sigma (x-\mu)^2}{N}}$$

$$\sigma = \sqrt{\frac{(3-12)^2+(6-12)^2+(9-12)^2+(15-12)^2+(27-12)^2}{5}}$$

$$\sigma = \sqrt{\frac{81+36+9+9+225}{5}} = \sqrt{\frac{360}{5}} = \sqrt{72} = 8.485$$

(4) Sampling means are

6 8 9 12 13 15
10 14 16
17

Mean of the sampling distribution of means

$$\mu_{\bar{x}} = \frac{6+8+9+12+13+15+10+14+16+17}{10}$$

$$= \frac{120}{10} = 12$$

(5) The standard deviation of the sampling distribution of means

$$\sigma_{\bar{x}} = \sqrt{\frac{(6-12)^2+(8-12)^2+(9-12)^2+(12-12)^2+....+(17-12)^2}{10}}$$

$$= \sqrt{\frac{36+16+9+1+9+4+4+16+25}{5}}$$

$$= \sqrt{\frac{120}{10}} = \sqrt{12} = 3.464$$

**Example 6** The mean of a certain normal population is equal to the standard error of mean of the sample of size 64 from the distribution. Find the probability that the mean of the sample size 36 will be negative.

**Solution** We have $n = 64$, $n_1 = 36$

Standard error of means $= \sigma_{\bar{x}} = \sigma/\sqrt{n} = \dfrac{\sigma}{\sqrt{64}} = \dfrac{\sigma}{8}$

Mean = Standard error $= \dfrac{\sigma}{8}$

i.e., $\mu = \dfrac{\sigma}{8}$

Since $z = \dfrac{\bar{x}-\mu}{\sigma/\sqrt{n_1}} = \dfrac{\bar{x}-\dfrac{\sigma}{8}}{\sigma/\sqrt{36}} = \dfrac{\bar{x}-\dfrac{\sigma}{8}}{\sigma/6} = \dfrac{6\bar{x}}{\sigma} - \dfrac{3}{4}$

where $\bar{x}$ is the mean of the sample size 36

$\bar{x}$ is negative when $z < 3/4$ i.e., 0.75

$$\therefore P(z < 0.75) = P(-\infty < z < 0.75)$$

$$= \int_{-\infty}^{0} \phi(z)\,dz + \int_{0}^{0.75} \phi(z)\,dz$$

$$= 0.5 + 0.2734$$

$$= 0.7734$$

**Example 7** A sample size of 400 is taken from a population where standard deviation is 16. Find the values of standard error and probable error.

**Solution** The standard deviation of the population = $\sigma = 16$

$n =$ size of the sample $= 400$

$$\text{Standard error} = \frac{\sigma}{\sqrt{n}} = \frac{16}{\sqrt{400}} = \frac{16}{20} = 0.8$$

Probable error $= 0.67449 \times 0.8 = 0.5395$

**Example 8** A normal population has a mean of 4.1 and standard deviation of 2%. Find the probability that the mean of a sample of size 900 will be negative.

**Solution** We have,

Mean of the population = $\mu = 0.1$

Standard deviation of the population = $\sigma = 2.1$

Size of the sample $n = 900$

$$z = \frac{\bar{x} - \mu}{\sigma/\sqrt{n_1}} = \frac{\bar{x} - 0.1}{2.1/\sqrt{900}} = \frac{\bar{x} - 0.1}{\left(\frac{2.1}{30}\right)}$$

i.e., $z = \dfrac{\bar{x} - 0.1}{0.07}$ or $\bar{x} = 0.1 + 0.07z$

where $z$ is the standard normal variate.

Hence $P(\bar{x} < 0) = P(0.1 + 0.07z < 0)$

$= P(0.07z < -0.1)$

$= P\left(Z < \dfrac{-0.1}{0.07}\right)$

$= P(z < -1.43)$

$= 0.5 - P(0 < Z < 1.43)$

$= 0.5 - 0.4236$

$= 0.0764$

**Example 9** A random sample of size 100 is taken from an infinite population having the mean $\mu = 76$ and the variance $\sigma^2 = 256$. What is the probability that $\bar{x}$ will be between 75 and 78.

**Solution** We have $n = 100$

$\mu$ = mean of the population = 76

Variance = $\sigma^2 = 256$

$\sigma = 16$

$$z = \frac{\overline{x} - \mu}{\sigma/\sqrt{n}}$$

$$\text{When } \overline{x} = 75,\ z = z_1(\text{say}) = \frac{75 - 76}{16/\sqrt{100}} = -0.625$$

$$\overline{x} = 78,\ z = z_2(\text{say}) = \frac{78 - 76}{16/\sqrt{100}} = 1.25$$

$$\therefore P(75 \le \overline{x} \le 78) = P(Z_1 \le Z \le Z_2)$$

$$= P(-0.625 \le Z \le 1.25)$$

$$= P(-0.625 \le Z \le 0) + P(0 \le Z \le 1.25)$$

$$= 0.2334 + 0.3944$$

$$= 0.628$$

**Example 10** A coin is tossed 400 times and it turns up head 212 times. Discuss whether the coin may be unbiased.

**Solution** If the coin is unbiased, then the expected number of heads in 400 tosses is

$$np = 400\left(\frac{1}{2}\right) = 200$$

The deviation of actual number of heads from the expected number of heads is therefore

$$212 - 200 = 12$$

The standard deviation $= \sqrt{npq} = \sqrt{400 . \frac{1}{2} . \frac{1}{2}} = \sqrt{100} = 10$

The deviation is therefore 1.2 times the standard deviation.

Hence it is the deviation likely to appear as a result of fluctuations of simple sampling.

Thus the coin is an unbiased one.

**Example 11** A cubical die was thrown 9000 times and a 2 or 3 were obtained 3286 times. On the assumption of a random throwing, do the data indicate an unbiased one?

**Solution** If the throwing of the die is unbiased, the chance of turning 2 or 3 is $\frac{2}{6}=\frac{1}{3}$

$$\therefore p=\frac{1}{3}, q=1-\frac{1}{3}=\frac{2}{3}$$

Expected number of successes $=np=9000\times\frac{1}{3}=3000$

The standard deviation of number of successes $=\sqrt{npq}=\sqrt{9000\times\frac{1}{3}\times\frac{2}{3}}=44.72$

The difference between the observed frequency and the expected frequency

$$=3286-3000=286$$

which is nearly 6.4 times the standard deviation i.e., more than $3\sqrt{npq}$

Hence, the die is a biased one.

**Example 12** A sample of 900 days is taken from the meteorological records of a certain district and 100 of them are found to be foggy. What are the probable limits to the percentage of foggy days in the district?

**Solution** We have

$$p=\text{probability of a foggy day}=\frac{100}{900}=\frac{1}{9}$$

$$q=1-p=1-\frac{1}{9}=\frac{8}{9}$$

$$n=900$$

$$\text{S.D}(\sigma)=\sqrt{npq}=\sqrt{900\left(\frac{1}{9}\right)\left(\frac{8}{9}\right)}=9.4$$

Probable limits are

$$100\pm 3\sigma=100\pm3(9.4)$$

$$=100\pm 28.2$$

i.e.,$(100-28.2, 100+28.2)=(71.8, 128.2)$

Therefore, the percentage of foggy days lies in range $\left(\frac{71.8}{9},\frac{128.2}{9}\right)$

i.e., the limits are 7.97% and 14.2%.

## 4.11 Standard Errors of Some Parameters

When the parent universe is assumed to be normal, some of the standard errors are given below:

| S.No | Statistic | Standard error |
|---|---|---|
| 1. | Mean | $\sigma/\sqrt{n}$ |
| 2. | Standard deviation | $\sqrt{q^2/2n}$ |
| 3. | Variance | $\sigma^2\sqrt{2/n}$ |
| 4. | Difference of two sample means $\bar{x}_1 - \bar{x}_2$ | $\sqrt{\sigma_1^2/n_1 + \sigma_2^2/n_2}$ |
| 5. | Difference of two samples standard deviations ($SD_1 - SD_2$) | $\sqrt{\sigma_1^2/2n_1 + \sigma_2^2/2n_2}$ |
| 6. | Median | $\sigma\sqrt{\pi/2n}$ |
| 7. | $\mu_3$ | $6\sigma^6/n$ |
| 8. | $\mu_4$ | $96\sigma^8/n$ |

### Exercises 4.1

1. Define (a) population (b) sample (c) sampling with replacement (d) sampling without replacement.

2. Define the terms

   (a) Random sampling

   (b) Simple (random) sampling

   (c) Stratified sampling

3. Define the terms (a) standard error (b) probable error (c) finite population correction multiplier

4. Find the finite population correction factor with $n = 10$ and $N = 1000$.

5. Find the value of correction factor when $n = 10$, $N = 400$.

6. A random sample of size 100 is taken from an infinite population having mean $\mu = 76$ and the variance $\sigma^2 = 256$. What is the probability that $\bar{x}$ will be between 75 and 78.

7. A research worker wishes to estimate mean of a population by using a sufficiently large sample. The probability is 95% that sample mean will not differ from the true mean by more than 25% of the standard deviation. How large a sample should be taken?

8. If the mean breaking strength of copper wire is 575 lbs, with a standard deviation of 8.3 lbs, how large a sample must be used in order that there will be one chance in 100 that the mean breaking strength of the sample is less than 572 lbs?

9. Samples of size 2 are taken from the population 1, 2, 3, 4, 5, 6 with replacement and without replacement. Find

   (a) Mean of the population.

   (b) Find the standard deviation of the population.

   (c) Calculate the mean of sampling distribution of means.

   (d) Find the standard deviation of sampling distributions of means.

   Verify that the mean of sampling distribution of means is equal to the mean of the population and the standard deviation of the sampling distribution of means is not equal to the standard deviation of the population.

10. A population consists of six numbers 4, 8, 12, 16, 20 and 24. If all samples of size 2 are drawn from this population without replacement, find

    (a) The mean of the population

    (b) The standard deviation of the population

    (c) The mean of the sampling distribution of means and

    (d) The standard deviation of the sampling distribution of means.

## Answers 4.1

(4) 0.99

(5) 0.98

(6) 0.628

(7) 62

(8) 42

(9) (a) 7/2 (b) $\sqrt{2.917}$=1.707 (c) 7/2 (d) 1.208, 1.08

(10) (a) 14 (b) 6.83 (c) 14 (d) 4.3209

## CHAPTER FIVE

# Theory of Estimation

## 5.1 Introduction

In this chapter, we shall introduce theory of estimation which was founded by Professor R. A Fisher. The process of estimation entails the data of a sample, some statistic that is referred to as the approximation of the corresponding parameter of the population from which the sample was drawn. The theory of estimation is also known as the theory of estimators. The theory seeks to investigate the conditions which an estimate should obey, what are the best estimates to employ in any given set of circumstances, and how good other estimates are in comparision. In this chapter, we also present some of the basic concepts essential for understanding hypothesis testing. We begin our study by introducing the types of estimation.

**Definition 5.1** Any procedure by which we reach a conclusion about a population on the basis of the information contained in the sample drawn from the population is known as statistical inference.

Statistical inference procedures allow us to make inferences about the population.

## 5.2 Point Estimate

A point estimate is a number obtained from computation on the observed values of the random variables which serve as an approximation to the parameters. It is defined as a particular value of a statistic which is used to estimate a given parameter. Point estimation is a single valued estimation and is also called the estimation of the parameter.

## 5.3 Interval Estimation

Interval estimation consists of two numerical values differing from the average of values with a specified degree of confidence. This will include the parameter estimated.

Interval estimation is the estimation of the population parameter by an interval around the point i.e., an interval estimation of the population parameter is the range of values used in making an estimation of the population parameter. It is defined as follows

"The interval estimation of a population parameter is the estimation of the parameter $\theta$ with the help of the interval $(t-s, t+s)$, where $t$ is the sample statistic i.e., $t-s \le \theta \le t+s$."

## 5.4 Criteria for a Good Estimator

A good estimator is one which is as close to the true value of the population parameter as possible. According to Fisher, the criteria for a good estimator are (i) consistency, (ii) efficiency, (iii) unbiasedness, (iv) sufficiency.

### 5.4.1 Consistency

Consistency ensures that the difference between the estimator and the parameter would become lesser and lesser in the probability sense as the size of the sample increases indefinitely. A consistent estimator is defined as follows.

**Definition 5.2** An estimator $t_n$ of the parameter $\theta$ is called consistent for the (universe) population parameter $\theta$, if it converges in probability to $\theta$ as the sample size $n$ tends to infinity.

In other words, the estimator $t_n$ is consistent if given two positive numbers, $\varepsilon$ and $\delta$ however small, we can find a positive integral $N$ depending on $\varepsilon$ and $\delta$, such that

$$p\{|t_n - \theta| > \varepsilon\} = 0 \text{ for all } \varepsilon > 0$$

or equivalently, if

$$\lim_{n\to\infty} p\{|t_n - \theta| < \varepsilon\} = 1 \text{ for all } \varepsilon > 0$$

**Solved examples**

**Example 1** Show that the sample median is a consistent estimator for the population mean of a normal distribution.

**Solution** Let $\mu$ denote the population mean of a normal distribution and $M_i$ be the median of the sample $x_1, x_2, \ldots, x_n$ taken from the normal population. Then for a large sample, the variate $t = \dfrac{(M_i - \mu)\sqrt{2n}}{\sigma\sqrt{\pi}}$ is a standard normal variate.

$$\therefore P\{|M_i - \mu| < \varepsilon\} = P\{|t| < \frac{\varepsilon\sqrt{2n}}{\sigma\sqrt{n}}\}$$

$$= \int_{-\varepsilon\frac{\sqrt{2n}}{\sigma\sqrt{n}}}^{\varepsilon\frac{\sqrt{2n}}{\sigma\sqrt{\pi}}} \frac{1}{\sqrt{2\pi}} e^{\frac{-1}{2}t^2} dt$$

$$= \int_{-\infty}^{\infty} \frac{1}{\sqrt{2\pi}} e^{\frac{-1}{2}t^2} dt \to 1 \text{ as } n \to \infty$$

Hence, the sample median is a consistent estimator for the population mean.

**Example 2** Show that in random sampling from a normal population, the sample mean is a consistent estimator.

**Solution** Let $x_1, x_2, \ldots, x_n$ be the random sample drawn from the normal population with mean $\mu$ and standard deviation $\sigma$. Then

$$\lim_{n\to\infty} P\{|\overline{x} - \mu| > \varepsilon\} = \lim_{n\to\infty} P\{|Z| > \varepsilon\}, \text{ where } \overline{z} = \overline{x} - \mu$$

$$= 1 - \lim_{n\to\infty} P\{|Z| > \varepsilon\}$$

$$= 1 - \lim_{n\to\infty} \int_{-\varepsilon}^{\varepsilon} \frac{\sqrt{n}}{\sigma\sqrt{2\pi}} e^{-\frac{nz^2}{2\sigma^2}} dz$$

$$= 1 - \lim_{n\to\infty} \int_{-\varepsilon\sqrt{n}|\sigma}^{\varepsilon\sqrt{n}|\sigma} \frac{1}{\sqrt{2\pi}} e^{-\frac{t^2}{2}} dt$$

$$= 1 - \int_{-\infty}^{\infty} \frac{1}{\sqrt{2\pi}} e^{-\frac{t^2}{2}} dt$$

$$= 1 - 1 = 0$$

Hence, the sample mean $\overline{x}$ is a consistent estimator of the population mean.

## 5.5 Unbiased Estimator

**Definition 5.3** An estimator $t_n$ of the parameter $\theta$ is said to be unbiased if there exists the relation $E(t_n) = \theta, n = 1, 2, 3, \ldots$

In other words, if the mean of the sampling distribution of a statistic is equal to the corresponding population parameter, the statistic is called an unbiased estimator of the parameter, otherwise it is called a biased estimator.

## Solved examples

**Example 1** Obtain an unbiased estimator of $\theta = \sigma^2$ for a normal distribution with mean $\mu$ and variance $\sigma^2$.

**Solution** The initial estimator $\theta = \sigma^2$

$$t_n = \frac{\Sigma(x_i - \bar{x})^2}{n}$$

Then, $t_n = nt_n - (n-1)t_{n-1}$

$$= n\frac{\sum\sum_{i<j}(x_i - x_j)^2}{n^2} - \frac{n-1}{n}\sum_{k=1}^{n}\frac{\sum\sum_{i<j}(x_i - x_j)^2}{(n-1)^2}$$

$$= \sum\sum_{i<j}(x_i - x_j)^2\left[\frac{1}{n} - \frac{1}{n(n-1)}\right]$$

$$= \frac{1}{n(n-1)}\sum\sum_{i<j}(x_i - x_j)^2$$

$$= \frac{1}{n-1}\sum_{i=1}^{n}(x_i - x_j)^2$$

which is exactly unbiased for $\theta$.

If $E(t_n) > \theta$, the estimator is said to be positively biased, and if $E(t_n) < \theta$, the estimator is said to be negatively unbiased.

An unbiased estimator may or may not be consistent depending on whether or not its distribution becomes concentrated near the mean as the sample size increases. If the sample size is finite, a consistent estimator may be biased, but as the size of the sample $n \to \infty$, the bias approaches zero.

**Example 2** If $x_1, x_2, \ldots, x_n$ be a random sample of size $n$ from a normal distribution, with mean zero and variance $\sigma^2$, show that $\sum_{i=1}^{n}\frac{x_i^2}{n}$ is an unbiased statistic for $\sigma^2$.

**Solution** Let $s^2 = \sum_{i=1}^{n}\frac{x_i^2}{n}$.

Then, $E(s^2) = E(\sum_{i=1}^{n} \frac{x_i^2}{n}) = \sigma^2$

Hence $s^2$ is an unbiased statistic for $\sigma^2$.

**Example 3** If $t$ is an unbiased estimator of the parameter $\theta$, show that $t^2$ is a biased estimator of $\theta^2$.

**Solution** We have $\text{var}(t) = E(t-\theta)^2$

$= E(t^2) - 2\theta E(t) + \theta^2$

$= E(t^2) - \theta^2$

Since $\text{Var}(t) \neq 0$, we have $E(t^2) - \theta^2 \neq 0$

i.e., $E(t^2) \neq 0$

Thus, $t^2$ is a biased estimator of $\theta^2$.

**Example 4** A population consists of four values 2, 4, 6 and 8. Draw all samples of size 2 with replacement. Verify that the sample mean is an unbiased estimator of the population mean.

**Solution** Population mean $= \mu = \dfrac{2+4+6+8}{4} = 5$

The samples of size 2 that can be drawn from the population 2, 4, 6 and 8, with replacement are

(2, 2), (2, 4), (2, 6), (2, 8)

(4, 2), (4, 4), (4, 6), (4, 8)

(6, 2), (6, 4), (6, 6), (6, 8)

(8, 2), (8, 4), (8, 6), (8, 8)

The sample means are:

2 3 4 5

3 4 5 6

4 5 6 7

5 6 7 8

For the above means we have the following distributions.

| Sample mean $(\overline{x_i})$ | Frequency $f_i(\overline{x})$ | $\overline{x_i} f_i(\overline{x})$ |
|---|---|---|
| 2 | 1 | 2 |
| 3 | 2 | 6 |
| 4 | 3 | 12 |
| 5 | 4 | 20 |
| 6 | 3 | 18 |
| 7 | 2 | 14 |
| 8 | 1 | 8 |
| | **16** | **80** |

$$\sum f_i(\overline{x}) = 16$$

$$\sum \overline{x_i} f_i(\overline{x}) = 80$$

$$E((\overline{X}) = \frac{80}{16} = 5$$

$E(\overline{X}) = \mu$, the sample mean is an unbiased estimator of population mean.

## 5.6 Efficiency

If $t_n$ and $t'_n$ are estimators of a parameter θ, such that Var($t_n$)<Var($t'_n$), for all $n$, then $t_n$ is said to be more efficient than $t'_n$.

The estimators which have the smallest limiting variances are called efficient estimators of the parameter θ. If the variance is small, the estimate is more reliable.

Efficiency $E$ of $t_n$ relative to $t'_n$ is defined as

$$E = \frac{\text{Var}(t_n)}{\text{Var}\,(t'_n)}$$

The value of $E$ is less than or equal to 1 i.e., $E \leq 1$.

A statistic with efficiency of 1 is said to be most efficient. Efficient estimators need not be consistent.

## 5.7 Sufficiency

An estimator $t_n$ is said to be sufficient if it contains all the information of the sample regarding the population parameter θ.

## 5.8 Likelihood Function

Let $f(x,\theta)$ be the frequency function of the random variable $X$, where $\theta$ is the parameter to be estimated and $x_1, x_2, \ldots x_n$ denote $n$ random variables corresponding to the $n$ samples of the variable $x$. Then the function given by $L(x_1, x_2, \ldots, x_n, \theta) = f(x_1,\theta)f(x_2,\theta)\ldots f(x_n,\theta)$ is known as the likelihood function.

If $L(x_1, x_2, \ldots, x_n, \theta)$ can be written as

$$L(x_1, x_2, \ldots, x_n, \theta) = L_1(t_n, \theta) L_2(x_1, x_2, \ldots, x_n)$$

where $L_2(x_1, x_2, \ldots, x_n)$ does not contain $\theta$, then $t_n$ is said to be sufficient for $\theta$.

**Example** Show that in estimating for the parameter in the Poisson distribution with general term $\frac{e^{-\mu}}{x!}\mu^x$, $\bar{x}$ is sufficient for $\mu$.

**Solution** We have $f(x,\mu) = \sum_{x=0}^{\infty} \frac{e^{-\mu}}{x!}\mu^x$ for the Poisson distribution.

Therefore, $L(x_1, x_2, \ldots, x_n, \mu) = \frac{e^{-\mu}}{x_1!}\mu^{x_1} \frac{e^{-\mu}}{x_2!}\mu^{x_2} \ldots \frac{e^{-\mu}}{x_2!}\mu^{x_n}$

$$= \frac{e^{-n\mu}.e^{x_1+x_2+\ldots+x_n}}{x_1!x_2!\ldots x_n!} \tag{1}$$

Since $\frac{x_1+x_2+\ldots+x_n}{n} = \mu$ i.e., $x_1+x_2+\ldots+x_n = n\mu$

From (1), we get $L(x_1, x_2, \ldots, x_n, \mu) = \left(e^{-n\mu}\mu^{nx}\right)\left(\frac{1}{x_1!x_2!x_3!\ldots x_n!}\right)$

$$L(x_1, x_2, \ldots, x_n, \mu) = L_1(\overline{x_1}, \mu) L_2(x_1, x_2, \ldots, x_n)$$

Hence $L$ is expressed as a product of two factors, i.e., $L$ is split up into two factors where the first factor contains the estimator $\bar{x}$ and the parameter $\mu$ and the second factor contains only $x_1$, $x_2, \ldots, x_n$.

Therefore, $\bar{x}$ is sufficient to estimate $\mu$.

## 5.9 Principle of Maximum Likelihood

A maximum likelihood estimator $t_n$ of the parameter $\theta$ in the frequency function $f(x, \theta)$ is an estimator that maximises the likelihood function $L(x_1, x_2, \ldots, x_n, \theta)$ as a function of $\theta$.

In other words, the principle of maximum likelihood states that $t_n$ is called the estimator of $\theta$ if there exists a statistic $t_n = t_n(x_1, x_2, \ldots, x_n)$ which maximises $L$ for a variation of $\theta$.

Thus, $t_n$ is a solution if any of

$$\frac{\partial L}{\partial \theta} = 0 \text{ or } \frac{\partial(\log L)}{\partial \theta} = 0$$

$$\frac{\partial^2 L}{\partial \theta^2} > 0 \text{ or } \frac{\partial^2(\log L)}{\partial \theta^2} < 0$$

### 5.9.1 Properties of the maximum likelihood estimator

Maximum likelihood estimators are (i) consistent, (ii) most efficient, and (iii) sufficient, if any exists.

**Example** For random sampling from a normal population, find the maximum likelihood estimators for

(a) the population mean, when the population variance is known.

(b) the population variance, when the population mean is known.

**Solution** The likelihood function is $L = \left(\frac{1}{\sigma\sqrt{2\pi}}\right)^n e^{\left[-\frac{1}{2\sigma^2}\sum_1^n (x_i-\mu)^2\right]}$

$$= \left(\frac{1}{2\pi\sigma^2}\right)^{\frac{1}{2}n} e^{-\frac{n}{2\sigma^2}\left(-\frac{n}{2\sigma^2}\right)(\bar{x}-\mu)^2+s^2}$$

Applying logarithms we get

$$\log L = \frac{n}{2}\log\frac{1}{2\pi} - \frac{1}{2}n\log\sigma^2 - \frac{n}{2\sigma^2}\left[(\bar{x}-\mu)^2 + s^2\right]$$

(a) If $\sigma^2$ is known, then in order to estimate for $\mu$, we have $\frac{\partial(\log L)}{\partial \mu} = 0$

which gives $(\bar{x}-\mu) = 0$ or $\bar{x} = \mu$

(b) If $\mu$ is known, then to estimate for $\sigma^2$, we have

$$\frac{\partial}{\partial\sigma^2}(\log L) = -\frac{n}{2}\frac{1}{\sigma^2} - \frac{n}{2\sigma^2}\left[(\bar{x}-\mu)^2 + s^2\right] = 0$$

$$\text{or } \sigma^2 = \sum_{i=1}^{n}\frac{(x_i-\bar{x})^2}{n}$$

## 5.10 Interval Estimator and Confidence Limits

The probability that we associate with an interval is called the confidence level. Higher the probability, the more is the confidence. In the interval estimation, the estimate for the parameter lies between two limits. These limits are known as the confidence limits.

The confidence limits are also known as the fiducial limits.

If $t$ is a sample statistic and $\theta$ is a population parameter, the interval estimation of $\theta$ is the estimation of the parameter $\theta$ with the help of the interval $(t-s, t+s)$. An interval estimate consists of two numerical values defining the range of values that with a specified degree of confidence, most likely includes the parameter estimated. In general, an interval estimate is expressed as:

Estimator $\neq$ (reliability coefficient) $\times$ standard error

A confidence interval is obtained from the formula

Estimator $\neq$ (reliability coefficient) $\times$ (standard error of the estimator)

If $t$ is the sample statistic used to estimate the corresponding population parameter $\theta$, then $(1-\alpha)$ 100% confidence limits for $\theta$ are $t\pm$ S.E.$(t).t_\alpha$, where $\alpha$ is the level of significance.

S.E.$(t)$ is the standard error of $t$ and $t_\alpha$ is the significant or critical value of $t$ at the significant level $\alpha$.

$t-$ S.E.$(t).t_\alpha$ and $t+$ S.E.$(t).t_\alpha$ are the limits of the confidence interval.

By central limit theorem, if $t$ is the sample statistic, then $z = \dfrac{t-E(t)}{\text{S.E.}(t)}$ is a standard normal variate with mean zero and variance $= 1$ as $n$ tends to $\infty$.

### 5.10.1 Confidence interval of the mean

(i) Consider a population. Let $\mu$ be the population mean and $\bar{x}$ be the sample mean of the sampling distribution. Assume that the sample is large. When the standard deviation $\sigma$ of the normal population is known, the S.E.$(\bar{x})$ is $\dfrac{\sigma}{\sqrt{n}}$ ($n$ is the large sample size). The confidence interval when the population is normal (or any population of large $n$) and $\sigma$ is $[\bar{x}-\text{S.E.}(\bar{x}).z_\alpha, \bar{x}+\text{S.E.}(\bar{x})\, z_\alpha]$

$$\text{i.e., } \left[\bar{x}-\frac{\sigma}{\sqrt{n}}.z_\alpha, \bar{x}+\frac{\sigma}{\sqrt{n}}z_\alpha\right]$$

$$\text{or } \left[\bar{x}-\frac{\sigma}{\sqrt{n}}.z_\alpha \le \mu \le \bar{x}+\frac{\sigma}{\sqrt{n}}z_\alpha\right]$$

(ii) When the standard deviation of the population is not known, the standard error of mean is S.E.$(\bar{x}) = \sigma_{\bar{x}} = \dfrac{s}{\sqrt{n-1}}$, where $n$ is size of the sample and $s$ is the standard deviation of the sampling distribution of the sample.

Therefore, the confidence interval is $\left[\bar{x} - \frac{s}{\sqrt{n-1}}.z_\alpha, \bar{x} + \frac{s}{\sqrt{n-1}} z_\alpha\right]$

i.e., $[\bar{x} - \sigma_{\bar{x}} z_\alpha, \bar{x} + \sigma_{\bar{x}} z_\alpha]$

or $\left[\bar{x} - \frac{\sigma}{\sqrt{n}} z_\alpha \leq \mu \leq \bar{x} + \frac{\sigma}{\sqrt{n}} z_\alpha\right]$

The value of $z_\alpha$ is selected from the following table.

| $\alpha$ | Confidence interval $(1-\alpha)$ or $(1-\alpha)100\%$ | Confidence coefficient $z_\alpha(or\ z)$ or $t_\alpha(or\ t)$ |
|---|---|---|
| 0.10 | 90% | 1.64 |
| 0.05 | 95% | 1.96 |
| 0.02 | 98% | 2.33 |
| 0.01 | 99% | 2.58 |
| Without any reference to the confidence level | -> | 3.00 |

$1-\alpha$ is called the degree of confidence.

We have the following table for selecting the values of $Z_\alpha$ and $Z_{\alpha/2}$.

| $\alpha$ | 0.400 | 0.300 | 0.200 | 0.100 | 0.050 | 0.025 | 0.020 | 0.0010 | 0.005 |
|---|---|---|---|---|---|---|---|---|---|
| $Z_\alpha$ | 0.253 | 0.524 | 0.842 | 1.282 | 1.645 | 1.960 | 2.054 | 2.326 | 2.576 |
| $Z_{a/2}$ | 0.842 | 1.036 | 1.282 | 1.645 | 1.960 | 2.240 | 2.326 | 2.576 | 2.807 |

**Theorem 5.1** If $\bar{x}$, the mean of random samples of size $n$ from a normal population with known variance $\sigma^2$, is to be used as an estimator of the mean of the population, the probability is $1-\alpha$ that the error will be less than $Z_{\alpha/2}\cdot\frac{\sigma}{\sqrt{n}}$.

Suppose that the mean of a random sample is to be used to estimate the mean of a normal population with the known variance $\sigma^2$, then the sampling distribution of $\bar{x}$ for random samples of size $n$ from a normal population with the mean $\mu$ and the variance $\sigma^2$ is a normal distribution with $\mu_{\bar{x}} = \mu$ and $\sigma_{\bar{x}}^2 = \frac{\sigma^2}{n}$ (i.e.,$\sigma_{\bar{x}} = \frac{\sigma}{\sqrt{n}}$).

Thus, we can write $P(|Z| < Z_{\alpha/2}) = 1-\alpha$

where $Z = \frac{\bar{x}-\mu}{\frac{\sigma}{\sqrt{n}}}$

and $Z_{\alpha/2}$ is such that the integral of the standard normal density from $Z_{\alpha/2}$ to $\infty$ equals $\alpha/2$. It follows that $P(|\bar{x}-\mu| < Z_{\alpha/2}\cdot\frac{\sigma}{\sqrt{n}}) = 1-\alpha$.

**Theorem 5.2** If $\bar{x}$ is the value of the mean of a random sample of size $n$ from a normal population with known variance $\sigma^2$, then $\left[\bar{x} - Z_{\alpha/2}\frac{\sigma}{\sqrt{n}} < \mu < \bar{x} + Z_{\alpha/2}\frac{\sigma}{\sqrt{n}}\right]$ is a $(1-\alpha)$ 100% confidence interval for the mean of the population.

Let $x_1, x_2, \ldots, x_n$ be a given sample from a normal distribution $N(\mu, \sigma^2)$. To construct a confidence interval for the unknown parameter, when variance $\sigma^2$ is known, for the probability $1-\alpha$, we can find a number $Z_{\alpha/2}$ (from the table) such that

$$P\left(-Z_{\alpha/2} \leq \frac{\bar{x}-\mu}{\sigma/\sqrt{n}} \leq Z_{\alpha/2}\right) = 1-\alpha \tag{1}$$

Since $\sigma > 0$, the following inequalities are equivalent.

$$-Z_{a/2} \leq \frac{\bar{x}-\mu}{\sigma/\sqrt{n}} \leq Z_{a/2},$$

$$-Z_{\alpha/2}\left(\frac{\sigma}{\sqrt{n}}\right) \leq \bar{x}-\mu \leq Z_{\alpha/2}\left(\frac{\sigma}{\sqrt{n}}\right)$$

$$-\bar{x}-Z_{\alpha/2}\left(\frac{\sigma}{\sqrt{n}}\right) \leq -\mu \leq -\bar{x}+Z_{\alpha/2}\left(\frac{\sigma}{\sqrt{n}}\right)$$

$$\bar{x}+Z_{\alpha/2}\left(\frac{\sigma}{\sqrt{n}}\right) \geq \mu \geq \bar{x}-Z_{\alpha/2}\left(\frac{\sigma}{\sqrt{n}}\right)$$

$$\bar{x}-Z_{\alpha/2}\left(\frac{\sigma}{\sqrt{n}}\right) \leq \mu \leq \bar{x}+Z_{\alpha/2}\left(\frac{\sigma}{\sqrt{n}}\right) \tag{2}$$

From (1) and (2), we have

$$P\left(\bar{x}-Z_{\alpha/2}\left(\frac{\sigma}{\sqrt{n}}\right) \leq \mu \leq \bar{x}+Z_{\alpha/2}\left(\frac{\sigma}{\sqrt{n}}\right)\right) = 1-\alpha$$

Thus, the probability that the random interval

$$\left[\bar{x}-Z_{\alpha/2}\left(\frac{\sigma}{\sqrt{n}}\right), \bar{x}+Z_{\alpha/2}\left(\frac{\sigma}{\sqrt{n}}\right)\right]$$

includes the unknown mean $\mu$ is $1-\alpha$.

Consider the probability $P(|\bar{x}-\mu| < Z_{\alpha/2} \cdot \frac{\sigma}{\sqrt{n}}) = 1-\alpha$. This can be written as $P\left[\bar{x}-Z_{\alpha/2}\frac{\sigma}{\sqrt{n}},\right.$ $\left.\bar{x}+Z_{\alpha/2}\frac{\sigma}{\sqrt{n}}\right] = 1-\alpha.$

The confidence interval formulas are not unique. The above confidence interval can be written as $\left[\mu < \bar{x} + Z_{\alpha}\frac{\sigma}{\sqrt{n}}\right]$ or it can be changed to a one-sided $(1-\alpha)100\%$ confidence interval formula written as $\left[\mu < \bar{x} + Z_{\alpha}\frac{\sigma}{\sqrt{n}}\right]$.

The maximum error of estimate for large samples is given below. For large $n$, the random variable $\frac{\bar{x}-\mu}{\frac{\sigma}{\sqrt{n}}}$ is the normal variate approximately.

If $\bar{x}$ is the mean of a random sample of size $n$ from the population with known variance $\sigma^2$, the $(1-\alpha)100\%$ confidence interval for $\mu$ is given by

$$\bar{x} - Z_{\alpha/2}\sigma/\sqrt{n} \leq \mu \leq \bar{x} + z_{\alpha/2}\frac{\sigma}{\sqrt{n}}$$

where $Z_{\alpha/2}$ is the $z$ value leaving an area of $\alpha/2$ to the right.

Hence, the maximum error of estimate $E$ with $(1-\alpha)$ probability is given by

$$E = z_{\alpha/2}\frac{\sigma}{\sqrt{n}}$$

## 5.11 Estimation of Difference Between Two Means

Let $\overline{x_1}$ and $\overline{x_2}$ denote the means of the independent random sample of sizes $n_1$ and $n_2$ respectively, drawn from two populations. Let $\mu_1$ and $\mu_2$ denote the means of the populations. If $\sigma_1^2$ and $\sigma_2^2$ denote the variances of these populations, then

$$Z = \frac{(\overline{X_1} - \overline{X_2}) - (\mu_1 - \mu_2)}{\sqrt{\frac{\sigma_1^2}{n_1} + \frac{\sigma_2^2}{n_2}}} \quad (\sigma_1^2, \sigma_2^2 \text{ are known})$$

has the normal distribution. The confidence interval formula for $\mu_1 - \mu_2$ is given by the following theorem.

**Theorem 5.3** If $\overline{x_1}$ and $\overline{x_2}$ are the values of the means of independent random samples of sizes $n_1$ and $n_2$ of normal populations with the known variances $\sigma_1^2$ and $\sigma_2^2$, then

$$(\overline{x_1} - \overline{x_2}) - z_{\alpha/2}\sqrt{\frac{\sigma_1^2}{n_1} + \frac{\sigma_2^2}{n_2}} < \mu_1 - \mu_2 < (\overline{x_1} - \overline{x_2}) + z_{\alpha/a}\sqrt{\frac{\sigma_1^2}{n_1} + \frac{\sigma_2^2}{n_2}}$$

is a $(1-\alpha)100\%$ confidence interval for the difference between population means. If $\sigma_1$ and $\sigma_2$ are unknown, we can assume $\sigma_1 = \sigma_2 = \sigma$.

Then we have $Z = \dfrac{(\overline{X_1} - \overline{X_2}) - (\mu_1 - \mu_2)}{\sigma\sqrt{\dfrac{1}{n_1} + \dfrac{1}{n_2}}}$

is a random variable having standard normal distribution.

## 5.12 The Estimator of Proportions

Let us assume that we are sampling a binomial population and our aim is to estimate the binomial parameter $p$. For a large $n$, we know that binomial distribution can be approximated with normal distribution

i.e., $Z = \dfrac{X - np}{\sqrt{np(1-p)}}$

can be treated as a random variable having approximately the standard normal distribution.

We have $P(-Z_{\alpha/2} < \dfrac{X - np}{\sqrt{np(1-p)}} < Z_{\alpha/2}) = 1 - \alpha$

The solution of the inequalities $-Z_{a/2} < \dfrac{X - np}{\sqrt{np(1-p)}}$ and $\dfrac{X = bp}{\sqrt{np(1-p)}} < Z_{a/2}$

yields a $(1-\alpha)100\%$ confidence interval for $p$, which is given by the following theorem.

**Theorem 5.4** If $X$ is a binomial random variable with parameters $n$ and $p$, $n$ large and $p = \dfrac{x}{n}$, then

$$p - Z_{\alpha/2}\sqrt{\frac{p(1-p)}{n}} < p < p + Z_{\alpha/2}\sqrt{\frac{p(1-p)}{n}}$$

is an approximate $(1-\alpha)100\%$ confidence interval for $p$.

If $p = \dfrac{x}{n}$ is used as an estimate of $p$, then we can assert with $(1-\alpha)100\%$ confidence that the error is less than $Z_{\alpha/2}\sqrt{\dfrac{p(1-p)}{n}}$.

Hence the maximum error for the proportion $p$ is given by $Z_{\alpha/2}\sqrt{\dfrac{p(1-p)}{n}}$ or $Z_{\alpha/2}\sqrt{\dfrac{pq}{n}}$

where $Q = 1 - P$ and the size of the sample is given by $n = p(1-p)\left(\dfrac{Z_{\alpha/2}}{E}\right)^2$

## 5.13 Bayesian Estimation

In the Bayesian estimation, the parameters are considered as a random variable having prior distributions. The main problem of a Bayesian estimation is that of combining prior feelings about a parameter with direct sample evidence. The concept introduced in the Bayesian method is subjective probability.

Let $\mu$ denote the population mean and $\sigma$ denote the standard deviation of the population. To estimate the value of $\mu$, we treat mean as a random variable whose distribution is indicative of the strong feeling's of a person about the possible value of $\mu$. Let $\mu_o$ and $\sigma_o$ denote the mean and standard deviation of some of subjective "prior distribution". Combining the prior feelings about the possible values of $\mu$ with direct sample evidence, the posterior distribution of $\mu$ is the Bayesian estimation approximated by normal distribution with $\mu_1 = \dfrac{n\bar{x}\sigma_0^2 + \mu_0\sigma^2}{n\sigma_0^2 + \sigma^2}$ and $\sigma_1 = \sqrt{\dfrac{\sigma_0^2\sigma^2}{n\sigma_0^2 + \sigma^2}}$,

where $\sigma^2$ is assumed to be known.

The value of $\mu_1$ and $\sigma_1$ are known as the mean and standard deviation of the Poisson distribution. In computing the value of $\mu_1$ and $\sigma_1$ when $\sigma^2$ is unknown and the sample size $n$ is large, it is replaced by $s^2$.

### Solved examples

**Example 1** A random sample of size 100 has mean 10, the population variance being 25. Find the interval estimate of the population mean with confidence level (i) 95% and (ii) 99%.

**Solution** We have $\bar{x} = 10$, $\sigma^2 = 25$ (given) and $n = 100$

$$\text{S.E.}(\bar{x}) = \sigma_{\bar{x}} = \frac{\sigma}{\sqrt{n}} = \frac{5}{\sqrt{100}} = 0.5$$

i. 95% confidence interval

We have $Z_\alpha = 1.96$, $\sigma_{\bar{x}} = 0.5$

The confidence interval is $\bar{x} \pm \sigma_{\bar{x}}.Z_\alpha = 10 \pm (0.5)(1.96) = 10 \pm 0.98$

$= (10 - 0.98, 10 + 0.98)$

$= (9.02, 10.98)$

ii. 99% confidence interval:

We have $Z_\alpha = 2.58$, $\sigma_{\bar{x}} = 0.5$

The confidence interval is $\bar{x} \pm \sigma_{\bar{x}}.Z_\alpha$

$= 10 \pm (0.5)(2.58) = 10 \pm 1.29$

$= (10 - 1.29, 10 + 1.29)$

$= (8.71, 11.29)$

**Example 2** A random sample of 50 items drawn from a particular population has a mean 30 with a standard deviation 28. Construct a 98% confidence interval estimate of the population mean.

**Solution** Since the population standard deviation is known, we compute an unbiased estimate of the standard error of mean.

We have $S$ = Standard deviation of the sample

$= 28$

$\bar{x} = 30,\ n = 50$

Confidence level = 98%

$$\text{S.E. } (\bar{x}) = \sigma_{\bar{x}} = \frac{s}{\sqrt{n-1}} = \frac{28}{\sqrt{50-1}} = 4$$

Hence, the confidence interval is $\bar{x} \pm \sigma_{\bar{x}}.Z_\alpha = 30 \pm (4)(2.33) = 30 \pm 9.32$

$= (30 - 9.32,\ 30 + 9.32)$

$= (20.68, 39.32)$

**Example 3** In a random selection of 50 of 600 road crossings in a town, the mean number of automobile accidents per year was found to be 3.8 and the sample standard deviation was 0.8. Construct a 95% confidence interval for the mean number of automobile accidents per crossing per year.

**Solution** We have $n = 50$, $N = 600$

Standard deviation of sample, $s = 0.8$ and $\bar{x} = 3.8$

Confidence level = 95%

$\therefore Z_\alpha = 1.96$

Since $\dfrac{n}{N} = \dfrac{50}{600} = 0.083$ is more than 0.05, the standard error is computed by the formula,

$$\text{S.E. } (\bar{x}) = \sigma_{\bar{x}} = \frac{s}{\sqrt{n-1}}\sqrt{\frac{N-n}{N-1}}$$

$$= \frac{0.8}{\sqrt{50-1}}\sqrt{\frac{600-50}{600-1}}$$

$$= \frac{0.8}{7}\sqrt{\frac{550}{599}} = 0.104$$

Required confidence interval is $\bar{x} \pm \sigma_{\bar{x}}.Z_{\alpha} = 3.8 \pm (0.104)(1.96) = 3.8 \pm 0.2046$

$$= (3.8 - 0.2046, 3.8 + 0.2046)$$

$$= (3.5954, 4.8046)$$

**Example 4** A population random variable has mean 100 and standard deviation 16. What are the values of mean and standard deviation of the sample mean for random sample of size 4 drawn with replacement?

**Solution** Since sampling is with replacement, $\sigma_{\bar{x}} = \dfrac{\sigma}{\sqrt{n}}$

It is given that mean of the population is $\mu = 100$

Standard deviation $\sigma = 16$

Size of the sample $n = 4$

$\therefore \mu_{\bar{x}} = \mu = 100$

$$\sigma_{\bar{x}} = \frac{\sigma}{\sqrt{n}} = \frac{16}{\sqrt{4}} = 8$$

**Example 5** The dean of a college wants to use the mean of a random sample to estimate the average amount of time that students take to get from one class to the next. She wants to be able to assert with 99% confidence that the error is almost 0.25 minute. If it can be presumed from experience that $\sigma = 1.40$ minutes, how large a sample will she have to take?

**Solution** We have $\sigma = 1.40$ mts

$E =$ maximum error $= 0.25$

$Z_{\alpha/2} = 2.575$

Therefore, $n = \left[\dfrac{Z_{\alpha/2}.\sigma}{E}\right]^2 = \left[\dfrac{(2575)(1.4)}{0.25}\right]^2 = 208$

Hence, the sample size $= 208$

**Example 6** A sample of 10 cam shafts intended for use in gasoline engines has an average eccentricity of 1.02 inches and a standard deviation of 0.044 inches. Assuming that the data may be treated as a random sample from a normal population, determine a 95% confidence interval for the actual mean eccentricity of the cam shaft.

**Solution** We have $n =$ size of the sample $= 10$

$\sigma =$ standard deviation $= 0.044$ inches

$\bar{x}$ = average eccentricity = 1.02 inches

95% confidence interval is $\bar{x} \pm \sigma_{\bar{x}}.Z_\alpha = 1.02 \pm (196)\left(\frac{0.044}{\sqrt{10}}\right) = 1.02 \pm 0.027$

(since the value of the confidence coefficient is 1.96 for 95% level of confidence)

$$= (1.02 - 0.027, 1.02 + 0.027)$$

$$= (0.993, 1.047)$$

**Example 7** It is desired to estimate the mean number of hours of continuous use until a certain computer will first require repairs. If it can be assumed that $\sigma$ = 48 hrs, how large a sample will be needed so that one will be able to assert with 90% confidence that the sample mean is off by at most 10 hours.

**Solution** Since standard deviation $\sigma$ = 48 hrs

Maximum error = $E$ = 10 hrs

The 90% value of confidence coefficient = 1.645

$$\text{Size of the sample} = n = \left(\frac{1.645.\sigma}{E}\right)^2 = \left(\frac{(1.645)(48)}{10}\right)^2$$

$$= 62.3$$

Hence $n = 62$

**Example 8** If a random sample of size $n$ = 20 from a normal population with variance $\sigma^2$ = 225 has a mean $\bar{x}$ = 64.3, construct a 95% confidence interval for the population mean $\mu$.

**Solution** We have $n = 20$, $\bar{x} = 64.3$, $\sigma = 15$ and $(1 - \alpha)100 = 95\%$

$$\alpha = 0.05$$

$$\text{or } \frac{\alpha}{2} = 0.025$$

$$Z_{\alpha/2} = 1.96$$

The required confidence interval is $\bar{x} - 1.96\frac{\sigma}{\sqrt{n}} < \mu < \bar{x} + 1.96\frac{\sigma}{\sqrt{n}}$

i.e., $64.3 - 1.96\frac{15}{\sqrt{20}} < \mu < 64.3 + 1.96\frac{15}{\sqrt{20}}$

or $57.7 < \mu < 70.9$

**Example 9** A random sample of size 81 was taken whose variance is 26.25 and mean is 32. Construct a 98% confidence interval.

**Solution** It is given that $n$ = size of the sample = 81

$$\bar{x} = 32$$

$$\sigma = \sqrt{20.25}$$

We have $Z_{\alpha/2} = 2.33$, S.E. $(\bar{x}) = \sigma_{\bar{x}} = \dfrac{\sigma}{\sqrt{n}} = \dfrac{\sqrt{20.25}}{\sqrt{81}} = \dfrac{4.5}{9} = 0.5$

Hence, the confidence interval is $\bar{x} \pm \dfrac{\sigma}{\sqrt{n}}.Z_\alpha = 32 \pm (0.5)(2.33) = 32 \pm 1.165$

$$= (31.835, 33.165)$$

**Example 10** If $X$ has the binomial distribution with parameters $n$ and $p$, show that the sample proportion, $\dfrac{X}{n}$ is an unbiased estimator of $p$.

**Solution** Since $E(X) = np$

We get $E(\dfrac{X}{n}) = \dfrac{1}{n}E(X)$

$$= \frac{1}{n}np = p$$

Hence $\dfrac{X}{n}$ is an unbiased estimator of $p$

**Example 11** Measurements of the weights of a random sample of 200 ball bearings made by a certain machine during one week showed a mean of 0.824 and a standard deviation of 0.042. Find 95% confidence limits for the mean weight of all the ball bearings.

**Solution** We have $\bar{x} = 0.824$

$\sigma = 0.042$ and $n = 200$

$$\text{S.E.}(\bar{x}) = \sigma_{\bar{x}} = \frac{\sigma}{\sqrt{n}} = \frac{0.042}{\sqrt{200}} = \frac{0.042}{14.142} = 0.00296$$

The 95% confidence interval is $\bar{x} \pm \sigma_{\bar{x}}.Z_\alpha = 0.824 \pm (0.002)(1.96) = 0.824 \pm 0.00582$

$$= (0.824 - 0.00582, 0.824 + 0.00582)$$

$= (0.81817, 0.82982)$

The confidence limits are 0.81817 and 0.82982.

**Example 12** If $x_1, x_2, \ldots, x_n$ be a random sample of size $n$ from a normal distribution with mean zero and variance $\sigma^2$, show that $\frac{1}{n}\sum_{i=1}^{n} x_i^2$ is an unbiased statistic for $\sigma^2$.

**Solution** Let $s^2 = \frac{1}{n}\sum_{i=1}^{n} x_i^2$

Then $E(s^2) = E(\frac{1}{n}\sum_{i=1}^{n} x_i^2) = \sigma^2$

Hence, $s^2$ is an unbiased statistic for $\sigma^2$.

## Exercise 5.1

1. Show that the sample median is a consistent estimator for the population mean of a normal distribution.
2. Define the terms

   (i) estimation, (ii) point estimation, (iii) unbiased estimator, (iv) consistent estimator
3. Define (i) consistency, (ii) efficiency, (iii) sufficiency
4. Show that the sample variance is a consistent estimator for the population variance of a normal distribution.
5. Show that for a random sampling from Cauchy population with density function

   $f(x, \mu) = \frac{1}{\pi} \cdot \frac{1}{1 - (x - M)^2}, -\infty \le x \le \infty$

   the sample mean is not a consistent estimator for the population.
6. Show that the sample mean is always a consistent estimator of the population mean provided that the population has a finite variance.
7. If $x_1, x_2, \ldots, x_n$ be a random sample of size $n$ from a normal distribution with $\bar{x}$ and variance $\sigma^2$, show that the statistic $s^2 = \frac{1}{n}\sum_{i=1}^{n}(x_i - \bar{x})^2$ is not an unbiased estimator of the parameter $\sigma^2$. Further, show that it is an asymptotically unbiased estimator of the parameter $\sigma^2$.
8. Show that $2\bar{x}$ is an unbiased estimate of $\theta$ for $f(x, \theta) = \frac{1}{\theta}, (0 < x < \theta)$.

9. Show that the sample mean $\overline{x}$ is an unbiased estimator of $\frac{1}{\theta}$ for the distribution $f(x,\theta) = (1-\theta)^{x-1}$, $x = 1,2,\ldots,(0<\theta<1)$.

10. Show that for samples of size $n$ from normal population with mean $\mu$ and variance $\sigma^2$, the statistic $t = \frac{1}{n+1}\sum_{i=1}^{n} x_i$ is most efficient for estimating $\mu$, though it is biased.

11. The mean and standard deviation of a population are 11,795 and 14,054 respectively. What can one assert with 95% confidence about the maximum error if $\overline{x} = 11.795$ and $n = 50$

12. A random sample of 400 items chosen from an infinite population is found to have a mean of 82 and a standard deviation of 18. Find 95% confidence limits for the mean of the population from which the sample is drawn.

13. Let $x_1, x_2, \ldots, x_n$ be a random sample drawn from a given population with mean $\mu$ and variance $\sigma^2$. Show that the sample mean $\overline{x}$ is an unbiased estimator of the population mean.

14. If $\theta_1$ and $\theta_2$ are two unbiased estimators of $\theta$, show that $w_1\theta_1 + w_2\theta_2$ is an unbiased estimator of $\theta$ (where $w_1 + w_2 = 1$).

15. Define maximum error of an estimate.

16. Define the terms (a) interval estimation (b) Bayesian estimation

17. What is the maximum error one can expect to make with probability 0.9, when using the mean of a random sample of size $n = 64$ to estimate the mean of a population with $\sigma^2 = 2.56$.

18. An industrial engineer intends to use the mean of a random sample of size $n = 150$ to estimate the average mechanical aptitude of assembly line works in a large industry. If on the basis of the experience, the engineer can assume that $\sigma = 6.2$ for such data, what can he assert with probability 0.99 about the maximum size of his error.

19. If the mean breaking strength of a copper wire is 505 lbs with a standard deviation of 15 lbs and the size of the sample is 49, construct the 95% confidence interval for the mean.

20. A population consists of four values 3, 7, 11 and 15. Draw all possible samples of size 2 with replacement. Verify that the sample mean is an unbiased estimator of the population mean.

## Answers 5.1

(11) 3896
(12) (80.236, 83.764)
(17) 0.329
(18) 1.3
(19) (500.8, 509.2)

CHAPTER SIX

# Hypothesis Testing

## 6.1 Introduction

In this chapter, we shall deal with the problem of hypothesis testing which enables us to make statements, about population parameters. Many experiments are carried out with the deliberate intention of testing a hypothesis. Hypothesis testing is a branch of statistics which helps us to arrive at a criterion of decision making. We now explain the meaning of hypothesis and some of the basic concepts essential for the better understanding of hypothesis testing methods.

## 6.2 Hypothesis

Hypothesis is an assumption or statement which may or may not be true. It is the conclusion which is tentatively drawn as logical basis. It is defined as follows.

**Definition 6.1** A hypothesis is a statement about a population parameter. It is an assumption made in order to arrive at a decision regarding the population through a sample of the population. A hypothesis should be specific, clear and precise. It should state as far as possible in mostly single terms so that the same is easily understood by all. It should state the relationship between variables. A hypothesis should be capable of testing.

## 6.3 Hypothesis Testing

It is a process of testing significance which concerns with the testing of some hypothesis regarding parameter of the population on the basis of a statistic from the population.

## 6.4 Types of Hypothesis

To verify our assumption which is based on a sample study, we collect data and find out the difference between the sample value and the population value. If there is no difference, or if the difference is very small then our hypothesised value is correct. In general, two types of hypothesis are constructed, namely, null hypothesis and alternative hypothesis.

### 6.4.1 Null hypothesis

In the process of statistical testing, the hypothesis concerning a population is rejected or accepted based on the sample drawn from the population. The statistician tests the hypothesis through observation and makes a probability statement. The hypothesis (i.e., statement) we have assumed is called the null hypothesis. It is defined as follows.

**Definition 6.2** The hypothesis formulated for the purpose of its rejection under the assumption that it is true is called the null hypothesis. Null hypothesis is denoted by $H_0$.

### 6.4.2 Alternative hypothesis

The negation of a null hypothesis is called an alternative hypothesis. It is denoted by $H_1$. If the null hypothesis is rejected, then the alternative hypothesis is accepted. The acceptance or rejection of a null hypothesis is based on the sample study. Suppose we want to test the hypothesis that the population mean is equal to the hypothesised mean 60. Symbolically $\mu$ can be expressed as

$H_0 : \mu = \mu_0 = 60$ (where $\mu_0 = \mu_{H_0}$)

The possible alternative hypothesis can be stated in one of the following forms.

$H_1 : \mu \neq \mu_0$ i.e $\mu > \mu_0$ or $\mu < \mu_0$ (two-tailed test)

$H_1 : \mu > \mu_0$ (one-tailed test)

$H_1 : \mu < \mu_0$ (one-tailed test)

In hypothesis testing, we usually proceed on the basis of a null hypothesis, the probability of rejecting the null hypothesis when it is true or the level of significance which is very small. A test hypothesis is the rule of rejecting $H_0$ and accepting $H_1$ or otherwise $H_0$.

## 6.5 Computation of a Test Statistic

A test statistic is computed after stating the null hypothesis. It is based on the appropriate probability distribution. The test statistic is used to test whether the null hypothesis $H_0$ should be accepted or rejected.

## 6.6 Level of Significance

The level of significance is an important concept in hypothesis testing. It is always some percentage. It should be chosen with great care, thought and reason. The level of significance is the maximum probability of rejecting a null hypothesis when it is true and is denoted by $\alpha$. The probability of making a correct decision is $1-\alpha$. The level of significance may be taken as 1% or 5% or 10% (i.e, $\alpha = 0.01$ or 0.05 or 0.1). If we fix the level of significance at 5%, then the probability of making a type I error is 0.05. This also means that we are 95% confident of making a correct decision. When no level of significance is mentioned, it is taken as $\alpha = 0.05$.

## 6.7 Critical Region

A region corresponding to a statistic which amounts to rejection of the null hypothesis $H_0$ is known as the critical region. It is also called as the region of rejection. The critical region is the region of the standard normal curve corresponding to a pre-determined level of significance $\alpha$. The region under the normal curve which is not covered by the normal curve is known as the acceptance region.

## 6.8 One-Tailed Test and Two-Tailed Tests

### 6.8.1 One-tailed test

A test of any statistical hypothesis where the alternative hypothesis is one-sided such as

$$H_0 : \mu = \mu_0 \quad \text{or} \quad H_0 : \mu = \mu_0 \quad \text{or} \quad H_0 : \mu \geq \mu_0 \quad H_0 : \mu \leq \mu_0$$

$$H_1 : \mu > \mu_0 \qquad\qquad H_1 : \mu < \mu_0 \qquad\qquad H_1 : \mu > \mu_0 \quad H_1 : \mu < \mu_0$$

is called a one-tailed test, where $\mu$ is the parameter to be tested and $H_0$ is the hypothesised value. When an alternative hypothesis is one-sided i.e., one-tailed, we apply the one-tailed test. There are two types of one-tailed tests. When it is described to see whether the population mean is as large as some specified value of the mean, we apply the right-tailed test. In the right-tailed test, the region of rejection (critical region) lies entirely on the right tail on the normal curve (Fig. 6.1).

When it is desired to see whether the population mean is at least as small as some specified value of the mean, we apply the left-tailed test. In the left-tailed test, the region of rejection (critical region) lies entirely on the left tail on the normal curve (Fig. 6.2).

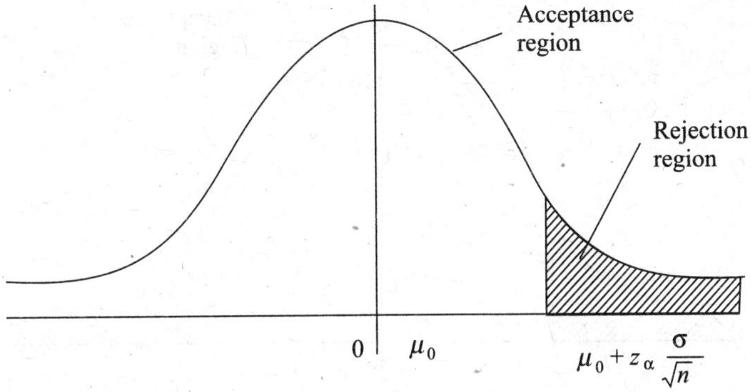


**Fig. 6.1** *Right-tailed test: Rejection and acceptance regions*

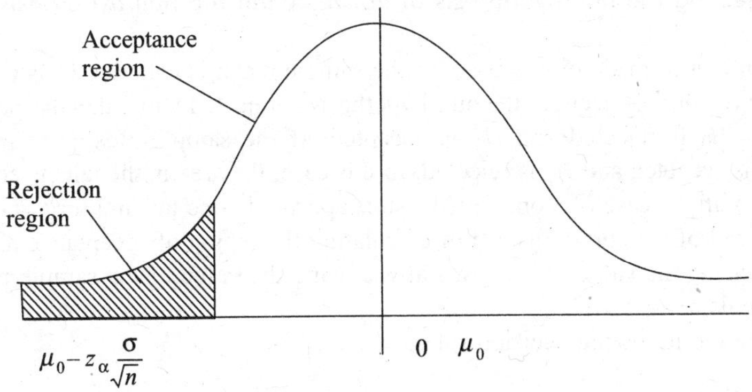


**Fig. 6.2** *Left-tailed test: Rejection and acceptance region*

## 6.8.2 Two-tailed test

A test of any statistical hypothesis where the alternative hypothesis is two-sided such as

$$H_0 : \mu = \mu_0$$

$$H_1 : \mu \neq \mu_0$$

is called a two-tailed test, where $\mu$ is the parameter to be tested and $\mu_0$ is the hypothesised value. In a two-tailed test, there are two rejection regions, one on each tail of the curve (Fig 6.3). If the significance level is 5% and the test applied is a two-tailed test, the probability of the region of rejection will be 0.05. It is equally split on both sides of the curve as 0.025. The region of acceptance in this case is 0.95.

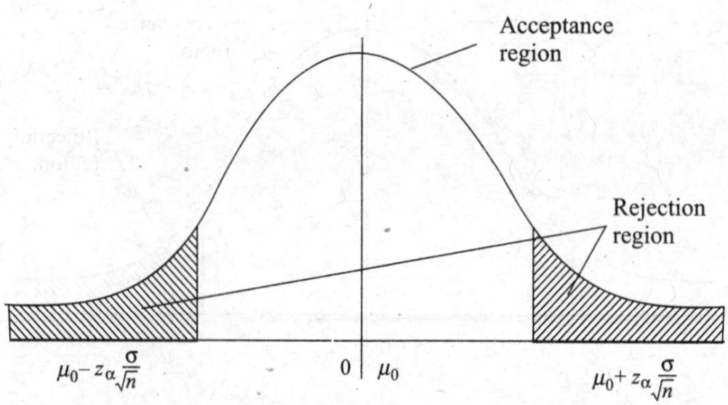


**Fig. 6.3** *Two-tailed test: Rejection and acceptance region*

### Decision of rejecting the null hypothesis or not rejecting the null hypothesis

The statistical inference made on the basis of the sample mean is called a decision. If the sample mean lies in the region of rejection, the null hypothesis is rejected while the alternative hypothesis is accepted i.e., $H_0$ is rejected and $H_1$ is accepted. If the sample mean lies in the region of acceptance, $H_0$ is accepted and $H_1$ is rejected. In this case, the area in the tails is equal to the level of significance $\alpha$. In the case of a one-tailed test, $\alpha$ appears in one tail and for two-tailed tests $\alpha/2$ appears in each tail of the curve. Instead of calculating the region of acceptance and rejection, we can directly calculate the value of the test statistic using the value of the sample mean calculated from the sample data.

Thus we have the following decision rules.

(a) $H_0 : \mu = \mu_0$ Accept $H_0$ if $\dfrac{|\bar{x} - \mu_0|}{\text{S.E. }(\bar{x})} \leq$ Table value of $Z_{\alpha/2}$ or $t_{\alpha/2}$

$H_1 : \mu \neq \mu_0$ Reject $H_0$ if $\dfrac{|\bar{x} - \mu_0|}{\text{S.E. }(\bar{x})} >$ Table value of $Z_{\alpha/2}$ or $t_{\alpha/2}$

(b) $H_0 : \mu \leq \mu_0$ Accept $H_0$ if $\dfrac{|\bar{x} - \mu_0|}{\text{S.E. }(\bar{x})} \leq$ Table value of $Z_\alpha$ or $t_\alpha$

$H_1 : \mu > \mu_0$ Reject $H_0$ if $\dfrac{|\bar{x} - \mu_0|}{\text{S.E. }(\bar{x})} >$ Table value of $Z_\alpha$ or $t_\alpha$

(c) $H_0 : \mu \geq \mu_0$ Accept $H_0$ if $\dfrac{|\bar{x} - \mu_0|}{\text{S.E. }(\bar{x})} \leq$ Table value of $Z_\alpha$ or $t_\alpha$

$H_1 : \mu < \mu_0$ Reject $H_0$ if $\dfrac{|\bar{x} - \mu_0|}{\text{S.E. }(\bar{x})} >$ Table value of $Z_\alpha$ or $t_\alpha$

### Remarks

(1) When the size of the sample is $n \geq 30$ i.e., the sample is large, we make use of the unbiased estimate of $\sigma$.

If the population S.D. σ is known and $n$ denotes the size of the sample, the test statistic

$$Z = \frac{\bar{x} - \mu}{\sigma/\sqrt{n}} \qquad \text{(normal)}$$

(2) When σ is unknown, and $n < 30$ (i.e., small sample), the statistic is $t = \dfrac{\bar{x} - \mu}{\hat{\sigma}/\sqrt{n}}$, ($\hat{\sigma}$ is the unbiased estimate of σ).

## 6.9 Errors

The decision regarding a null hypothesis may or may not be correct. There are two types of errors we can make

| | Accept $H_0$ | Reject $H_0$ |
|---|---|---|
| $H_0$ (true) | Correct decision | Wrong decision |
| | Type I error | Type II error |
| $H_0$ (false) | Wrong decision | Correct decision |

The probability of rejecting $H_0$ when $H_0$ is true is denoted by α and the probability of accepting $H_0$ when $H_1$ is true is denoted by β.

i.e., Prob (Type I error ) $= \alpha$

Prob (Type II error) $= \beta$

An increase in the sample size $n$ will reduce α.

The probability that we will reject a false null hypothesis is given by $1-\beta$. The quantity $1-\beta$ is called the power of a test. For a given test, we may specify any number of possible values of the parameter of interest and for each compute the value of $1-\beta$. The result is called a power function and the corresponding curve is called a power curve.

By adjusting the values of α, we can reduce the size of the rejection region. Probability of making one type of error can be reduced by allowing an increase in the probability of the other type of error. Hence, a decrease in the probability of the other.

## 6.10 Procedure for Hypothesis Testing

The main question in hypothesis testing is whether to accept the null hypothesis or not to accept the null hypothesis. The following tests are involved in hypothesis testing.

1. **Stating of the hypothesis**

   The null hypothesis $H_0$, and the alternative hypothesis $H_1$ are constructed in this step. It is of the form given below.

| Right-tailed | | Left-tailed | | Two-tailed |
|---|---|---|---|---|
| $H_0 : \mu \leq \mu_0$ | or | $H_0 : \mu > \mu_0$ | or | $H_0 : \mu = \mu_0$ |
| $H_1 : \mu > \mu_0$ | | $H_1 : \mu < \mu_0$ | | $H_1 : \mu \neq \mu_0$ |

where $\mu$ is the population mean and $\mu_0$ is its hypothesis mean.

2. **Identification of test stastic**
For large samples ($n \geq 30$), when population $n$'s standard deviation is known, the test statistic is $Z = \dfrac{\bar{x} - \mu}{\sigma/\sqrt{n}}$. The corresponding distribution is normal. For small samples (i.e., $n < 30$), we use the $t$-test.
3. **Specify the level of significance**
A very important concept in hypothesis testing is the level of significance. It is always some percentage. It may be 1% or 5% or 10%.The level of significance is the maximum value of the probability of rejecting the null hypothesis when it is true and is usually determined in advance before testing the hypothesis.
4. Determine the value of the test stastic.
5. Check whether the probability value is less than or equal to $\alpha$ and accordingly reject the null hypothesis, or accept it.

## 6.11 Important Tests of Hypothesis

For the purpose of testing a hypothesis, several tests of hypothesis were developed. They can be classified as

(1) Parametric test

(2) Non-parametric test

Parametric tests are also known as the standard distribution free tests of hypothesis. In this section, we consider only parametric tests. The important parametric tests are

(1) $Z$-test

(2) $t$-test

(3) $\chi^2$-test

(4) $F$-test

The parametric tests are based on the assumption of normality.

$Z$-test is used for comparing the mean of a sample to some hypothesised mean for the population in case of a large sample. It is also used when the population variance is known. $Z$-test is also used for comparing the sample proportion to a theoretical value of population proportion.

$t$-test is used in the case of small samples. It is based on the $t$-distribution and is considered an appropriate test for judging the significance of a sample mean or for judging the significance of the difference between the means of two samples.

Chi-square test is used for comparing a sample variance to a theoretical population variance. It is based on chi-square distribution.

$F$-test is used to compare the variances of the two independent samples. It is based on the $F$-distribution.

## 6.12 Critical Values

The value of the test statistic which separates the critical region and the acceptance region is called critical value. It is also called the significant value and is dependent on

(a) the level of significance and

(b) the alternative hypothesis.

For large samples corresponding to the statistic $t$, the variable $Z = \frac{t - E(t)}{\text{S.E.}(t)}$ is normally distributed with mean 0 and variance 1. If $\alpha$ is the level of significance, the value $Z_\alpha$ of the test statistic for a two-tailed test is given by

$$P(|Z| > Z_\alpha) = \alpha$$

i.e., $Z_\alpha$ is the value of $Z$ so that the total area of the critical region and the two tails is $\alpha$. Since the normal curve is a symmetrical curve,

$$P(|Z| > Z_\alpha) = \alpha$$

$$P(Z > Z_\alpha) + P(Z < -Z_\alpha) = \alpha$$

or $2\,P(Z > Z_\alpha) = \alpha$

i.e., $P(Z > Z_\alpha) = \frac{\alpha}{2}$

Hence the area in each tail is $\frac{\alpha}{2}$. The following table gives us the critical values of Z.

| Critical value ($Z_\alpha$) | Level of significance | | |
|---|---|---|---|
| | 1% | 5% | 10% |
| Two-tailed test | $\lvert Z_\alpha \rvert = 2.58$ | $\lvert Z_\alpha \rvert = 1.96$ | $\lvert Z_\alpha \rvert = 1.645$ |
| Right-tailed test | $Z_\alpha = 2.33$ | $Z_\alpha = 1.645$ | $Z_\alpha = 1.28$ |
| Left-tailed test | $Z_\alpha = -2.33$ | $Z_\alpha = -1.645$ | $Z_\alpha = -1.28$ |

## 6.13 Test of Significance: Large Samples

### 6.13.1 Test of significance for single mean

Let $x_1, x_2, \ldots, x_n$ be a random sample of size $n$, drawn from a large population with mean $\mu$ and variance $\sigma^2$. Let $x$ denote the mean of the sample and $s$ denote the variance of the sample.

We know that $\bar{x} \sim N\left(\mu, \dfrac{\sigma^2}{n}\right)$.

The standard normal variate corresponding to $\bar{x}$ is

$$Z = \frac{\bar{x}-\mu}{\text{S.E.}(\bar{x})}, \text{ where S.E.}(\bar{x}) = \frac{\sigma}{\sqrt{n}}$$

We set up the null hypothesis that there is no difference between the sample mean and the population mean. The test statistic is

$$Z = \frac{\bar{x}-\mu}{\sigma/\sqrt{n}} \quad (\sigma \text{ is known and S.E.}(\bar{x}) = \frac{\sigma}{\sqrt{n}})$$

If $\sigma$ is not known

$$Z = \frac{\bar{x}-\mu}{s/\sqrt{n}}$$

where $s$ is the standard deviation of the sample.

**Solved examples**

**Example 1** The heights of college students in a city are normally distributed with S.D. 6 cms. A sample of 100 students have mean height 158 cms. Test the hypothesis that the mean height of college students in the college is 160 cms.

**Solution** We have

$\bar{x} = 158$ ( Mean of the sample)

$\mu = 160$ (Mean of the population)

$\sigma = 6$ (S.D)

$n = 100$ (size of the sample)

Level of significance: 5% and 1%

$H_0 : \mu = 160$ i.e., difference is not significant

$H_1 : \mu \neq 160$

We apply the two-tailed test

Test statistic is

$$Z = \frac{\bar{x} - \mu}{\sigma/\sqrt{n}} = \frac{158 - 160}{6/\sqrt{100}} = \frac{-2}{6/10} = -3.333$$

$\therefore |Z| = 3.333$

Table value of Z at 5% level of significance = 1.96

Since calculated value of Z at 5% level of significance is greater than the table value of Z, we reject $H_0$ at 5% level of significance.

The table value of Z at 1% level of significance is 2.58. The calculated value at 1% level of significance is greater than the table value of Z. Hence we reject the null hypothesis $H_0$ at 1% i.e., $H_0$ is rejected both at 1% and 5% level of significance.

**Example 2** A sample of 400 items is taken from a population whose standard deviation is 10. The mean of the sample is 40. Test whether the sample has come from the population with mean 38. Also calculate 95% confidence interval for the population mean.

**Solution** Here, we have $\bar{x} = 40$ (Mean of the sample)

$\mu = 38$ (Mean of the population)

$\sigma = 10$ (S.D.)

$n = 400$

Null hypothesis: $H_0 : \mu = 38$

i.e., the sample is from the population whose mean = 38

Alternative hypothesis: $H_1 : \mu \neq 38$

Level of significance: $\alpha = 0.05$

Table value of $Z = Z_\alpha = 1.96$

Test statistic:

$$Z = \frac{\bar{x} - \mu}{\sigma/\sqrt{n}} = \frac{40 - 38}{10/\sqrt{400}} = \frac{2}{10/20} = \frac{40}{10} = 4$$

Calculated value of $Z = 4$ is greater than the table value of Z at 0.05 level of significance. Hence, null hypothesis is rejected, i.e., confidence interval for the population whose mean is 38. 95%

confidence interval for the population mean is given by

$$\bar{x} \pm 1.96.\frac{\sigma}{\sqrt{n}} = 40 \pm 1.96.\frac{10}{\sqrt{400}} = 40 \pm 0.98$$

or $[40 - 0.98, 40 + 0.98] = [39.02, 40.98]$

**Example 3** The mean and standard deviation of a population are 11,795 and 14,054 respectively. If $n = 30$, find 95% confidence interval for the mean.

**Solution** It is given that $\bar{x} = 11795$, $\sigma = 14054$, $n = 50$.

Therefore, $\dfrac{\sigma}{\sqrt{n}} = \dfrac{14054}{\sqrt{50}} = \dfrac{14054}{7.071} = 1987.55$

and $1.96.\dfrac{\sigma}{\sqrt{n}} = (1.96)(1987.55) = 3895.607$

Hence the 95% confidence interval is

$$\bar{x} \pm 1.96.\frac{\sigma}{\sqrt{n}} = [11795 - 3895.607, 1175 + 3895.607]$$

$$= [7899.39, 15690.60]$$

**Example 4** It is claimed that a random sample of 49 tyres has a mean life of 15200 kms. Is the sample drawn from a population whose mean is 15,150 kms and whose standard deviation is 1200 kms? Test the significance at 0.05 levels.

**Solution** We have $\bar{x} = 15200, \mu = 15150, \sigma = 1200$ and $n = 49$

Null hypothesis : $H_0$: $\mu = 15150$

Alternative hypothesis: $H_1$: $\mu \neq 15150$ (two-tailed test)

Level of significance $= \alpha = 0.05$

Table value of $Z = 1.96$

Test statistic is

$$Z = \frac{\bar{x} - \mu}{\sigma/\sqrt{n}} = \frac{15200 - 15150}{1200/\sqrt{49}} = \frac{50}{171.42} = 0.29168$$

Since the calculated value of $Z$ is less than the table value at 5% level, we accept the null hypothesis.

**Example 5** A sample size of 400 was drawn and the sample mean was found to be 98. Test whether this sample could have come from a normal population with mean 100 and standard deviation 8 at 5% level of significance.

**Solution** Here we have $\bar{x} = 98, \mu = 100, \sigma = 8$ and $n = 400$

Null hypothesis: $H_0 : \mu = 100$

i.e., the sample comes from a normal population with mean 100.

Alternative hypothesis $H_1 : \mu \neq 100$ (two-tailed test)

Level of significance $= \alpha = 0.05$

Table value of $Z$ at 5% level of significance $= 1.96$

Test statistic:

$$Z = \frac{\bar{x} - \mu}{\sigma/\sqrt{n}} = \frac{98 - 100}{8/\sqrt{400}} = \frac{-2}{8/20} = \frac{-10}{2} = -5$$

$$|Z| = |-5| = 5 > 1.96$$

Since the calculated value of $Z$ is greater than the table value of $Z$ at 5% level of significance, we reject $H_0$. That is the sample was not drawn from the normal population with mean 100 and standard deviation 8.

**Example 6** The average marks in mathematics of a sample of 100 students was 51 with a standard deviation of 6 marks. Could this have been a random sample from a population with average marks of 50.

**Solution** Here we have $\bar{x} = 51,\ s = 6, \mu = 50, n = 100$

Null hypothesis: $H_0 : 50$.

The sample is from a population with an average of 50 marks, i.e., $\mu = 50$

Alternate hypothesis : $H_1 : \mu \neq 50$

Let the level of significance be taken as 5%

Table value of $Z$ for 5% level of significance is $|Z| = 1.96$

$$Z = \frac{\bar{x} - \mu}{s/\sqrt{n}} = \frac{51 - 50}{6/\sqrt{100}} = \frac{1}{6/10} = \frac{10}{6} = 1.666$$

Since the calculated value of $Z$ is less than the table value of $Z$ at 5% level of significance, we accept $H_0$. That is, the sample is drawn from a population with a mean of 50 marks. The difference is not significant.

**Example 7** The mean of a certain production process is known to be 50 with a standard deviation of 2.5. The production manager may welcome any change in the mean value towards the higher side but would like to safeguard against decreasing values of mean. He takes a sample of 12 items that gives a mean value of 46.5. What inference should the manager take for the production process on the basis of sample results. Use 5% level of significance for the purpose.

**Solution** $\bar{x} = 46.5, \mu = 50, \sigma = 2.5$

Null hypothesis $H_0 : \mu = 50$

Alternative hypothesis $H_1 : \mu < 50$ (one-sided test)

Level of significance $\alpha = 5\%$

Table value of $Z = 1.645$

$$Z = \frac{\bar{x} - \mu}{\sigma/\sqrt{n}} = \frac{46.5 - 50}{2.5/\sqrt{12}} = \frac{-3.5}{2.5/3.464} = \frac{-3.5}{0.721} = -4.854$$

$$|Z| = |-4.854| = 4.854$$

i.e., calculated value of $Z$ is greater than the table value of $Z$.

Hence, $H_0$ is rejected. Therefore, the production process shows a mean which is significantly less than the population mean and this calls for taking same corrective measures concerning the production process.

**Example 8** A random sample of 100 students gave a mean height of 58 kgs, with standard deviation 4 kgs. Test the hypothesis that the mean weight in the population is 60 kgs.

**Solution** Here $\bar{x} = 58, S = 4, \mu = 60, n = 100$

$H_0 : \mu = 60$ kgs

$H_1 : \mu \neq 60$ kgs

Level of significance $\alpha = 5\%$

Table value of $Z_\alpha = 1.96$

$$Z = \frac{\bar{x} - \mu}{S/\sqrt{n}} = \frac{58 - 60}{4/\sqrt{100}} = -5$$

$$|Z| = |-5| = 5$$

Since the calculated value of $Z$ is greater than the table value at 5% level of significance, $H_0$ is rejected.

$\therefore \mu \neq 60$

### 6.13.2 Test of significance for difference of means of two large samples

Let $\overline{x_1}$ be the mean of an independent random sample of size $n_1$ from a population with mean $\mu_1$ and variance $\sigma_1^2$. Let $\overline{x_2}$ be the mean of an independent random sample of size $n_2$ from a population with mean $\mu_2$ and variance $\sigma_2^2$, where $n_1$ and $n_2$ are large.

Clearly, $\overline{x_1} \sim N\left(\mu_1, \dfrac{\sigma_1^2}{n_1}\right)$ and $\overline{x_2} \sim N\left(\mu_2, \dfrac{\sigma_2^2}{n_2}\right)$

and $Z = \dfrac{(\overline{x_1} - \overline{x_2}) - E\left(\overline{X_1} - \overline{X_2}\right)}{\text{S.E.}(\overline{x_1} - \overline{x_2})} \sim N(0,1)$

i.e., $Z$ is a standard normal variate.

Setting up the null hypothesis, we have $E(\overline{x_1} - \overline{x_2}) = E(\overline{x_1}) - E(\overline{x_2})$

$$= \mu_1 - \mu_2 = 0$$

Since $\overline{x_1}$ and $\overline{x_2}$ are independent, the covariance terms vanish. We have

$$\text{Var}\ (\overline{x_1} - \overline{x_2}) = \text{Var}\ (\overline{x_1}) + \text{Var}(\overline{x_2})$$

$$= \frac{\sigma_1^2}{n_1} + \frac{\sigma_2^2}{n_2}$$

Hence, under the null hypothesis $H_0 : \mu_1 = \mu_2$, the test statistic is

$$Z = \frac{\overline{x_1} - \overline{x_2}}{\sqrt{\dfrac{\sigma_1^2}{n_1} + \dfrac{\sigma_2^2}{n_2}}} \qquad (1)$$

If $\sigma_1 = \sigma_2 = \sigma$, then the test statistic is

$$Z = \frac{\overline{x_1} - \overline{x_2}}{\sqrt{\sigma^2\left(\dfrac{1}{n_1} + \dfrac{1}{n_2}\right)}} = \frac{\overline{x_1} - \overline{x_2}}{\sigma\sqrt{\left(\dfrac{1}{n_1} + \dfrac{1}{n_2}\right)}} \qquad (2)$$

If $\sigma_1$ and $\sigma_2$ are not known, the test statistic is

$$Z = \frac{\overline{x_1} - \overline{x_2}}{\sqrt{\frac{S_1^2}{n_1} + \frac{S_2^2}{n_2}}}$$

where $\sigma_1 \neq \sigma_2$

If $\sigma_1 = \sigma_2$ and $\sigma$ is not known, we compute $\sigma^2$ by using the formula,

$$\sigma^2 = \frac{n_1 s_1^2 + n_2 s_2^2}{n_1 + n_2}$$

In this case, the test statistic is

$$Z = \frac{\overline{x_1} - \overline{x_2}}{\sqrt{\frac{n_1 s_1^2 + n_2 s_2^2}{n_1 + n_2}\left(\frac{1}{n_1} + \frac{1}{n_2}\right)}}$$

Substitute $s_1$ for $\sigma_1$ and $s_2$ for $\sigma_2$ provided $n_1$ and $n_2$ are large.

$$\text{i.e., } Z = \frac{\overline{x_1} - \overline{x_2}}{\sqrt{\frac{n_1 s_1^2 + n_2 s_2^2}{n_1 n_2}}} \tag{4}$$

If we want to test the null hypothesis $\mu_1 - \mu_2 = \delta$, where $\delta$ is a specified constant against one of the alternatives

$$\mu_1 - \mu_2 \neq \delta$$

$$\mu_1 - \mu_2 < \delta$$

$$\mu_1 - \mu_2 > \delta$$

We apply likelihood ratio technique and at the test based on $\overline{x_1} - \overline{x_2}$, the corresponding critical regions for the test can be written as

$|Z| \geq Z_{\alpha/2}, Z \geq Z_\alpha$ and $Z \leq Z_\alpha$, where

$$Z = \frac{\overline{X_1} - \overline{X_2} - \delta}{\sqrt{\frac{\sigma_1^2}{n_1} + \frac{\sigma_2^2}{n_2}}}$$

When we deal with independent random samples drawn from populations with unknown variances, that may not be normal. We substitute $S_1$ for $\sigma_1$ and $S_2$ for $\sigma_2$ provided $n_1$ and $n_2$ are large.

When the sizes of the independent random samples i.e., $n_1$ and $n_2$ are small and $\sigma_1$ and $\sigma_2$ are unknown, the above test cannot be used. For independent random samples from two normal population having the same unknown variable, we use the test

$$Z = \frac{\overline{X_1} - \overline{X_2} - \delta}{S\sqrt{\frac{1}{n_1} + \frac{1}{n_2}}}$$

$$\text{where } S = \frac{(n_1 - 1)s_1^2 + (n_2 - 1)s_2^2}{n_1 + n_2 - 2}$$

## Solved examples

**Example 1** Random samples drawn from two places gave the following data relating to the heights of children

| | Place A | Place B |
|---|---|---|
| Mean height | 68.50 | 68.58 |
| Standard deviation | 2.5 | 3.0 |
| Number of items in the sample | 1200 | 1500 |

Test at 5% level that the mean height is the same for the children at two places.

**Solution** Here we have $\overline{x_1} = 68.50, \quad \overline{x_2} = 68.58$

$$n_1 = 1200, \quad n_2 = 1500$$

$$\sigma_1 = 2.5, \quad \sigma_2 = 3.0$$

Level of significance = 0.05

$$Z_{0.05} = 1.96$$

$$\text{S.E.}(\overline{x_1} - \overline{x_2}) = \sqrt{\frac{\sigma_1^2}{n_1} + \frac{\sigma_2^2}{n_2}}$$

$$= \sqrt{\frac{(2.5)^2}{1200} + \frac{3^2}{1500}}$$

$$= \sqrt{\frac{6.25}{1200} + \frac{9}{1500}} = 0.1058$$

Null hypothesis $H_0 : \mu_1 = \mu_2$

Alternative hypothesis $H_1 : \mu_1 \neq \mu_2$ (two-tailed test )

$$Z_{0.05} = 1.96 \text{ (table value)}$$

$$Z = \frac{\overline{X_1} - \overline{X_2}}{\text{S.E.}(\overline{x_1} - \overline{x_2})} = \frac{68.50 - 68.58}{0.1058} = -0.756$$

$|Z| = 0.756 < 1.96$ i.e., calculated value of $Z$ is less than the table value. Hence, null hypothesis is accepted.

**Example 2** The mean produce of a sample of 100 fields is 200 lbs per acre with a standard deviation of 10 lbs. Another sample of 150 fields gives the mean of 220 lbs with a standard deviation 12 lbs. Can the two samples be considered to have been taken from the same population whose standard deviation is 11 lbs? Use 5% level of significance.

**Solution** It is given that

$\overline{x_1} = 200$ lbs, $\overline{x_2} = 220$ lbs

$n_1 = 100$, $n_2 = 150$

$\sigma_1 = 10$ lbs, $\sigma_2 = 12$ lbs

$\sigma =$ S.D. of the population $= 11$ lbs

$H_0 : \mu_1 = \mu_2$

$H_1 : \mu_1 \neq \mu_2$ (two-tailed test)

Level of significance = 5%

$Z_\alpha = 1.96$

$$\text{Test statistic: } Z = \frac{\overline{x_1} - \overline{x_2}}{\sqrt{\sigma^2\left(\dfrac{1}{n_1} + \dfrac{1}{n_2}\right)}} = \frac{200 - 220}{\sqrt{(11)^2\left[\dfrac{1}{100} + \dfrac{1}{150}\right]}}$$

$$= \frac{-20}{\sqrt{(121)\left(\dfrac{250}{(100)(150)}\right)}} = \frac{-20}{1.42} = -14.08$$

$\therefore |Z| = |-14.08| = 14.08$

The calculated value of $Z$ is greater than the table value at 5% level of significance.

Hence, we reject the null hypothesis $H_0$. Therefore, the difference between the means of the two samples is significant and it is not due to sampling fluctuations.

**Example 3** Two sets of 100 students each were taught to read by two different methods. After the instructions were over, a reading test given to them revealed

$\overline{x_1} = 73.4, \ \overline{x_2} = 70.3.$

$S_1 = 8$ and $S_2 = 10.$

Test the hypothesis that $\mu_1 = \mu_2$.

**Solution** It is given that

$\overline{x_1} = 73.4, \overline{x_2} = 70.3$

$S_1 = 8, S_2 = 10$

$n_1 = n_2 = 100$

Null hypothesis $H_0 : \mu_1 = \mu_2$

Alternative hypothesis: $H_1 : \mu_1 \neq \mu_2$

$$Z = \frac{\overline{x_1} - \overline{x_2}}{\sqrt{\dfrac{S_1^2}{n_1} + \dfrac{S_2^2}{n_2}}} = \frac{73.4 - 70.3}{\sqrt{\dfrac{64}{100} + \dfrac{100}{100}}}$$

$$= \frac{3.1}{\sqrt{164}} \times \sqrt{100} = \frac{31}{12.806} = 2.4207$$

$|Z| = 2.4207 > 1.96$, therefore at 5% level of significance $H_0$ is rejected.

$|Z| = 2.4207 < 2.58$, therefore at 1% level of significance $H_0$ is accepted.

**Example 4** Samples of students were drawn from two universities and from their weights in kg, the standard deviations are calculated. Make a large sample test the significance of the difference between the means.

| | Mean | S.D. | Size of the sample |
|---|---|---|---|
| University A | 55 | 10 | 400 |
| University B | 57 | 15 | 100 |

**Solution** We have

$\overline{x_1} = 55, \ \overline{x_2} = 57$

$\sigma_1 = 10, \quad \sigma_2 = 15$

$n_1 = 400, \ n_2 = 100$

Null hypothesis $H_0 : \mu_1 = \mu_2$

Alternative hypothesis: $H_1 : \mu_1 \neq \mu_2$

Level of significance = 0.05

Table value of Z = 1.96

$$\text{Test statistic: } Z = \frac{\overline{X_1} - \overline{X_2}}{\sqrt{\frac{\sigma_1^2}{n_1} + \frac{\sigma_2^2}{n_2}}} = \frac{55 - 57}{\sqrt{\frac{100}{400} + \frac{225}{100}}}$$

$$= \frac{-2}{\sqrt{\frac{100 + 900}{400}}} = \frac{-4}{\sqrt{10}} = -1.26$$

$$\therefore |Z| = 1.26 < 1.96$$

i.e., calculated value of $Z$ is less than the table value of $Z$ at 5% level of significance. Hence $H_0$ is accepted.

**Example 5** In a certain factory there are two independent processes for manufacturing the same item. The average weight in a sample of 250 items produced from one process is found to be 120 gms with S.D. of 12 gms, while the corresponding figures in a sample of 400 items from the other processes are 124 gms and 14 gms. Is the difference between mean weights significant at 10% level of significance?

**Solution** The given values are

$\overline{x_1} = 120, \quad \overline{x_2} = 124$

$S_1 = 12, \quad S_2 = 14$

$n_1 = 250, \quad n_2 = 400$

Level of significance = 10%

Null hypothesis $H_0 : \mu_1 = \mu_2$

Alternative hypothesis: $H_1 : \mu_1 \neq \mu_2$ (two-tailed test)

Z-value = 1.645 at 10% level

$$\text{Test statistic is } Z = \frac{\overline{x_1} - \overline{x_2}}{\sqrt{\frac{S_1^2}{n_1} + \frac{S_2^2}{n_2}}} = \frac{120 - 124}{\sqrt{\frac{12^2}{250} + \frac{14^2}{400}}}$$

$$= \frac{-4}{1.0325} = -3.87$$

$$|Z| = 3.87 > 1.645 \text{ (i.e., Z at 10\% level)}$$

Since, the calculated value of Z is greater than the table value of Z, we reject the null hypothesis. Therefore, there is a significant difference between the two sample mean weights.

### 6.13.3 Test of significance for the difference of standard deviations of two large samples

If $s_1$ and $s_2$ are the standard deviations of two independent samples and $H_0 : \sigma_1 = \sigma_2$ is the null hypothesis, the test statistic is

$$Z = \frac{s_1 - s_2}{\text{S.E.}(S_1 - S_2)} \sim N(0,1)$$

If $\sigma_1, \sigma_2$ are known, then for large samples

$$Z = \frac{s_1 - s_2}{\sqrt{\frac{\sigma_1^2}{2n_1} + \frac{\sigma_2^2}{2n_2}}}$$

If $\sigma_1, \sigma_2$ are not known, the test statistic is

$$Z = \frac{s_1 - s_2}{\sqrt{\frac{s_1^2}{2n_1} + \frac{s_2^2}{2n_2}}}$$

where $\sqrt{\frac{s_1^2}{2n_1} + \frac{s_2^2}{2n_2}}$ is the S.E. $(s_1 - s_2)$.

#### Solved examples

**Example 1** Intelligence test of two groups of boys and girls gave the following results

Girls: $\overline{x_1} = 84$, S.D. of $s_1 = 10$, $n_1 = 121$

Boys: $\overline{x_2} = 80$, S.D. of $s_2 = 14$, $n_2 = 81$

Is the difference between the standard deviations significant?

**Solution** Null hypothesis $H_0 : \sigma_1 = \sigma_2$

Alternative hypothesis: $H_1 : \sigma_1 \neq \sigma_2$ (two-tailed test)

Level of significance = 5%

$Z = 1.96$ at 5% level of significance

We have $Z = \dfrac{s_1 - s_2}{\sqrt{\dfrac{s_1^2}{2n_1} + \dfrac{s_2^2}{2n_2}}} = \dfrac{10 - 14}{\sqrt{\dfrac{10^2}{242} + \dfrac{14^2}{162}}}$.

$$= \frac{-4}{\sqrt{0.4132 + 1.209}} = \frac{-4}{\sqrt{1.623}}$$

$$= \frac{-4}{1.274} = -3.139$$

$\therefore |Z| = |-3.139| = 3.139 > 1.96$

Since the calculated value of $Z$ is greater than the table value of $Z$ at 5% level of significance, $H_0$ is rejected. Therefore, the difference between two standard deviations is significant.

**Example 2** The mean yield of two sets of plots and their variability are as given below. Examine whether the difference in the variances in yields are significant.

| | Set of 40 plots | Set of 60 plots |
|---|---|---|
| Mean yield per plot | 1258 kg | 1243 kg |
| S.D. per plot | 34 | 28 |

**Solution** We have $\overline{x_1} = 1258$ kgs, $\overline{x_2} = 1248$ kgs

$s_1 = 34,\ s_2 = 28$

$n_1 = 40,\ n_2 = 60$

$\sigma_1, \sigma_2$ unknown.

Null hypothesis $H_0 : \sigma_1 = \sigma_2$

Alternative hypothesis: $H_1 : \sigma_1 \neq \sigma_2$ (two-tailed test)

Level of significance $\alpha = 5\%$

$Z_\alpha = 1.96$

Test statistic: $Z = \dfrac{s_1 - s_2}{\sqrt{\dfrac{s_1^2}{2n_1} + \dfrac{s_2^2}{2n_2}}} = \dfrac{34 - 28}{\sqrt{\dfrac{34^2}{80} + \dfrac{28^2}{120}}}$

$$= \frac{6}{\sqrt{4.45 + 6.63}} = \frac{6}{\sqrt{11.08}} = 1.31$$

$Z = 1.31 < 1.96$ i.e., table value of $Z$ at 5% level of significance. Hence, we accept null hypothesis, i.e., there is insignificant difference between the two sample mean weights.

**Example 3** In a study of the effect of chemicals on the labourers of a chemical unit, the following results were obtained on their systolic blood pressures. Examine whether the differences in blood pressures are significant.

| | Males | | Females | |
|---|---|---|---|---|
| | Exposed group | Controlled group | Exposed group | Controlled group |
| Number | 250 | 55 | 105 | 50 |
| Mean | 117.5 | 121.0 | 111 | 112 |
| S.D. | 10.50 | 11 | 19.4 | 9.5 |

**Solution** For the males, we have

$\overline{x_1} = 117.5,\ \overline{x_2} = 121$

$s_1 = 10.5,\ s_2 = 11$

$n_1 = 250,\ n_2 = 55$

Null hypothesis $H_0$: Assume that there is no difference in their means

Alternative hypothesis $H_1$: The difference is significant since the samples are from the same population

$$\sigma^2 = \frac{n_1 s_1^2 + n_2 s_2^2}{n_1 + n_2} = \frac{(250)(10.5)^2 + (55)(11)^2}{250 + 55} = 112.188$$

$\therefore \sigma = 10.59$

$$\text{Test statistic: } Z = \frac{\bar{x}_1 - \bar{x}_2}{\sqrt{\sigma^2\left(\frac{1}{n_1} + \frac{1}{n_2}\right)}} = \frac{117.5 - 121}{10.59\sqrt{\frac{1}{250} + \frac{1}{55}}}.$$

$$= \frac{-3.5}{(10.5)(0.1489)} = \frac{-3.5}{1.5772} = -2.219$$

$\therefore |Z| = 2.219 > 1.96$

Hence the difference is significant at 5% level since $|Z| = 2.219 < 2.58$; the difference is not significant at 1% level.

For females, we have

$\overline{x_1} = 111, \overline{x_2} = 112$

$s_1 = 9.4,\ s_2 = 9.5$

$n_1 = 105,\ n_2 = 50$

$H_0$: The difference is not significant.

$H_1$: The difference is significant.

$$\sigma^2 = \frac{n_1 s_1^2 + n_2 s_2^2}{n_1 + n_2} = \frac{(105)(9.4)^2 + (50)(9.5)^2}{105 + 50}$$

$$\sigma = \sqrt{88.9646} = 9.432$$

$$Z = \frac{\overline{x_1} - \overline{x_2}}{\sqrt{\sigma^2\left(\frac{1}{n_1} + \frac{1}{n_2}\right)}} = \frac{111 - 112}{9.432\sqrt{\frac{1}{105} + \frac{1}{50}}}$$

$$= \frac{-1}{9.432\sqrt{\frac{155}{105 \times 50}}} = \frac{-1}{9.432(0.1778)} = -0.6172$$

$\therefore |Z| = 0.6172 < 1.96$

Hence, $H_0$ is accepted at 5% level of significance and at 1% level of significance. Therefore, the difference is not significant with respect to the females.

**Example 4** An experiment is performed to determine whether the average nicotine content of one kind of cigarette exceeds that of another kind by 0.22 milligram. If $n_1 = 50$, cigarettes of the first kind had an average nicotine content of $\overline{x_1} = 2.62$ milligrams with a S.D. of $s_1 = 0.12$ mgm, whereas $n_2 = 40$ cigarettes of the other kind had an average nicotine content of $\overline{x_2} = 2.40$ mgm with a S.D. of $s_2 = 0.16$ mgm. Test the null hypothesis that $\mu_1 - \mu_2 = 0.22$ against the alternative hypothesis $\mu_1 - \mu_2 \neq 0.22$.

**Solution** The given values are

$\overline{x_1} = 2.66,\ \overline{x_2} = 2.40$

$s_1 = 0.12$ mgm, $s_2 = 0.16$ mgm

$n_1 = 60,\ n_2 = 2.40$

Null hypothesis $H_0 : \mu_1 - \mu_2 = 0.22$

Alternative hypothesis $H_1 : \mu_1 - \mu_2 \neq 0.22$

The test statistic is $Z = \dfrac{\overline{X_1} - \overline{X_2} - \delta}{\sqrt{\dfrac{\sigma_1^2}{n_1} + \dfrac{\sigma_2^2}{n_2}}}$

Substituting $s_1$ for $\sigma_1$ and $s_2$ for $\sigma_2$, we have

$$Z = \frac{\overline{X_1} - \overline{X_2} - \delta}{\sqrt{\dfrac{s_1^2}{n_1} + \dfrac{s_2^2}{n_2}}} = \frac{2.66 - 2.40 - 0.22}{\sqrt{\dfrac{(0.12)^2}{60} + \dfrac{(0.16)^2}{40}}}$$

$$= \frac{0.04}{\sqrt{\dfrac{0.0144}{60} + \dfrac{0.0256}{40}}} = \frac{0.04}{\sqrt{0.00024 + 0.00064}}$$

$$= \frac{0.04}{0.0296} = 1.351$$

The corresponding probability value is $2(0.5000 - 0.4115) = 0.0885$. Since 0.0855 exceeds 0.05, the null hypothesis cannot be rejected.

## Exercise 6.1

1. Define the terms

   (a) Hypothesis

   (b) Null hypothesis

   (c) Alternative hypothesis

2. Define

   (i) Level of significance

   (ii) Hypothesis testing

3. Define

   (a) Type I error

   (b) Type II error

   (c) Confidence interval

4. Explain the procedure for hypothesis testing.

5. A random sample of 50 items drawn from a particular population has a mean of 30 with an S.D. of 28. Construct a 98% confidence interval estimate of the population mean.

6. For a given sample of 200 items drawn from a large population, the mean is 65 and the standard deviation is 8. Find the 95% confidence limits for the population mean.

7. A sample of size 400 was drawn and the sample mean was found to be 99. Test whether this sample could have come from a normal population with mean 100 and standard deviation 8 at 5% level of significance.

8. A stenographer claims that she can type at the rate of 120 words per minute. Can we reject her claim on the basis of 100 trials in which she demonstrates a mean of 116 words with a standard deviation of 15 words? Use 5% level of significance.

9. A sample of 400 items is taken from a normal population whose mean is 4 and whose variance is also 4. If the sample mean is 4.45, can the sample be regarded as a truly random sample.

10. A sample of 900 members has a claim of 3.4 cms and S.D. 2.61 cms. Can the sample be regarded as one drawn from a population with mean 3.25 cms? Using the level of significance as 0.05, test whether the claim is acceptable.

11. According to the norms established for a mechanical aptitude test, persons who are 18 years old have an average height of 73.2" with an S.D. of 8.6. If 45 randomly selected persons of that age averaged 76.7", test the null hypothesis $\mu = 73.2$ against the alternative hypothesis $\mu \neq 73.2$ at 0.01 level of significance.

12. In a study of an automobile insurance company, a random sample of 80 body repair costs had a mean of Rs. 473.36 and a standard deviation of Rs. 62.35. If $\overline{x}$ is used as a point estimate to the true average repair costs, with what confidence can we assert that the maximum error does not exceed Rs. 107.

13. Random samples drawn from two places gave the following data relating to the heights of children.

| | Place A | Place B |
|---|---|---|
| Mean height (cm) | 68.50 | 68.58 |
| Standard deviation (cm) | 2.5 | 3.0 |
| No. of items in sample | 1200 | 1500 |

Test at 5% level that the mean height is the same for children at the two places

14. A random sample of 50 male employees is taken at the end of a year and the mean number of hours of absenteeism for the year is found to be 63 hours. A similar sample of 50 female employees has a mean of 66 hours. Could these samples be drawn from a population with the same mean S.D. of 10 hours? State clearly the assumption made by you.

15. The standard deviation of the height of students of a college is 4.0 cm. Two samples are taken. The standard deviation of 100 B.Com (honours) students is 3.5 cm and 50 B.A economics students is 4.5 cm. Test the significance of the difference of standard deviation of the samples.

16. Intelligence test of two groups of boys and girls gave the following results.

Girls: 84; Standard deviation $S_1 = 10$; $\overline{x}_1 = 121$

Boys: 81; Standard deviation $S_2 = 12$; $\overline{x}_2 = 81$

Is the difference between the means significant?

17. For two samples, the following data are given

$n_1 = 1000, \quad \overline{x_1} = 67.42 \quad S_1 = 2.58$
$n_2 = 1200, \quad \overline{x_2} = 67.25 \quad S_2 = 2.50$

Is the difference between the standard deviations significant?

18. The following data are from an investigation

| | No. of cases | Mean wages | S.D. of the wage |
|---|---|---|---|
| Sample I | 400 | Rs. 47.4 | Rs. 3.1 |
| Sample II | 900 | Rs. 50.4 | Rs. 3.3 |

Find out whether the two mean wages differ significantly.

19. From the data given below, compute the standard error of the difference of the two sample means and find out if the two means significantly differ at 5% level of significance.

| | No. of items | Mean | S.D. of the wage |
|---|---|---|---|
| Group I | 50 | 181 | 3.0 |
| Group II | 75 | 179 | 3.6 |

20. In a study of the effect of chemicals on the labourers in a chemical unit, the following results were obtained on their systolic blood pressures. Examine whether differences in the blood pressures are significant.

| | Males | | Females | |
|---|---|---|---|---|
| | Exposed group | Controlled group | Exposed group | Controlled group |
| Number | 250 | 55 | 103 | 50 |
| Mean | 117.5 | 121.6 | 111.7 | 112.5 |
| S.D. | 10.58 | 10.82 | 9.33 | 9.38 |

21. A sample of 400 male students is found to have a mean height of 67.47 units. Can it be reasonably regarded as a sample from a large population with mean 67.39 units and S.D. 1.30 units.

22. A random sample of 100 students gave a mean weight of 58 kg with a S.D. of 4 kg. Test the hypothesis that the mean weight in the population is 60 kg.

23. A random sample of 200 measurements from a large population gave a mean value 50 and an S.D. of 9. Determine the 95% confidence interval for the mean of population.

24. According to the norms established for a mechanical aptitude test, persons who are 18 years old have an average height of 73.2" with a standard deviation of 8.6 cm. If 4 randomly selected persons of that age averaged 76.7", test the hypothesis $\mu = 73.2$ against the alternative hypothesis $\mu > 73.2$ at 0.01 level of significance.

25. An ambulance service claims that it takes, on an average, less than 10 minutes to reach its destination in emergency calls. A sample of 36 calls has a mean of 11 minutes and a variance of 16 minutes. Test the significance at 0.05 level.

26. The means of two large samples of sizes 1000 and 2000 members are 67.5 inches and 68.0 inches respectively. Can the samples be rejected as drawn from the same population with an S.D. of 2.5 inches?

27. In a city, 250 men out of 750 were found to be smokers. Does this information support the conclusion that the majority of men in the city are smokers.

28. Given the following information relating to two places A and B, test whether there is any significant difference between their mean wages.

| | A | B |
|---|---|---|
| Mean wages | 47 | 49 |
| S.D. | 28 | 40 |
| No. of workers | 1000 | 1500 |

29. The number of accidents per day were studied for 144 days in town A and for 100 days in town B and the following information was obtained

| | Town A | Town B |
|---|---|---|
| Mean number of accidents | 4.5 | 5.4 |
| Standard deviation | 1.2 | 1.5 |

Is the difference between the mean accidents of the two towns statistically significant?

30. A normal population has a mean of 0.1 and a standard deviation of 2.1. Find the probability that mean of a sample of 900 will be negative.

[Hint: we have $z = \dfrac{\bar{x}-\mu}{\sigma/\sigma_n} = \dfrac{\bar{x}-0.1}{2-1/\sqrt{900}} = \dfrac{\bar{x}-0.1}{0.07}$

$$\text{or } \bar{x} = 0.1 + 0.07z$$

$$\begin{aligned}\therefore P(\bar{x} < 0) &= P(0.1 + 0.07z < 0) \\ &= P\left(z < \frac{-0.1}{0.07}\right) \\ &= P(z < -1.4286) \\ &= P(z > 1.4286) \\ &= 0.5 - P(0 < z < 1.4286) \\ &= 0.5 - 0.4236 \\ &= 0.0764]\end{aligned}$$

31. The mean produce of wheat of a sample of 100 fields comes to 200 kg per acre and another sample of 150 fields gives a mean of 220 kg. Assuming the standard deviation of the yield to be 11 kgs for the universe test, is there a significant difference between the means of the samples?

32. An oceanographer wants to check whether the average depth of the ocean in a certain region is 574 fathoms as had previously been recorded. What can he conclude at the level of significance $\alpha = 0.05$, if the surroundings taken at 40 random locations in the given region yielded a mean of 59.1 fathoms with a standard deviation of 5.2 fathoms?

33. In a random sample of 400 men, the mean is found to be 40. In another sample of 50 men, the mean is 45. Can these samples be drawn from the population whose S.D. is 5?

34. It was found that the average income of 36 workers is Rs 12,000 per annum. Was it taken from a population whose mean is Rs 11,800 per annum and S.D. 800?

35. Test whether the means of the samples are significantly different, given the following data.

| | Size | Mean | S.D. |
|---|---|---|---|
| Sample I | 100 | 110 | 13 |
| Sample II | 100 | 112 | 16 |

A random sample of 400 items has an average length of 15 cms. Can this be regarded as a sample from a large population with mean 16 cms and S.D. 5 cms.

36. A random sample of 400 items has an average length of 15 cms. Can this be regarded as a sample from a large population with mean 16 cms and S.D. 5 cms.

## Answers 6.1

(5) 20.68, 39.32

(6) 65 ± 1.109

(7) $H_0$ rejected

(8) $H_0$ rejected i.e., stenographer claim is rejected

(9) $H_0$ rejected i.e., the sample cannot be regarded as having been drawn from the same population with mean 4

(10) The sample is drawn from a population with mean 3.25 cms

(11) $H_0$ rejected

(12) 85%

(13) Null hypothesis $H_0$ is accepted

(14) $H_0$ accepted. The two samples have been drawn from the same population

(15) $H_0$ : rejected

(16) The difference between the means is not significant

(17) The sample standard deviations do not differ significantly

(18) The difference is significant

(20) The difference is significant for males. The difference is not significant for females

(21) $H_0$ accepted

(22) $H_0$ rejected, mean weight cannot be 60 kg

(23) 48.8, 51.2

(24) Null hypothesis is accepted

(25) Null hypothesis is accepted

(26) The samples are not drawn from the same normal population of standard deviation of 2.5 inches

(27) Null hypothesis is rejected

(28) $H_0$ accepted

(29) $|Z|= 5$, significant

(30) 0.0764

(31) The difference is significant

(32) Null hypothesis rejected

(33) Not significant

(34) Significant (35) Significant
(36) Significant

## 6.14 Test of Significance for a Single Proportion

Consider a sample which is drawn from a population with a population proportion (percentage) $P$. If $n$ denotes the sample and $p$ is the sample proportion or percentage, then we have

$$\text{Standard error percentage} = \sqrt{\frac{P(100-P)}{n}}$$

$P =$ Population percentage

$$E(p) = P, \text{ Var }(p) = \frac{PQ}{n}$$

$$\text{S.E.}(p) = \sqrt{\frac{PQ}{n}}.$$

$$Z = \frac{p - E(p)}{\text{S.E.}(p)} = \frac{p-P}{\sqrt{\frac{PQ}{n}}} \sim N(0,1)$$

$Z$ is called the test statistic and is used as a test for a single proportion.

$$\text{If } |Z| = \left|\frac{p-P}{\sqrt{\frac{PQ}{n}}}\right| < Z_\alpha \qquad (1)$$

The difference of proportions $(p-P)$ is not significant at the level of significance $\alpha$.

From (1), we have $-Z_\alpha < \dfrac{p-P}{\sqrt{\frac{PQ}{n}}} < Z_\alpha$

$$\text{i.e., } P - Z_\alpha\sqrt{\frac{PQ}{n}} < p < P + Z_\alpha\sqrt{\frac{PQ}{n}}$$

(i) The probable limits for observed proportion of success i.e., sample proportion are $P \pm Z_\alpha\sqrt{\frac{PQ}{n}}$

(ii) If $P$ is unknown, the confidence limits for the proportion in the population are $p \pm Z_\alpha\sqrt{\frac{pq}{n}}$, where $q = 1-p$

(iii) If the proportion is not given, the 99% confidence limits for $p$ are $P \pm 3\sqrt{\frac{PQ}{n}}$ and the 99% confidence limits for P are $p \pm 3\sqrt{\frac{pq}{n}}$

(iv) If $\alpha$ is the level of significance under a two-tailed test, then

$$\int_{-Z_\alpha}^{Z_\alpha} \phi(t)\,dt = 1 - \alpha$$

i.e, $$\int_{0}^{Z_\alpha} \phi(t)\,dt = 0.5 - \alpha/2$$

where $\alpha$ is the level of significance.

## Solved examples

**Example 1** A random sample of 600 toys was taken from a large consignment and 80 were found to be defective. Find the percentage of defective toys in the consignment

**Solution** Here we have

$n = 600$, no. of toys $= 80$

$p =$ Proportion of defective toys in the sample

$$= \frac{80}{600} = \frac{8}{60} = 0.133$$

$$q = 1 - 0.133 = 0.867$$

The limits for population proportion $P$ are given by $p \pm 3\sqrt{\frac{pq}{n}}$

$$= 0.133 \pm 3\sqrt{\frac{0.133 \times 0.867}{600}}$$

$$= 0.133 \pm 0.041589$$

$$= 0.17458 \text{ and } 0.091411$$

Therefore, the percentage of defective toys in the consignment lies between 17.458 and 9.1411.

**Example 2** Out of a simple sample of 1000 individuals from the inhabitants of a country, we find 36% of them have blue eyes, the remaining have eyes of some other colour. What can be inferred about the proportion of blue eyed individuals in the whole population?

**Solution** Here $n = 1000$, $p =$ population of individuals having blue eyes

$$= \frac{36}{100} = 0.36$$

$$q = 1 - p = 1 - 0.36 = 0.64$$

Probable limits are $p \pm 3\sqrt{\frac{pq}{n}}$

$$= 0.36 \pm 3\sqrt{\frac{0.36 \times 0.64}{1000}}$$

$$= 0.36 \pm 3\,(0.01517)$$

$$= 0.36 \pm 0.0455$$

$$= 0.4055 \text{ and } 0.3145$$

Hence, the limits in percentage are 40.55% and 31.45%.

**Example 3** In a locality of 20,000 families, a sample of 800 families was selected. Of these 800 families, 260 families were found to have a monthly income of Rs. 150 or less. It is desired to estimate how many out of 20,000 families have a monthly income of Rs. 150 or less. Within what limits would you place your estimate?

**Solution** Here we have $n = 800$, $p = \frac{260}{800} = \frac{26}{80} = 0.325$

$$q = 1 - p = 0.675$$

Standard error of the proportion of families having monthly income of Rs. 150 or less

$$= \sqrt{\frac{pq}{n}} = \sqrt{\frac{0.325 \times 0.675}{800}}$$

$$= 0.0165 \text{ i.e., } 1.65\%$$

Hence, the required limits in percentage are 32.5% ± (3 × 1.65)%

= 32.5% ± 4.95% = 37.45% and 27.55%

Therefore, between 5510 and 7490 families have a monthly income of Rs. 150 or less.

**Example 4** If we can assert with 95% confidence that the maximum error is 0.5 and $P$ is given as 0.2. Find the size of the sample.

**Solution** We have $P = 0.2$, $Q = 1 - P = 1 - 0.2 = 0.8$

$Z_{\alpha/2} = 1.96$

Substituting in $E = Z_{\alpha/2}\sqrt{\dfrac{PQ}{n}}$, we get

$$0.05 = 1.96\sqrt{\frac{0.2 \times 0.8}{n}}$$

$$\text{or } n = \frac{(1.96)^2}{(0.05)^2}(0.16) = 245.86$$

i.e., $n = 246$ (approx)

**Example 5** If out of 80 patients treated with an antibiotic, 59 got cured, find 99% confidence limits to the true population of the cured.

**Solution** We have $n = 80$, $P = \dfrac{x}{n} = \dfrac{59}{80} = 0.7375$

$$Q = 1 - P = 1 - 0.7375 = 0.2625$$

The required confidence interval is

$$= (0.7375 - 3\sqrt{\frac{PQ}{n}}, 0.7375 + 3\sqrt{\frac{PQ}{n}})$$

$$= (0.7375 - 3\sqrt{\frac{0.7375 \times 0.2625}{80}}, 0.7375 + 3\sqrt{\frac{0.7375 \times 0.2625}{80}})$$

$$= (0.7375 - 3(0.049), 0.7375 + 3(0.049))$$

$$= (0.59, 0.88)$$

**Example 6** In a random sample of 400 industrial accidents, it was found that 231 were due at least partially to unsafe working conditions. Construct a 99% confidence interval for the corresponding true proportion.

**Solution** We have, $n = 400$, $x = 231$

$$P = \frac{x}{n} = \frac{231}{400} = 0.5775$$

$$Q = 1 - P = 1 - 0.5775 = 0.4225$$

The required confidence interval is $\left(P - 3\sqrt{\dfrac{PQ}{n}}, P + 3\sqrt{\dfrac{PQ}{n}}\right)$

$$= \left(0.5775 - 3\sqrt{\frac{0.5775 \times 0.4225}{400}}, 0.5775 + 3\sqrt{\frac{0.5775 \times 0.4225}{400}}\right)$$

$$= (0.5775 - 3(0.0247), 0.5775 + 3(0.0247))$$

$$= (0.5034, 0.6516)$$

**Example 7** Among 900 people in a state, 90 were found to be chapathi eaters. Construct a 99% confidence interval for the true proportion.

**Solution:** Here we have $x = 90$, $n = 900$

$$P = \frac{x}{n} = \frac{90}{900} = \frac{1}{10}$$

$$Q = 1 - P = 1 - \frac{1}{10} = \frac{9}{10}$$

$$\sqrt{\frac{PQ}{n}} = \sqrt{\frac{\left(\frac{1}{10}\right)\left(\frac{9}{10}\right)}{900}} = \sqrt{\frac{1}{(100)(100)}}$$

$$= \frac{1}{100} = 0.01$$

The confidence interval is $\left(P - 3\sqrt{\frac{PQ}{n}}, P + 3\sqrt{\frac{PQ}{n}}\right)$

$$= (0.1 - 3 \times 0.01, 0.1 + 3 \times 0.01)$$

$$= (0.07, 0.13)$$

**Remarks** The confidence limits can also be computed by using

$$\left(P - Z_{\alpha/2}\sqrt{\frac{PQ}{n}}, P + Z_{\alpha/2}\sqrt{\frac{PQ}{n}}\right)$$

**Example 8** A sample size of two persons selected at random from large city show that the percentage of matrix in the sample is 52%. It is believed that the male to total population ratio is half. Test whether this belief is confirmed by the observation.

**Solution** It is given that $P = 50$, $Q = 100 - 50 = 50$

$$p = 52$$

Null hypothesis: $H_0 : P = 50\%$

Alternative hypothesis: $H_1 : P \neq 50\%$

Level of significance $\alpha = 0.05$, $Z_\alpha = 1.96$

$$\text{S.E.}(p) = \sqrt{\frac{PQ}{n}} = \sqrt{\frac{P(100-P)}{n}} = \sqrt{\frac{50(100-50)}{600}} = 2.04$$

Test statistic: $Z = \dfrac{p-P}{\text{S.E.}(p)} = \dfrac{52-50}{2.04} = 0.9803$

Since calculated value of $Z$ is less than the table value of $Z$ at 5% level of significance, null hypothesis $H_0$ is accepted i.e., male population is 50%.

**Example 9** If 20 people were attacked by a disease and only 18 survived, will you reject the hypothesis that the survival rate if attacked by this disease is 85% in favour of the hypothesis that is more at 5% level?

**Solution** Number of people survived $= x = 18$

Size of the sample $= n = 20$

$p =$ Population of the people survived $= \dfrac{x}{n} = \dfrac{18}{20} = \dfrac{9}{10} = 0.9$

It is given that $P = 85\% = 0.85$

$Q = 1 - P = 1 - 0.85 = 0.15$

Null hypothesis $H_0 : P = 0.85$

Alternative hypothesis: $H_1 = P > 0.85$ (one- tailed test)

Level of significance $\alpha = 0.05$

Table value of $Z = 1.645$

Test statistic:

$$Z = \frac{p-P}{\text{S.E.}(p)} = \frac{0.9-0.85}{\sqrt{\frac{(0.85)(0.15)}{20}}} = \frac{0.05}{0.0798} = 0.6265$$

Calculated value of Z is less than the table value of Z at 5% level of significance. Null hypothesis $H_0$ is accepted.

Hence, the proportion of the survived people is 0.85.

**Example 10** In a random sample of 125 cola drinkers, 68 say that, they prefer Thumps up to Pepsi. Test the null hypothesis $P = 0.5$ against the alternative hypothesis $P > 0.5$.

**Solution** Here, we have

$$n = 125, x = 68, p = \frac{x}{n} = \frac{68}{125} = 0.544$$

Null hypothesis $H_0 : P = 0.5$

Alternative hypothesis $H_1 : P > 0.5$

Assume that the level of significance $\alpha = 0.05$, $Z_\alpha = 1.645$

$$\text{The test statistic} = \frac{p-P}{\text{S.E.}(p)} = \frac{0.544-0.5}{\sqrt{\frac{(0.5)(0.5)}{125}}} = \frac{0.044}{0.045} = 0.9839$$

Calculated value of Z is less than the table value of Z at 5% level of significance. Therefore, null hypothesis $H_0$ is accepted.

## Exercise 6.2

1. A dice was thrown 400 times and "six" resulted 80 times. Do the data justify the hypothesis of an unbiased dice.
2. A dice was thrown 9000 times. Of these, 3220 yielded 3 or 4. Is this consistent with the hypothesis that the die was unbiased?
3. In a sample of 400 parts manufactured by a factory, the number of defective parts was found to be 30. The company however claimed that only 5% of their product is defective. Is the claim tenable?
4. Experience has shown that 20% of a manufactured product is of top quality. In one day's production of 400 articles, only 50 are of the top quality. Test the hypothesis at 0.05 level.
5. A manufacturer claims that only 4% of his products are defective. A random sample of 500 were taken among which 100 were defective. Test the hypothesis at 5% level.
6. In a random sample, 160 workers were exposed to a certain amount of radiation. 24 experienced some ill effects. Construct a 99% confidence interval for the corresponding true percentage.
7. Experiences have shown that 10% of a manufactured product is of top quality. What can you say about the maximum error with 95% confidence for 100 items?
8. 500 apples are taken from a large consignment and 50 are found to be bad. Estimate the percentage of bad apples in the consignment and assign the 95% confidence limits within which the percentage probability lies.

9. In a city, 250 men out of 750 men were found to be smokers. Does this information support the conclusion that the majority of men in this city are smokers.

10. In a sample of 500 people from a village in Rajasthan, 280 are found to be rice eaters and the rest wheat eaters. Can we assume that both food practices are equally popular?

11. A coin is tossed 400 times and it turns up head 216 times. Discuss whether the coin may be an unbiased one, and explain briefly the theoretical principals you would use on this purpose.

12. A cultivator of bananas claims that only 3 out of 100 bananas supplied by him are defective. A random sample of 700 bananas contained 45 defective bananas. Test whether the claim of the cultivator is correct.

13. A wholesaler in apples claims that only 4% of the apples supplied by him are defective. A random sample of 600 apples contained 36 defective apples. Test the claim of the wholesaler.

## Answers 6.2

(1) The die is unbiased
(2) The die is unbiased
(3) The claim is rejected
(4) Null hypothesis is rejected
(5) Null hypothesis is rejected
(6) 0.065, 0.234
(7) 0.0588
(8) 0.075, 0.125
(9) Majority are smokers
(10) Both food eaters are not equally popular
(11) Coin is unbiased
(12) Cultivator's claim cannot be accepted
(13) Claim cannot be accepted

## 6.15 Testing of Significance for Difference of Proportions

Suppose two samples of sizes $n_1$ and $n_2$ are drawn from two different populations. To test the significance of difference between the two proportions, we consider the following cases.

**Case I** When the population proportions $P_1$ and $P_2$ are known:

In this case $Q_1 = 1 - P_1$ and $Q_2 = 1 - P_2$

The test statistic is $Z = \dfrac{P_1 - P_2}{\sqrt{\dfrac{P_1 Q_1}{n_1} + \dfrac{P_2 Q_2}{n_2}}}$

**Case II** When the population proportions $P_1$ and $P_2$ are not known but sample proportions $p_1$ and $p_2$ are known:

The test statistic is $Z = \dfrac{p_1 - p_2}{\sqrt{\dfrac{p_1 q_1}{n_1} + \dfrac{p_2 q_2}{n_2}}}$

**Case III** Method of pooling

In this method, the sample proportions $P_1$ and $P_2$ are pooled into a single proportion $P$, by using the formula

$$= \frac{n_1 P_1 + n_2 P_2}{n_1 + n_2}$$

So that $Q = 1 - P$

The test statistic in this case is $Z = \dfrac{p_1 - p_1}{\sqrt{PQ\left(\dfrac{1}{n_1} + \dfrac{1}{n_1}\right)}}$

## Solved examples

**Example 1** A machine puts out 16 imperfect articles in a sample of 500. After the machine is overhauled, it puts out 3 imperfect articles in a batch of 100. Has the machine improved?

**Solution** Here we have

$$p_1 = \frac{16}{500} = 0.032$$

$$p_2 = \frac{3}{100} = 0.03$$

$$q_1 = 1 - p_1 = 1 - 0.032 = 0.968$$

$$q_2 = 1 - p_2 = 1 - 0.03 = 0.970$$

Null hypothesis $H_0 : P_1 = P_2$

Alternative hypothesis $H_1 : P_1 < P_2$ (one-tailed test)

Level of significance $= 0.05$, value of $Z = 1.645$

Standard error of difference $= \sqrt{\dfrac{p_1 q_1}{n_1} + \dfrac{p_2 q_2}{n_2}}$

$$= \sqrt{\frac{(0.032)(0.968)}{500} + \frac{(0.03)(0.97)}{100}}$$

$$= 0.0187$$

The test statistic is $Z = \dfrac{p_1 - p_2}{\sqrt{\dfrac{p_1q_1}{n_1} + \dfrac{p_2q_2}{n_2}}} = \dfrac{0.032 - 0.030}{0.0187}$

$= 0.107$

Since the calculated value of Z at 5% level of significance is less than the table value of Z, we accept the null hypothesis i.e., the machine has not improved after overhauling.

**Example 2** During a country wide investigation, the incidence of T.B was found to be 1%. In a college of 400 strong, 5 were affected, whereas in another 1200 strong, 10 were affected. Does this indicate any significance difference?

**Solution** Here, we have

$$p_1 = \frac{5}{500} = 0.0125, \; p_2 = \frac{10}{1200} = 0.008333$$

$$P = 1\% = \frac{1}{100} = 0.01, \; Q = 0.99$$

Null hypothesis $H_0 : P_1 = P_2$ i.e., there is no significant difference

Alternative hypothesis $H_1 : P_1 \neq P_2$ (two-tailed test)

Level of significance $= 0.05$, table value of $Z = 1.96$

The test statistic is $Z = \dfrac{p_1 - p_2}{\sqrt{PQ\left(\dfrac{1}{n_1} + \dfrac{1}{n_2}\right)}} = \dfrac{0.0125 - 0.008333}{\sqrt{(0.01)(0.99)\left(\dfrac{1}{400} + \dfrac{1}{1200}\right)}}$

$$= \frac{0.0041667}{\sqrt{\dfrac{1}{100} \times \dfrac{99}{100}\left(\dfrac{3+1}{1200}\right)}} = \frac{0.0041667}{\sqrt{\dfrac{99}{10^2}}\sqrt{\dfrac{1}{300}}}$$

$$= \frac{0.0041667 \times 10^3}{\sqrt{99}\sqrt{\dfrac{1}{3}}}$$

$$= \frac{4.1667}{\sqrt{33}} = \frac{4.1667}{5.744} = 0.7254$$

Since the calculated value of Z at 5% level of significance is less than the table value of Z, we accept null hypothesis i.e., the difference is not significant.

**Example 3** In a random sample of 1000 persons from a town A, 400 are found to be consumers of wheat. In a sample of 800 from town B, 400 are found to be consumers of wheat. Do these data reveal a significant difference between town A and town B so far as the proportion of wheat consumers is concerned?

**Solution** We have $p_1 = \dfrac{400}{1000} = 0.4,\ n_1 = 1000$

$$p_2 = \frac{400}{800} = 0.5,\ n_2 = 800$$

$$P = \frac{n_1 p_1 + n_2 p_2}{n_1 + n_2} = \frac{(1000)(0.4) + (800)(0.5)}{1000 + 800} = \frac{4}{9}$$

$$Q = 1 - P = 1 - \frac{4}{9} = \frac{5}{9}$$

Null hypothesis $H_0 : P_1 = P_2$

Alternative hypothesis $H_1 : P_1 \neq P_2$ (two-tailed test)

Level of significance = 1%

Table value of Z = 2.58

The test statistic is $Z = \dfrac{p_1 - p_2}{\sqrt{PQ\left(\dfrac{1}{n_1} + \dfrac{1}{n_2}\right)}} = \dfrac{0.4 - 0.5}{\sqrt{\dfrac{4}{9} \times \dfrac{5}{9}\left(\dfrac{1}{100} + \dfrac{1}{800}\right)}}$

$$= \frac{-0.1}{0.024} = -4.17$$

$\therefore |Z| = 4.17 > 2.58$

Since the difference is more than 2.58 at 1% level of significance, we reject null hypothesis.

Therefore, the data reveal a significant difference between town A and town B so far as the proportion of wheat consumers is concerned.

**Example 4** Random samples of 400 men and 600 women were asked whether they would like to have a school near their residence. 200 men and 350 women were in favour of the proposal. Test the hypothesis that the proportions of men and women in favour of the proposal are the same at 5% level of significance.

**Solution** Size of the sample of men $= n_1 = 400$

Size of the sample of women $= n_2 = 600$

Proportion of men $= p_1 = \dfrac{200}{400} = 0.5$

Proportion of women $= p_2 = \dfrac{350}{600} = 0.5833$

Null hypothesis $H_0 : P_1 = P_2$

i.e., there is no significant difference between the attitude of men and women as far as the proposal is concerned.

Alternative hypothesis $H_1 : P_1 \neq P_2$ (two-tailed)

Level of significance = 0.05 (5%)

Table value of $Z = 1.96$

The test statistic is $Z = \dfrac{p_1 - p_2}{\sqrt{PQ\left(\dfrac{1}{n_1} + \dfrac{1}{n_2}\right)}}$

We have $P = \dfrac{n_1 p_1 + n_2 p_2}{n_1 + n_2} = \dfrac{(400)(0.5) + (600)(0.5833)}{400 + 600}$

$= 0.5499$

$$Q = 1 - P = 1 - 0.5499 = 0.4501$$

$$\sqrt{PQ\left(\frac{1}{n_1} + \frac{1}{n_2}\right)} = \sqrt{(0.5499)(0.4501)\left(\frac{1}{400} + \frac{1}{600}\right)}$$

$$= 0.03207$$

$$|Z| = \left|\frac{p_1 - p_2}{\sqrt{PQ\left(\dfrac{1}{n_1} + \dfrac{1}{n_2}\right)}}\right| = \left|\frac{0.5 - 0.5833}{0.03207}\right| = 2.597$$

Since $|Z| = 2.597 > 1.96$, the null hypothesis is rejected significantly in their attitude. Men and women differ.

**Example 5** In two large populations, there are 30% and 25% respectively of fair haired people. Is this difference likely to be hidden in samples of 1200 and 900 respectively from the two populations?

**Solution** Here, we have

$P_1$ : Population of fair haired people in the first population = 30% = 0.3

$P_2$ : Population of fair hared people in another population = 25% = 0.25

$$Q_1 = 1 - P_1 = 1 - 0.3 = 0.7$$

$$Q_2 = 1 - P_2 = 1 - 0.25 = 0.75$$

Null hypothesis $H_0 : P_1 = P_2$

i.e., The difference in population proportions is likely to be hidden in sampling

$$H_1 : P_1 \neq P_2$$

Level of significance = 0.05

$$Z = \frac{P_1 - P_2}{\sqrt{\frac{P_1Q_1}{n_1} + \frac{P_2Q_2}{n_2}}} = \frac{0.30 - 0.25}{\sqrt{\frac{(0.3)(0.7)}{1200} + \frac{(0.25)(0.75)}{900}}}$$

$$= \frac{0.05}{\frac{1}{10}\sqrt{\frac{0.21}{12} + \frac{0.1875}{9}}}$$

$$= \frac{0.05}{\sqrt{0.0175 + 0.0208}}$$

$$= 2.5537 > 1.96$$

At 5% level of significance, calculated value of Z is greater than the table value. Hence $H_0$ is rejected. Since $|Z| < 2.58$, $H_0$ is accepted at 1% level.

## Exercise 6.3

1. One thousand articles from a factory are examined and found to be 3% defective. Fifteen hundred similar articles from a second factory are found to be only 2% defective. Can it be reasonably concluded that the product of the first factory is inferior to the second.

2. Out of a sample of 100 persons, 800 persons were found to be coffee drinkers. Subsequently, the excise duty on coffee was increased. After the increase in excise duty of coffee seeds, 800 people were found to take coffee out of a sample of 8200. Test whether there is any significant increase in the consumption of coffee after the increase in excise duty.

3. A machine produced 20 defective units in a sample of 400. After overhauling the machine, it produced 10 defective items in a batch of 300. Has the machine improved due to over hauling?

4. In a random sample of 400 students of the university teaching departments, it was found that 300 students failed in an examination. In another random sample of 500 students of the affiliated colleges, the number failed in the examination was found to be 300. Find out whether the population of failures in the university teaching department is significantly greater than the proportion of failures in the university teaching department and affiliated colleges taken together.($H_0$ rejected)

5. In village A, out of a random sample of 1000 persons, 100 were found to be vegetarians, while in another village B, out of 1500 persons, 180 were found to be vegetarians. Do you find a significant difference in the food habits of the people of the two villages.

6. In a simple random sample of 600 men taken from a big city , 400 are found to be smokers. In another simple random sample of 900 taken from another city 450 are smokers. Do the data indicate that there is a significant difference in the habit of smoking in the cities

7. In a sample of 600 students, 400 were found to use pens. In another college from a sample of 900 students, 450 were found to use pens. Test whether the two colleges are significantly different with regard to using of pens.

8. In town A, there were 956 births of which 52.5% were male births, while in towns A and B, this population in a total of 1406 births was 0.496. Is there any significant difference in the proportion of male births in the two towns.

## Answers 6.3

(1) The difference is not significant

(2) $H_0$ is rejected: There is a significant difference in the consumption of coffee duty after the increase in excise duty

(3) $H_0$ is accepted

(4) $H_0$ rejected

(5) There is no significant difference in the food habits of the people of the two villages

(6) There is a significant difference in the habit of smoking of the two cities

(7) $H_0$ is rejected

(8) There is a significant difference in male births in the two towns

## CHAPTER SEVEN

# $\chi^2$-Distribution

## 7.1 Introduction

In this chapter, we introduce chi-square distribution, the measure of which enables us to find the degree of discrepancy between the observed and expected frequencies and then to determine whether the discrepancy between the observed and expected frequencies is due to error of sampling or due to chance.

The chi-square is denoted by the symbol $\chi^2$. It was discovered by Helmert in 1875 and was rediscovered by Karl Pearson in 1900. $\chi^2$ is always positive. The value of chi-square lies between 0 and $\infty$.

Since chi-square is not derived from the observation in a population, it is not a parameter. The chi-square test is not a parametric test.

The chi-square distribution is used for testing the goodness of fit. It is used for finding associations and relations between attributes. It is also used to test the homogeneity of independent estimates of population. Chi-square is computed on the basis of frequencies in a sample and the value of chi-square so obtained is a statistic.

### $\chi^2$-Test

$\chi^2$-test is defined as $\chi^2 = \sum \frac{(O_i - E_i)^2}{E_i}$

where $O_i$ = Observed frequencies

$E_i$ = Expected frequencies

## 7.2 Contingency Table

A classification table containing $r$ rows and $c$ columns with observed frequencies is called a contingency table. A $2 \times 2$ contingency table is of the form

| | |
|---|---|
| $a$ | $b$ |
| $c$ | $d$ |

.

If the data is divided into $m$ classes, $A_1, A_2, \ldots, A_m$ according to an attribute $A$ and $n$ classes, $B_1, B_2, \ldots, B_n$ according to another attribute $B$, the $m \times n$ contingency table can be formed as follows.

| Attributes | $B_1$ | $B_2$ | ........ | $B_j$ | ........ | $B_n$ |
|---|---|---|---|---|---|---|
| $A_1$ | $O_{11}$ | $O_{12}$ | ........ | $O_{1j}$ | ........ | $O_{1n}$ |
| $A_2$ | $O_{21}$ | $O_{22}$ | ........ | $O_{2j}$ | ........ | $O_{2n}$ |
| ........ | | | | | | |
| ........ | | | | | | |
| $A_i$ | $O_{i1}$ | $O_{i2}$ | ........ | $O_{ij}$ | ........ | $O_{in}$ |
| ........ | | | | | | |
| ........ | | | | | | |
| $A_m$ | $O_{m1}$ | $O_{m2}$ | ........ | $O_{mj}$ | ........ | $O_{mn}$ |

where $O_{ij}$ denotes the $i$th row $j$th column frequency of the cell belonging to both the classes $A_i$ and $B_j$.

## 7.3 Calculation of Expected Frequencies

Consider the $2 \times 2$ contingency table shown in Table 7.1

**Table 7.1** $2 \times 2$ *contingency table*

| | | | Total |
|---|---|---|---|
| | $a$ | $b$ | $a+b$ |
| | $c$ | $d$ | $c+d$ |
| Total | $a+c$ | $b+d$ | $N$ |

where $N = a+b+c+d$.

The expected frequencies corresponding to the cell frequencies $a, b, c$ and $d$ can be expressed in the form of Table 7.2.

**Table 7.2** *Expected frequency table for* $2 \times 2$ *contingency table*

| | |
|---|---|
| $(a+c)(a+b)/N$ | $(b+d)(a+b)/N$ |
| $(a+c)(c+d)/N$ | $(b+d)(c+d)/N$ |

This method can be extended to compute the expected frequencies of a $m \times n$ contingency table.

## 7.4 $\chi^2$-Distribution

Let samples of size $n$ be drawn from a normal population with standard deviation $\sigma$. If for each sample we calculate $\chi^2$ a sampling distribution of $\chi^2$ can be obtained. It is given by

$$y = f(\chi^2) = y_0 e^{-\frac{1}{2}\chi^2} (\chi^2)^{\frac{1}{2}(\nu-2)}$$

$$= y_0 e^{-\frac{1}{2}\chi^2} (\chi)^{(\nu-2)}$$

where $\nu = n - 1$ is the number of degrees of freedom and $y_0$ is a constant such that the total area under the curve is one. The chi-square distribution corresponding to various values of $\nu$ are shown in Fig. 7.1

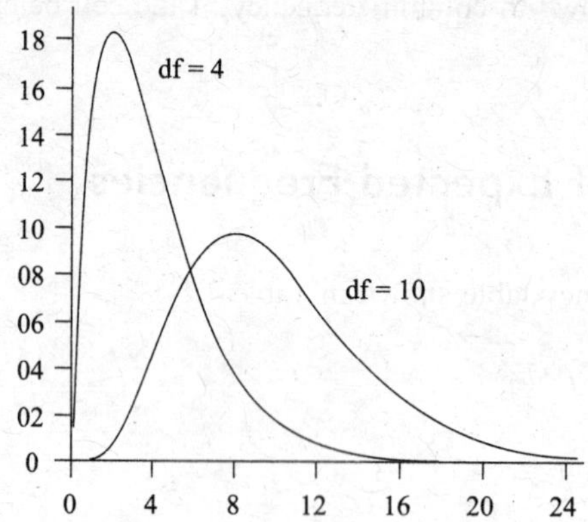


**Fig. 7.1** *Chi-square distribution*

### Maximum value of $y$

We have $y = y_0 e^{-\frac{1}{2}\chi^2} (\chi^2)^{\frac{1}{2}(\nu-2)}$ (1)

Differentiating (1) with respect to $\chi^2$ we get

$$\frac{dy}{d\chi^2} = y_0 \left[\frac{1}{2}(\nu-2)(\chi^2)^{\frac{1}{2}(\nu-4)} e^{\frac{-1}{2}\chi^2} - \frac{1}{2} e^{-\frac{\chi^2}{2}} (\chi^2)^{\frac{1}{2}(\nu-2)}\right]$$

Taking $\frac{dy}{d\chi^2} = 0$ for maximum value of $y$, we get

$$\chi^2 = \nu - 2 \text{ for } \nu \geq 2 \text{ (since } \chi^2 \text{ cannot be negative)}$$

The constant $y_0$ is related in such a way that the area $\chi^2 = 0$ to $\chi^2 = \infty$ is unity.

In this case

$$y_0 = \frac{1}{2^{\nu/2}\Gamma(\nu/v)}$$

### 7.4.1 Characteristic function of $\chi^2$-distribution

We have

$$\phi(\chi^2(t)) = E\left(e^{it\chi^2}\right)$$

$$= \int_0^\infty e^{it\chi^2} f(\chi^2)\, d\chi^2$$

$$= \frac{1}{2^{\nu/2}\Gamma\nu/2}\int_0^\infty e^{-\frac{1}{2}(1-2it)\chi^2} (\chi^2)^{\frac{1}{2}n-1}\, d\chi^2$$

$$= (1-2it)^{-n/2}$$

## 7.5 Mean and Variance of $\chi^2$

The moment generating function of $\chi^2$ with respect to the origin is

$$M_0(t) = \int_0^\infty e^{t\chi^2} \frac{1}{2^{\nu/2}\Gamma(\nu/2)} e^{-\frac{1}{2}\chi^2}\left(\frac{1}{2}\chi^2\right)^{\frac{1}{2}(\nu-1)} d(\chi^2)$$

$$= \frac{1}{2^{\nu/2}\Gamma(\nu/2)}\int_0^\infty \exp\left\{-\left(\frac{1-2t}{2}\right)\chi^2\right\}(\chi^2)^{\frac{\nu}{2}-1} d(\chi^2)$$

$$= \frac{1}{2^{\nu/2}\Gamma(\nu/2)} \cdot \frac{\Gamma\left(\frac{1}{2}\nu\right)}{\left\{\frac{1}{2}(1-2t)\right\}^{\nu/2}}$$

$$= (1-2t)^{-\nu/2}, (|2t| < 1)$$

$$= 1 + \frac{\nu}{2}(2t) + \frac{1}{2}\frac{\nu\left(\frac{1}{2}\nu+1\right)}{2!}(2t)^2 + \ldots + \frac{\frac{1}{2}\nu\left(\frac{1}{2}\nu+1\right)\ldots\left(\frac{1}{2}\nu+r-1\right)}{r!}(2t)^r + \ldots$$

$$\mu_r' = \text{coefficient } \frac{t^r}{r!}$$

$$= 2^r \frac{1}{2}\nu\left(\frac{1}{2}\nu+1\right)\left(\frac{1}{2}\nu+2\right)\ldots\left(\frac{1}{2}\nu+r-1\right)$$

$$= \nu(\nu+2)(\nu+4)\ldots(\nu+2r-2)$$

Hence, the mean value of $\chi^2$ is $\nu$ and variance is $\nu(\nu+2)-\nu^2 = 2\nu$

Thus, $\chi^2 - \dfrac{\nu}{\sqrt{2\nu}}$ is a standard variate.

## 7.6 Additive Property of an Independent Chi-square Variate

**Theorem 7.1** If $\chi_1^2$ and $\chi_2^2$ are independent $\chi^2$ variates with $n_1$ and $n_2$ degrees of freedom, then $\chi_1^2+\chi_2^2$ is a $\chi^2$ variate with $n_1+n_2$ degrees of freedom.

*Proof* The moment generating function of $(\chi_1^2+\chi_2^2)$ = (m.g.f of $\chi_1^2$) (m.g.f of $\chi_2^2$)

$$= (1-2t)^{-n_1/2}(1-2t)^{-n_2/2}$$

$$= (1-2t)^{-(n_2+n_2)/2}$$

which is the moment generating function of $\chi^2$ variate with $n_1+n_2$ degrees of freedom. Hence, proved.

**Theorem 7.2** The chi-square distribution tends to a normal distribution as $n$ tends to infinity.

*Proof* We have

$$M(t) = e^{-\frac{nt}{\sqrt{2n}}} M_o\left(\frac{t}{\sqrt{2n}}\right)$$

$$= e^{-\frac{nt}{\sqrt{2n}}}\left(1-\frac{2t}{\sqrt{2n}}\right)^{\frac{-n}{2}}.$$

Therefore, $\log_e M(t) = \dfrac{-nt}{\sqrt{2n}} - \dfrac{n}{2}\log_e\left(1-\dfrac{2t}{\sqrt{2n}}\right)$

or $\log_e M(t) = \dfrac{-nt}{\sqrt{2n}} + \dfrac{n}{2}\left(\dfrac{2t}{\sqrt{2n}} + \dfrac{1}{2}\left(\dfrac{2t}{\sqrt{2n}}\right)^2 + \ldots\right)$

or $\log_e M(t) = \dfrac{1}{2}t^2 + O\left(\dfrac{1}{n}\right)$

Further, as $n \to \infty \log_e M(t) \to \frac{1}{2}t^2$

or $M(t) \to e^{\frac{1}{2}t^2}$

Hence, the result.

## Solved examples

**Example 1** Show that the value of $\chi^2$ for the contingency table

| Classes | $A$ | $A^1$ | Total |
|---|---|---|---|
| $B$ | $a$ | b | $a+b$ |
| $B^1$ | c | $d$ | $c+d$ |
| Total | $a+c$ | $b+d$ | $N=a+b+c+d$ |

($a$, $b$, $c$, $d$ are cell frequencies)

calculated from the independent frequencies is

$$\chi^2 = \frac{N(ad-bc)^2}{(a+b)(c+d)(b+d)(a+c)}, \quad (N=a+b+c+d)$$

**Solution** Since the marginal totals are fixed, the probability for a member belonging to class A is $\frac{a+c}{N}$ and is a constant.

Further, the attributes being independent, the probability for it to belong to both classes A and B is $\left(\frac{a+b}{N}\right)\left(\frac{a+c}{N}\right)$.

The expected frequency of the class $A$, denoted by $E(a)$ is given by

$$E(a) = N.\left(\frac{a+b}{N}\right)\left(\frac{a+c}{N}\right) = \frac{(a+b)(a+c)}{N}$$

Similarly, we get

$$E(b) = \frac{(a+b)(b+d)}{N}$$

$$E(c) = \frac{(c+d)(a+c)}{N}$$

$$E(d) = \frac{(c+d)(b+d)}{N}$$

By the definition of $\chi^2$, we have

$$\chi^2 = \frac{[a-E(a)]^2}{E(a)} + \frac{[b-E(b)]^2}{E(b)} + \frac{[c-E(c)]^2}{E(c)} + \frac{[d-E(d)]^2}{E(d)}$$

$$= \frac{\left[a - \frac{(a+b)(a+c)}{a+b+c+d}\right]^2}{\frac{(a+b)(a+c)}{a+b+c+d}} + \frac{\left[b - \frac{(a+b)(b+d)}{a+b+c+d}\right]^2}{\frac{(a+b)(b+d)}{a+b+c+d}} + \frac{\left[c - \frac{(c+d)(a+c)}{a+b+c+d}\right]^2}{\frac{(c+d)(a+c)}{a+b+c+d}} + \frac{\left[d - \frac{(c+d)(b+d)}{a+b+c+d}\right]^2}{\frac{(c+d)(b+d)}{a+b+c+d}}$$

$$= \frac{(ad-bc)^2}{a+b+c+d}\left[\frac{1}{(a+b)(a+c)} + \frac{1}{(a+b)(b+d)} + \frac{1}{(c+d)(a+c)} + \frac{1}{(c+d)(b+d)}\right]$$

$$= \frac{(ad-bc)^2}{a+b+c+d}\left[\frac{a+b+c+d}{(a+b)(a+c)(b+d)} + \frac{a+b+c+d}{(a+c)(c+d)(b+d)}\right]$$

$$= (ad-bc)^2\left[\frac{1}{(a+b)(a+c)(b+d)} + \frac{1}{(a+c)(c+d)(b+d)}\right]$$

$$= (ad-bc)^2\left[\frac{c+d+a+b}{(a+b)(a+c)(b+d)(c+d)}\right]$$

$$= \left[\frac{(c+d+a+b)(ad-bc)^2}{(a+b)(a+c)(b+d)(c+d)}\right]$$

$$= \frac{N(ad-bc)^2}{(a+b)(c+d)(b+d)(a+c)}$$

**Example 2** Show that for 2 degrees of freedom, the probability $P$ of a value of $\chi^2$ greater than $\chi_0^2$ is $e^{-\frac{1}{2}\chi_0^2}$ and hence show that $\chi_0^2 = 2\log_e\left(\frac{1}{P}\right)$. Deduce the value of $\chi^2$ when $P = 0.05$.

**Solution** We know that

$$P(\chi^2) = \frac{1}{2^{\nu-2/2}} \int_{\chi_0}^{\infty} e^{-\frac{\chi^2}{2}} \chi^{\nu-1} d(\chi)$$

When $\nu = 2$,

$$P(\chi^2) = \frac{1}{2^0\Gamma\left(2/2\right)} \int_{\chi_0}^{\infty} e^{-\frac{\chi^2}{2}} \chi d(\chi)$$

$$= \int_{\chi_0}^{\infty} e^{-\frac{\chi^2}{2}} \chi d(\chi)$$

$$= \left[-e^{-\frac{\chi^2}{2}}\right]_{\chi_0}^{\infty} = e^{-\frac{1}{2}\chi_0^2}$$

$$\therefore e^{-\frac{1}{2}\chi_0^2} = P \text{ or } e^{\frac{1}{2}\chi_0^2} = \left(\frac{1}{P}\right)$$

$$\text{or } \frac{\chi_0^2}{2} = \log_e\left(\frac{1}{P}\right)$$

$$\text{or } \chi_0^2 = 2\log_e\left(\frac{1}{P}\right).$$

When $P = 0.05$, we have

$$\chi_0^2 = 2\log_e\left(\frac{1}{0.05}\right)$$

$$\chi_0^2 = 2\log_e\left(\frac{1}{\frac{1}{20}}\right)$$

$$\chi_0^2 = 2\log_e(20) = 3.012$$

**Example 3** Prove that for a chi-square distribution with $n$ degrees of freedom

$$\mu_{r+1} = 2r(\mu_r + n\mu_{r-1}), \ (\nu > 0)$$

**Solution** The moment generating function of $\chi^2$ distribution with $n$ degrees of freedom about the mean is

$$\mu(t) = e^{-nt}(1-2t)^{n/2}$$

Applying logarithms on both sides and differentiating, we get

$$\frac{\mu'(t)}{\mu(t)} = -n + \frac{n}{2}\left(\frac{2}{1-2t}\right)$$

i.e., $(1-2t)\mu'(t) = 2nt\mu(t)$

Using Leibnitz theorem and differentiating with respect to $t$, we get

$$(1-2t)\ \mu^{r+1}(t) + 2(-2)\mu^{r}(t) = 2nt\mu^{r}(t) + 2nr\mu^{r-1}(t)$$

Putting $t = 0$ and using the relation

$$\mu_r = \left[\frac{d}{dt}\mu^r(t)\right]_{t=0} = \mu^r(\nu)$$

We get $\mu_{r+1} = 2r(\mu_r + n\mu_{r-1})$

## 7.7 Degrees of Freedom

It is the number of values in a set which may be arbitrarily assigned. The number of independent variables is usually called the degree of freedom. It is denoted by the symbol $\nu$. If these are normalised variables subject to $k$ independent linear constraints, then the degrees of freedom is $\nu = n - k$.

If the data is given in the form of a row containing $n$ observations, then the degrees of freedom is $n-1$. Similarly if the data is given in the form of a column, the degrees of freedom is $n-1$, where $n$ is the number of observation, in the column. If there are $r$ rows and $c$ columns, the degrees of freedom is given by $(r-1)(c-1)$.

## 7.8 Conditions for Using $\chi^2$-Test

1. The total number of observations used in this test must be large (i.e., $n \geq 50$)
2. Each of the observations making up the sample for the $\chi^2$ test should be independent of each other.
3. The test is wholly dependent on the degrees of freedom.
4. The frequencies used in a $\chi^2$ test should be absolute and not relative in terms.
5. The expected frequency of any item or cell should not be less than 5. If it is less than 5, then the frequencies from the adjacent items or cells should be pooled together in order to make it 5 or more than 5 (preferably not less than 10).
6. The observations collected for $\chi^2$ must be based on the method of random sampling

   The constraints on the cell frequencies if any must be linear.

## 7.9 Uses of $\chi^2$-Test

$\chi^2$-test is an important test. We require only the degrees of freedom for using this test. It is a powerful test and is used

(a) as a test of goodness of fit

(b) as a test of independence of attributes, and

(c) as a test of homogeneity

### 7.9.1 $\chi^2$-Test as a test of goodness of fit

$\chi^2$-test is applied as a test of goodness of fit to determine whether the actual (i.e., observed) frequencies are close to the expected (i.e., theoretical) frequencies. The degrees of freedom in this case are $\nu = n - 1$, where $n$ is the number of observations.

## Solved examples

**Example 1** In 90 throws of a die, face 1 turned 9 times, face 2 or 3 turned 24 times, face 4 or 5 turned 36 times and 6 turned 18 times. Test at 10% level if the dice is honest, it being given that $\chi^2$ for 3 df $= 6.25$ at 10% level of significance.

**Solution** $H_0$: The die is honest

$H_1$ : The die is not honest

Expected frequencies for each face $= 90 \times \dfrac{1}{6} = 15$

Level of significance $=$ 10% (i.e., 0.1)

Degrees of freedom $= 4 - 1 = 3$

$\chi^2$-value for 3 df at 10% level of significance $= 6.25$

We have the following table

| Face turned | Observed $O_i$ | Expected $E_i$ | $(O_i - E_i)$ | $\dfrac{(O_i - E_i)^2}{E_i}$ |
|---|---|---|---|---|
| 1 | 9 | 15 | −6 | 2.4 |
| 2 or 3 | 27 | 30 | −3 | 0.3 |
| 4 or 5 | 36 | 30 | 6 | 1.2 |
| 6 | 18 | 15 | 3 | 0.6 |
| Total | 90 | 90 | | 4.5 |

$$\chi^2 = \frac{\Sigma(O_i - E_i)^2}{E_i} = 4.5$$

Since the calculated value of $\chi^2$ is less than the table value at 10% level of significance and for 3 df, we accept the null hypothesis and conclude that the die honest.

**Example 2** Find whether or not the following observed distribution of phenotypes in a sample of 384 drosophila flies have a significant goodness of fit with the proposed Mendelian 9:3:3:1 distribution (test at 5% level of significance)

| Phenotypes: | AB | Ab | aB | ab | Total |
|---|---|---|---|---|---|
| Number of: animals | 232 | 76 | 58 | 18 | 384 |

**Solution** Degrees of freedom $= 4 - 1 = 3$

Null hypothesis: $H_0$ : 9:3:3:1

i.e., variation is not significant.

Alternative hypothesis $H_1$ : 1:1:1:1

Level of significance $= \alpha = 0.05$

Table value i.e critical value of $\chi^2 = 7.82$ for 3 df

| Observed $O_i$ | Expected $E_i$ | $(O_i - E_i)$ | $(O_i - E_i)^2$ | $(O_i - E_i)^2/E_i$ |
|---|---|---|---|---|
| 232 | $384/16 \times 9 = 216$ | 16 | 256 | 1.185 |
| 76 | $384/16 \times 3 = 72$ | 4 | 16 | 0.223 |
| 58 | $384/16 \times 3 = 72$ | −14 | 196 | 2.723 |
| 18 | $384/16 \times 124$ | −6 | 36 | 2.25 |
| Total = 384 | | | | 6.381 |

The calculated value of $\chi^2$ for 3 df at 5% level of significance is less than the table value. Hence we accept $H_0$, i.e., variation is not significant.

Hence, the result is a good fit for the 9 : 3 : 3 : 1 hypothesis.

**Example 3** Among 64 offsprings of a certain cross between guinea pigs, 32 were red, 10 were black and 22 were white. According to the genetic model, these numbers should be in the ratio 9:3:4. Are the data consistent with the model at 5% level?

**Solution** $H_0$: Data are consistent with the model

$H_1$ : Data are not consistent with the model

Level of significance = 5%

Degrees of freedom $= n - 1 = 3 - 1 = 2$

Table value of $\chi^2$ for 2 df at 5% level = 5.991

Expected frequencies are in the ratio 9:3:4

i.e., $64.\left(\frac{9}{16}\right) = 36,\ 64.\left(\frac{3}{16}\right) = 12,\ 64.\left(\frac{4}{16}\right) = 16$

| Observed $O_i$ | Expected $E_i$ | $(O_i - E_i)^2$ | $\frac{(O_i - E_i)^2}{E_i}$ |
|---|---|---|---|
| 32 | 36 | 16 | 0.4444 |
| 10 | 12 | 4 | 0.3333 |
| 22 | 16 | 36 | 2.25 |
| Total | | | 3.0277 |

$$\chi^2 = \frac{\Sigma(O_i - E_i)^2}{E_i} = 3.0277$$

The calculated value of $\chi^2$ for 2 df at 5% level of significance is less than the table value. Hence we accept $H_0$ i.e., data are consistent with the model.

**Example 4** A sample analysis of examination results of 500 students was made. It was found that 230 students had failed. 160 had secured a third class, 80 were placed in second class and 30 got first class. Do these figures commensurate with the general examination results which is in the ratio 4:3:2:1 for various categories respectively?

**Solution**

Null hypothesis: $H_0$ : The observed results commensurate with the general examination results

Alternative hypothesis: $H_1$: It is not true that the observed results commensurate with the general examination results.

Level of significance = 5%

Degrees of freedom $= n - 1 = 4 - 1 = 3$

The total frequency $= N = 500$

Table value $\chi^2$ for 3 df at 5% level of significance = 7.81 for 3 df

Dividing 500 in the ratio 4:3:2:1, we get 200, 150, 100 and 50.

Therefore, the expected frequencies are 200, 150, 100 and 50 corresponding to the observed frequencies 230, 160, 80 and 30.

| Class/division | Observed $O_i$ | Expected $E_i$ | $(O_i - E_i)$ | $\frac{(O_i - E_i)^2}{E_i}$ |
|---|---|---|---|---|
| Failed | 230 | 200 | 30 | 4.500 |
| Third | 160 | 150 | 10 | 0.6666 |
| Second | 80 | 100 | −20 | 4.0000 |
| First | 30 | 50 | −20 | 8.00 |

$$\chi^2 = \frac{\Sigma(O_i - E_i)^2}{E_i} = 17.1666$$

Since the calculated value $\chi^2$ is greater than the table value of $\chi^2$, the null hypothesis is rejected.

**Example 5** The table below gives the number of aircraft accidents that occurred during the various days of the week. Test whether the accidents are uniformly distributed over the week.

| Days : | Mon | Tue | Wed | Thu | Fri | Sat |
|---|---|---|---|---|---|---|
| No. of accidents: | 14 | 18 | 12 | 11 | 15 | 14 |

**Solution**

Null hypothesis : $H_0$ : The accidents are uniformly distributed over the week.

Alternative hypothesis: The accidents are not uniformly distributed over the week

Total number of accidents $= 14+18+12+11+15+14=84$

Level of significance $= 5\%$

Degrees of freedom $= 6-1=5$

Table value of $\chi^2 = 11.07$

The expected frequency of accidents each day is $= \dfrac{84}{6} = 14$

| Day | Observed $O_i$ | Expected $E_i$ | $(O_i - E_i)$ | $(O_i - E_i)^2$ | $\dfrac{(O_i - E_i)^2}{E_i}$ |
|---|---|---|---|---|---|
| Mon | 14 | 14 | 0 | 0 | 0 |
| Tue | 18 | 14 | 4 | 16 | 1.1428 |
| Wed | 12 | 14 | −2 | 4 | 0.2857 |
| Thu | 11 | 14 | −3 | 9 | 0.6428 |
| Fri | 15 | 14 | 1 | 1 | 0.0714 |
| Sat | 14 | 14 | 0 | 0 | 0.000 |
| Total | 84 | 84 | | | 2.1427 |

Since $2.1427 < 11.07$, the calculated value of $\chi^2$ at 5% level, for 5 df is less than the table value of $\chi^2$ the null hypothesis is accepted. That is, the accidents are uniformly distributed over the week.

**Example 6** 4 coins are tossed 160 times and the following results were obtained.

| Number of heads: | 0 | 1 | 2 | 3 | 4 |
|---|---|---|---|---|---|
| Observed frequencies: | 17 | 52 | 54 | 31 | 6 |

Under the assumption that coins are balanced, find the expected frequencies of getting 0, 1, 2, 3 or 4 heads and test the goodness of fit.

**Solution** Null hypothesis: $H_0$: The coins are balanced.

Alternative hypothesis: The coins are not balanced.

Level of significance = 5%

Degrees of freedom = 5 − 1 = 4

Table value of $\chi^2 = 9.488$

We have $N = 160,\ p = \frac{1}{2},\ q = \frac{1}{2}$

The expected frequencies of 0, 1, 2, 3 or 4 successes are

$$160 \times {}^4C_0 \left(\frac{1}{2}\right)^0 \left(\frac{1}{2}\right)^4 = 10, \quad 160 \times {}^4C_1 \left(\frac{1}{2}\right)^1 \left(\frac{1}{2}\right)^3 = 40,$$

$$160 \times {}^4C_2 \left(\frac{1}{2}\right)^2 \left(\frac{1}{2}\right)^2 = 60, \quad 160 \times {}^4C_3 \left(\frac{1}{2}\right)^3 \left(\frac{1}{2}\right)^1 = 40,$$

$$160 \times {}^4C_4 \left(\frac{1}{2}\right)^4 \left(\frac{1}{2}\right)^0 = 10$$

| No of heads | Observed $O_i$ | Expected $E_i$ | $(O_i - E_i)$ | $(O_i - E_i)^2$ | $\frac{(O_i - E_i)^2}{E_i}$ |
|---|---|---|---|---|---|
| 0 | 17 | 10 | 7 | 49 | 4.900 |
| 1 | 52 | 40 | 12 | 144 | 3.600 |
| 2 | 54 | 60 | −6 | 36 | 0.600 |
| 3 | 31 | 40 | −9 | 81 | 2.025 |
| 4 | 6 | 10 | −4 | 16 | 1.600 |
| Total | 160 | 160 | | | 12.725 |

$$\chi^2 = \frac{\Sigma(O_i - E_i)^2}{E_i} = 12.725$$

The calculated value of $\chi^2$ is greater than the table value of $\chi^2$ at 5% level of significance and 4 df. Therefore, null hypothesis is rejected. The coins are not balanced. Hence the fit is poor.

**Example 7** Fit a Poisson distribution to the following data and test the goodness of fit

| $x$: | 0 | 1 | 2 | 3 | 4 | 5 | 6 |
|---|---|---|---|---|---|---|---|
| $f$: | 275 | 72 | 30 | 7 | 5 | 2 | 1 |

**Solution**

Null hypothesis : $H_0$ : The Poisson fit is good for the given data.

Alternative hypothesis: The Poisson fit is not a good fit for the given data

Level of significance = 5%

$$\text{Mean of the distribution} = \frac{\sum f_i x_i}{\sum f_i}$$

$$= \frac{0+72+60+21+20+10+6}{275+72+30+7+5+2+1}$$

$$= \frac{189}{392} = 0.482$$

The (expected) frequency of 0, 1, 2, 3, 4, 5 and 6 successes by using recurrence formula for Poisson distribution are

$$N(0) = 392e^{-0.482} = 242.1$$

$$N(1) = 392e^{-0.482}(0.482) = 116.7$$

$$N(2) = 116.7.\left(\frac{0.482}{2}\right) = 28.12$$

$$N(3) = 28.12.\left(\frac{0.482}{3}\right) = 4.52$$

$$N(4) = 4.52.\left(\frac{0.482}{4}\right) = 0.54$$

$$N(5) = 0.54.\left(\frac{0.482}{5}\right) = 0.052$$

$$N(6) = 0.052.\left(\frac{0.482}{6}\right) = 0.004$$

The frequency table is

| $x$: | 0 | 1 | 2 | 3 | 4 | 5 | 6 |
|---|---|---|---|---|---|---|---|
| Observed frequency: | 275 | 72 | 30 | 7 | 5 | 2 | 1 |
| Expected frequency: | 242.1 | 116.7 | 28.1 | 4.5 | 0.5 | 0.1 | 0 |

Since the last four frequencies are small we regroup them and obtain the following table.

Calculation of $\chi^2$

| Observed $O_i$ | Expected $E_i$ | $(O_i - E_i)$ | $(O_i - E_i)^2$ | $\frac{(O_i - E_i)^2}{E_i}$ |
|---|---|---|---|---|
| 275 | 242.1 | 32.9 | 1082.41 | 4.4709 |
| 72 | 116.7 | 44.7 | 1998.09 | 17.1216 |
| 30 | 28.1 | 1.9 | 3.61 | 0.1285 |
| 15 | 5.1 | 9.9 | 98.01 | 19.217 |
| Total | | | | 40.938 |

$$\chi^2 = \frac{\Sigma (O_i - E_i)^2}{E_i} = 40.938$$

Degrees of freedom for the Poisson fit $= n - 2 = 4 - 2 = 2$

Table value of $\chi^2$ at 5% for 2 df $= 5.991$

The calculated value of $\chi^2$ is 40.938 which is greater than the table value.

We reject null hypothesis and conclude that the Poisson fit is not a good fit for the given data.

## 7.9.2 Test for independence of attributes

The chi-square test can also be applied to test the association between attributes such as honesty, smoking, drinking etc, when the sample data is presented in the form of a contingency table with any number of rows and columns.

Data relating to attributes or to continuous variates or to a mixture of both, will in practice be arranged in compartments called cells.The frequency falling in a cell will be referred to as the cell frequency. A $m \times n$ continuous table will have $mn$ cells.

### Solved examples

**Example 1** The following table gives the classification of 150 workers according to sex and nature of work. Test whether the nature of work is independent of the sex of the work.

| | Stable | Unstable | Total |
|---|---|---|---|
| Males | 60 | 30 | 90 |
| Females | 15 | 45 | 60 |
| Total | 75 | 75 | 150 |

**Solution**

Null hypothesis $H_0$: The nature of the work is independent of the sex of the worker.

Alternative hypothesis $H_1$: The nature of the work is not independent of the sex of the worker.

Degrees of freedom $= (r-1)(c-1) = (2-1)(2-1) = 1$

Level of significance = 5%

Table value of $\chi^2 = 3.84$

Expected frequencies are given in the following table

Table expected frequencies:

| | Stable | Unstable |
|---|---|---|
| Males | $75 \times 90/150 = 45$ | $75 \times 90/150 = 45$ |
| Females | $75 \times 60/150 = 30$ | $75 \times 60/150 = 30$ |

Calculation of $\chi^2$

| Observed $O_i$ | Expected $E_i$ | $(O_i - E_i)$ | $(O_i - E_i)^2$ | $\frac{(O_i - E_i)^2}{E_i}$ |
|---|---|---|---|---|
| 60 | 45 | 15 | 225 | 5.00 |
| 15 | 30 | −15 | 225 | 7.50 |
| 30 | 45 | −15 | 225 | 5.00 |
| 45 | 30 | 15 | 225 | 7.50 |
| Total | | | | 25 |

$$\chi^2 = \frac{\Sigma (O_i - E_i)^2}{E_i} = 25$$

The calculated value of $\chi^2$ is greater than the table value at 5% and 1 df. Hence we reject the null hypothesis.

We conclude that the nature of work is not independent of the sex of the worker.

**Example 2** In a certain sample of 2000 families, 1400 families are consumers of tea. Out of 1800 Hindu families, 1,236 families consume tea. Use the $\chi^2$-test and state whether there is any significant difference between consumption of tea among Hindu and non-Hindu families.

**Solution** From the given data, the 2 × 2 contingency table that can be formed is given below.

| | Hindu | Non-Hindu | Total |
|---|---|---|---|
| Consuming tea | 1236 | 164 | 1400 |
| Not consuming tea | 564 | 36 | 600 |
| Total | 1800 | 200 | 200 |

Null hypothesis $H_0$: The attributes are independent i.e., there is no significant difference between the communities as far as consuming of tea is concerned.

Alternative hypothesis $H_1$: The attributes are not independent

Level of significance: 5%

df = 1, table value of $\chi^2$ at 5% for 1 df = 3.841

The expected frequencies corresponding to the given observed frequencies are given below.

| | |
|---|---|
| $1800 \times 1400/2000 = 1264$ | $200 \times 1400/2000 = 140$ |
| $1800 \times 600/2000 = 540$ | $200 \times 600/2000 = 60$ |

Computation of $\chi^2$

| Observed $O_i$ | Expected $E_i$ | $(O_i - E_i)$ | $(O_i - E_i)^2$ | $\frac{(O_i - E_i)^2}{E_i}$ |
|---|---|---|---|---|
| 1236 | 1260 | −24 | 576 | 0.457 |
| 564 | 540 | 24 | 576 | 1.068 |
| 164 | 140 | 24 | 576 | 4.114 |
| 36 | 60 | −24 | 576 | 9.600 |
| Total | | | | 15.239 |

$$\chi^2 = \frac{\Sigma (O_i - E_i)^2}{E_i} = 15.239$$

The calculated value of $\chi^2$ is higher than the table value of $\chi^2$ at 5% and 1 df. Therefore, the null hypothesis is rejected.

The two communities differ significantly as far as consumption of tea is concerned.

**Example 3** A tobacco company claims that there is no relationship between smoking and lung ailments. To investigate the claim, a random sample of 300 males in the age group of 40 and 50 are given a medical test. The observed sample results are tabulated below

| | Lung ailment | Non-lung ailment | Total |
|---|---|---|---|
| Smokers | 75 | 105 | 180 |
| Non-smokers | 25 | 95 | 120 |
| Total | 100 | 200 | 300 |

On the basis of this information, can it be concluded that smoking and lung ailments are independent (given $\chi^2_{0.05} = 3.841$ for 1 df)?

**Solution** Null hypothesis $H_0$ : Smoking and lung ailments are not associated

Alternative hypothesis $H_1$ : Smoking and lung ailments are associated

Level of significance = 5%

Table value of $\chi^2$ for 1 df = 3.841

The expected frequencies are

| | |
|---|---|
| 100 × 180/300 = 60 | 200 × 180/300 = 120 |
| 100 × 120/300 = 40 | 200 × 120/300 = 80 |

Computation of $\chi^2$

| Observed $O_i$ | Expected $E_i$ | $(O_i - E_i)$ | $(O_i - E_i)^2$ | $\frac{(O_i - E_i)^2}{E_i}$ |
|---|---|---|---|---|
| 75 | 60 | 15 | 225 | 3.750 |
| 25 | 40 | −15 | 225 | 5.625 |
| 105 | 120 | −15 | 225 | 1.875 |
| 95 | 80 | 15 | 225 | 2.813 |
| Total | | | | 14.063 |

$$\chi^2 = \frac{\Sigma(O_i - E_i)^2}{E_i} = 14.063$$

For 1 df at 5% level of significance, the calculated value of $\chi^2$ is more than the table value. Hence, we reject null hypothesis.

Therefore, smoking and lung ailments are not independent.

**Example 4** Given the following contingency table for hair colour and eye colour, find the value of $\chi^2$. Is there a good association between the two?

| | | Hair colour | | | |
|---|---|---|---|---|---|
| | | Fair | Brown | Black | Total |
| Eye colour | Blue | 15 | 5 | 20 | 40 |
| | Grey | 20 | 10 | 20 | 50 |
| | Brown | 25 | 15 | 20 | 60 |
| | Total | 60 | 30 | 60 | 150 |

**Solution** Null hypothesis $H_0$ : The two attributes, hair colour and eye colour, are independent.

Alternative hypothesis: $H_1$ : The two attributes, colour of the hair and colour of the eyes, are not independent.

Level of significance = 9.488 for 4 df at 5% level of significance.

Expected frequencies are

| | | |
|---|---|---|
| $60 \times 40/150 = 16$ | $30 \times 40/150 = 8$ | $60 \times 40/150 = 16$ |
| $60 \times 50/150 = 20$ | $30 \times 50/150 = 10$ | $60 \times 50/150 = 20$ |
| $60 \times 60/150 = 24$ | $30 \times 60/150 = 12$ | $60 \times 60/150 = 24$ |

Computation of $\chi^2$

| Observed $O_i$ | Expected $E_i$ | $(O_i - E_i)$ | $(O_i - E_i)^2$ | $\frac{(O_i - E_i)^2}{E_i}$ |
|---|---|---|---|---|
| 15 | 16 | −1 | 1 | 0.0625 |
| 5 | 8 | −3 | 9 | 1.125 |
| 20 | 16 | 4 | 16 | 1 |
| 20 | 20 | 0 | 0 | 0 |
| 10 | 10 | 0 | 0 | 0 |
| 20 | 20 | 0 | 0 | 0 |
| 25 | 24 | 1 | 1 | 0.042 |
| 15 | 12 | 3 | 9 | 0.75 |
| 20 | 24 | −4 | 16 | 0.666 |
| Total | | | | 3.6458 |

$$\sum = \frac{(O_i - E_i)^2}{E_i} = 3.6458$$

The calculated value of $\chi^2$ for $(3-1)(3-1) = 2 \times 2 = 4$ df at 5% level, is less than the table value of 9.488. Hence we accept the null hypothesis i.e., the hair colour and eye colour are independent.

**Example 5** The following table shows the result of an experiment to investigate the effect of vaccination induced on the animals against a particular disease. Use the chi-square test to test the hypothesis that the vaccinated and unvaccinated groups are similarly prone to the disease i.e., vaccination and the disease are independent.

| Vaccinated | Got disease | Did not get disease |
|---|---|---|
| | 9 | 42 |
| Not vaccinated | 17 | 28 |

(Value of $\chi^2$ for 1 df at 5% level is equal to 3.841)

**Solution** We have $a = 9$, $b = 42$, $c = 17$, $d = 28$

$$N = a + b + c + d = 9 + 42 + 17 + 28 = 96$$

Null hypothesis $H_0$ : The vaccination and disease are independent.

Table value of $\chi^2$ for 1 df at 5% level = 3.841

$$\chi^2 = \frac{N(ad-bc)^2}{(a+b)(c+d)(b+d)(a+c)} = \frac{96(9\times 28 - 17\times 42)^2}{(51)(26)(70)(45)}$$

$$= \frac{96(-462)^2}{4176900} = 4.906$$

The calculated value of $\chi^2$ is more than the table value of $\chi^2$ at 5% level and 1 df. Therefore, we reject $H_0$ and conclude that the disease and vaccination are not independent.

## Yate's Correction

In a $2\times 2$ contingency table, if the cell frequency is small, Yate's correction is necessary. If the expected value or frequency in any observation is less than 5 in one df, we apply Yate's correction. We subtract 0.5 from the absolute difference between observed and expected frequencies. By making Yate's correction, $\chi^2$ becomes continuous and the formula after making Yate's correction is

$$\chi^2 = \Sigma \frac{(|O_i - E_i| - 0.5)^2}{E_i} \qquad \text{(0.5 is Yate's correction)}$$

**Example 6** Two batches each of 12 animals are taken to test the effectiveness of an inoculation. One batch was inoculated and the others batch was not inoculated. The frequencies of the dead and surviving animals are given below in both cases. Can the inoculation be regarded as effective against the disease?

| | Dead | Survived | Total |
|---|---|---|---|
| Inoculated | 2 | 10 | 12 |
| Not inoculated | 8 | 4 | 12 |
| Total | 10 | 14 | |

($\chi^2_{0.05}$ for 1 df = 3.841)

## Solution

Null hypothesis $H_0$ : Inoculation and the disease are independent.

Alternative hypothesis $H_1$ : Inoculation and the disease are not independent.

Df $= (r-1)(c-1) = (2-1)(2-1) = 1$

Level of significance = 5%

Table value of $\chi^2$ for 1 df at 5% level = 3.841

Frequencies in the cells are small ($a$ and $d$ are small) so we make Yate's correction. The expected frequencies are given below.

Expected frequency table

| | |
|---|---|
| $(10 \times 12)/24 = 5$ | $(14 \times 12)/24 = 7$ |
| $(10 \times 12)/24 = 5$ | $(14 \times 12)/24 = 7$ |

Computation of $\chi^2$

| Observed $O_i$ | Expected $E_i$ | $\lvert O_i - E_i\rvert$ | $\lvert O_i - E_i\rvert - 0.5$ | $(\lvert O_i - E_i\rvert - 0.5)^2$ | $\frac{(\lvert O_i - E_i\rvert - 0.5)^2}{E_i}$ |
|---|---|---|---|---|---|
| 2 | 5 | −3 | 2.5 | 6.25 | 1.25 |
| 10 | 7 | 3 | 2.5 | 6.25 | 0.89285 |
| 8 | 5 | 3 | 2.5 | 6.25 | 1.25 |
| 4 | 7 | −3 | 2.5 | 6.25 | 0.89285 |
| Total | | | | | 4.2857 |

$$\chi^2 = \sum \frac{(|O_i - E_i| -- 0.5)^2}{E_i} = 4.2857$$

For 1 df at 5% level of significance, the calculated value of $\chi^2$ is greater than the table value. Hence, $H_0$ is rejected. The inoculation and the disease are not independent and we conclude that the inoculation is effective against the disease.

### 7.9.3 $\chi^2$-test as a test for homogeneity

The $\chi^2$-test may be used to test the homogeneity of the attributes in respect to particular characteristics. It is performed to decide whether separate samples are sufficiently uniform to be added together. The $\chi^2$-test may also be used to test the population variance.

### 7.9.4 $\chi^2$-distribution of sample variance

Let $\sigma^2$ denote the population variance and $S^2$ denote the sample variance. The sampling distribution $(n-1)\frac{S^2}{\sigma^2}$ has $\chi^2$- distribution with $n-1$ degrees of freedom. It is very useful in making an inference about the population variance $\sigma^2$ by using sample variance $S^2$. Also, it is used in making interval estimates of the population variance which is given by

$$\frac{(n-1)S^2}{\chi^2_{\alpha/2}} \leq \alpha \leq \frac{(n-1)S}{\chi^2_{1-\alpha/2}}$$

where $\alpha$ is the level of significance, and $\chi^2_\alpha$ is the value of the chi-square distribution giving an area to the right of $\chi^2_\alpha$ .

The test may also be used as a hypothesis test for the value of the population variance.

### 7.9.5 Testing a hypothesis about the variance of a normally distributed population: Decision rule

Let $\sigma_0^2$ denote the hypothesised value of the population variance. Then the decision rule for accepting or rejecting $H_0$ is as follows.

| | | |
|---|---|---|
| **1.** | **Two-tailed test** | **Decision rule** |
| | $H_0 : \sigma^2 = \sigma_0^2$ | Accept $H_0$ if computed value of $\chi^2 \leq \chi^2_{\alpha/2}$ (table value) |
| | $H_1 : \sigma^2 \neq \sigma_0^2$ | Reject $H_0$ if computed value of $\chi^2 > \chi^2_{\alpha/2}$ (table value) |
| **2.** | **One-tailed test** | |
| | $H_0 : \sigma^2 \geq \sigma_0^2$ | Accept $H_0$ if computed value of $\chi^2 <$ table value of $\chi^2_\alpha$ |
| | $H_1 : \sigma^2 < \sigma_0^2$ | Reject $H_0$ if computed value of $\chi^2 >$ table value of $\chi^2_\alpha$ |

## Solved examples

**Example 1** According to the census report of 2004, the number of women to every thousand men in the following 5 states/union territories are as follows:

In Andaman 846, in Delhi 821, in Chandigarh 777, in Daman and Diu 710, in Dadar and Nagar Haveli 812. By an appropriate statistical method, determine whether a 1:1 sex ratio among human population can be attributed to 5 regions ($\chi^2_{0.05}$ for 4 df = 7.88).

**Solution** Null hypothesis $H_0$ : Sex ratio is 1:1 among the human population of 5 states

Alternative hypothesis $H_0$ : Sex ratio is not 1:1 among the human population of 5 states

Table value of chi-square for 4 df at 5% level = 7.88

Computation of $\chi^2$ for 5 states

| | Observed $O_i$ | Expected $E_i$ | $\lvert O_i - E_i \rvert$ | $\frac{(\lvert O_i - E_i \rvert)^2}{E_i}$ | $\chi^2$ | df |
|---|---|---|---|---|---|---|
| Andaman | Men = 1000 | 923 | 77 | 6.42 | 12.84 | 1 |
| | Women = 846 | 923 | 77 | 6.42 | | |
| Delhi | Men = 1000 | 910.5 | 89.5 | 8.79 | 17.58 | 1 |
| | Women = 821 | 910.5 | 89.5 | 8.79 | | |
| Chandigarh | Men = 1000 | 888.5 | 111.5 | 13.99 | 27.98 | 1 |
| | Women = 777 | 888.5 | 111.5 | 19.99 | | |
| Damn and Diu | Men = 1000 | 855 | 145 | 24.59 | 49.18 | 1 |
| | Women = 710 | 855 | 145 | 24.59 | | |
| Dadar and Nagar Haveli | Men = 1000 | 906 | 94 | 9.75 | 19.5 | 1 |
| | Women = 812 | 906 | 94 | 9.75 | | |
| Men | 5000 | | | | | |
| Women | 3966 | | Total | | 127.08 | 5 |

$\chi^2$ for summed data

| Observed | Expected | $\lvert O_i - E_i \rvert$ | $\frac{(\lvert O_i - E_i \rvert)^2}{E_i}$ | df |
|---|---|---|---|---|
| Men = 5000 | 4483 | 517 | 59.62 | 2 − 1 = |
| Women = 3966 | 4483 | 517 | 59.62 | 1 |

| | Chi-square | df |
|---|---|---|
| Total | 127.08 | 5 |
| Summed | 119.2 | 1 |
| Homogeneity | 7.88 | 4 |

The calculated value of $\chi^2$ at 5% level of significance for 4 df is less than the table value of $\chi^2$. We accept $H_0$.

**Example 2** Heights in cms of 10 students are given below

61, 65, 67, 66, 68, 70, 64, 65, 68, 71

Can we say that the variance of the distribution heights of all students from which the above sample of 10 students was drawn is equal to 50?

**Solution** We have

$$\text{Mean of the sample} = \frac{61+65+62+66+68+70+64+65+68+71}{10}$$

$$= \frac{660}{10} = 66$$

Null hypothesis $H_0 : \sigma^2 = \sigma_0^2 = 50$

Alternative hypothesis $H_1 : \sigma^2 \neq \sigma_0^2$, i.e., $\sigma^2 \neq 50$ (Two-tailed test)

Level of significance : 5% $(\alpha = 0.05)\chi^2_{1-\alpha/2} = \chi^2_{0.975} = 2.70$

Computation of $\chi^2$

| $x_i$ | $x_i - \bar{x}$ | $(x_i - \bar{x})^2$ |
|---|---|---|
| 61 | −5 | 25 |
| 65 | −1 | 1 |
| 62 | −4 | 16 |
| 66 | 0 | 0 |
| 68 | 2 | 4 |
| 70 | 4 | 16 |
| 64 | −2 | 4 |
| 65 | −1 | 1 |
| 68 | 2 | 4 |
| 71 | 5 | 25 |

Test statistic is

$$\chi^2 = (n-1)\frac{S^2}{\sigma_0^2}$$

$$= \frac{(n-1)}{\sigma_0^2} \cdot \frac{\Sigma (x_i - \bar{x})^2}{n-1}$$

$$= \frac{\Sigma (x_i - \bar{x})^2}{\sigma_0^2} = \frac{96}{50} = 1.92$$

Since calculated value of $\chi^2$ for 9 df at 5% level of significance is less than the value of $\chi^2_{1-\alpha/2}$ at 5% level, we reject null hypothesis and conclude that population variance is not 50.

## Exercise 7.1

1. The following figures show the distribution of digits in numbers chosen at random from a telephone directory

| Digits: | 0 | 1 | 2 | 3 | 4 | 5 | 6 | 7 | 8 | 9 |
|---|---|---|---|---|---|---|---|---|---|---|
| Frequency: | 1026 | 1107 | 997 | 996 | 1075 | 993 | 1107 | 972 | 964 | 853 |

   Test whether the digits may be taken to occur equally frequently in the directory ($\chi^2_{0.05}$ = 16.919 for 9 df)

   [Hint: Expected frequency of each observation $= \frac{10000}{10} = 1000$]

2. A die is thrown 264 times with the following results. Show that the die is biased.

| No. appeared on the die: | 1 | 2 | 3 | 4 | 5 | 6 |
|---|---|---|---|---|---|---|
| Frequency: | 40 | 32 | 28 | 58 | 54 | 60 |

   (given $\chi^2$ for 5 df at 5% level = 11.07)

3. On the basis of information given below about the treatment of 200 patients suffering from a disease, state whether the new treatment is comparatively superior to the conventional treatment.

| | Favourable | Not favourable | Total |
|---|---|---|---|
| New treatment | 60 | 30 | 90 |
| Conventional treatment | 40 | 70 | 110 |

   ($\chi^2_{0.05}$ = 3.841 for 1 df)

4. 200 digits were chosen at random from a set of tables. The frequencies of the digits were

| Digit: | 0 | 1 | 2 | 3 | 4 | 5 | 6 | 7 | 8 | 9 |
|---|---|---|---|---|---|---|---|---|---|---|
| Frequency: | 18 | 19 | 23 | 21 | 16 | 25 | 22 | 20 | 21 | 15 |

   Use chi-square test to assess the correctness of the hypothesis that the digits were distributed in equal number in the tables from which these were chosen. (For d.f at 5% level of significance $\chi^2$ value is 16.919)

5. In a sample of owls, it is found that red males are 35 in number, red females are 70, gray males are 50, gray females are 45. Colouration is due to the plumage. Is the colouration independent of the sex of the sample?

6. The following table gives the frequencies of occurrence of the digits 0, 1, 2,. . . 9, in the last place of the four-figure logarithm of numbers. Examine if there is any peculiarity.

| Digit: | 0 | 1 | 2 | 3 | 4 | 5 | 6 | 7 | 8 | 9 |
|---|---|---|---|---|---|---|---|---|---|---|
| Frequency: | 6 | 16 | 15 | 10 | 12 | 12 | 3 | 2 | 9 | 5 |

($\chi^2$ value for 5 df at 5% level of significance is 11.07)

7. A die was thrown 498 times. Denoting $x$ to be the number appearing on the top face of it, the observed frequency of $x$ is given below

| $x$: | 1 | 2 | 3 | 4 | 5 | 6 |
|---|---|---|---|---|---|---|
| $t$: | 69 | 78 | 85 | 82 | 86 | 98 |

What opinion would you form for the accuracy of the die?

($\chi^2$ value for 5 df at 5% level of significance is 11.07)

8. Among 64 offsprings of a certain cross between guinea pigs, 34 were red, 10 were black and 30 were white. According to the genetic model, these numbers should be in the ratio 9:3:4. Are the data consistent with the model at 5% level?

9. In an investigation into the health and nutrition of two groups of children from different social states, the following results are obtained.

| Social status / Health | Proof | Rich | total |
|---|---|---|---|
| Below normal | 130 | 20 | 150 |
| 'Normal | 102 | 108 | 210 |
| Above normal | 24 | 96 | 120 |

Discuss the relation between the health and their social status.

10. The following table gives a classification of a sample of 160 plants by their colour and flatness of leaf.

| | Flat leaves | Curled leaves | Total |
|---|---|---|---|
| White flower | 99 | 36 | 135 |
| Red flower | 20 | 5 | 25 |
| Total | 119 | 41 | 160 |

Test whether the flower colour is independent of the flatness of the leaf.

11. From the following information, state whether the two attributes viz., condition of house and condition of child are independent.

| Condition of the child | Condition of the house | | Total |
|---|---|---|---|
| | Clean | Dirty | |
| Clean | 69 | 51 | 120 |
| Fairly clean | 81 | 20 | 101 |
| Dirty | 35 | 44 | 79 |
| Total | 185 | 115 | 300 |

($\chi^2$ at 5% level for 2 df = 5.991)

12. A certain drug was administered to 500 people out of a total of 800 included in the sample to test its efficiency against typhoid. The results are given below.

| | Typhoid | No typhoid | Total |
|---|---|---|---|
| Drug | 200 | 300 | 500 |
| No drug | 280 | 20 | 300 |
| Total | 480 | 320 | 800 |

On the basis of the data can we say that the drug is effective in preventing typhoid.

13. In an experiment on the immunisation of goats from anthrax, the following results were obtained. Derive your inference on the vaccine.

| | Died of anthrax | Survived | Total |
|---|---|---|---|
| Inoculated with vaccine | 2 | 10 | 12 |
| Not inoculated | 6 | 6 | 12 |
| Total | 8 | 16 | |

14. 50 students are selected at random from 500 students enrolled in a computer programme. They were classified according to age and grade points giving the following data.

| | Age | | |
|---|---|---|---|
| Grade points | 20 and under | Age 21to 30 | Above 30 |
| Up to 5.0 | 3 | 5 | 2 |
| 5.1 to 7.5 | 8 | 7 | 5 |
| 7.5 to 10.0 | 4 | 8 | 8 |

Test at 5 % level of significance the hypothesis that age and grade points are independent ($\chi^2$ at 5% level for 2 df = 5.991)

15. The following data relates to the sales in a time of trade depression of a certain commodity demand.

| District where sales are | District not hit by depression | District hit by depression | Total |
|---|---|---|---|
| Satisfactory | 350 | 80 | 330 |
| Not satisfactory | 140 | 30 | 170 |
| Total | 390 | 110 | 500 |

Do these data suggest that the sales are significantly affected by depression?
($\chi^2_{0.05}$ = 3.841 for 1 df)

16. A survey of 200 families having three children selected at random gave the following results

| Male births: | 0 | 1 | 2 | 3 |
|---|---|---|---|---|
| No. of families: | 40 | 58 | 62 | 40 |

Test the hypothesis that male and female births are equally likely at 5% level of significance ($\chi^2_{0.05}$ = 7.82 for 3 df)

17. Two researchers adopted different sampling techniques. While investigating the same group of students to find the number of students falling into different intelligence levels, the results are as follows

| Researchers | Below average | Average | Above average | genius | Total |
|---|---|---|---|---|---|
| X | 86 | 60 | 44 | 10 | 200 |
| Y | 40 | 33 | 25 | 2 | 100 |
| Total | 126 | 93 | 69 | 12 | 300 |

Would you say that the sampling techniques adopted by the two researchers are significantly different.

($\chi^2_{0.05}$ for 2 df and 3 df are 5.991 and 7.82 respectively)

18. From the following data, find whether there is any significant liking in the habit of drinking soft drinks among categories of employees.

| Soft drinks | Clerks | Teachers | Officers |
|---|---|---|---|
| Pepsi | 10 | 25 | 65 |
| Thums up | 15 | 30 | 65 |
| Fanta | 50 | 60 | 30 |

($\chi^2$ for 4 df at 5% level of significance = 9.4888)

19. A firm manufacturing rivets wants to limit variations to their length as much as possible. The lengths (in cms) of 10 rivets manufactured by a new process are

| | | | | |
|---|---|---|---|---|
| 21.5 | 1.99 | 2.05 | 2.12 | 2.17 |
| 2.01 | 1.98 | 2.03 | 2.25 | 1.92 |

Examine whether the new process can be considered superior to the old if the population has a standard deviation of 0.145 cm.

($\chi^2_{0.05}$= 16.919 for 9 df )

20. 4 coins where tossed 160 times and the following results were obtained.

| No of heads : | 0 | 1 | 2 | 3 | 4 |
|---|---|---|---|---|---|
| Observed frequencies: | 17 | 52 | 54 | 31 | 6 |

Under the assumption that wins are balanced, find the expected frequencies of getting 0, 1, 2, 3 or 4 heads and test the goodness of fit.

($\chi^2_{0.05}$ value is 9.488)

21. Out of 8000 graduates in a city, 800 are males. Out of 1600 graduate employees, 120 are females. Use chi-square test to determine if any distinction is made in the appointment process on the basis of sex. ($\chi^2_{0.05}$ for 1 df = 3.841)

22. In the accounting department of a bank, 100 accounts are selected at random and examined for errors. The following results have been obtained.

| No of errors : | 0 | 1 | 2 | 3 | 4 | 5 | 6 |
|---|---|---|---|---|---|---|---|
| No of accounts: | 36 | 40 | 19 | 2 | 0 | 2 | 1 |

Does this information verify that errors are distributed according to the Poisson probability ($\chi^2_{0.05}$ = 7.815 for 3 df)?

## Answers 7.1

(1) Null hypothesis is rejected.

(3) Null hypothesis is rejected. The new treatment is superior to the conventional one.

(5) The colouration of the individuals are not independent of sex.

(6) $H_0$ accepted. There is a peculiarity.

(7) Die is unbiased.

(8) The data is consistent with the model.

(9) Health and social status are associated i.e., $H_0$ is rejected.

(10) $H_0$ is rejected. Flower colour is independent of the flatness of leaf.

(11) $H_0$ rejected. There is an association between the condition of the child and condition of the house).

(12) The drug is effective.

(13) $H_0$ is accepted. The vaccine is ineffective in controlling the disease.

(14) $H_0$ accepted. Grade points and age are independent of each other.

(15) $H_0$ is accepted. The sales are not significantly affected by depression.

(16) $H_0$ is rejected.

(18) $H_0$ rejected. Habit of drinking soft drinks depends on the category.

(19) $H_0$ accepted. The new process cannot be considered superior to the old process.

(20) The fit is poor.

(21) $H_0$ rejected. There is a distinction.

(22) $H_0$ accepted.

CHAPTER EIGHT

# Test of Significance: Small Samples

## 8.1 Introduction

In this chapter, we discuss tests of significance for small samples. The important tests for small samples are

(i) Chi-square test

(ii) $t$-test

(iii) $F$-test

$\chi^2$-distribution was already introduced in the previous chapter. We now introduce $t$- and $F$-distribution in this chapter.

### t-Distribution

When the population standard deviation is not known and the size of the sample is less than or equal to thirty, we use a $t$-test. The parameter $t$ was introduced by W.S. Gosset in the year 1908, but $t$-distribution was established by Fisher in the year 1920. $t$-distribution is also known as students $t$-distribution. Let $x_1$, $x_2, \ldots$, $x_n$ be the members of a random sample drawn from a normal population with mean $\mu$ and variance $\sigma^2$. We define test statistic as

$$t = \frac{\overline{x} - \mu}{S/\sqrt{n}}$$

where $\overline{x}$ is mean of the sample given by $\frac{\Sigma x_i}{n}$.

If we take $(n-1)S^2 = \sum_{i=1}^{n} (x_i - \overline{x})^2 = nS^2$, then we get

$$\frac{t^2}{\nu} = \frac{t^2}{n-1} = \frac{(\bar{x}-\mu)^2 n}{(n-1)S^2} = \frac{(\bar{x}-\mu)^2 \Big/ \frac{\sigma^2}{n}}{nS^2 \big/ \sigma^2} (\nu = n-1 \text{ is degrees of freedom})$$

Since $x_i$, $i = 1, 2, 3, \ldots, n$ is a random sample from the normal population with mean $\mu$ and variance $\sigma^2$, $\dfrac{\bar{x}-\mu}{\sigma/\sqrt{n}} \sim N(0,1)$.

$\dfrac{(\bar{x}-\mu)^2 n}{\sigma^2}$ is the square of a standard normal variate and it is distributed as a chi-square variate, with one degree freedom. $nS^2/\sigma^2$ is also distributed as a $\chi^2$- variate with $n-1$ degrees of freedom. Hence $\dfrac{t^2}{\nu}$ is the ratio of two independent $\chi^2$- variates distributed with 1 and $\nu$ degrees of freedom respectively. Therefore, $\dfrac{t^2}{\nu}$ is a $\beta_2\left(\dfrac{1}{2}, \dfrac{\nu}{2}\right)$ variate. For different samples, it was shown by Fisher that its distribution is

$$y = \frac{y_0}{\left(1+\frac{t^2}{\nu}\right)^{\frac{\nu+1}{2}}}, -\infty < x < \infty \tag{1}$$

The constant $y_0$ is chosen in such a way that

$$\int_{-\infty}^{\infty} y dt = 1$$

(i.e., area of the curve given by (1) is unity)

$$\text{i.e., } y_0 \int_{-\infty}^{\infty} \frac{y_0}{\left(1+\frac{t^2}{\nu}\right)^{\frac{\nu+1}{2}}} dt = 1$$

$$\text{i.e., } y_0 \int_{0}^{\infty} \frac{\left(\frac{t^2}{\nu}\right)^{-1/2}}{\left(1+\frac{t^2}{\nu}\right)^{\frac{1}{2}\left(\frac{\nu+1}{2}\right)}} d\left(\frac{t^2}{\nu}\right) = 1$$

$$\text{or} \quad y_0 \nu^{1/2} \beta\left(\frac{1}{2}, \frac{\nu}{2}\right) = 1$$

$$\text{or} \quad y_0 = \frac{1}{\sqrt{\nu}\beta\left(\frac{1}{2}, \frac{\nu}{2}\right)}$$

$$y = f(t) = \frac{1}{\sqrt{\nu}\beta\left(\frac{1}{2},\frac{\nu}{2}\right)\left(1+\frac{t^2}{\nu}\right)^{\frac{\nu+1}{2}}} \tag{2}$$

Equation (2) is known as a $t$-distribution. It is called the equation of $t$-probability curve.

The probability that we get a value of $t$ between the limits $t_1$ and $t_2$ is given by

$$P = \int_{t_1}^{t_2} \frac{y_0}{\left(1+\frac{t^2}{\nu}\right)^{\frac{\nu+1}{2}}} dt$$

where $y_0 = \dfrac{1}{\sqrt{\nu}\beta\left(\frac{1}{2},\frac{\nu}{2}\right)}$

## 8.2 Moments About Mean

The origin is mean for a $t$-distribution. All the moments of odd order about the origin vanish. The moment of even order about the origin is

$$\mu'_{2r} = 2\int_0^\infty \frac{t^{2r}}{\sqrt{\nu}\beta\left(\frac{\nu}{2},\frac{1}{2}\right)} \cdot \frac{dt}{\left(1+\frac{t^2}{\nu}\right)^{\frac{\nu+1}{2}}}$$

$$= 2\int_0^\infty \frac{t^{2r}\left(\frac{t^2}{\nu}\right)^{\frac{-1}{2}}}{\beta\left(\frac{\nu}{2},\frac{1}{2}\right)} \frac{d\left(\frac{t^2}{\nu}\right)}{\left(1+\frac{t^2}{\nu}\right)^{\frac{\nu+1}{2}}}$$

$$= \frac{\nu^2}{\beta\left(\frac{\nu}{2},\frac{1}{2}\right)} \int_0^\infty \frac{\left(\frac{t^2}{\nu}\right)^{r+\frac{1}{2}-1}}{\left(1+\frac{t^2}{\nu}\right)^{\frac{\nu+1}{2}}} d\left(\frac{t^2}{\nu}\right)$$

Substituting, $1+\dfrac{t^2}{\nu} = y$, we get $\dfrac{t^2}{\nu} = \dfrac{1-y}{y}$

and $\mu_{2r}' = \dfrac{\nu}{\beta\left(\dfrac{\nu}{2}, \dfrac{1}{2}\right)} \beta\left(r + \dfrac{1}{2}, \dfrac{\nu}{2} - r\right), r < \dfrac{\nu}{2}$

Therefore, $\mu_{2r} = \dfrac{(2r-1)(2r-3)\ldots 3.1}{(\nu-2)(\nu-4)\ldots(\nu-2r)} . \nu^r, n < \dfrac{\nu}{2}$

In particular $\beta_1 = \dfrac{\mu_3^2}{\mu_2^3} = 0$

$$\beta_2 = \frac{\mu_4}{\mu_3} = \frac{3(\nu-2)}{(\nu-4)}$$

$$= \frac{3\left(1 - \frac{2}{\nu}\right)}{\left(1 - \frac{4}{\nu}\right)}$$

As $\nu \to \infty$, we get $\beta_2 = \lim\limits_{\nu \to \infty} \dfrac{3\left(1 - \frac{2}{\nu}\right)}{\left(1 - \frac{4}{\nu}\right)} = 3$

## 8.3 Properties of *t*-Probability Curve

1. The equation of $t$-probability contains only even powers of $t$, therefore, the curve is symmetrical about the line $t = 0$ (Fig. 8.1).
2. $t$-distribution extends to infinity on either side and is asymptotic to the $t$-axis at each end.
3. $t$-distribution has a greater spread than the normal distribution and its graph is similar to that of a normal distribution.
4. Sampling distribution of $t$ is independent of the population parameters $\mu$ and $\sigma^2$. It depends only on the size of the sample.
5. The curve has a maximum ordinate at $t = 0$, the mean and mode coincide at $t = 0$.
6. As $\nu \to \infty$, $y \to e^{-\frac{1}{2}t^{\nu}}$, hence $t$ is distributed normally.

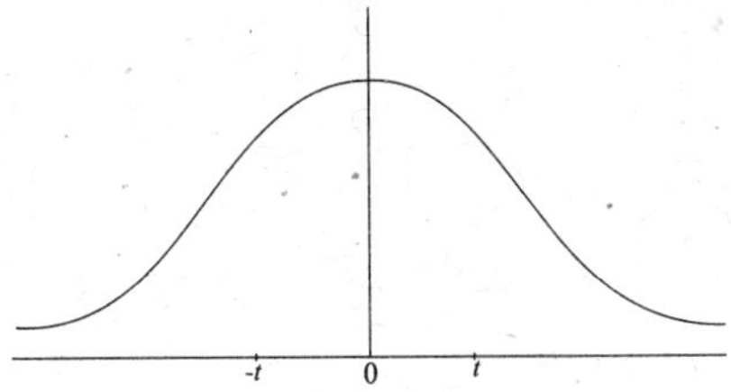


**Fig. 8.1** *t-distribution*

## 8.4 Assumptions for *t*-Test

The $t$-test is applied under the following assumptions.

(a) The samples are drawn from a normal population and are random.

(b) The population standard deviation may not be known.

(c) For testing the equality of two population means, the population variances are regarded as equal.

## 8.5 Uses of *t*-Distribution

The $t$-distribution is used to test the significance of

(i) a mean for small samples when the population variance is not known.

(ii) the difference between two means for small samples.

The values of $t$ for different degrees of freedom are given at the end of this book.

## 8.6 Interval Estimate of Population Mean

Let $\overline{x}$ denote the sample mean and $n$ denote the size of a sample. Then the interval estimate of the population mean $\mu$ at 5% level of significance, is given by $\overline{x} \pm t_{0.05}\frac{s}{\sqrt{n}}$.

$$\overline{x} \pm t_{\alpha}\frac{S}{\sqrt{n-1}} \text{ (where } \frac{s^2}{n} = \frac{s^2}{n-1})$$

$$S = \sqrt{\frac{\sum (x_i - \overline{x})^2}{n}} \text{ and } \frac{s}{n-1} \text{ is the standard error mean.}$$

## 8.7 Types of *t*-Test

There are two types of *t*-test

(i) unpaired *t*-test

(ii) paired *t*-test

## 8.8 Significant Values of *t*

The significant value of $t$ at $\alpha$ level of significance and degrees of freedom $\nu$, for a two-tailed test is given by (Fig. 8.2)

$$P\{|t| > t_\nu(\alpha)\} = \alpha$$

$$P\{|t| \leq t_\nu(\alpha)\} = 1 - \alpha$$

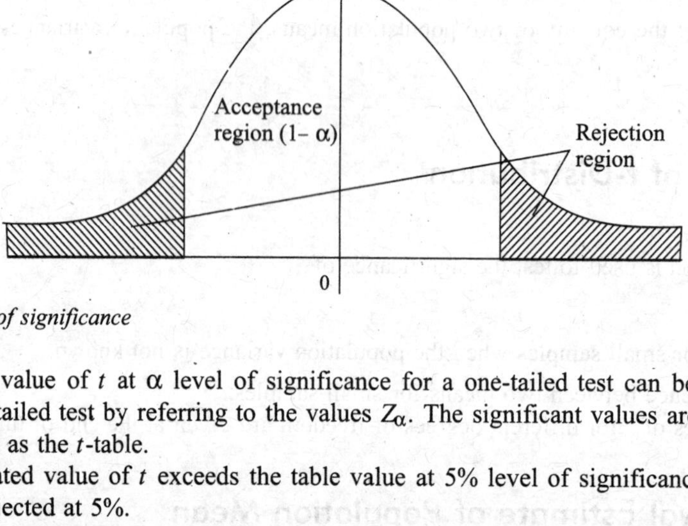


**Fig. 8.2** *Level of significance*

The significant value of $t$ at $\alpha$ level of significance for a one-tailed test can be obtained from those of a two-tailed test by referring to the values $Z_\alpha$. The significant values are given by $\alpha$ in the table known as the $t$-table.

If the calculated value of $t$ exceeds the table value at 5% level of significance, then the null hypothesis is rejected at 5%.

Similarly for 1% level of significance, we can reject the null hypothesis at 1% if the calculated value of $t$ exceeds the table value.

## 8.9 Test of Significance of a Single Mean

To test the significance of a mean of a small sample the test statistic is

$$t = \frac{\bar{x} - \mu}{\left(\dfrac{S}{\sqrt{n-1}}\right)}$$

where $\bar{x} = \dfrac{\sum x_i}{n}$

$\dfrac{S}{\sqrt{n-1}}$ is the standard error of mean

$n =$ size of the sample ($n \leq 30$)

$$S = \text{standard deviation} = \sqrt{\frac{\sum (x_i - \bar{x})^2}{n-1}}$$

If $\bar{x}$ is a fraction, we compute $S$ by using the formulae

$$\bar{x} = A + \frac{\sum d}{n}, \; d = x - A$$

$$S^2 = \frac{1}{n-1}\left[\sum d^2 - \frac{(\sum d)^2}{n}\right]$$

where $A$ is the assumed mean.

If the calculated value of $|t|$ is less than the table value of $t_\alpha$ ($\alpha =$ level of significance), we accept null hypothesis for the degrees of freedom $\nu$.

If the calculated value of $|t|$ is greater than the table value of $t_\alpha$ at the level of significance $\alpha$, for the degrees of freedom $\nu$, we reject the null hypothesis and accept the alternative hypothesis.

### Solved examples

**Example 1** A random sample of size 7 from a normal population gave a mean of 977.51 and a standard deviation of 4.42. Find 95% confidence interval for the population mean.

**Solution** Here we have $n = 7$, $\bar{x} = 977.51$, df $= n - 1 = 6$ and $S = 4.42$

$$\text{S.E.} = \frac{s}{\sqrt{n-1}} = \frac{4.42}{\sqrt{7-1}} = \frac{4.42}{\sqrt{6}} = \frac{4.42}{2.4494} = 1.8045$$

95% confidence limits for $\mu$ are

$$\bar{x} - t_{0.05}\frac{S}{\sqrt{n-1}} < \mu < \bar{x} + t_{0.05}\frac{S}{\sqrt{n-1}}$$

i.e., $977.51 - (2.447)(1.8045) < \mu < 977.51 + (2.447)(1.8045)$

i.e., $977.51 - 4.4156 < \mu < 977.51 + 4.4156$

i.e., $973.094 < \mu < 981.9256$

The required confidence interval is (973.094, 981.9256)

**Example 2** A sample of 10 cam shafts intended for use in a gasoline engine has an average eccentricity of 1.02 and a standard deviation of 0.044 inches. The data may be treated as a random sample from a normal population. Determine a 95% confidence interval for the actual mean eccentricity of the cam shaft.

**Solution** It is given that

$n =$ size of the sample $= 10$

$S =$ standard deviation $= 0.044$ inches

$\bar{x} = 1.02$ inches

Degrees of freedom $\nu = n - 1 = 10 - 1 = 9$

$$\frac{S}{\sqrt{n}} = \frac{0.044}{\sqrt{10}} = \frac{0.044}{3.1622} = 0.0139$$

The confidence interval is

$$\bar{x} \pm t_{\alpha/2}\frac{S}{\sqrt{n}} = 1.02 \pm 2.262 \times 0.0139$$

$$= 1.02 \pm 0.0314$$

i.e., $(1.02 - 0.0314, 1.02 + 0.0314)$

$$= (0.9886, 1.0514)$$

**Example 3** Find the students $t$ for the following variable values in a sample of eight. −4, −2, −2, 0, 2, 2, 3, 3, taking the mean of the universe to be zero.

**Solution** Table for computing mean $\bar{x}$ and $S$

| $x$ | $x-\bar{x}$ | $(x-\bar{x})^2$ |
|---|---|---|
| −4 | −4.25 | 18.0625 |
| −2 | −2.25 | 5.0625 |
| −2 | −2.25 | 5.0625 |
| 0 | −0.25 | 0.0625 |
| 2 | 1.75 | 3.0625 |
| 2 | 1.75 | 3.0625 |
| 3 | 2.75 | 7.5625 |
| 3 | 2.75 | 7.5625 |
| $\sum x_i = 2$ | | 49.5080 |

$$\bar{x} = \frac{\sum x_i}{n} = \frac{2}{8} = \frac{1}{4} = 0.25$$

$$S = \sqrt{\frac{\sum (x_i - \bar{x})^2}{n-1}} = \sqrt{\frac{49.5080}{8-1}} = \sqrt{7.071428}$$

$$= 2.6592$$

$$\text{Test statistic } t = \frac{\bar{x}-\mu}{\left(\frac{S}{\sqrt{n}}\right)} = \frac{0.25-0}{2.6592} \times \sqrt{8}$$

$$= \frac{(0.25)(2.8284)}{2.6592}$$

$$= 0.2659$$

**Example 4** A certain stimulus administered to each of 12 patients resulted in the following increases of blood pressures.

5, 2, 8, −1, 3, 0, 6, −2, 1, 5, 0, 4

Can it be calculated that the stimulus will be in general accomplished by an increase in blood pressure, given that for 11 degrees of freedom, the value of is 2.201?

**Solution** For 11 df = 2.201 (given)

$$\text{Mean of the sample } \bar{x} = \frac{\sum x_i}{n} = \frac{5+2+8-1+3+0+6-2+1+5+0+4}{12}$$

$$= \frac{31}{12} = 2.581$$

Calculation of $S$

| $x$ | $x-\bar{x}$ | $(x-\bar{x})^2$ |
|---|---|---|
| 5 | 2.42 | 5.8564 |
| 2 | −0.58 | 0.3364 |
| 8 | 5.42 | 29.3764 |
| −1 | −3.58 | 12.8164 |
| 3 | 0.42 | 0.1764 |
| 0 | −2.58 | 6.6564 |
| 6 | 3.42 | 11.6964 |
| −2 | −4.58 | 20.9764 |
| 1 | −1.58 | 2.4964 |
| 5 | 2.42 | 5.8564 |
| 0 | −2.58 | 6.6564 |
| 4 | 1.42 | 2.0614 |
| | | 104.9268 |

$$S^2 = \frac{\sum (x_i - \bar{x})^2}{n-1} = \frac{104.9268}{12-1} = 9.5389$$

$$\therefore S = \sqrt{9.5389} = 3.0885$$

Assuming that mean of the universe is zero, we get

$$t = \frac{\bar{x}-\mu}{\left(\frac{S}{\sqrt{n}}\right)} = \frac{2.58-0}{3.0885} \times \sqrt{12}$$

$$= \frac{(2.58)(3.464)}{3.0885} = 2.8936$$

Since the calculated value of $t$ for 11 df at 5% level of significance is greater than the table value of $t$, we reject the null hypothesis and conclude that the stimulus will be in general accompanied by an increase in blood pressure.

**Example 5** A random blood sample for test of fasting sugar for 10 boys gave the following data in ms/dl.

70, 120, 110, 101, 88, 83, 95, 107, 100, 98

Does this support the assumption of population mean of 100 mg/dl? Find a reasonable range in which most of the mean fasting sugar test of the 10 boys lie.

**Solution** We have $n = 10$

$$\sum x_i = 70 + 120 + 110 + 101 + 88 + 83 + 95 + 107 + 100 + 91$$

$$= 972$$

$$\bar{x} = \frac{\sum x_i}{n} = \frac{972}{10} = 97.2$$

Null hypothesis $H_0 : \mu = 100$

Alternative hypothesis $H_1 : \mu \neq 100$ (two-tailed test)

Degrees of freedom $\nu = n - 1 = 10 - 1 = 9$

Level of significance $\alpha = 0.05$

Table value of $t$ for 9 df at 5% level of significance = 2.262

| $x$ | $x - \bar{x}$ | $(x - \bar{x})^2$ |
|---|---|---|
| 70 | −27.2 | 739.84 |
| 120 | 22.8 | 519.84 |
| 110 | 12.8 | 163.84 |
| 101 | 3.8 | 14.44 |
| 88 | −9.2 | 84.64 |
| 83 | −14.2 | 201.64 |
| 95 | −2.2 | 4.84 |
| 107 | 9.8 | 96.04 |
| 100 | 2.8 | 7.84 |
| 98 | 0.8 | 0.64 |
| | | $\sum (x_i - \bar{x})^2 = 1833.60$ |

$$S^2 = \frac{\sum (x_i - \bar{x})^2}{n-1} = \frac{1833.60}{10-1} = \frac{1833.60}{9} = 203.74$$

$$\text{Standard error of mean} = \frac{S}{\sqrt{n}} = \frac{\sqrt{203.74}}{\sqrt{10}} = \sqrt{20.374}$$

$$\text{Test statistic: } t = \frac{\bar{x} - \mu}{\left(\frac{S}{\sqrt{n}}\right)} = \frac{-2.8}{\sqrt{20.374}} = \frac{-2.8}{4.5137} = -0.6203$$

$\therefore |t| = 0.6203 < 2.262$

i.e., calculated value of $t$ at 5% level of significance for 9 df is less than the table value.

Hence we accept null hypothesis.

The 95% confidence limits for $\mu$ are

$$\bar{x} \pm t_{0.05} \frac{S}{\sqrt{n}} = 97.2 \pm 2.262(4.51)$$

i.e., [97.2 – 10.20, 97.2 + 10.20]

i.e., [87.0, 107.40]

i.e., the 95% confidence interval for $\mu$ is [87.0, 107.40].

**Example 6** A sample of 26 bulbs gives a mean life of 990 hours with an S.D. of 20 hours. The manufacturer claims that the mean life of the bulbs is 1000 hours. Is the sample not up to standard?

**Solution** We have $n = 26$

$\bar{x} = 990$

$\mu = 100$

S.D. = 20

df $= n - 1 = 26 - 1 = 25$

Null hypothesis $H_0$ : The sample is up to standard

Alternative hypothesis $H_1$ : The sample is not up to standard

$$t = \frac{\bar{x} - \mu}{\left(\frac{S}{\sqrt{n-1}}\right)} = \frac{990 - 1000}{\frac{20}{\sqrt{26-1}}} = \frac{-10}{\left(\frac{20}{5}\right)}$$

$$= \frac{-50}{20} = -2.5$$

$\therefore |t| = 2.5$

Table value of $t$ for 25 df at 5% level is 1.708. Since the calculated value of $t$ is greater than the table value of $t$, we reject the null hypothesis.

Therefore, the sample is not up to standard.

**Example 7** The average breaking strength of steel rods is specified to be 18.5 thousand pounds. To test this, a sample of 14 rods was tested. The mean and standard deviations obtained were 17.85 and 1.955 respectively. Is the result of the experiment significant?

**Solution** Here we have $n = 14$, $\bar{x} = 17.85$, $\mu = 18.5$

$S =$ SD $= 1.955$

Degrees of freedom $= n - 1 = 14 - 1 = 13$

Table value of $t$ for $13\nu$ df at 5% level of significance = 2.116 (i.e., critical value)

Null hypothesis $H_0$ : The result of the experiment is not significant

Alternative hypothesis $H_1 : \mu \neq 18.5$

$$\text{The test statistic } t = \frac{\bar{x}-\mu}{\left(\dfrac{S}{\sqrt{n-1}}\right)} = \frac{17.85-18.5}{\dfrac{1.955}{\sqrt{13}}} = \frac{-0.65}{0.542} = -1.199$$

Therefore, $|t| = 1.199 < 2.166$

The calculated value of $t$ is less than the table value of $t$. Hence we accept the null hypothesis i.e., the result of the experiment is not significant.

**Example 8** A machine is designed to produce insulating washers for electrical devices of average thickness 0.025 cm. A random sample of 10 washers was found to have a thickness 0.024 cm, with S.D. of 0.002 cm. Test the significance of the deviation (value of $t$ for 9 degrees of freedom at 5% level of significance is 2.262).

**Solution** Size of the given sample = 10 (small sample)

Sample mean $= \bar{x} = 0.024$ cm

Population mean $= \mu = 0.025$ cm

$S =$ Standard deviation $= 0.002$ cm

Degrees of freedom $= n-1 = 10-1 = 9$

Table value of $t$ for 9 df at 5% level of significance = 2.262

Null hypothesis $H_0$ : The difference between $\bar{x}$ and $\mu$ is not significant.

Alternative hypothesis $H_1 : \mu \neq 0.025$

$$\text{The test statistic } t = \frac{\bar{x}-\mu}{\left(\dfrac{S}{\sqrt{n-1}}\right)} = \frac{0.024-0.025}{\dfrac{0.002}{\sqrt{10-1}}}$$

$$= \frac{-0.01}{0.002} \times 3 = -1.5$$

$\therefore |t| = 1.5 < 2.262$

The calculated value of $t$ is less than the table value of $t$ at 5% level of significance and 9 df. We accept null hypothesis and conclude that the difference between $\bar{x}$ and $\mu$ is not significant.

**Example 9** A sample of 20 items has a mean 42 units and S.D. 5 units. Test the hypothesis that it is a random sample from a normal population with mean 46 units.

**Solution** We have $n = 20$ (small sample), $\bar{x} = 46$, $\mu = 46$, $S = 5$

Null hypothesis $H_0 : \mu = 46$

Alternative hypothesis $H_1 : \mu \neq 46$

Degrees of freedom $= n - 1 = 20 - 1 = 19$

Table value of $t$ for 19 df at 5% level of significance $= 2.09$

$$\text{The test statistic } t = \frac{\bar{x} - \mu}{\left(\dfrac{S}{\sqrt{n-1}}\right)} = \frac{42 - 46}{\dfrac{5}{\sqrt{19}}}$$

$$= \frac{-4\sqrt{19}}{5} = \frac{-4 \times 4.3588}{5}$$

$$= -3.4871$$

Therefore, $|t| = 3.4871 > 2.09$

The calculated value of $t$ is greater than the table value at 5% level for 19 df. Hence, we reject the null hypothesis $H_0 : \mu = 46$ i.e., the difference between the sample mean and population mean is significant.

## 8.10 Student's *t*-Test for Difference of Means

Suppose we want to test if two independent samples $x_1, x_2, x_3, \ldots, x_{n_1}$ and $y_1, y_2, y_3, \ldots, y_{n_2}$ of sizes $n_1$ and $n_2$ have been drawn from the normal population with means $\mu_1$ and $\mu_2$ respectively.

Assume that the population variances are equal.

$$\text{The statistic } t = \frac{(\bar{x} - \bar{y}) - (\mu_1 - \mu_2)}{S\sqrt{\dfrac{1}{n_1} + \dfrac{1}{n_2}}}$$

$$\text{where } \bar{x} = \frac{\Sigma x_i}{n_1}, \ \bar{y} = \frac{\Sigma y_i}{n_2}$$

$$S^2 = \frac{1}{n_1 + n_2 - 2}\left[\sum_1^n (x_i - \bar{x})^2 + \sum_1^n (y_i - \bar{y})^2\right]$$

is an unbiased estimate of the population variance $\sigma^2$.

$t$ follows $t$-distribution with degrees of freedom $n_1 + n_2 - 2$.

Under the null hypothesis $H_0$, that

(i) Samples have been drawn from the populations with the same means i.e., $\mu_1 = \mu_2$

(ii) Sample means $\bar{x}$ and $\bar{y}$ do not differ significantly,

we compute the test statistic

$$t = \frac{(\bar{x} - \bar{y})}{S\sqrt{\dfrac{1}{n_1} + \dfrac{1}{n_2}}}$$

The degrees of freedom is $n_1 + n_2 - 2$.

If $|t|$ is less than the table value of $t_\alpha$ for a given level of significance $\alpha, n_1 + n_2 - 2$ df, we accept the null hypothesis $H_0$. Otherwise, we reject the null hypothesis.

The confidence interval for $\mu_1 - \mu_2$ is a (100, $1 - \alpha$)% interval given by

$$\bar{x} - \bar{y} \pm t_{\alpha/2}\sqrt{\frac{\Sigma(x_i - \bar{x})^2 + \Sigma(y_i - \bar{y})^2}{n_1 + n_2 - 2}\left[\frac{1}{n_1} + \frac{1}{n_2}\right]}$$

$$\text{i.e., } \bar{x} - \bar{y} \pm t_{\alpha/2} S\sqrt{\frac{1}{n_1} + \frac{1}{n_2}}$$

where the degrees of freedom is $\nu = n_1 + n_2 - 2$.

## Solved examples

**Example 1** The nicotine content in milligrams of the samples of tobacco were found as follows

Sample A: 24 27 26 21 25
Sample B: 27 30 28 31 22 36
Can it be said that the two samples come from normal populations with the same mean?

**Solution** We have $\bar{x} = \dfrac{24 + 27 + 26 + 21 + 25}{5} = \dfrac{123}{5} = 24.6$

$$\bar{y} = \frac{27 + 30 + 28 + 31 + 22 + 36}{6} = \frac{174}{6} = 29$$

Calculation of sample S.D.s

| $x$ | $x-\bar{x}$ | $(x-\bar{x})^2$ | $y$ | $y-\bar{y}$ | $(y-\bar{y})^2$ |
|---|---|---|---|---|---|
| 24 | −0.6 | 0.36 | 27 | −2 | 4 |
| 27 | 2.4 | 5.76 | 30 | 1 | 1 |
| 26 | 1.4 | 1.96 | 28 | −1 | 1 |
| 21 | −3.6 | 12.96 | 31 | 2 | 4 |
| 25 | 0.4 | 0.16 | 22 | −7 | 49 |
| — | — | — | 36 | 7 | 49 |
| 123 | 0 | 2.12 | 174 | 0 | 108 |

$\therefore \Sigma(x-\bar{x})^2 = 21.2,\ \Sigma(y-\bar{y})^2 = 108$

$$S^2 = \frac{1}{n_1+n_2-2}\left[\sum_1^n (x_i-\bar{x})^2 + \sum_1^n (y_i-\bar{y})^2\right]$$

$$= \frac{1}{5+6-1}[21.2+108] = 14.35$$

$\therefore S = \sqrt{14.35} = 3.78$

Null hypothesis $H_0 : \mu_1 = \mu_2$ i.e., the two samples have been drawn from normal populations with the same mean.

Alternative hypothesis $H_1 : \mu_1 \neq \mu_2$

Level of significance = 5%

Degrees of freedom = $n_1+n_2-2 = 5+6-2 = 9$

$t_{0.05}$ for 9 df = 2.262

Test statistic is

$$t = \frac{(\bar{x}-\bar{y})}{S\sqrt{\frac{1}{n_1}+\frac{1}{n_2}}} = \frac{24.6-29}{3.78\sqrt{\frac{1}{5}+\frac{1}{6}}} = -\ 1.92$$

$\therefore |t| = 1.92 < 2.262$

The calculated value of $t$ is less than the table value of $t$ at 5% level of significance for 9 df. Therefore, we accept $H_0$.

The samples are drawn from normal populations with the same mean.

**Example 2** A group of 7 week old chicken reared on a high protein diet weigh 12, 15, 11, 16, 14, 14 and 16 ounces. A second group of five chickens similarly treated, except that they receive a low protein diet, weigh 8, 10, 14, 10 and 13 ounces. Test whether there is evidence that additional protein has increased the weight of the chicken. (The table value of $t$ for $\nu$ = 10 at 5% level of significance is 2.33)

**Solution** Let the weights of chicken reared on high protein diet be denoted by $x_i$ and the weights of chicken reared on low protein diet be denoted by $y_i$.

$$\text{We have } \bar{x} = \frac{\sum x_i}{n_1} = \frac{12+15+11+16+14+14+16}{7} = \frac{98}{7} = 14$$

$$\bar{y} = \frac{\sum y_i}{n_2} = \frac{8+10+14+10+13}{5} = \frac{55}{5} = 11$$

Null hypothesis $H_0 : \mu_1 = \mu_2$

i.e., additional protein has not increased the weight of the chickens

Alternative hypothesis $H_1 : \mu_1 > \mu_2$ (one-tailed test)

Computation of standard deviation

| $x$ | $x-\bar{x}$ | $(x-\bar{x})^2$ | $y$ | $y-\bar{y}$ | $(y-\bar{y})^2$ |
|---|---|---|---|---|---|
| 12 | −2 | 4 | | | |
| 15 | 1 | 1 | 8 | −3 | 9 |
| 11 | −3 | 9 | 10 | −1 | 1 |
| 16 | 2 | 4 | 14 | 3 | 9 |
| 14 | 0 | 0 | 10 | −1 | 1 |
| 14 | 0 | 0 | 13 | 2 | 4 |
| 16 | 2 | 4 | | | |
| 98 | | 22 | 55 | | 24 |

$$S = \sqrt{\frac{\sum (x_i - \bar{x})^2 + \sum (y_i - \bar{y})^2}{n_1 + n_2 - 2}} = \sqrt{\frac{22+24}{7+5-2}} = \sqrt{\frac{46}{10}} = \sqrt{4.6} = 2.1447$$

Level of significance = 5% ( $\alpha$ = 0.05)

Critical value i.e., table value of $t$ for 10 df at 5% level = 1.81

$$\text{Test statistic is } t = \frac{(\bar{x}-\bar{y})}{S\sqrt{\frac{1}{n_1}+\frac{1}{n_2}}} = \frac{14-11}{2.1447\sqrt{\frac{1}{7}+\frac{1}{5}}}$$

$$= \frac{3}{2.1447} \times \frac{\sqrt{25}}{\sqrt{5+7}} = \frac{3 \times 5.916}{2.1447 \times 3.4641}$$

$$= \frac{17.748}{7.4294} = 2.3888$$

Since the calculated value of $t$ is greater than the table value of $t$, at 5% level of significance for 10 degrees of freedom, we reject $H_0$ and conclude that additional protein has increased the weight of the chicken.

**Example 3** Two salesmen A and B work in a certain district from a sampling survey conducted by the head office. The following results were obtained. State whether there is any significant difference in the average sales between the two salesmen.

| | A | B |
|---|---|---|
| No. of sales: | 20 | 18 |
| Average sales (in Rs): | 170 | 203 |
| Standard deviation (in Rs): | 20 | 25 |

**Solution** We have

$$n_1 = 20, \quad n_2 = 18$$
$$\overline{x_1} = 170, \quad \overline{x_2} = 205$$
$$S_1 = 20, \quad S_2 = 25$$

Degrees of freedom $= n_1 + n_2 - 2 = 20 + 16 - 2$

Level of significance $\alpha = 0.05$

Table value of $t = 1.9$ (for 36 df at 5%)

Standard error of difference of means

$$= \text{S.E.}(\overline{x_1} - \overline{x_2})$$

$$= S\sqrt{\frac{1}{n_1} + \frac{1}{n_2}}$$

$$\text{where } S = \sqrt{\frac{n_1 s_1^2 + n_2 s_2^2}{n_1 + n_2 - 2}} = \sqrt{\frac{(20)(20)^2 + (18)(25)^2}{20 + 18 - 2}}$$

$$= \sqrt{\frac{800 + 11250}{36}} = 23.1240$$

$$\text{Therefore, S.E.}(\overline{x_1} - \overline{x_2}) = 23.1240\sqrt{\frac{1}{20} + \frac{1}{18}} = 23.1240\sqrt{\frac{9 + 10}{180}}$$

$$= 23.1420\sqrt{\frac{19}{180}} = 23.1420 \times 0.3248$$

$$= 7.5128$$

$$\text{Test statistic is } t = \frac{(\overline{x_1} - \overline{x_2})}{S\sqrt{\dfrac{1}{n_1} + \dfrac{1}{n_2}}} = \frac{170 - 205}{7.5128} = \frac{-35}{7.5128} = -4.6587$$

$\therefore |t| = 4.6587 > 1.96$

i.e., the calculated value of $t$ is greater than the table value of $t$ at 5% level of significance, for 36 df.

Hence, we reject null hypothesis and conclude that there is significant difference in the average sales between the two sales men.

**Example 4** A group of five patients treated with medicine A weigh 42, 39, 48, 60 and 41 kgs. A second group of 7 patients from the same hospital with medicine B weigh 38, 42, 56, 64, 68, 69 and 62 kgs. Do you agree with the claim that the medicine B increases the weight significantly (The value of $t$ at 5% level of significance for 10 df is 2.228).

**Solution** Let $x_i$ denote the weights of patients treated with medicine A, and $y_i$ denote the weights of patients treated with medicine B.

We have $n_1 = 5$, $n_2 = 7$, df $= n_1 + n_2 - 2 = 5 + 7 - 2 = 10$

Null hypothesis $H_0 : \mu_A = \mu_B$

i.e., there is no significant difference between the medicines A and B regarding their effect on increase in weight.

Alternative hypothesis $H_1 : \mu_A \neq \mu_B$ (two-tailed test)

Table value of $t$ at 5% level of significance for 10 df = 2.228

$$\bar{x} = \frac{\Sigma x_i}{n_1} = \frac{42+39+48+60+41}{5} = \frac{230}{5} = 46$$

$$\bar{y} = \frac{\Sigma y_i}{n_2} = \frac{38+42+56+64+68+69+62}{7} = 59$$

| $x$ | $x-\bar{x}$ | $(x-\bar{x})^2$ | $y$ | $y-\bar{y}$ | $(y-\bar{y})^2$ |
|---|---|---|---|---|---|
| 42 | −4 | 16 | 38 | −19 | 361 |
| 39 | −7 | 49 | 42 | −15 | 225 |
| 48 | 2 | 4 | 56 | −1 | 1 |
| 60 | 14 | 196 | 64 | 7 | 49 |
| 41 | −5 | 25 | 68 | 11 | 121 |
| — | — | — | 69 | 12 | 144 |
| — | — | — | 62 | 5 | 25 |
| 230 | — | 290 | 399 | — | 926 |

$$S = \sqrt{\frac{\Sigma(x_i-\bar{x})^2 + \Sigma(y_i-\bar{y})^2}{n_1+n_2-2}} = \sqrt{\frac{290+926}{5+7-2}}$$

$$= \sqrt{\frac{1216}{10}} = \sqrt{121.6} = 11.027$$

Test statistic is $t = \dfrac{(\bar{x} - \bar{y})}{S\sqrt{\dfrac{1}{n_1} + \dfrac{1}{n_2}}}$

$$= \frac{46 - 57}{11.027\sqrt{\frac{1}{5} + \frac{1}{7}}} = \frac{-11 \times \sqrt{35}}{(11.027)\sqrt{12}}$$

$$= \frac{-11 \times 5.916}{(11.027) \times 3.464}$$

$$= \frac{-65.076}{38.197} = -1.703$$

$\therefore |t| = 1.703$

The calculated value of $t$ is less than the table value of $t$ at 5% level of significance, for 10 df. Therefore, we accept null hypothesis $H_0$.

There is no significant difference between medicines A and B as regards the increase in weights.

**Example 5** The students of two schools were measured for their heights. One school was in the east coast, another was in the west coast, where there is slight difference in weather. The sampling results are as follows.

| | | | | | | | | | |
|---|---|---|---|---|---|---|---|---|---|
| East coast: | 43 | 45 | 48 | 49 | 51 | 51 | – | – | – |
| West coast: | 47 | 49 | 51 | 53 | 54 | 55 | 55 | 56 | 57 |

Find whether there is any impact of weather on height, taking other variables as constant (given $t_{0.05}$ for 13 df = 2.16).

**Solution** Here we have $n_1 = 6$, $n_2 = 9$

Let the east coast student height be denoted by $x$ and the west coast student height be denoted by $y$.

$$\text{Then } \bar{x} = \frac{\sum x_i}{n_1} = \frac{43 + 45 + 48 + 49 + 51 + 52}{6} = \frac{288}{6} = 48$$

$$\bar{y} = \frac{\sum y_i}{n_2} = \frac{47 + 49 + 51 + 53 + 54 + 55 + 55 + 56 + 57}{9} = \frac{477}{9} = 53$$

Null hypothesis $H_0$ : Weather has no impact on the heights of the students.

Alternative hypothesis $H_1$ : Weather has an impact on the heights of the students.

Degrees of freedom $= 6+9-2 = 13$

Level of significance $= 5\%$

Table value of $t$ for 13 df $= 2.16$

$$S^2 = \frac{1}{n_1+n_2-2}\left[\sum_1^n (x_i-\bar{x})^2 + \sum_1^n (y_i-\bar{y})^2\right]$$

$$= \frac{60+90}{6+9-2} = \frac{150}{13} = 11.5$$

| East coast ($x$) | $x-\bar{x}$ | $(x-\bar{x})^2$ | West coast ($y$) | $y-\bar{y}$ | $(y-\bar{y})^2$ |
|---|---|---|---|---|---|
| 43 | −5 | 25 | 47 | −6 | 36 |
| 45 | −3 | 9 | 49 | −4 | 16 |
| 48 | 0 | 0 | 51 | −2 | 4 |
| 49 | 1 | 1 | 53 | 0 | 0 |
| 51 | 3 | 9 | 54 | 1 | 1 |
| 52 | 4 | 16 | 55 | 2 | 4 |
| – | – | – | 55 | 2 | 4 |
| – | – | – | 56 | 3 | 9 |
| – | – | – | 57 | 4 | 16 |
| 288 | | 60 | 477 | | 90 |

$\therefore S = \sqrt{11.5} = 3.39$

Test statistic is $t = \dfrac{(\bar{x}-\bar{y})}{S\sqrt{\dfrac{1}{n_1}+\dfrac{1}{n_2}}} = \dfrac{48-53}{3.39\sqrt{\dfrac{1}{6}+\dfrac{1}{9}}}$

$$= \frac{-5}{3.39\sqrt{\dfrac{5}{18}}} = \frac{-5}{1.76} = -2.84$$

$\therefore |t| = 2.84$

The calculated value of $t$ is greater than the table value of $t$ at 5% level of significance, for 13 df.

Hence, we reject null hypothesis and accept the alternative hypothesis.

## Exercise 8.1

1. Define $t$-distribution and write the properties of the $t$-probability curve.
2. Ten students are selected at random from a college and their heights are found to be 100, 104, 108, 110, 118, 120, 122, 124, 126 and 118 cms. In the height of these data, discuss

the suggestion that the mean height of the students of the college is 110 cms. Use 5% level of significance.

3. A machine which produces mica insulating washers having a thickness of 10 miles (1 mile = 0.001 inch). A sample of 10 washers has an average thickness of 9.52 miles with a standard deviation of 0.60 miles. Find the value of $t$.

4. Ten cartons are taken at random from an automatic machine. The mean net weight of the 10 cartons is 11.8 kg and standard deviation is 0.15 kg. Do the sample means differ significantly from the intended height of 12 kg ($t_{0.05}$ for 9 df = 2.26)?

5. A fertilizer mixing machine is set to give 12 kg of nitrate for every quintal bag of fertilizer. Ten 100 kgs bags are examined. The percentages of nitrate are as follows: 11, 14, 13, 12, 13, 12, 1,3, 14, 11, 12. Is there reason to believe that the machine is effective ($t_{0.05}$ for 9 df = 2.262)?

6. Two samples of sizes 6 and 5 gave the following data

| | Sample I | Sample II |
|---|---|---|
| Mean | 40 | 50 |
| S.D. | 8 | 10 |

Is the difference of the mean significant? The value of $t$ for $t$ degrees of freedom at 5% level of significance is 2.26.

7. Samples of two types of sodium vapour lamps were tested for length of life and the following data were obtained.

| | Type I | Type II |
|---|---|---|
| Sample no. | 8 | 7 |
| Sample mean | 1234 hrs | 1036 hrs |
| Sample S.D | 36 hrs | 40 hrs |

Is the difference in the mean sufficient to warrant an inference that type I is superior to type II regarding length of life (type I is superior to type II)?

8. Sandal powder is packed into packets by a machine. A random sample of 12 packets is drawn and their weights are found to be (in kgs) 0.49, 0.48, 0.47, 0.48, 0.49, 0.50, 0.51, 0.49, 0.50, 0.51, 0.48. Test if the average packing can be taken as 0.5 kg ($t_{0.05}$ for 11 df = 2.20)

9. The following results are obtained from a sample of 10 boxes of biscuits. Mean weight of the content = 490 gm. Standard deviation of the weights = 9 gm. Could the sample come from a population having a mean of 500 gms ($t_{0.05}$ for 9 df = 2.26)?

10. The wages of 10 workers taken at random from a factory are given as follows. Wages (in Rs): 578, 572, 570, 568, 572, 578, 570, 572, 596, 584. Is it possible that the mean height of all workers of this factory could be Rs 580 (given at 9 df $t_{0.05}$ = 2.26)?

11. The 9 items of a sample have the following values: 45, 47, 50, 52, 48, 47, 49, 53, 51. Does the mean values differ significantly from the assumed mean 47.5 (for 8 df , $t_{0.05}$ = 2.31)?

12. Write short notes on: (a) Students $t$-distribution (b) Properties of $t$-distribution.

13. Test whether the sample having values 63, 63, 64, 55, 66, 69, 70, 70 and 71 has been chosen from a population with mean of 65 at 5% level of significance ($t_{0.05}$ = 2.31 for 8 df).

14. A certain medicine administered to each of 10 patients resulted in the following increases in the blood pressure: 8, 8, 7, 5, 4, 1, 0, 0, –1, –1. Can it be concluded that the medicine was responsible for the increase in blood pressure?

15. The weight of 10 people of a locality are found to be 70, 67, 62, 68, 61, 68, 70, 64, 64, 66 kgs. Is it reasonable to believe that the average weights of the people of locality are greater than 64 kgs. Test at 5% level of significance (Given $t_{0.05}$ = 1.83 for 9 df )

16. The following table contains the data resulting from a sample of managers trained under different programmes.

| Programme sampled | Mean sensitivity of the programme | Number of managers observed | Estimated S.D. for sensitivity after the programme |
|---|---|---|---|
| Formal | 92% | 12 | 15% |
| Informal | 84% | 15 | 19% |

Test at 0.05 level of significance whether the sensitivity achieved by the formal programme is significantly higher than that achieved under the informal programme.

17. A random sample of 10 tins of oil filled in by an automatic machine gave the following weights in kg: 2.05, 2.01, 2.04, 1.91, 1.96, 2.01, 1.97, 1.97, 2.04, 2.02. Can we accept at 5% level of significance, the claim that the average weight of the tin is 2 kg?

18. A random sample of size 16 has 53 as its mean. The sum of the squares of the deviations taken from the mean is 150. Can this sample be regarded as taken from the population having 56 as its mean. Obtain 95% and 99% confidence limits of the mean population (for $\nu = 15$, $t_{0.01}$ = 2.95, $t_{0.05}$ = 2.131)

19. A sample of 20 items has mean 42 units and S.D. 5 units. Test the hypothesis that it is a random sample from a normal population with mean 45 units.

20. The means of two random samples of size 9 and 7 are 196.42 and 198.82 respectively. The sum of the squares of the deviations from the mean is 26.94 and 18.73 respectively. Can the sample be considered to have been drawn from the same normal population.

21. Two horses A and B were tested according to the time (in seconds) to run a particular track, with the following results.

| | | | | | | | |
|---|---|---|---|---|---|---|---|
| Horse A : | 28 | 30 | 32 | 33 | 33 | 29 | 34 |
| Horse B: | 29 | 30 | 30 | 24 | 27 | 27 | – |

Test whether you can discriminate between the two horses. You can use the fact that at 5% level of significance, value of $t$ for 11 df is 2.2.

22. A group of 5 patients treated with medicine A weighed 42, 39, 48, 60 and 41 kgs. A second group of 7 patients from the same hospital treated with medicine B weighed 38, 42, 56, 64, 68, 69 and 62 kgs. Do you agree with the claim that medicine B increases the weight significantly?

23. The memory capacity of 10 students was tested before and after training. State whether the training was effective or not from the following scores.

| Before : | 12 | 14 | 11 | 8 | 7 | 10 | 3 | 0 | 5 | 6 |
|---|---|---|---|---|---|---|---|---|---|---|
| After: | 15 | 16 | 10 | 7 | 5 | 13 | 10 | 2 | 8 | 8 |

24. An I.Q test was administered on 5 persons before and after they were trained. The results are given below.

| I.Q before training | 110 | 120 | 123 | 132 | 125 |
|---|---|---|---|---|---|
| I.Q after training | 120 | 118 | 125 | 136 | 121 |

Test whether there is any change in I.Q after the training programme.

25. Measurements of a sample of weights were determined as 8.3, 10.6, 9.7, 8.8, 10.2, 9.4. Determine unbiased and efficient estimates of (a) the population mean and (b) the population variance.

26. Ten students were given intensive coaching for a month in mathematics. The scores obtained in tests 1 and 5 are given below.

| Serial no. of students: | 1 | 2 | 3 | 4 | 5 | 6 | 7 | 8 | 9 | 10 |
|---|---|---|---|---|---|---|---|---|---|---|
| Marks in first test: | 50 | 52 | 53 | 60 | 65 | 67 | 48 | 69 | 72 | 80 |
| Marks in second test: | 65 | 55 | 65 | 65 | 60 | 67 | 49 | 82 | 74 | 86 |

Do the scores from test 1 to test 5 show an improvement? Test at 5% level of significance ($t_{0.05}$ for 9 df = 1.833 and 2-tailed test $t_{0.05}$ for 9 df = 2.262)

27. A certain stimulus administered to each of 12 patients resulted in the following changes in the blood pressure 5, 2, 8, −1 3, 0, −2, 1, 5, 0, 4, 6. Can it be concluded that the stimulus in general be accompanied by an increase in blood pressure.

28. Experience shows that a fixed dose of a certain drug causes an average increase of pulse rate by 10 beats per minute with S.D. of 4. A group of 9 patients given the same dose showed the following increase: 13, 15, 14, 10, 8, 12, 16, 9, 20. Test 5% level of significance whether this group is different in response to the drug.

## Answers 8.1

2. Null hypothesis is accepted i.e., mean height of the students is 110 cm.

4. $H_0$ rejected. The sample means differ.

5. There is no reason to believe that the machine is defective.

6. The difference of means is not significant.

8. $H_0$ is rejected.

9. $H_0$ is rejected.

10. $H_0$ is accepted.

11. $H_0$ is accepted.

13. The difference is significant.

14. The difference is significant. The medicine is responsible for the increase in blood pressure.

15. $H_0$ is rejected. Average weight is greater than 64 kg.

16. $H_0$ rejected: The sensitivity achieved by the formal programme is higher than that achieved under the informal programme.

17. That the average weight of the tin is 2 kg is accepted.

18. $H_0$ is rejected, [50.316, 54.684], [50.67, 55.33].

19. The sample could not have come from the given population.

20. The two samples are not drawn from the same population.

21. There is significant difference between the two horses and they can be discriminated.

22. The medicines A ands B do not differ significantly as regards their effect on the increase of weight.

23. Not significant.

24. No change in I.Q.

25. 9.5 kgs, 0.736.

26. For 1-tailed test, null hypothesis is rejected.

27. $H_0$ rejected. The stimulus in general is accompanied by an increase in blood pressure.

28. $H_0$ rejected: The given group is different in response to the drug.

## 8.11 Paired *t*-Test

In this case the size of two samples are equal i.e., $n_1 = n_2 = n$

Let $x_1, x_2, \ldots, x_n$ and $y_1, y_2, \ldots, y_n$ be two samples. Let $d_1, d_2, \ldots, d_n$ be the differences between the corresponding members of the two samples.

If $\bar{d} = \dfrac{\Sigma d}{n}$ and $S = \dfrac{\Sigma (d - \bar{d})^2}{n-1}$, then the $t$-statistic is defined as

$$t = \frac{\bar{d} - 0}{s/\sqrt{n}}$$

$t$-statistic is applied for $n - 1$ degrees of freedom.

## Solved examples

**Example 1** A certain diet was newly introduced to each of 12 pigs resulting in the following increase in body weight.

6, 3, 8, –2, 3, 0, –1, 1, 6, 0, 5 and 4

Can you conclude that the diet is effective in increasing the weight of pigs (given $t_{0.05} = 2.20$ for 11 df)?

**Solution** Here we have $n = 12$

$$d_1 = 6, d_2 = 3, d_3 = 8, d_4 = -2, d_5 = 3, d_6 = 0,$$

$$d_7 = -1, d_8 = 1, d_9 = 6, d_{10} = 0, d_{11} = 5, d_{12} = 4$$

where $d_i$ denotes the increase in weights of the pigs.

Null hypothesis: $H_0$ : There is no significant difference in the body weights of the pigs before and after introduction of the new diet. i.e., $\mu_x = \mu_y$

Alternative hypothesis $H_1 : \mu_x \neq \mu_y$

Degrees of freedom $= n - 1 = 12 - 1 = 11$

Level of significance $= 5\%, t_{0.05}$ for 11 df $= 2.20$

$$\Sigma d = 6 + 3 + 8 + (-2) + 3 + 0 + (-1) + 1 + 6 + 0 + 5 + 4 = 33$$

$$\bar{d} = \frac{\Sigma d}{n} = \frac{33}{12} = 2.75$$

$$\Sigma d^2 = 6^2 + 3^2 + 8^2 + (-2)^2 + 3^2 + 0^2 + (-1)^2 + 1^2 + 6^2 + 0^2 + 5^2 + 4^2 = 201$$

$$\therefore S^2 = \frac{1}{n-1}\left[\Sigma d^2 - \left(\frac{(\Sigma d)^2}{n}\right)\right] = \frac{1}{12-1}\left[201 - \frac{(33)^2}{12}\right]$$

$$= \frac{1}{11}[201 - 90.75] = 10.0227$$

$$S = \sqrt{10.0227} = 3.1658$$

Test statistic $t$ is

$$t = \frac{\bar{d} - 0}{S/\sqrt{n}} = \frac{2.75}{3.1658/\sqrt{12}} = \frac{2.75 \times 3.464}{3.1658} = 3.0091$$

Since the calculated value of $t$ is greater than the table value of $t$ at 5% level of significance for 11 df, we reject null hypothesis. Therefore, the new diet is effective in increasing the body weight of the pigs.

## 8.12 *F*-Distribution

In this section, we introduce $F$-distribution and another statistic $F$

$F$ statistic is defined by

$$F = \frac{S_1^2}{S_2^2}, \text{ where } S_1^2 > S_2^2$$

and $S_1^2 = \dfrac{1}{n_1 - 1} \sum (x - \bar{x})^2$

$$S_2^2 = \frac{1}{n_2 - 1} \sum (y - \bar{y})^2$$

For the degrees of freedom $\nu_1 = n_1 - 1, \nu_2 = n_2 - 1$, the value of $F$ is always greater than 1. $F$ follows $F$-distribution with $(\nu_1, \nu_2)$ df.

If $S_1^2 < S_2^2$, (i.e., $S_2^2 > S_1^2$), $F$ is given by

$$F = \frac{S_2^2}{S_1^2}$$

So that the value of $F$ is greater than 1. In this case, $F$ follows $F$-distribution with $(\nu_2, \nu_1)$ df.

$F$ test is based on $F$-distribution which is defined as the ratio of two independent chi-square variates. $F$-distribution with $(\nu_1, \nu_2)$ degrees of freedom is given by the probability function

$$dp = \frac{\nu_1^{\frac{1}{2}\nu_1} \nu_2^{\frac{1}{2}\nu_2} F^{\frac{\nu_1}{2} - 2}}{\beta\left(\frac{\nu_1}{2}, \frac{\nu_2}{2}\right) (\nu_1 F + \nu_2)^{\frac{1}{2}(\nu_1 + \nu_2)}} dF, 0 < F < \infty$$

From the definition of $F$, we have

$$\frac{\nu_1}{\nu_2}F = \frac{(n_1-1)S_1^2/\sigma^2}{(n_2-1)S_2^2/\sigma^2}$$

The numerator and the denominator of the second member are independent $\chi^2$ variates with $\nu_1$ and $\nu_2$ degrees of freedom respectively. Hence $\frac{\nu_1}{\nu_2}F$ is a $\beta_2\left(\frac{\nu_1}{2}, \frac{\nu_2}{2}\right)$ variate so that the probability that a random value of $F$ will fall in the interval $dF$ is given by

$$dp = \frac{\nu_1^{\frac{1}{2}\nu_1}\nu_2^{\frac{1}{2}\nu_2}F^{\frac{\nu_1}{2}-2}}{\beta\left(\frac{\nu_1}{2}, \frac{\nu_2}{2}\right)(\nu_1 F+\nu_2)^{\frac{1}{2}(\nu_1+\nu_2)}}dF, 0 < F < \infty$$

$F$-probability curve is J-shaped. The curve is bell shaped $\nu_1 > 2$.

The distribution of $F$ is independent of the population variance and depends on $\nu_1$ and $\nu_2$ only.

**F-test is used to test**

(a) whether two independent samples have been drawn from normal populations with the same variance.

(b) whether the two independent estimates of the population variances are homogeneous or not.

**F-test is based on the following assumptions.**

(i) The distribution in each group should be normally distributed.

(ii) Error should be independent of each observed value.

(iii) Variance within each group should be equal for all groups.

When the sample sizes are large, the assumption of normality is not necessary.

## Solved examples

**Example 1** Two horses A and B were tested according to the time (in seconds) to run a particular track, with the following results.

| | | | | | | | |
|---|---|---|---|---|---|---|---|
| Horse A: | 28 | 30 | 32 | 33 | 33 | 29 | 34 |
| Horse B: | 29 | 30 | 30 | 24 | 27 | 29 | – |

Test whether the two horses have the same running capacity.

**Solution** We have $n_1 = 7, n_2 = 6$

Then $\bar{x} = \frac{\Sigma x_i}{n_1} = \frac{219}{7} = 31.28, \bar{y} = \frac{\Sigma y_i}{n_2} = \frac{169}{6} = 28.2$

Let the time taken by horse A be denoted by $x$ and the time taken by horse B be denoted by $y$.

Degrees of freedom $= n_1 - 1,\ n_2 - 1 = 6,\ 5$

Critical value i.e. level of significance = 5%

| $x$ | $x-\bar{x}$ | $(x-\bar{x})^2$ | $y$ | $y-\bar{y}$ | $(y-\bar{y})^2$ |
|---|---|---|---|---|---|
| 28 | −3.28 | 10.75 | 28 | 0. | 0.64 |
| 30 | −1.28 | 1.64 | 30 | 1.8 | 3.24 |
| 32 | 0.72 | 0.52 | 30 | 1.8 | 3.24 |
| 33 | 1.72 | 2.96 | 24 | −4.2 | 17.64 |
| 33 | 1.72 | 2.96 | 27 | −1.2 | 1.44 |
| 29 | −2.28 | 5.20 | 29 | 0.8 | 0.64 |
| 34 | 2.72 | 7.40 | – | – | – |
| 219 | | 31.43 | 169 | | 26.84 |

$$S_1^2 = \frac{1}{n_1 - 1}\sum (x-\bar{x})^2 = \frac{31.43}{6} = 5.224$$

$$S_2^2 = \frac{1}{n_2 - 1}\sum (y-\bar{y})^2 = \frac{26.84}{5} = 5.368$$

Null hypothesis $H_0$ : The two horses have the same running capacity i.e., $\sigma_1^2 = \sigma_2^2$

Alternative hypothesis $H_1 : \sigma_1^2 \neq \sigma_2^2$

Table value of $F$ for (5, 6) degrees of freedom at 5% level of significance is 4.39.

Test statistic is $F = \dfrac{S_2^2}{S_1^2} = \dfrac{5.368}{5.224} = 1.02$ (since $S_2^2 > S_1^2$)

Since the calculated value of $F$ is less than the table value of $F$ for (5, 6) df at 5% level of significance, we accept $H_0$.

The two horses have the same running capacity.

**Example 2** In a sample of 8 observations, the sum of squares of deviations of the sample values from the sample mean was 94.5 and in another sample of 10 observations it was found to be 101.7. Test whether the difference is significant.

**Solution** It is given that $\sum (x-\bar{x})^2 = 94.5, n_1 = 8$

$$\sum (y-\bar{y})^2 = 101.7,\ n_2 = 10$$

$$S_1^2 = \frac{1}{n_1 - 1}\sum (x - \bar{x})^2 = \frac{94.5}{8 - 1} = 13.5$$

$$S_2^2 = \frac{1}{n_2 - 1}\sum (y - \bar{y})^2 = \frac{1010.7}{10 - 1} = 11.3$$

Null hypothesis $H_0 : \sigma_1^2 = \sigma_2^2$ i.e., the difference is not significant

Alternative hypothesis $H_1 : \sigma_1^2 \neq \sigma_2^2$

Level of significance: 5%

Critical value i.e., table value of $F$ for $(8 - 1,\ 10 - 1) = (7, 9)$ df is 3.29

$$F = \frac{S_1^2}{S_2^2} = \frac{13.5}{11.3} = 1.195$$

Since the calculated value of $F$ is greater than the table value of $F$ at 5% level of significance for (7, 9) df, we accept the null hypothesis and conclude that the difference is not significant.

**Example 3** In a test given to two groups of students drawn from two normal populations, the marks were as follows

| | | | | | | | | | |
|---|---|---|---|---|---|---|---|---|---|
| Sample A: | 18 | 20 | 36 | 50 | 49 | 36 | 34 | 49 | 41 |
| Sample B: | 29 | 28 | 26 | 35 | 30 | 44 | 46 | | |

Examine 5% level whether the two populations have the same variance.

**Solution** Null hypothesis $H_0 :\ \sigma_A^2 = \sigma_B^2$

Alternative hypothesis $H_1 :\ \sigma_A^2 \neq \sigma_B^2$

Level of significance 5%,

We have $\bar{x} = \dfrac{\sum x_i}{n_1} = \dfrac{18 + 20 + 36 + 50 + 49 + 36 + 34 + 49 + 41}{9} = \dfrac{333}{9} = 37,$

$$\bar{y} = \frac{\sum y_i}{n_2} = \frac{29 + 28 + 26 + 35 + 30 + 44 + 46}{7} = \frac{238}{7} = 34$$

Calculation of variances

| Sample A | | | Sample B | | |
|---|---|---|---|---|---|
| $x$ | $x-\overline{x}$ | $(x-\overline{x})^2$ | $y$ | $y-\overline{y}$ | $(y-\overline{y})^2$ |
| 18 | –19 | 361 | | | |
| 20 | –7 | 289 | 29 | 5 | 25 |
| 36 | –1 | 1 | 28 | 6 | 36 |
| 50 | 13 | 169 | 26 | 8 | 64 |
| 49 | 12 | 144 | 35 | 1 | 1 |
| 36 | –1 | 1 | 30 | 4 | 16 |
| 34 | –3 | 9 | 44 | 10 | 100 |
| 49 | 12 | 144 | 46 | 12 | 144 |
| 41 | 4 | 16 | | | |
| | | 1134 | | | 238 |

$$S_A^2 = \frac{1}{n_1 - 1}\sum(x-\overline{x})^2 = \frac{1134}{9-1} = 141.75$$

$$S_B^2 = \frac{1}{n_2 - 1}\sum(y-\overline{y})^2 = \frac{386}{7-1} = 64.33$$

We have $S_A^2 > S_B^2$

$$F = \frac{S_A^2}{S_B^2} = \frac{141.75}{64.33} = 2.203$$

Degrees of freedom $= (\nu_1\ , \nu_2) = (n_1 - 1,\ n_2 - 1)$

$$= (9-1, 7-1)$$

$$= (8, 6)$$

Table value of $F$ at 5% for (8,6) df $= 4.15$

Calculated value of $F$ is less than the table value of $F$ at 5% level (8, 6) df. Hence we accept $H_0$ and conclude that the populations from where the samples have been drawn have the same variance.

**Example 4** Two samples of sizes 9 and 8 gave the sum of the squares of deviations from their respective means equal to 160 and 91 respectively. Can they be regarded as drawn from the same normal population?

**Solution** It is given that $\sum(x-\overline{x})^2 = 160, n_1 = 9$

$$\sum(y-\overline{y})^2 = 91,\ n_2 = 8$$

$$S_1^2 = \frac{1}{n_1 - 1} \sum (x - \bar{x})^2 = \frac{160}{9 - 1} = 20$$

$$S_2^2 = \frac{1}{n_2 - 1} \sum (y - \bar{y})^2 = \frac{91}{8 - 1} = 13$$

Null hypothesis $H_0$ : Both the samples have come from the same normal population

Alternative hypothesis $H_1$ : The samples are not from the same normal population

Since $S_1^2 > S_2^2$, we have

Degrees of freedom $(n_1 - 1, n_2 - 1) = (8, 7)$

Level of significance = 5%

Table value of $F$ at 5% level of significance for (8,7) df = 3.73

Test statistic:

$$F = \frac{S_1^2}{S_2^2} = \frac{20}{13} = 1.5385$$

Calculated value of F is less than the table value of $F$ at 5% level (8, 6) df. Hence we accept null hypothesis.

The two samples can be regarded as drawn from the same normal population.

**Example 5** Two samples are drawn from two normal populations. From the following data, test whether the two samples have the same variances at 5% level.

| Sample I | 60 | 65 | 71 | 74 | 76 | 82 | 85 | 87 | | |
|---|---|---|---|---|---|---|---|---|---|---|
| Sample II | 64 | 66 | 67 | 85 | 78 | 88 | 86 | 85 | 63 | 91 |

**Solution** Null hypothesis $H_0 : \sigma_1^2 = \sigma_2^2$ i.e., the two samples have same variances

Alternative hypothesis $H_1 : \sigma_1^2 \neq \sigma_2^2$ (two-tailed test)

$$\bar{x} = \frac{\sum x_i}{n_1} = \frac{60 + 65 + 71 + 74 + 76 + 82 + 85 + 87}{8} = \frac{600}{8} = 75$$

$$\bar{y} = \frac{\sum y_i}{n_2} = \frac{64 + 66 + 67 + 85 + 78 + 88 + 86 + 85 + 63 + 91}{10} = \frac{770}{10} = 77$$

| Sample A | | | Sample B | | |
|---|---|---|---|---|---|
| $x$ | $x-\bar{x}$ | $(x-\bar{x})^2$ | $y$ | $y-\bar{y}$ | $(y-\bar{y})^2$ |
| | | | 61 | −16 | 256 |
| 60 | −15 | 225 | 66 | −19 | 121 |
| 65 | −10 | 100 | 67 | 10 | 100 |
| 71 | −4 | 16 | 85 | 8 | 64 |
| 74 | −1 | 1 | 78 | 1 | 1 |
| 76 | 1 | 1 | 88 | 11 | 121 |
| 82 | 7 | 49 | 86 | 9 | 81 |
| 85 | 10 | 100 | 85 | 8 | 64 |
| 87 | 12 | 144 | 63 | −14 | 196 |
| | | | 91 | 14 | 196 |
| 600 | | 636 | 770 | | 1200 |

$$S_1^2 = \frac{1}{n_1-1}\sum(x-\bar{x})^2 = \frac{636}{8-1} = 90.857$$

$$S_2^2 = \frac{1}{n_2-1}\sum(y-\bar{y})^2 = \frac{1200}{10-1} = 133.33$$

Since $S_2^2 > S_1^2$

$$F = \frac{S_2^2}{S_1^2} = \frac{133.33}{90.851} = 1.467$$

Calculated value of $F$ for $(n_1 - 1, n_2 - 1) = (9, 7)$ df at 5% level = 3.68

Therefore, the calculated value of $F$ is less than the table value of $F$ at 5% level of significance for (9, 7) df.

We accept the null hypothesis. The samples 1 and 11 have the same variance.

**Example 6** In a laboratory experiment, two samples of blood gave the following results.

| Sample | Size | Sample mean | Sum of squares of deviations from mean |
|---|---|---|---|
| 1 | 10 | 15 | 90 |
| 2 | 12 | 14 | 108 |

Test the equality of sample variances at 5% level of significance.

**Solution** Here we have $n_1 = 10$, $n_2 = 12$

$\bar{x} = 15, \bar{y} = 14$

$\Sigma(x-\bar{x})^2 = 90, \Sigma(y-\bar{y})^2 = 108$

$$S_1^2 = \frac{1}{n_1 - 1}\sum(x-\bar{x})^2 = \frac{90}{10-1} = 10$$

$$S_2^2 = \frac{1}{n_2 - 1}\sum(y-\bar{y})^2 = \frac{108}{12-1} = 9.82$$

Null hypothesis $H_0 : \sigma_1^2 = \sigma_2^2$

Alternative hypothesis $H_1 : \sigma_1^2 \neq \sigma_2^2$

Test statistic: $F = \dfrac{S_1^2}{S_2^2} = \dfrac{10}{9.82} = 1.018$

Table value of $F$ at 5% level of significance for $(10-1, 12-1) = (9, 11)$ df is 2.90.

Therefore, the calculated value of $F$ is less than the table value. Hence we accept null hypothesis. The two programmes have the same variance.

## Exercise 8.2

1. In one sample of 10 observations, the sum of the squares of the deviations of the sample values from sample mean was 120 and in another sample of 12 observations it was $\frac{3}{4}$. Test whether this difference is significant at 5% level of significance.

2. Two samples of sizes 9 and 8 gave the sum of the squares of deviations from their respective means equal to 160 and 91 respectively. Can they be regarded as drawn from the same normal population?

3. In one sample of 10 observations, the sum of squares of the deviations of the sample values from the sample mean was 120 and in another sample of 12 observations it was 314. Test whether the difference is significant at 5% level of significance.

4. Write short notes on $F$-distribution and $F$-test.

5. Mention the applications of $F$-test.

6. Two independent samples are of sizes 8 and 10. The sum of squares of deviations of sample values from their respective sample means were 84.4 and 102.6. Test whether the difference of variances of the populations is significant or not.

7. Two random samples of sizes 8 and 7 had the following values of variances.

   | | | | | | | | | |
   |---|---|---|---|---|---|---|---|---|
   | Sample A : | 9 | 11 | 13 | 11 | 15 | 9 | 12 | 14 |
   | Sample B : | 10 | 12 | 10 | 14 | 9 | 8 | 10 | |

   Do these estimates of population variance differ significantly?

8. Two independent samples of sizes 9 and 7 from a normal population have the following values of variances.

| | | | | | | | | | |
|---|---|---|---|---|---|---|---|---|---|
| Sample I : | 18 | 13 | 12 | 15 | 12 | 14 | 16 | 14 | 15 |
| Sample II: | 16 | 19 | 13 | 16 | 18 | 13 | 15 | | |

Do these estimates of population variance differ significantly?

## Answers 8.2

1. The sample might have come from two populations having the same variances.
2. The two samples can be regarded as drawn from two normal populations with the same variance. $H_0$ accepted.
3. $H_0$ accepted i.e., $\sigma_1^2 = \sigma_2^2$.
6. $\sigma_1^2 = \sigma_2^2$ i.e., accept $H_0$.
7. Not significant.
8. $H_0$ accepted. The difference is not significant.

## CHAPTER NINE

# Curve Fitting

## 9.1 Introduction

In this chapter, we consider the problems of fitting an equation or a curve to data involving paired values. An approximate non-mathematical relationship between the two variables can be established by a diagram called scatter diagram. The extra mathematical relationship between the two variables is given by a simple algebraic expression called curve fitting. Though there are infinite variate of curves in mathematics, the curves used for the purpose of curve fitting are relatively limited in type. The straight line is the simplest and one of the most important curves used.

## 9.2 Straight Line

The equation

$$y = a + bx \tag{1}$$

is an equation of the first degree in $x$ and $y$. It represents a straight line. The difference

$$y_i = a + bx_i \tag{2}$$

is zero if and only if the point $(x_i, y_i)$ lies on the line given by (1).

## 9.3 Fitting a Straight Line

Usually, fitting a straight line means finding the values of the parameters $a$ and $b$ of the straight line given by (1). As well as actually constructing the lines itself, the graphic methods of least squares are two useful methods for fitting a straight line.

### 9.3.1 The graphical method

This method can be used whenever the given formula can be plotted as a straight line either directly or after suitable transformation. The straight line is drawn after a careful visual estimate of its position has been made with the aid of a ruler. The coordinates of any two points on the line not too near together are then measured and substituted in equation (1). The resulting equations in $a$ and $b$ are then solved of their parameters. Graphical method, whatever its theoretical attractions, suffers from the disadvantage that it is difficult to apply in practice except for the straight line. This method will give fairly good results when a finely divided coordinate paper is used, but in general it is not recommended.

An alternative method which is in almost universal use at the present time is known as the method of least squares. We proceed to discuss it at length.

### 9.3.2 The method of least squares

This method of curve fitting was suggested earlier in the 19th century by the French mathematician Adrian Legendre. Let $(x_r, y_r)$ denote a set of observed data points. Our problem is to find the estimating curve of prescribed form which minimises the sum of squared $y$-deviations between observed and estimated values. The method of least squares assumes that the best fitting line is the curve for which the sum of the squares of the vertical distance of the point $(x_r, y_r)$, $r = 1, 2, ..., n$, from the line is minimum. This method is more accurate than the graphical method.

Let $\quad y = a + bx \quad$ [Refer to (1)]

be the equation of the line (Fig. 1). The ordinate of any point $Q_r$ on the line vertically above or below a given point $P_r$ can be found by substituting the abscissa $x_r$ in the right-hand side of (1). The two coordinates of $Q_r$ will be (See Fig. 9.1) $(x_r, a + bx_r)$. The vertical distance $e_r$ from the line of any point $P_r$ with coordinates $(x_r, y_r)$ will therefore be given by the equation.

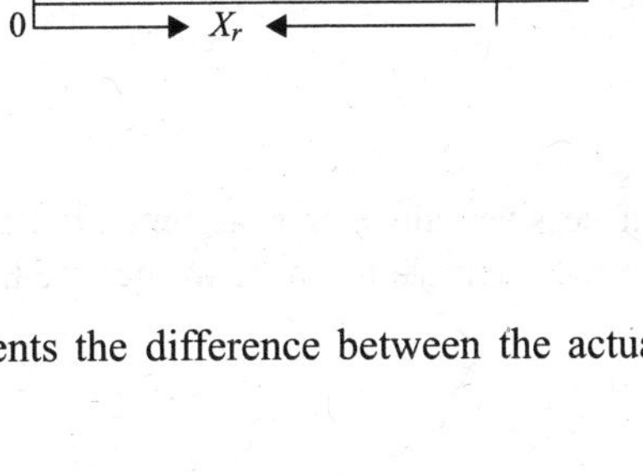


**Fig. 9.1**

$$e_r = y_r - (a + bx_r) \tag{3}$$

We may say that $e_r$ represents the difference between the actual ordinate $y_r$ of a point and its theoretical ordinate $a + bx_r$.

Let $q = \sum_{r=1}^{n} e_r^2 = \sum_{r=1}^{n} (y_r - (a + bx_r))^2$ (4)

The best fitting line is that line for which the sum of the squares $q = \sum_{r=1}^{n} e_r^2$ is a minimum. We find the values of $a$ and $b$ which gives minimum $q$ as follows. Differentiating $q$ partially with respect to $a$ and $b$ and equating this partially to zero, we obtain

$$\frac{\partial q}{\partial a} = \sum_{r=1}^{n}(-2)\left[y_y - (a + bx_r)\right] = 0$$

and $$\frac{\partial q}{\partial b} = \sum_{r=1}^{n}(-2)x_r\left[y_y - (a + bx_r)\right] = 0$$

which yield the so-called system of normal equations

$$\sum_{r=1}^{n} y_r = an + b\sum_{r=1}^{n} x_r \tag{5}$$

and $$\sum_{r=1}^{n} y_r x_r = a\sum_{r=1}^{n} x_r + b\sum_{r=1}^{n} x_r^2 \tag{6}$$

Solving this system of equations, we get

$$a = \frac{\sum_{r=1}^{n} x_r^2 \sum_{r=1}^{n} y_r - \sum_{r=1}^{n} x_r \sum_{r=1}^{n} x_r y_r}{n\sum_{r=1}^{n} x_r^2 - \left(\sum_{r=1}^{n} x_r^2\right)^2} \tag{7}$$

$$b = \frac{n\left(\sum_{r=1}^{n} x_r y_r\right) - \left(\sum_{r=1}^{n} x_r\right)\left(\sum_{r=1}^{n} y_r^2\right)}{n\left(\sum_{r=1}^{n} x_r^2\right) - \left(\sum_{r=1}^{n} x_r\right)^2} \tag{8}$$

where $n$ is the number of points $(x_r, y_r)$.

To simplify the formula, we get

$$t_r = \frac{x_r - \bar{x}}{h}$$

where $h$ is the interval between successive values of $x$, assumed to be equally spaced. Shifting the origin to mean $\bar{x}$ and compressing the intervals to unity, we get the transformed equation referred to the new $(t,\ y)$ axis as

$$\bar{y} = \bar{a} + \bar{b}t \tag{10}$$

where $\overline{a}$ and $\overline{b}$ are the parameters in the new equation. In terms of the new coordinate formulae, (7) and (8) becomes

$$\overline{a} = \frac{\sum_{r=1}^{n} t_r^2 \sum_{r=1}^{n} y_r - \sum_{r=1}^{n} t_r \sum_{r=1}^{n} t_r y_r}{n \sum_{r=1}^{n} t_r^2 - \left(\sum_{r=1}^{n} t_r^2\right)^2} \tag{11}$$

$$\overline{b} = \frac{n\left(\sum_{r=1}^{n} t_r y_r\right) - \left(\sum_{r=1}^{n} t_r\right)\left(\sum_{r=1}^{n} y_r^2\right)}{n\left(\sum_{r=1}^{n} t_r^2\right) - \left(\sum_{r=1}^{n} t_r\right)^2} \tag{12}$$

Transformation (9) replaces the $x_r$s by unit deviations $t_r$ from the mean $\overline{x}$.

Since $\sum_1^n t_r = 0$, the formulas (11) and (12) reduce to

$$\overline{a} = \frac{\sum_{r=1}^{n} t_r^2 \sum_{r=1}^{n} y_r - 0}{n \sum_{r=1}^{n} t_r^2 - 0}$$

and $\overline{b} = \dfrac{n\left(\sum_{r=1}^{n} t_r y_r\right) - 0}{n\left(\sum_{r=1}^{n} t_r^2\right) - 0}$

They will further simplify to the form

$$\overline{a} = \frac{1}{n}\sum_1^n y_r = \overline{y} \tag{13}$$

$$\overline{b} = \frac{\left(\sum_{r=1}^{n} t_r y_r\right)}{\left(\sum_{r=1}^{n} t_r^2\right)} \tag{14}$$

## 9.4 Fitting of a Parabola

When a set of points exhibits a parabolic trend, the fitting of a quadratic function

$$y = a + bx + cx^2$$

to data may be carried out by the method of least squares, which leads to the three equations.

$$na + b\sum_{1}^{n} x_r + c\sum_{1}^{n} x_r^2 = \sum_{1}^{n} y_r$$

$$a\sum_{1}^{n} x_r + b\sum_{1}^{n} x_r^2 + c\sum_{1}^{n} x_r^3 = \sum_{1}^{n} x_r y_r$$

$$a\sum_{1}^{n} x_r^2 + b\sum_{1}^{n} x_r^3 + c\sum_{1}^{n} x_r^4 = \sum_{1}^{n} x_r^2 y_r$$

The above equations can be solved for *a*, *b*, *c*.

## 9.5 Exponential Function $y = ae^{bx}$

Transforming the exponential equation $y = ae^{bx}$ by taking logarithms on both sides, we get

$$\log y = \log a + bx \log e$$

If we replace log *y* by *Y* and the constant log *a* and *b* log e by *d* and *b*′ respectively, we obtain

$$Y' = a' + b'x$$

which defines a straight line.

### Solved examples

**Example 1** Find the least square line $y = a + bx$ for the data points (–1, 10), (0, 9), (1, 7), (2, 5), (3,4), (4,3) , (5,0) and (6,–1).

**Solution** Here

| $x_r$ | $y_r$ | $x_r^2$ | $x_r y_r$ |
|---|---|---|---|
| –1 | 10 | 1 | –10 |
| 0 | 9 | 0 | 0 |
| 1 | 7 | 1 | 7 |
| 2 | 5 | 4 | 10 |
| 3 | 4 | 9 | 12 |
| 4 | 3 | 16 | 12 |
| 5 | 0 | 25 | 0 |
| 6 | –1 | 36 | –6 |
| 20 | 37 | 92 | 25 |

From the table we have,

The normal equations are

$$\sum_{r=1}^{n} y_r = an + b\sum_{r=1}^{n} x_r$$

and $\sum_{r=1}^{n} y_r x_r = a\sum_{r=1}^{n} x_r + b\sum_{r=1}^{n} x_r^2$

Putting the value in normal equations

$$8a + 20b = 37$$

$$20a + 97b = 25$$

And solving this equations, we get

$$a = 8.6428571$$

$$b = -1.6071429$$

Therefore, the least square line is

$$y = 8.6428571 + (-1)\,1.6071429x$$

$$y = -1.6071429x + 8.6428571$$

**Example 2** Find the least square line $y = a + bx$ for the data

| $x_r$ | −2 | −1 | 0 | 1 | 2 |
|---|---|---|---|---|---|
| $y_r$ | 1 | 2 | 3 | 3 | 4 |

**Solution**

| $x_r$ | $y_r$ | $x_r^2$ | $x_r y_r$ |
|---|---|---|---|
| −2 | 1 | 4 | −2 |
| −1 | 2 | 1 | −2 |
| 0 | 3 | 0 | 0 |
| 1 | 3 | 1 | 3 |
| 2 | 4 | 4 | 8 |
| 0 | 12 | 10 | 7 |

In this case $n = 5, \sum x_r = 0, \sum y_r = 13, \sum x_r^2 = 10$, and $\sum x_r y_r = 7$

The normal equations are

$$\sum_{r=1}^{n} y_r = an + b\sum_{r=1}^{n} x_r$$

and $$\sum_{r=1}^{n} y_r x_r = a\sum_{r=1}^{n} x_r + b\sum_{r=1}^{n} x_r^2$$

Putting the values of $n, \sum x_r, \sum y_r, \sum x_r^2$ and $\sum x_r y_r$ in the above equations, we get

$$5a = 13, 10b = 7$$

Solving, $a = 2.6, b = 0.7$

The required line of fit is $y = 2.6 + (0.7)x$.

**Example 3** Find a formula for a line of the form $y = a + bx + cx^2$ which will fit the following data.

| $x$ | 0 | 0.1 | 0.2 | 0.3 | 0.4 | 0.5 | 0.6 | 0.7 | 0.8 | 0.9 |
|---|---|---|---|---|---|---|---|---|---|---|
| $y$ | 3.1950 | 3.2299 | 3.2532 | 3.2611 | 3.2516 | 3.2282 | 3.1807 | 3.1266 | 3.059 | 2.975 |

**Solution** The normal equations are

$$na + b\sum_1^n x_r + c\sum_1^n x_r^2 = \sum_1^n y_r$$

$$a\sum_1^n x_r + b\sum_1^n x_r^2 + c\sum_1^n x_r^3 = \sum_1^n x_r y_r$$

$$a\sum_1^n x_r^2 + b\sum_1^n x_r^3 + c\sum_1^n x_r^4 = \sum_1^n x_r^2 y_r$$

Substituting the values of $\sum x_r, \sum x_r^2, \sum x_r^3, \sum x_r y_r, \sum x_r^2 y_r, \sum y_r$ and $n$, we get

$$10a + 4.5b + 2.85c = 31.7616$$

$$4.5a + 2.58b + 2.025c = 14.0896$$

$$2.85a + 2.025b + 1.5333c = 8.82881$$

And solving these equations we obtain

$$a = 3.1951, b = 0.44254, c = -0.76531$$

The required equation is $y = 3.1951 + 0.44254x - 0.76531x^2$.

**Example 4** Fit a second degree parabola to the following data

| $x$ | 0 | 1 | 2 | 3 | 4 |
|---|---|---|---|---|---|
| $y$ | 0 | 1.8 | 1.3 | 2.5 | 6.3 |

**Solution** The values of $x$ are 0, 1, 2, 3, 4.

The numbers of values are odd. Shifting the origin to the middle value 2 of $x$, and making the substitutions $u = x - 2, v = y$, the curve of fit has $v = a + bu + cu^2$

We obtain

| $x$ | $y$ | $u$ | $v$ | $uv$ | $u^2$ | $u^2v$ | $u^3$ | $u^4$ |
|---|---|---|---|---|---|---|---|---|
| 0 | 1 | −1 | 1 | −2 | 4 | 4 | −8 | 16 |
| 1 | 1.8 | −2 | 1.8 | 1.6 | 1 | 1.8 | −1 | 1 |
| 2 | 1.2 | 0 | 1.3 | 0 | 0 | 0 | 0 | 0 |
| 3 | 2.5 | 1 | 2.5 | 2.5 | 1 | 2.5 | 1 | 1 |
| 4 | 6.3 | 2 | 6.3 | 12.6 | 4 | 25.2 | 8 | 16 |
| – | – | 0 | 12.9 | 11.3 | 10 | 33.5 | 0 | 34 |

The normal equations are

$$\sum v = na + b\sum u + c\sum u^2$$

$$\sum uv = a\sum u + b\sum u^2 + c\sum u^3$$

$$\sum u^2v = a\sum u^2 + b\sum u^3 + c\sum u^4$$

Putting the values of $n, \sum u, \sum v, \sum uv, \ldots$etc.,

$$12.9 = 5a + 10c$$

$$11.3 = 10b$$

$$33.5 = 10a + 34c$$

Solving these normal equations, we get

$$a = 1.48, b = 1.13, c = 0.550$$

The required equation is

$$v = 1.48 + 1.13u + 0.55u^2$$

Substituting $u = x - 2, v = y$ in the above equation, we get

$$y = 1.42 + 1.13\,(x - 2) + 0.55\,(x - 2)^2$$

The required curve of fit is the parabola

$$y = 1.42 - 1.07x + 0.55x^2$$

**Example 5** Obtain normal equations for fitting a curve of the form $y = ax + \frac{b}{x}$ for a point $(x_r, y_r)$, $r = 1, 2, .., x$.

**Solution** The curve $y = ax + \frac{b}{x}$ (1)

passes through the points $(x_r, y_r)$, $r = 1, 2, \ldots., x$

Therefore, we have

$$y_1 = ax_1 + \frac{b}{x_1} \tag{2}$$

$$y_2 = ax_2 + \frac{b}{x_2}$$

.

.

.

$$y_n = ax_n + \frac{b}{x_n}$$

Putting $\frac{1}{x} = z$ in (1), we get

$$y = ax + bz$$

or $S = \Sigma(y_r - ax_r - bz_r)^2$ (4)

Differentiating partially with respect to $a$ and $b$, we get

$$\frac{\partial S}{\partial a} = -2\sum x_r (y_r - ax_r - bz_r) = -2\sum (x_r y_r - ax_r^2 - bx_r z_r)$$

and $\frac{\partial S}{\partial b} = -2\sum z_r (y_r - ax_r - bz_r) = -2\sum (z_r y_r - ax_r z_r - bz_r^2)$

For $S$ to be minimum, we have

$$\frac{\partial S}{\partial a} = 0 \text{ and } \frac{\partial S}{\partial b} = 0$$

$$\frac{\partial S}{\partial a} = 0 \text{ or } -2\sum (x_r y_r - ax_r^2 - bx_r z_r) = 0$$

or $\sum x_r y_r = a\sum x_r^2 + b\sum x_r z_r$ (5)

$$\frac{\partial S}{\partial b} = 0 \text{ or } -2\sum (z_r y_r - ax_r z_r - bz_r^2) = 0$$

or $\sum y_r z_r = a\sum x_r z_r + b\sum z_r^2$ (6)

But $z = \dfrac{1}{x}$

From (5) and (6), the required normal equations can be written as

$$\sum x_r y_r = a\sum x_r^2 + b\sum x_r \frac{1}{x_r}$$

i.e., $nb + a\sum x_r^2 = \sum x_r y_r$ (7)

and $\sum y_r z_r = a\sum x_r z_r + b\sum z_r^2$

i.e., $na + b\sum \dfrac{1}{x_r^2} = \sum \dfrac{y_r}{x_r}$ (8)

**Example 6** Given the following data

| $v$ (ft/min) | 350 | 400 | 500 | 600 |
|---|---|---|---|---|
| $t$ (min) | 61 | 26 | 7 | 2.6 |

If $v$ and $t$ are connected by the relation $v = at^b$, find the best possible values of $a$ and $b$.

**Solution** We have $v = at^b$ (1)

Therefore, $\log_{10} v = \log_{10} a + b\log_{10} t$

Substituting, $x = \log_{10} t, y = \log_{10} v, a' = \log_{10} a$ in (1) we get

$$y = a' + bx$$

The normal equations can be written as

$$4a' + b\sum x_r = \sum y_r \tag{2}$$

$$a'\sum x_r + b\sum x_r^2 = \sum x_r y_r \tag{3}$$

| $v$ | $t$ | $x$ | $y$ | $xy$ | $x^2$ |
|---|---|---|---|---|---|
| 350 | 61 | 1.7858 | 2.5441 | 4.542 | 3.187 |
| 400 | 26 | 1.4150 | 2.6021 | 3.682 | 2.002 |
| 500 | 7 | 0.8451 | 2.6990 | 2.281 | 0.174 |
| 600 | 2.6 | 0.4150 | 2.7782 | 1.153 | 0.172 |
| | | 4.4604 | 10.6234 | 11.658 | 6.075 |

Substituting the above values in (2) and (3), we get

$$4a' + 4.4604b = 10.623 \tag{4}$$

$$4.4604a' + 6.075b = 11.658 \tag{5}$$

Solving (4) and (5), we get $a' = 2.845, b = -0.167$

$a =$ anti $\log a' =$ anti log $2.845 = 699.8$

**Example 7** Using the method of least squares, fit a curve of the form $y = ab^x$ to the following data

| $x$ | 2 | 3 | 4 | 5 | 6 |
|---|---|---|---|---|---|
| $y$ | 8.3 | 15.4 | 33.1 | 65.2 | 127.4 |

**Solution** Here we have $n = 5$ (number of observations)

Consider $y = ab^x$ (1)

Applying logarithms on both sides, we get

$$\log_{10} y = \log_{10} a + x \log_{10} b \quad (2)$$

Taking $\log_{10} y = Y$, equation (2) can be written as

$$Y = a' + b'x \quad (3)$$

where $a' = \log_{10} a, b' = \log_{10} b$

Equation (3) is linear in $x$ and $y$, the normal equations are

$$na' + b'\sum x = \sum y$$

$$a'\sum x + b'\sum x^2 = \sum xy$$

We have

| $x$ | $y$ | $Y = \log_{10} y$ | $XY$ | $x^2$ |
|---|---|---|---|---|
| 2 | 8.3 | 0.9191 | 1.8382 | 4 |
| 3 | 15.4 | 1.1875 | 3.5625 | 9 |
| 4 | 33.1 | 1.5198 | 6.0792 | 16 |
| 5 | 65.2 | 1.1842 | 9.0710 | 25 |
| 6 | 127.4 | 2.1052 | 12.6312 | 36 |
| 20 | | 7.5458 | 33.1821 | 90 |

The normal equations are

$$5a' + 20b' = 7.5458 \quad (4)$$

$$20a' + 90b' = 33.1821 \quad (5)$$

Solving (4) and (5), we get

$$a' = 0.3096, b' = 0.2995$$

Since $a' = 0.3096$ we have $\log_{10} a = 0.3096$ or $a = 2.0399$.

Since $b' = 0.2995$, we get $\log_{10} b = 0.2999$ or $b = 1.9948$.

The least square curve is $y = 2.0399\,(1.9948)^x$.

**Example 8** Fit a least square curve of the form $y = ae^{bx}$ $(a > 0)$ to the data given below.

**Solution** Consider $y = ae^{bx}$

Applying logarithms (with base 10) on both sides, we get

$$\log_{10} y = \log_{10} a + bx \log_{10} e \tag{1}$$

Taking $\log_{10} y = Y$, the equation (1) can be written as

$$Y' = a' + b'x \tag{2}$$

where $a' = \log_{10} a, b' = b \log_{10} e$

Equation (2) is a linear equation in $x$ and $Y$, the normal equations are

$$na' + b'\sum x = \sum Y$$

$$a'\sum x + b'\sum x^2 = \sum xY$$

The normal equation can be written as

$$4a' + 10b' = 2.1684 \tag{3}$$

$$10a' + 30b' = 6.5051 \tag{4}$$

Solving equations (3) and (4), we get

$$a' = 0.0001, b' = 0.2168$$

Now $a' = 0.0001 \Rightarrow \log_{10} a = 0.0001 \Rightarrow a = 1.0002$

$$b' = 0.2168 \Rightarrow b \log_{10} e = 0.2168$$

or $b = \dfrac{0.2168}{\log_{10} e} = \dfrac{0.2169}{0.4343}$

or $b = 0.4992$

Therefore, the required curve is $y = (1.0002)\,e^{0.4992x}$

## Exercise 9.1

1. Find the least square line $y = a + bx$ for the data

| $x_r$ | $-4$ | $-2$ | 0 | 2 | 4 |
|---|---|---|---|---|---|
| $y_r$ | 1.2 | 2.8 | 6.2 | 7.8 | 13.2 |

2. Fit a straight line to the following data regarding $x$ as the independent variable.

| $x$ | 0 | 1 | 2 | 3 | 4 |
|---|---|---|---|---|---|
| $y$ | 1 | 1.8 | 3.3 | 4.5 | 6.3 |

3. Find the least square line $y = a + bx$ for the following data.

| $x$ | $-2$ | 0 | 2 | 4 | 6 |
|---|---|---|---|---|---|
| $y$ | 1 | 3 | 6 | 8 | 13 |

4. Find the least squares parabolic fit $y = a + bx + cx^2$ for the data

| $x$ | $-3$ | $-1$ | 1 | 3 |
|---|---|---|---|---|
| $y$ | 15 | 5 | 1 | 5 |

5. Find the least square parabola for the points (−3, 3), (0, 1), (2, 1), (4, 3).

6. The profit of a certain company in the $x$th year of its life are given by

| $x$ | 1 | 2 | 3 | 4 |
|---|---|---|---|---|
| $y$ | 1250 | 1400 | 1950 | 2300 |

Taking $u = x - 3$ and $v = \dfrac{y - 1650}{50}$, show that the parabola of the second degree of $v$ on $u$ is $v + 0.086 = 5.30u + 0.643u^2$ and deduce that the parabola of the second degree of $y$ on $x$ is $y = 114 + 72x + 32.15x^2$.

7. Find the least square line $y = a_0 + a_1x$ for the data

| $x$ | 1 | 2 | 3 | 4 |
|---|---|---|---|---|
| $y$ | 0 | 1 | 1 | 2 |

8. Find the least square fit straight line (of the form $y = a_0 + a_1x$) for the data of fertiliser application on the yield of a plant.

| Fertiliser applied (gm/week/plant) | 0 | 10 | 20 | 30 | 40 | 50 |
|---|---|---|---|---|---|---|
| Yield (kg) | 0.8 | 0.8 | 1.3 | 1.6 | 1.7 | 1.8 |

9. Find the normal equations that arise from fitting by the least squares method, an equation of the form $y = a_0a_1 \sin x$ to the set of points (0,0), $\left(\frac{\pi}{6}, 1\right)$, $\left(\frac{\pi}{2}\right)$ and $\left(5\frac{\pi}{6}, 2\right)$. Solve for $a_0$ and $a_1$.

10. The following data relates the percentage of allowing elements to the compressive strength of an alloy:

| % alloy element | 10 | 15 | 20 | 25 | 30 |
|---|---|---|---|---|---|
| Compressive strength | 27.066 | 29.57 | 31.166 | 31.366 | 31.0 |

11. Find the least squares parabolic fit for the following data

| $x$ | $-3$ | $-1$ | 1 | 3 |
|---|---|---|---|---|
| $y$ | 15 | 5 | 1 | 5 |

12. Find the polynomial of degree 2 that best fits the following data in the least square sense.

| $x$ | $-2$ | $-1$ | 0 | 1 | 2 |
|---|---|---|---|---|---|
| $y$ | $-3.150$ | $-1.390$ | 0.620 | 2.880 | 5.378 |

13. If $p$ is the pull required to lift a load $W$ by means of a pulley block, find a linear law of the form $p = mW + c$, connecting $p$ and $W$ using the given data.

| $p$ | 12 | 15 | 21 | 25 |
|---|---|---|---|---|
| $W$ | 50 | 70 | 100 | 120 |

14. In some determination of the volume $v$ of carbon dioxide dissolved in a given volume of water at different temperatures, the following pairs of values were obtained:

| $\theta$ | 0 | 5 | 10 | 15 |
|---|---|---|---|---|
| $v$ | 1.80 | 1.45 | 1.80 | 1.00 |

Obtain by the method of least squares, a relation of the form $v = a + b\theta$ which best fits these observations.

15. The observations from an experiment are as given below:

| $y$ | 2 | 10 | 26 | 61 |
|---|---|---|---|---|
| $x$ | 600 | 500 | 400 | 350 |

It is known that a relation of the type $y = ae^{bx}$ exists. Find the best possible values of $a$ and $b$.

16. Fit the least square curve $y = ax^b$ to the following data:

| $x$ | 1 | 2 | 3 | 4 | 5 |
|---|---|---|---|---|---|
| $y$ | 0.5 | 2 | 4.5 | 8 | 12.5 |

17. Using the method of least squares, fit a relation of the form $y = ab^x$ to the following data:

| $x$ | 2 | 3 | 4 | 5 | 6 |
|---|---|---|---|---|---|
| $y$ | 144 | 172.8 | 207.4 | 248.8 | 298.5 |

18. Using the method of least squares, fit a curve of the form $y = ab^x$ to the following data:

| $x$ | 1 | 2 | 3 | 4 |
|---|---|---|---|---|
| $y$ | 4 | 11 | 35 | 100 |

19. Fit a least square curve of the form $y = ax^b$ for the following data, where $a$ and $b$ are constants:

| $x$ | 61 | 26 | 7 | 2.6 |
|---|---|---|---|---|
| $y$ | 350 | 400 | 500 | 600 |

20. The pressure and volume of a gas are related by the equation $Pv^{\lambda} = k$($\lambda$ and k are constants). Fit this equation for the data given below:

| $P$ | 0.5 | 1.0 | 1.5 | 2.0 | 2.5 | 3.0 |
|---|---|---|---|---|---|---|
| $v$ | 1.62 | 1.00 | 0.75 | 0.62 | 0.52 | 0.046 |

## Answers 9.1

1. $y = 6.24 + 1.45x$
2. $y = 0.72 + 1.33x$
3. $y = 3.3 + 1.45x$
4. $y = 2.125 - 1.70x + 0.875x^2$
5. $y = 0.850519 - 0.192495x + 0.178462x^2$
7. $y = -\frac{1}{2} + \frac{3}{5}x$
8. $y = 0.7575 + 0.0229x$
9. $a_0 = 0, a_1 = 3$
10. $y = 18.988 + 1.0125x + 0.02048x^2$
11. $y = \frac{7}{8}x^2 - \frac{17}{10}x + \frac{17}{8}$
12. $y = 0.621 + 2.1326x + 0.1233x^2$
13. $2.2759 + 0.1879W$, 30.4635 kg
14. $v = 1.758 - 0.053\theta$
16. $y = 0.5012x^{1.9977}$
17. $y = 9986x^{1.2}$
18. $y = (1.3268)(2.9845)^x$
19. $y = 701.94x^{0.1708}$
20. $pv^{1.4225} = 0.997$

## CHAPTER TEN

# Correlation and Regression

## 10.1 Introduction

In this chapter, we discuss the concepts of correlation and regression. In many problems of science we may be dealing with two variables which may have a relationship or association between themselves. Correlation measures the strength of the relationship between the two variables. The correlation in a sample is measured by the sample coefficient of correlation which is denoted by $r$ and the coefficient of correlation in a population is measured by the population coefficient correlation which is denoted by $\rho$(rho).

Regression measures the nature and extent of a correlation. Regression is used to predict one variable with respect to another variable.

## 10.2 Correlation

Correlation is a statistical measure for finding out the degree of association between two or more variables.

**Definition 10.1** The relationship between two variables such that a change in one variable results in the positive or negative change in the other is called correlation. A greater change in one variable resulting in a corresponding greater or smaller change in the other variable is also known as correlation.

### 10.2.1 Types of correlation

Correlation is classified into many types

(i) **Positive or negative** When two variables tend to move together in the same direction then the correlation is called positive correlation, otherwise negative correlation.

**Examples** Height and weight, demand and supply

(ii) **Simple and multiple** When we study only two variables say $X$ and $Y$, then the relationship is described as a simple correlation.

If we study more than two variables simultaneously, then the correlation is called multiple correlations.

**Examples** Relationship of price, demand and supply.

(iii) **Partial and total** If the study of the variables excludes some other variables, then it is called partial correlation.

If all the facts are taken into account, then the correlation is called total correlation. The amount of change in the other variables is called non-linear correlation.

## 10.3 Coefficient of Correlation

The extent or degree of relationship between two variables measured in terms of another parameter is called the coefficient of correlation.

It is denoted by $r$.

Let $X$ and $Y$ denote two variables and $r$ denote the coefficient of correlation between $X$ and $Y$. Depending on the value of $r$, we can classify correlation as follows

(a) If $r = 1$, both the variables $X$ and $Y$ increase or decrease in the same proportion. In this case we say that there is perfect positive correlation.

(b) If $r = -1$, both the variables $X$ and $Y$ are inversely proportional to each other. We say that there is perfect negative correlation.

(c) If $r = 0$, we say that there is no relation between $X$ and $Y$.

(d) If $0 < r < 1$, there is moderate (partial) positive correlation between $X$ and $Y$.

(e) If $-1 < r < 0$, there is moderate (partial) negative correlation between $X$ and $Y$.

### 10.3.1 Properties of coefficient of correlation

1. The correlation is a measure of the relationship between two variables.
2. The value of the coefficient of correlation lies between $-1$ and $1$ i.e., $-1 \leq r \leq 1$.
3. If $r = 0$, the variables are said to be independent.
4. If $r = \pm 1$, there is perfect correlation coefficient.

## 10.4 Methods of Finding the Coefficient of Correlation

Coefficient of correlation may be computed by any one or more of the following methods.

(a) Scatter diagram
(b) Correlationogram
(c) Direct method (Karl Pearson's method)
(d) Two way frequency table method
(e) Concurrent deviation method

In this chapter, we discuss some of these methods.

## 10.5 Scatter Diagram

When the pair of values $(x_1, y_1), (x_2, y_2), (x_3, y_3), \ldots, (x_n, y_n)$ are plotted on a graph paper, the points show the pattern in which they lie, such a diagram is called a scatter diagram.

Consider the points$(x_1, y_1), (x_2, y_2), (x_3, y_3), \ldots, (x_n, y_n)$. In a scatter diagram, the variable $x$ is shown along the $x$-axis (horizontal axis) and the variable $y$ is shown along the $y$-axis (vertical axis) and all the pairs of values of $x$ and $y$ are shown by points (or dots) on the graph paper. The scatter diagram of these points reveals the nature and strength of correlation between these variables $x$ and $y$. We observe the following.

If the points plotted lie on a straight line rising from lower left to upper right, then there is a perfect positive correlation between the variables $x$ and $y$ (Fig. 8.1 (a)). If all the points do not lie on a straight line, but their tendency is to rise from the lower left to the upper right then there is a positive correlation between the variable $x$ and $y$ (Fig. 8.1(b)). In these cases, the two variables $x$ and $y$ are in the same direction and the association between the variables is direct.

If the movements of the variables $x$ and $y$ are opposite in direction and the scatter diagram is a straight line, the correlation is said to be negative (Fig. 8.2) i.e., association between the variables is said to be indirect.

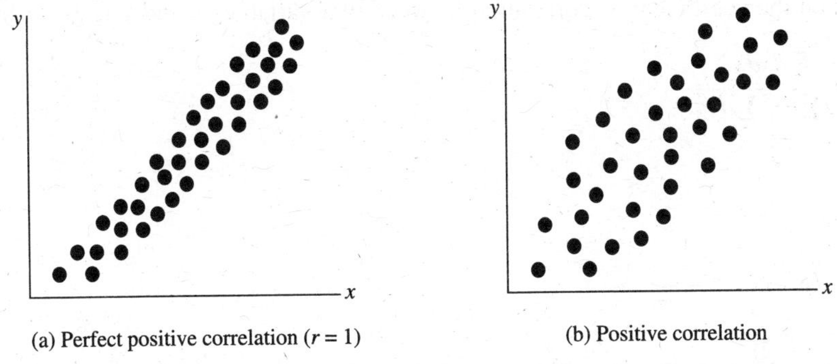


(a) Perfect positive correlation ($r = 1$)    (b) Positive correlation

**Fig. 8.1** *Scatter diagram of a positive correlation*

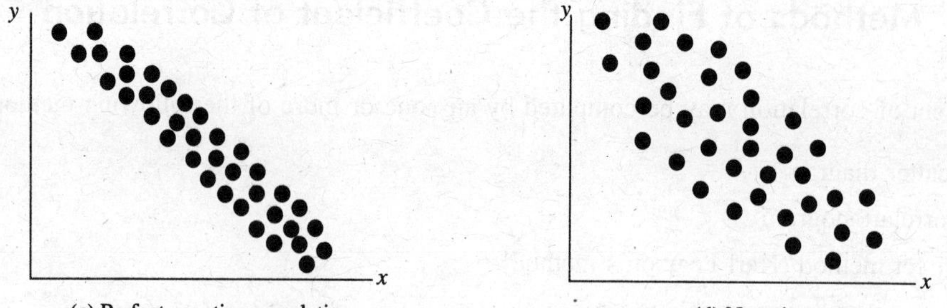


(c) Perfect negative correlation (d) Negative

**Fig. 8.2** *Scatter diagram of a negative correlation*

**Example** Draw a scatter diagram for the following data.

| $x$ | 3 | 5 | 7 | 9 | 11 | 13 | 15 |
|---|---|---|---|---|---|---|---|
| $y$ | 5 | 8 | 11 | 13 | 15 | 17 | 19 |

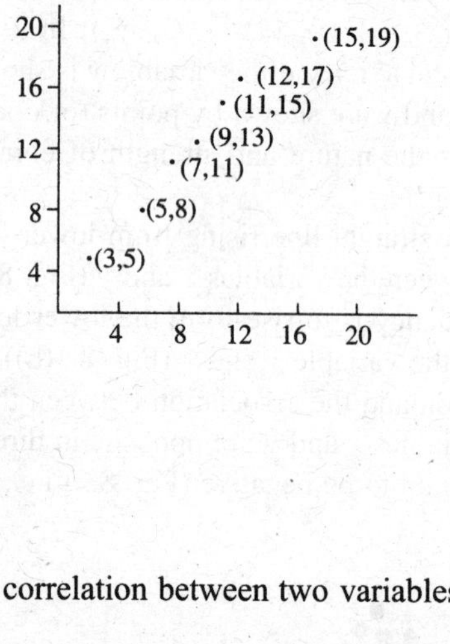


## 10.6 Direct Method

In this method the coefficient of correlation between two variables $x$ and $y$ is given by

$$r = \frac{\sum dxdy}{\sqrt{\sum dx^2 \sum dy^2}}$$

where $dx = x - \bar{x}$

$$dy = y - \bar{y}$$

$$dx^2 = (x - \bar{x})^2$$

$$dy^2 = (y - \bar{y})^2$$

$\bar{x}$ and $\bar{y}$ are the means of the variates $X$ and $Y$ respectively.

The above formula can also be written as

$$r = \frac{\sum dxdy}{n\sigma_x\sigma_y} \text{ or } r = \frac{\text{covariance}\,(XY)}{\sigma_x\sigma_y}$$

where,

$\sigma_x$= Standard deviation of $X$

$\sigma_y$ = Standard deviation of $Y$

$n$ = number of observations in $X$ or $Y$ series

This method is known as Karl Pearson's method. It is a popular mathematical method.

## Solved examples

**Example 1** Find the coefficient of correlation between $X$ and $Y$ for the following data

| $X$ | 1 | 2 | 3 | 4 | 5 | 6 | 7 | 8 | 9 |
|---|---|---|---|---|---|---|---|---|---|
| $Y$ | 10 | 11 | 12 | 14 | 13 | 15 | 16 | 17 | 18 |

**Solution** We have

$$\bar{x} = \frac{\sum x}{n} = \frac{1+2+3+4+5+6+7+8+9}{9} = \frac{45}{9} = 5$$

$$\bar{y} = \frac{\sum y}{n} = \frac{10+11+12+13+14+15+16+17+18}{9} = \frac{126}{9} = 14$$

Computation table

| X | $dx = x-5$ | $dx^2$ | Y | $dy = y-14$ | $dy^2$ | $dxdy$ |
|---|---|---|---|---|---|---|
| 1 | –4 | 16 | 10 | –4 | 16 | 16 |
| 2 | –3 | 9 | 11 | –3 | 9 | 9 |
| 3 | –2 | 4 | 12 | –2 | 4 | 4 |
| 4 | –1 | 1 | 14 | 0 | 0 | 0 |
| 5 | 0 | 0 | 13 | –1 | 1 | 0 |
| 6 | 1 | 1 | 15 | 1 | 1 | 1 |
| 7 | 2 | 4 | 16 | 2 | 4 | 4 |
| 8 | 3 | 9 | 17 | 3 | 9 | 9 |
| 9 | 4 | 16 | 18 | 4 | 16 | 16 |
| $\sum x = 45$ | 0 | $\sum dx^2 = 60$ | $\sum Y = 136$ | 0 | $\sum dy^2 = 60$ | $\sum dxdy = 59$ |

$$\text{Coefficient of correlation } r = \frac{\sum dxdy}{\sqrt{\sum dx^2 \sum dy^2}} = \frac{59}{\sqrt{60 \times 60}} = \frac{59}{60} = 0.9833$$

**Example 2** Calculate the coefficient of correlation between the variables $X$ and $Y$ from the following data

| $X$: | 65 | 66 | 67 | 67 | 68 | 69 | 70 | 72 |
|---|---|---|---|---|---|---|---|---|
| $Y$: | 67 | 68 | 65 | 68 | 72 | 72 | 69 | 71 |

**Solution** We have $\bar{x} = \dfrac{\sum x}{n} = \dfrac{65+66+67+67+68+69+70+72}{8} = \dfrac{544}{8} = 68$

$$\bar{y} = \frac{\sum y}{n} = \frac{67+68+65+68+72+72+69+71}{8} = \frac{552}{8} = 69$$

| X | $dx = x-68$ | $dx^2$ | Y | $dy = y-69$ | $dy^2$ | $dxdy$ |
|---|---|---|---|---|---|---|
| 65 | –3 | 9 | 67 | –2 | 4 | 6 |
| 66 | –2 | 4 | 68 | –1 | 1 | 2 |
| 67 | –1 | 1 | 65 | –4 | 16 | 4 |
| 67 | –1 | 1 | 68 | –1 | 1 | 1 |
| 68 | 0 | 0 | 72 | 3 | 9 | 0 |
| 69 | 1 | 1 | 72 | 3 | 9 | 3 |
| 70 | 2 | 4 | 69 | 0 | 0 | 0 |
| 72 | 4 | 16 | 71 | 4 | 16 | 8 |
| 544 | | 36 | 552 | | 44 | 24 |

$$r = \frac{\sum dxdy}{\sqrt{\sum dx^2 \sum dy^2}} = \frac{24}{\sqrt{36 \times 44}} = 0.603$$

**Example 3** Find the coefficient of correlation between the heights of fathers and sons from the following data.

| Height of father | 64 | 65 | 66 | 67 | 68 | 69 | 70 |
|---|---|---|---|---|---|---|---|
| Height of son | 66 | 67 | 68 | 69 | 70 | 71 | 72 |

**Solution** Let $X$ denote the height of the father and $Y$ denote the height of the son.

We have $\bar{x} = \dfrac{\sum x}{n} = \dfrac{64+65+66+67+68+69+70}{7} = \dfrac{469}{7} = 67$

$$\bar{y} = \frac{\sum y}{n} = \frac{66+67+68+69+70+71+72}{7} = \frac{483}{7} = 69$$

| X | $dx = x - 67$ | $dx^2$ | Y | $dy = y - 69$ | $dy^2$ | $dxdy$ |
|---|---|---|---|---|---|---|
| 64 | –3 | 9 | 66 | –3 | 9 | 9 |
| 65 | –2 | 4 | 67 | –2 | 4 | 4 |
| 66 | –1 | 1 | 68 | –1 | 1 | 1 |
| 67 | 0 | 0 | 69 | 0 | 0 | 0 |
| 68 | 1 | 1 | 70 | 1 | 1 | 1 |
| 69 | 2 | 4 | 71 | 2 | 4 | 4 |
| 70 | 3 | 9 | 72 | 3 | 9 | 9 |
| 469 | | 28 | 483 | | 28 | 28 |

$$r = \frac{\sum dxdy}{\sqrt{\sum dx^2 \sum dy^2}} = \frac{28}{\sqrt{28 \times 28}} = 1$$

**Example 4** Find coefficient of correlation for the following data and comment

| Fertilisers used: | 15 | 18 | 20 | 24 | 30 | 35 | 40 | 50 |
|---|---|---|---|---|---|---|---|---|
| Productivity: | 85 | 93 | 95 | 105 | 120 | 130 | 150 | 160 |

**Solution** Let $X$ denote the fertiliser used and $Y$ denote the productivity.

Computation of coefficient of correlation.

Assume $A = 29$, and $B = 119$

| Fertiliser used | | | Productivity | | | $dxdy$ |
|---|---|---|---|---|---|---|
| $X$ | $dx = x - 29$ | $dx^2$ | $Y$ | $dy = y - 119$ | $dy^2$ | $dxdy$ |
| 15 | –14 | 196 | 85 | –34 | 1156 | 476 |
| 18 | –11 | 121 | 93 | –26 | 676 | 286 |
| 20 | –9 | 81 | 95 | –24 | 576 | 216 |
| 24 | –5 | 25 | 105 | –14 | 196 | 70 |
| 30 | 1 | 1 | 120 | 1 | 1 | 1 |
| 35 | 6 | 36 | 130 | 11 | 121 | 66 |
| 40 | 11 | 121 | 150 | 31 | 961 | 341 |
| 50 | 21 | 441 | 160 | 41 | 1681 | 861 |
| | 0 | 1022 | | –14 | 5368 | 2317 |

$$\text{Coefficient of correlation } r = \frac{\sum dxdy - \frac{\sum dx \times \sum dy}{N}}{\sqrt{\sum dx^2 - \frac{(\sum dx)^2}{N}}\sqrt{\sum dy^2 - \frac{(\sum dy)^2}{N}}}$$

$$= \frac{2317 - \dfrac{0 \times (-14)}{8}}{\sqrt{1022 - \dfrac{0^2}{8}}\sqrt{5368 - \dfrac{(-14)^2}{8}}}$$

$$= \frac{2317}{\sqrt{1022}\sqrt{5368 - \dfrac{196}{8}}}$$

$$= \frac{2317}{\sqrt{1032 \times 5343.51}} = 0.99$$

There is a high degree of correlation between fertiliser used and productivity.

## 10.7 Spearman's Rank Correlation Coefficient

This method is based on rank and is useful in measuring characteristics such as beauty, intelligence, character etc. It is applicable only to individual observations. It is defined as follows

$$r = 1 - \frac{6\sum D^2}{n(n^2 - 1)}$$

where $n$ = number of paired observations

$D^2$ = sum of squares of the difference of two ranks

$r$ = rank correlation coefficient

The rank correlation coefficient is also denoted by $P$.

### Solved examples

**Example 1** A random sample of 5 college students is selected and their grades in mathematics and statistics are found to be

| Mathematics: | 85 | 60 | 73 | 40 | 90 |
|---|---|---|---|---|---|
| Statistics: | 93 | 75 | 65 | 50 | 80 |

Calculate rank correlation coefficient.

**Solution** Calculation of rank correlation coefficient.

| Marks in mathematics $X$ | Ranks of $X$ | Marks in statistics $Y$ | Ranks of $Y$ | Rank difference $D$ | $D^2$ |
|---|---|---|---|---|---|
| 85 | 2 | 93 | 1 | 1 | 1 |
| 60 | 4 | 75 | 3 | 1 | 1 |
| 73 | 3 | 65 | 4 | –1 | –1 |
| 40 | 5 | 50 | 5 | 0 | 0 |
| 90 | 1 | 80 | 2 | –1 | 1 |

$$r = 1 - \frac{6\sum D^2}{n\,(n^2-1)} = 1 - \frac{(6)(4)}{5\,(5^2-1)} = \frac{4}{5} = 0.8$$

**Rank correlation coefficient when the ranks are tied**

When two or more values are equal it is customary that values are given the average of the ranks they would have received. In this case the formula for computing rank correlation coefficient takes the form

$$r = 1 - \frac{6\,(\sum D^2 + T_x + T_y)}{n\,(n^2-1)}$$

$$T_x = \sum \frac{1}{12}\,(t_i^3 - t_i)$$

($i$ denotes the $i$th tie. $t_i$ denotes the number of ranks tied in the $i$th tie in the variable $X$)

$$T_y = \sum \frac{1}{12}\,(t_j^3 - t_j)$$

($j$ denotes the $j$th tie, $t_j$ denotes the number of ranks tied in the $j$th tie in the variable $Y$)

**Example 2** Compute the rank correlation coefficient for the following data

| $x$ | 68 | 64 | 75 | 50 | 64 | 80 | 75 | 40 | 55 | 64 |
|---|---|---|---|---|---|---|---|---|---|---|
| $y$ | 62 | 58 | 68 | 45 | 81 | 60 | 68 | 48 | 50 | 70 |

**Solution**

In the first ranking there are two sets of ties, and in the second ranking, there is one set of ties.

Therefore,

$$T_x = \frac{1}{12}\{(2^3-2)+(3^3-3)\}$$

$$T_y = \frac{1}{12}\{(2^3-2\}$$

| $X$ | Rank of $X$ | $Y$ | Rank of $Y$ | $D = X - Y$ | $D^2$ |
|---|---|---|---|---|---|
| 68 | 4 | 62 | 5 | –1 | 1 |
| 64 | 6 | 58 | 7 | –1 | 1 |
| 75 | 2.5 | 68 | 3.5 | –1 | 1 |
| 50 | 9 | 45 | 10 | –1 | 1 |
| 64 | 6 | 81 | 1 | 5 | 25 |
| 80 | 1 | 60 | 6 | –5 | 25 |
| 75 | 2.5 | 68 | 3.5 | –1 | 1 |
| 40 | 10 | 48 | 9 | 1 | 1 |
| 55 | 8 | 50 | 8 | 0 | 0 |
| 64 | 6 | 70 | 2 | 4 | 16 |

$$r = 1 - \frac{6\left(\Sigma D^2 + \frac{1\,(t^3 - t)}{12} + \ldots\right)}{n\,(n^2 - 1)}$$

$$= 1 - \frac{6\left[72 + \frac{1}{12}(2^3 - 2) + \frac{1}{12}(3^3 - 3) + \frac{1}{12}(2^3 - 2)\right]}{10\,(10^2 - 1)}$$

$$= 1 - \frac{6\,[72 + 0.5 + 2 + 0.5]}{990} = 0.545$$

## Exercise 10.1

1. Find the coefficient of correlation for the data

| Height of father | 66 | 67 | 67 | 68 | 69 | 71 | 73 |
|---|---|---|---|---|---|---|---|
| Height of son | 68 | 64 | 68 | 72 | 70 | 69 | 70 |

2. Calculate the coefficient of correlation from the following data

| $X$ | 12 | 9 | 8 | 10 | 11 | 13 | 7 |
|---|---|---|---|---|---|---|---|
| $Y$ | 14 | 8 | 6 | 9 | 11 | 12 | 3 |

3. Ten students get the following percentage of marks in mathematics and physics. Find the coefficient of rank correlation.

| Mathematics $X$ | 8 | 36 | 98 | 25 | 75 | 82 | 92 | 62 | 35 |
|---|---|---|---|---|---|---|---|---|---|
| Physics $Y$ | 84 | 51 | 91 | 60 | 68 | 62 | 86 | 58 | 49 |

4. Find the coefficient of correlation between $x$ and $y$ and interpret.

| $X$: | 1 | 3 | 4 | 5 | 7 | 8 | 10 |
|---|---|---|---|---|---|---|---|
| $Y$: | 2 | 6 | 8 | 10 | 14 | 16 | 20 |

5. Find the correlation of correlation for the data given below.

| $X$ | 78 | 36 | 98 | 25 | 75 | 82 | 90 | 62 | 35 | 39 |
|---|---|---|---|---|---|---|---|---|---|---|
| $Y$ | 84 | 51 | 91 | 60 | 68 | 62 | 86 | 58 | 53 | 47 |

6. Compute the coefficient of correlation between $A$ and $B$

| $A$ | 5 | 10 | 5 | 11 | 12 | 4 | 3 | 2 | 7 | 1 |
|---|---|---|---|---|---|---|---|---|---|---|
| $B$ | 1 | 6 | 2 | 8 | 5 | 1 | 4 | 6 | 5 | 2 |

7. Find the coefficient of correlation.

| $X$ | 1 | 2 | 3 | 4 | 5 | 6 | 7 | 8 | 9 |
|---|---|---|---|---|---|---|---|---|---|
| $Y$ | 9 | 8 | 10 | 12 | 11 | 13 | 14 | 16 | 15 |

8. Calculate the Pearson's coefficient of correlation from the following data using 44 and 26 respectively as the origins of $X$ and $Y$.

| $X$ | 43 | 44 | 46 | 40 | 44 | 42 | 45 | 42 | 38 | 40 | 42 | 57 |
|---|---|---|---|---|---|---|---|---|---|---|---|---|
| $Y$ | 29 | 31 | 19 | 18 | 19 | 27 | 27 | 29 | 41 | 30 | 26 | 10 |

9. Calculate the Karl Pearson's correlation coefficient between $X$ and $Y$ for the following data.

| $X$ | 43 | 54 | 59 | 68 | 76 |
|---|---|---|---|---|---|
| $Y$ | 105 | 98 | 84 | 63 | 56 |

10. Calculate the coefficient of correlation between $X$ and $Y$.

| $X$ | 92 | 89 | 87 | 86 | 83 | 77 | 71 | 63 | 53 | 50 |
|---|---|---|---|---|---|---|---|---|---|---|
| $Y$ | 86 | 83 | 91 | 77 | 68 | 85 | 52 | 82 | 37 | 57 |

11. Calculate the coefficient of correlation between $X$ and $Y$.

| $X$ | 1 | 2 | 3 | 4 | 5 | 6 | 7 | 8 | 9 |
|---|---|---|---|---|---|---|---|---|---|
| $Y$ | 12 | 11 | 13 | 15 | 14 | 17 | 16 | 19 | 10 |

### Answers 10.1

1. $r = 0.472$
2. $r = 0.95$
3. 0.4551
4. Perfect positive correlation
5. $r = 0.78$
6. $r = +0.58$
7. $r = 0.95$
10. $r = 0.73$
11. $r = 0.933$

## 10.8 Calculation of *r* (Correlation Coefficient) by Karl Pearson's Formula

If $(x_1, y_1), (x_2, y_2), (x_3, y_3), \ldots, (x_n, y_n)$ be $n$ paired observations, then

$$r = \frac{\Sigma[(x_i - \bar{x})(y_i - \bar{y})]}{n\sigma_x\sigma_y}$$

or simply $r = \dfrac{\sum[(x_i - \bar{x})(y_i - \bar{y})]}{\sqrt{\sum(x_i - \bar{x})^2}\sqrt{\sum(y_i - \bar{y})^2}}$

where $\sigma_x$= standard deviation of $x_1, x_2, \ldots, x_n$

$\sigma_y$= standard deviation of $y_1, y_2, \ldots, y_n$

$$\bar{x} = \frac{\sum x_i}{n}, \bar{y} = \frac{\sum y_i}{n}$$

and $\sigma_x = \sqrt{\dfrac{\sum X_i}{n}}$, $\sigma_y = \sqrt{\dfrac{\sum Y_i}{n}}$

If $X_i = x_i - \bar{x}$ and $Y_i = y_i - \bar{y}$, then

$$r = \frac{\sum X_i Y_i}{n\sigma_x\sigma_y}$$

$$r = \frac{\sum x_i y_i - \frac{(\sum x_i)(\sum y_i)}{n}}{\sqrt{\sum x_i^2 - \frac{(\sum x_i)^2}{n}}\sqrt{\sum y_i^2 - \frac{(\sum y_i)^2}{n}}}$$

Karl Pearson's formula is a direct method of computing $r$. It can be proved mathematically that $-1 \leq r \leq 1$. The Karl Pearson's coefficient $r$, is also denoted by $\rho$ (rho) and is also called Karl Pearson's moment correlation coefficient.

## 10.9 Regression

Correlation methods are used to know, how two or more variables are interrelated. Correlation cannot be used to estimate or perfect the most likely values of one variable for specified values of the other variable. The term "regression" was coined by Sir Francis Galton (while studying the linear relation between two variables).

The independent variable in regression analysis is called the "predictor" or "regressor" and the dependent variable is called the regressed variable.

**Definition 10.2** Regression is the measure of the average relationship between two or more variables in terms of the original units of data.

## 10.10 Regression Equation

The functional relationship of a dependent variable with one or more independent variables is called a regression equation. It is also called a prediction equation (or estimating equation).

## 10.11 Curve of Regression

The graph of the regression equation is called the curve of regression. If the curve is a straight line, then it is called the line of regression.

## 10.12 Types of Regression

If there are only two variables under consideration, then the regression is called simple regression.

**Example**

(i) Study of regression between heights and age for a group of persons.

(ii) The study of regression between income and expenditure for a group of persons. In this case the relationship is linear.

If there are more than two variables under consideration then the regression is called multiple regressions.

If there are more than two variables under considerations and only the relation between two variables is established, after excluding the effect of the remaining variables, then the regression is called partial regression.

If the relationship between $x$ and $y$ is non-linear, then the regression is a curvilinear regression. In some cases, polynomials are selected to predict or estimate. This is called polynomial regression.

## 10.13 Regression Equations (Linear Fit)

### 10.13.1 Linear regression equation of $y$ on $x$

In linear regression if we fit a straight line of the form $y = a + bx$ to the given data by the method of least squares, we obtain the regression of $y$ on $x$.

Let $(x_1, y_1), (x_2, y_2), (x_3, y_3), \ldots, (x_n, y_n)$ denote $n$ pairs of observations and let the corresponding straight line to be fitted to these data points be

$$y = a + bx \tag{1}$$

Applying the method of least squares, we get the following normal equations

$$na + b\sum x_i = \sum y_i \tag{2}$$

$$a\sum x_i + b\sum x_i^2 = \sum x_i y_i \tag{3}$$

Dividing equation (2) by $n$, we get

$$a+b\frac{\sum x_i}{n}=\frac{\sum y_i}{n}$$

or $a+b.\bar{x}=\bar{y}$ (4)

Subtracting (4) from (1), we obtain

$$(y_i-\bar{y})=b\,(x_i-\bar{x}) \tag{5}$$

Multiplying (2) by $\sum x_i$ and (3) by $n$ and then subtracting, we get

$$\left(\sum x_i\right)\left(\sum y_i\right)-n\sum x_iy_i=b\left(\sum x_i\right)^2-nb\sum x_i^2$$

or $b\left[n\sum x_i^2-\left(\sum x_i\right)^2\right]=n\sum x_iy_i-\sum x_i\sum y_i$

or $b=\dfrac{n\sum x_iy_i-\sum x_i\sum y_i}{\left[n\sum x_i^2-\left(\sum x_i\right)^2\right]}$

or $b=\dfrac{\dfrac{\sum x_iy_i}{n}-\dfrac{\sum x_i}{n}.\dfrac{\sum y_i}{n}}{\dfrac{\sum x_i^2}{n}-\dfrac{\left(\sum x_i\right)^2}{n}}$ (6)

Replacing $b$ by $b_{yx}$ in (5), we get

$$(y_i-\bar{y})=b_{yx}\,(x_i-\bar{x}) \tag{7}$$

where $b_{yx}=\dfrac{\dfrac{\sum x_iy_i}{n}-\dfrac{\sum x_i}{n}.\dfrac{\sum y_i}{n}}{\dfrac{\sum x_i^2}{n}-\dfrac{\left(\sum x_i\right)^2}{n}}$

Equation (7) is called the regression equation of $y$ on $x$, and is used to estimate the values of $y$ for given values of $x$.

$b_{yx}$ is also given by $b_{yx}=r\dfrac{\sigma_y}{\sigma_x}$

and it is called the regression coefficient of $y$ on $x$.

Regression coefficient is an absolute figure. It explains that the decrease in one variable is associated with the increase in the other variable. The regression coefficients $b_{yx}$ and $b_{xy}$ will have the same sign and are independent of change of origin but not of scale.

### 10.13.2 Regression equation of $x$ on $y$

It is the best equation of the best fitted line of the type

$$x=a'+b'y$$

to the given data.

Applying the principle of least squares, we get the following two normal equations

$$na' + b'\sum y = \sum x \quad \text{(i)}$$

$$a'\sum y + b'\sum y^2 = \sum xy \quad \text{(ii)}$$

Solving (i) and (ii) and proceeding as before, we obtain the regression equation of $x$ on $y$ as follows,

$$(x - \bar{x}) = b_{xy}(y - \bar{y})$$

where $b_{xy} = \dfrac{\dfrac{\sum x_i y_i}{n} - \dfrac{\sum x_i}{n} \cdot \dfrac{\sum y_i}{n}}{\dfrac{\sum y_i^2}{n} - \dfrac{(\sum y_i)^2}{n}}$

$b_{xy}$ is also given by $b_{xy} = r\dfrac{\sigma_x}{\sigma_y}$

and is called the regression coefficient of $x$ on $y$.

Since $b_{yx} = r\dfrac{\sigma_y}{\sigma_x}$ and $b_{xy} = r\dfrac{\sigma_x}{\sigma_y}$

we have $b_{yx} . b_{xy} = r^2$

**Note**

1. If $X_i = x_i - \bar{x}$, $Y_i = y_i - \bar{y}$, then

$$b_{yx} = \frac{\dfrac{\sum X_i Y_i}{n}}{\dfrac{\sum X_i^2}{n}} = \frac{\sum X_i Y_i}{\sum X_i^2} \left(\because \sum X_i = 0, \sum Y_i = 0\right)$$

Similarly $b_{xy} = \dfrac{\sum X_i Y_i}{\sum Y_i^2}$

The two regression lines

$$(x_i - \bar{x}) = b_{xy}(y_i - \bar{y}) \quad \text{(iv)}$$

and

$$(y_i - \bar{y}) = b_{yx}(x_i - \bar{x}) \quad \text{(v)}$$

are identical if

$$b_{yx} \times b_{xy} = 1$$

or $b_{yx} = \dfrac{1}{b_{xy}}$ or $r^2 = 1$

i.e., the lines (iv) and (v) are identical if $r^2 = 1$ i.e., $r = \pm 1$.

2. The two regression lines always intersect at $(\overline{x}, \overline{y})$.

## 10.14 Angle Between two Lines of Regression

Consider the regression lines

$$(y - \overline{y}) = b_{yx}(x - \overline{x}) \tag{1}$$

$$(x - \overline{x}) = b_{xy}(y - \overline{y}) \tag{2}$$

Equation (2) can be written as $(y - \overline{y}) = \dfrac{1}{b_{xy}}(x - \overline{x})$ (3)

Let $\theta$ be the angle between the regression lines, then the slopes of the lines (1) and (3) are

$$m_1 = b_{yx} = r\frac{\sigma_y}{\sigma_x}$$

and $m_2 = \dfrac{1}{b_{xy}} = \dfrac{\sigma_y}{r\sigma_x}$

We have $\tan\theta = \pm\dfrac{m_2 - m_1}{1 + m_2 m_1}$

$$= \pm\frac{\dfrac{\sigma_y}{r\sigma_x} - \dfrac{r\sigma_y}{\sigma_x}}{1 + \dfrac{\sigma_y}{r\sigma_x}\cdot\dfrac{r\sigma_x}{\sigma_y}} = \pm\frac{\dfrac{\sigma_y}{\sigma_x}\left(\dfrac{1}{r} - r\right)}{1 + \dfrac{\sigma_y^2}{\sigma_x^2}}$$

$$= \pm\left(\frac{1 - r^2}{r}\right)\frac{\sigma_y}{\sigma_x}\cdot\frac{\sigma_x^2}{\sigma_x^2 + \sigma_y^2}$$

$$= \pm\left[\left(\frac{1 - r^2}{r}\right)\frac{\sigma_x \sigma_y}{\sigma_x^2 + \sigma_y^2}\right]$$

Since $r^2 \leq 1$ and $\sigma_x, \sigma_y$ are positive, the positive sign gives the acute angle between the lines and the negative sign gives the obtuse angle between the lines.

If $\theta_1$ denotes the acute angle and $\theta_2$ denotes the obtuse angle between the regression lines, then

$$\theta_1 = \tan^{-1}\left[\left(\frac{1-r^2}{r}\right)\frac{\sigma_x\sigma_y}{\sigma_x^2+\sigma_y^2}\right]$$

and $$\theta_2 = \tan^{-1}\left[\left(\frac{r^2-1}{r}\right)\frac{\sigma_x\sigma_y}{\sigma_x^2+\sigma_y^2}\right]$$

If $r = 0$, then $\tan\theta = \infty$ and $\theta = \frac{\pi}{2}$. In this case, $x$ and $y$ are uncorrelated and lines of regression are perpendicular to each other.

If $r = \pm 1$, then $\tan\theta = 0$, and $\theta = 0$. In this case, there is a perfect correlation (positive or negative) between $x$ and $y$. The two lines of regression coincide, but are not parallel since the lines pass through the point $(\bar{x}, \bar{y})$.

**Note** If $r = 0$, then from the equation of lines of regression, we have

$$(y-\bar{y}) = r\frac{\sigma_y}{\sigma_x}(x_i-\bar{x}) = 0$$

$$(y-\bar{y}) = 0 \text{ or } y = \bar{y}$$

and $$(x-\bar{x}) = r\frac{\sigma_x}{\sigma_y}(y-\bar{y}) = 0$$

$$x-\bar{x} = 0 \text{ or } x = \bar{x}$$

When $r = 0$, the equations of the lines of regression are $x = \bar{x}$ and $y = \bar{y}$ which are the equations of the lines parallel to the axes.

If two variables (i.e $x$ and $y$) are independent, their regression lines are straight and at right angles to each other.

## Solved examples

**Example 1** For the following data, find the regression line (by applying the method of least squares).

| $x$ | 5 | 10 | 15 | 20 | 25 |
|---|---|---|---|---|---|
| $y$ | 20 | 40 | 30 | 60 | 50 |

**Solution** We have

| $x$ | $y$ | $x^2$ | $y^2$ | $xy$ |
|---|---|---|---|---|
| 5 | 20 | 25 | 400 | 100 |
| 10 | 30 | 100 | 900 | 400 |
| 15 | 40 | 225 | 1600 | 450 |
| 20 | 60 | 400 | 3600 | 1200 |
| 25 | 50 | 625 | 2500 | 1250 |
| 75 | 200 | 1375 | 9000 | 3400 |

$\therefore \sum x_i = 75, \sum y_i = 200, \sum x_i^2 = 1375, \sum y_i^2 = 9000, \sum x_i y_i = 3400$

**Regression equation of $y$ on $x$**

The normal equations are

$$na + b\sum x_i = \sum y_i$$

$$a\sum x_i + b\sum x_i^2 = \sum x_i y_i$$

i.e., $5a + 75b = 200$ (1)

$$75a + 375b = 3400 \quad (2)$$

Solving these equations, we get

$$a = 16 \text{ and } b = 1.6$$

Thee regression equation of $y$ on $x$ is

$$y = 16 + 1.6x$$

**Regression equation of $x$ on $y$**

The normal equations are

$$na' + b'\sum y_i = \sum x_i$$

$$a'\sum y_i + b'\sum y_i^2 = \sum x_i y_i$$

i.e., $5a + 200b = 75$ (3)

$$200a + 9000b = 3400 \quad (4)$$

Solving equations (3) and (4), we get

$a = -1$ and $b = 0.4$

The regression equation of $x$ on $y$ is

$$x = -1 + 0.4y$$

The regression equations are

$$y = 16 + 1.6x$$

and $x = -1 + 0.4y$

**Example 2** For the following data, find the regression line of $y$ on $x$.

| $x$ | 1 | 2 | 3 | 4 | 5 | 8 | 10 |
|---|---|---|---|---|---|---|---|
| $y$ | 9 | 8 | 10 | 12 | 14 | 16 | 15 |

**Solution** We have

| $x$ | $y$ | $xy$ | $x^2$ |
|---|---|---|---|
| 1 | 9 | 9 | 1 |
| 2 | 8 | 16 | 4 |
| 3 | 10 | 30 | 9 |
| 4 | 12 | 48 | 16 |
| 5 | 14 | 70 | 25 |
| 8 | 16 | 128 | 64 |
| 10 | 15 | 150 | 100 |

Total, $\sum x_i = 33, \sum y_i = 84, \sum x_i^2 = 219, \sum x_i y_i = 451$

$$\bar{x} = \frac{\sum x_i}{n} = \frac{33}{7} = 4.714$$

$$\bar{y} = \frac{\sum y_i}{n} = \frac{84}{7} = 12$$

$$\text{and } b = \frac{n\sum x_i y_i - \sum x_i \sum y_i}{\left[n\sum x_i^2 - (\sum x_i)^2\right]} = \frac{7(451) - (33)(84)}{7(2119) - (33)^2}$$

$$= 0.867$$

The regression equation of $y$ on $x$ is

$$(y_i - \bar{y}) = b_{yx}(x_i - \bar{x})$$

i.e., $y - 12 = 0.867(x - 4.714)$

or $y = 0.867x + 7.9129$

**Example 3** From the following data, fit two regression equations by finding actual means (of $x$ and $y$) i.e., by the actual means method.

| $x$ | 1 | 2 | 3 | 4 | 5 | 6 | 7 |
|---|---|---|---|---|---|---|---|
| $y$ | 2 | 4 | 7 | 6 | 5 | 6 | 5 |

**Solution** We change the origin and find the regression equations as follows.

We have

$$\bar{x} = \frac{\sum x_i}{n} = \frac{1+2+3+4+5+6+7}{7} = \frac{28}{7} = 4$$

$$\bar{y} = \frac{\sum y_i}{n} = \frac{2+4+7+6+5+6+5}{7} = \frac{35}{7} = 5$$

| $x$ | $y$ | $X = x - \bar{x}$ | $Y = y - \bar{y}$ | $X^2$ | $Y^2$ | $XY$ |
|---|---|---|---|---|---|---|
| 1 | 2 | –3 | –3 | 9 | 9 | 9 |
| 2 | 4 | –2 | –1 | 4 | 4 | 2 |
| 3 | 7 | –1 | 2 | 1 | 1 | –2 |
| 4 | 6 | 0 | 1 | 0 | 0 | 0 |
| 5 | 5 | 1 | 0 | 1 | 1 | 0 |
| 6 | 6 | 2 | 1 | 4 | 4 | 2 |
| 7 | 5 | 3 | 0 | 9 | 9 | 0 |
| 28 | 35 | 0 | 0 | 28 | 16 | 11 |

We have

$$\sum x_i = 28, \sum y_i = 35, \sum X_i = 0, \sum Y_i = 0, \sum X_i^2 = 28, \sum Y_i^2 = 16, \sum X_i Y_i = 11$$

$$b_{yx} = \frac{\sum X_i Y_i}{\sum X_i^2} = \frac{11}{28} = 0.3928 = 0.393 \text{ (approximately)}$$

$$b_{xy} = \frac{\sum X_i Y_i}{\sum Y_i^2} = \frac{11}{16} = 0.6875 = 0.688 \text{ (approximately)}$$

The regression equation of $y$ on $x$ is

$$(y - \bar{y}) = b_{yx}(x - \bar{x})$$

$$(y - 5) = 0.393\,(x - 4)$$

or $y = 0.393x + 3.428$

And the regression equation of $x$ on $y$ is

$$(x-\bar{x}) = b_{xy}\,(y-\bar{y})$$

$$(x-4) = 0.688\,(y-5)$$

or $x = 0.688y + 0.56$

The required equations are

$$y = 0.393x + 3.428$$

$$x = 0.688y + 0.56$$

**Example 4** From the following results, obtain the two regression equations and estimate the yield of crops when the rainfall is 29 cms and the rainfall when the yield is 600 kg.

| | $Y$(yield in kgs) | $X$ (rainfall in cms) |
|---|---|---|
| Mean | 508.4 | 26.7 |
| S.D | 36.8 | 4.6 |

Coefficient of correlation between yield and rainfall is 0.52.

**Solution** We have

$$\bar{x} = 26.7, \bar{y} = 508.4$$

$$\sigma_x = 4.6, \sigma_y = 36.8$$

and $r = 0.52$

$$b_{yx} = r\frac{\sigma_y}{\sigma_x} = (0.52)\frac{36.8}{4.6} = 4.16$$

$$b_{xy} = r\frac{\sigma_x}{\sigma_y} = (0.52)\frac{4.6}{36.8} = 0.065$$

Regression equation of $y$ on $x$

$$(y_i-\bar{y}) = b_{yx}\,(x_i-\bar{x})$$

i.e $y = 508.4 = 4.16\,(x-26.7)$

or $y = 397.328 + 4.16x$

When $x = 29$, we have $y = 397.328 + 4.16\,(29) = 517.968$ kg

Regression equation of $x$ on $y$

$$(x_i - \bar{x}) = b_{xy}(y_i - \bar{y})$$

or $x - 26.7 = -0.065\,(y - 508.4)$

or $x = -6.346 + 0.065x$

When $y = 600$ kg, $x = -6.346 + 0.065\,(-600)$

$$= 32.654 \text{ cms}$$

The regression equations are

$$y = 397.328 + 4.16x$$

$$x = -6.346 + 0.065y$$

When the rainfall is 29 cms, the yield of the crop is 517.968 kg, and when the yield is 600 kg, the rainfall is 32.654 cms.

**Example 5** Find the most likely price of a commodity in Bombay corresponding to the price of Rs.70 at Calcutta from the following:

| | Calcutta | Bombay |
|---|---|---|
| Average prize | 65 | 67 |
| Standard deviation | 2.5 | 3.5 |

Correlation coefficient between the prices of commodity in the two cities is 0.8.

**Solution** We have

$$\bar{x} = 65, \bar{y} = 67$$

$$\sigma_x = 2.5, \sigma_y = 3.5$$

and $r = 0.8$

$$(y - \bar{y}) = r\frac{\sigma_y}{\sigma_x}(x - \bar{x}) = 0$$

$$\Rightarrow y - 67 = (0.8).\left(\frac{3.5}{2.5}\right)(x - 65)$$

$$\Rightarrow y = 67 + 1.12x - 72.8$$

$$\Rightarrow y = -5.8 + 1.12x$$

When $x = 70$,

$$\Rightarrow y = -5.8 + 1.12 \times 70 = -5.8 + 78.4$$

$$\Rightarrow y = 72.60$$

The price of the commodity in Bombay corresponding to Rs.70 at Calcutta is 72.60.

**Example 6** The regression equation calculated from a given set of observations

$$x = -0.4y + 6.4$$

and $\quad y = -0.6x + 4.6$

Calculate $\bar{x}, \bar{y}$ and $r_{xy}$

**Solution**

We have

$$x = -0.4y + 6.4 \tag{1}$$

and

$$y = -0.6x + 4.6 \tag{2}$$

From (2), we have

$$y = -0.6(-0.4y + 6.4) + 4.6$$

$$\Rightarrow y = 0.24y - 3.84 + 4.6$$

$$\Rightarrow 0.76y = 0.76$$

$$\Rightarrow y = 1$$

From (1), we have $x = -0.4 \times 1 + 6.4 = 6.0$

But $(\bar{x}, \bar{y})$ is the point of intersection of (1) and (2).

Hence $(\bar{x}, \bar{y}) = (6, 1)$

$$\bar{x} = 6, \bar{y} = 1$$

Clearly, equation (1) is the regression equation $x$ on $y$ and equation (2) is the regression equation $y$ on $x$.

We have, $b_{yx} = -0.6, b_{xy} = -0.4$

and $r^2 = (-0.4)(-0.6) = 0.24$

$$r = r_{xy} = \pm\sqrt{0.24}$$

Since $b_{yx}$ and $b_{xy}$ are both negative, $r = r_{xy}$ is negative.

$$r_{xy} = -\sqrt{0.24}$$

**Example 7** Show that the coefficient of correlation is the geometric mean (G.M.) of the coefficient of regression.

**Solution** The coefficients of regression are $r\frac{\sigma_x}{\sigma_y}$ and $r\frac{\sigma_y}{\sigma_x}$.

Geometric mean of the regression coefficients is

$$\sqrt{r\frac{\sigma_x}{\sigma_y}.r\frac{\sigma_y}{\sigma_x}} = \sqrt{r^2} = r$$

$$= \text{coefficient of correlation}$$

**Example 8** In a partially destroyed laboratory record of an analysis of correlation data, the following results are legible.

Variance of $x = 9$

Regression equation: $8x - 10y + 66 = 0,\ 40x - 18y = 214$

What were (i) the mean values of $x$ and $y$?

(ii) the standard deviation of $y$?

(iii) the coefficient of correlation between $x$ and $y$.

**Solution** Variance of $x = 9$

i.e., $\sigma_x^2 = 9 \Rightarrow \sigma_x = 3$

Solving the regression equations,

$$8x - 10y + 66 = 0 \quad (1)$$

$$40x - 18y = 214 \quad (2)$$

We obtain $x = 13, y = 17$

Since the pair of intersection of the regression lines is $(\bar{x}, \bar{y})$, we have

$$(\bar{x}, \bar{y}) = (x, y) = (13, 17)$$

$\therefore \bar{x} = 13, \bar{y} = 17$

The regression (1) and (2) can be written as

$$y = 0.8x + 6.6 \tag{3}$$

$$x = 0.45y + 5.35 \tag{4}$$

The regression coefficient of $y$ on $x$ is $r\dfrac{\sigma_y}{\sigma_x} = 0.8$ (5)

and the regression coefficient of $x$ on $y$ is $r\dfrac{\sigma_x}{\sigma_y} = 0.45$ (6)

Multiplying (5) and (6), we get

$$r^2 = 0.45 \times 0.8$$

$$\Rightarrow r^2 = 0.36$$

$$\Rightarrow r = 0.6$$

Putting the values of $r$ and $\sigma_x$ in (5), we get the value of $\sigma_y$ as follows

$$r\frac{\sigma_y}{\sigma_x} = 0.8$$

$$\Rightarrow (0.6)\frac{\sigma_y}{3} = 0.8$$

$$\Rightarrow \sigma_y = \frac{0.8}{0.2} = 4$$

**Example 9** If one of the regression coefficients is greater than unity, show that the other regression coefficient is less than unity.

**Solution** Let one of the regression coefficients, say $b_{yx} > 1$

Then $b_{yx} > 1 \Rightarrow \dfrac{1}{b_{yx}} < 1$

Since, $b_{yx}.b_{xy} = r^2 \leq 1$

We have $b_{xy} \leq \dfrac{1}{b_{yx}}$

Hence, $b_{xy} < 1$ $\left(\because \dfrac{1}{b_{yx}} < 1\right)$

**Example 10** Show that the arithmetic mean of the regression coefficients is greater than the correlation coefficient.

**Solution** We have to show that $\frac{b_{xy}+b_{yx}}{2} > r$

Consider $(\sigma_y - \sigma_x)^2 > 0$

Clearly $(\sigma_y - \sigma_x)^2 > 0$

(Since square of two real qualities is always >0)

$$\sigma_x^2 + \sigma_y^2 - 2\sigma_x\sigma_y > 0$$

$$\text{or } \frac{\sigma_x^2}{\sigma_y\sigma_x} + \frac{\sigma_y^2}{\sigma_y\sigma_x} > 2$$

$$\text{or } \frac{\sigma_x}{\sigma_y} + \frac{\sigma_y}{\sigma_x} > 2$$

$$\text{or } r\frac{\sigma_x}{\sigma_y} + r\frac{\sigma_y}{\sigma_x} > 2r$$

$$\text{or } b_{xy} + b_{yx} > 2r$$

$$\text{or } \frac{b_{xy}+b_{yx}}{2} > r$$

Hence proved.

**Example 11** Given that $x = 4y+5$ and $y = kx+4$ are two lines of regression, show that $0 \leq k \leq \frac{1}{4}$. If $k = \frac{1}{8}$, find the means of the variables, and ratio of their variances.

**Solution** $x = 4y+5 \Rightarrow b_{xy} = 4$

$$y = kx+4 \Rightarrow b_{yx} = k$$

$$r^2 = b_{xy}.b_{yx} = 4.k$$

But $-1 \leq r \leq 1$

$$\Rightarrow 0 \leq r^2 \leq 1$$

$$\Rightarrow 0 \leq 4k \leq 1$$

$$\Rightarrow 0 \leq k \leq \frac{1}{4}$$

When $k = \frac{1}{8}$, $r^2 = 4.\frac{1}{8} = \frac{1}{2} \Rightarrow r = 0.7071$

$$y = kx + 4$$

or $$y = \frac{1}{8}x + 4$$

or $$8y = x + 32$$

or $$8y = 4y + 5 + 32$$

or $$4y = 37 \Rightarrow y = 9.25$$

Now $x = 4y + 5 \Rightarrow x = 4\,(9.25) + 5$

or $x = 42$

Therefore, $(\bar{x}, \bar{y}) = (x, y)$ (the point of intersection is $(\bar{x}, \bar{y})$ )

$$= (42, 9.25)$$

i.e., $\bar{x} = 42, \bar{y} = 9.25$

Hence, $$\frac{b_{xy}}{b_{yx}} = \frac{r\dfrac{\sigma_x}{\sigma_y}}{r\dfrac{\sigma_y}{\sigma_x}} = \frac{\sigma_x.\sigma_x}{\sigma_y.\sigma_y} = \frac{4}{\left(\dfrac{1}{8}\right)} = 32$$

i.e., $$\frac{(\sigma_x)^2}{(\sigma_y)^2} = 32$$

The ratio of the variances is 32:1

## 10.15 Multi Linear Regression

In some cases, the value of a variate may not depend only on a single variable. It may happen that there are several variables, which when taken jointly, will serve as a satisfactory basis for estimating the desired variable. If $x_1, x_2, \ldots, x_k$ represent the independent variables, $y$ is the variable which is to be predicted, and represents the regression equation.

$$y' = a_0 + a_1x_1 + a_2x_2 + \ldots. + a_kx_k$$

The unknown coefficients $a_0, a_1, a_2, \ldots, a_k$ will be estimated by the method of least squares. To obtain the values of the variables, we have $n$ set of values of $(k+1)$ variables. Geometrically, the problem is one of finding the equation of the plane which best fits in the sense of least squares of $n$ points in $(K+1)$ dimension. The normal equations are

$$na_0 + a_1\sum x_1 + a_2\sum x_2 + .... + a_k\sum x_k = \sum y$$

$$a_0\sum x_1 + a_1\sum x_1^2 + a_2\sum x_1x_2 + ... + a_k\sum x_1x_k = \sum x_1y$$

$$a_0\sum x_k + a_1\sum x_1x_k + a_2\sum x_kx_2 + ... + a_k\sum x_k^2 = \sum x_ky$$

If there are two independent variables say $x_1$ and $x_2$, the normal equations are

$$na_0 + a_1\sum x_1 + a_2\sum x_2 = \sum y$$

$$a_0\sum x_1 + a_1\sum x_1^2 + a_2\sum x_1x_2 = \sum x_1y$$

$$a_0\sum x_2 + a_1\sum x_1x_2 + a_2\sum x_2^2 = \sum x_2y$$

and the regression equation is

$$y = a_0 + a_1x_1 + a_2x_2$$

## 10.16 Uses of Regression Analysis

There are many uses of regression analysis. In many situations, the dependent variable $y$ is such that it cannot be measured directly. In such cases, some auxiliary variables are taken as independent variables in a regression to estimate the value of $y$. Regression equation is often used as a prediction equation. Regression analysis is used in predicting yield of a crop, for different doses of a fertiliser, and in predicting future demand of food. Regression analysis is also used to estimate the height of at person at a given age, by finding the regression of height on age.

### Exercise 10.2

1. Heights of fathers and sons are given below in inches:

| Height of father | 65 | 66 | 67 | 67 | 68 | 69 | 71 | 73 |
|---|---|---|---|---|---|---|---|---|
| Height of son | 67 | 68 | 64 | 68 | 72 | 70 | 69 | 70 |

Form the lines of regression and calculate the average height of the son when the height of the father is 67.5 inches.

2. For the following data, determine the regression lines:

| $x$ | 6 | 2 | 10 | 4 | 8 |
|---|---|---|---|---|---|
| $y$ | 9 | 11 | 5 | 8 | 7 |

3. Find the regression equations for the following data:

| Age of husband ($x$) | 36 | 23 | 27 | 28 | 28 | 29 | 30 | 31 | 33 | 35 |
|---|---|---|---|---|---|---|---|---|---|---|
| Age of wife ($y$) | 29 | 18 | 20 | 22 | 27 | 21 | 29 | 27 | 29 | 28 |

4. By the method of least squares, find the regression of $y$ and $x$. Find the value of $y$ when $x = 4$. Also find the regression equation of $x$ on $y$ and find the value of $x$ when $y = 24$. Use the table given below:

| $x$ | 1 | 3 | 5 | 7 | 9 |
|---|---|---|---|---|---|
| $y$ | 15 | 18 | 21 | 23 | 22 |

5. Using the method of least squares, find the two regression equation for the data given below:

| $x$ | 5 | 10 | 15 | 20 | 25 |
|---|---|---|---|---|---|
| $y$ | 20 | 40 | 30 | 60 | 50 |

6. Define regression and find the regression equation of $y$ on $x$, given

| $x$ | 2 | 6 | 4 | 3 | 2 | 2 | 8 | 4 |
|---|---|---|---|---|---|---|---|---|
| $y$ | 7 | 2 | 1 | 1 | 2 | 3 | 2 | 6 |

7. From the following data, obtain the two regression equations.

| Sales | 91 | 97 | 108 | 121 | 67 | 124 | 51 |
|---|---|---|---|---|---|---|---|
| Purchase | 71 | 75 | 69 | 97 | 70 | 91 | 39 |

8. Find the equation of regression lines for the following pairs for the variables $x$ and $y$. (1, 2), (2, 5), (3, 3),(4, 8), (5, 7).

9. From the following data, find the yield of wheat in kg per unit area when the rainfall is 9 inches.

| | Means | S.D. |
|---|---|---|
| Yield of wheat (in kg) | 10 | 8 |
| Annual rainfall (in inches) | 8 | 2 |

10. Show that the regression coefficients are independent of the change of origin but not the change of scale.

11. Given $\sum x_i = 60, \sum y_i = 40, \sum x_i^2 = 4160, \sum y_i^2 = 1, \sum x_i y_i = 1150, x = 10$, find the regression equation of $x$ or $y$. Also find $r$.

12. Using the data given below, find the demand when the price of the quantity is Rs.12.50.

| | Price | Demand |
|---|---|---|
| Means | 10 | 35 |
| Standard deviation | 2 | 5 |

Coefficient of correlation $(r) = 0.8$

13. Find the mean of $x_i$ and $y_i$. Also find the coefficient of correlation given

$$2y - x - 50 = 0$$
$$2y - 2x - 10 = 0$$

14. From the following data, obtain the two regression equations and calculate the correlation coefficient.

| $X$ | 1 | 2 | 3 | 4 | 5 | 6 | 7 | 8 | 9 |
|---|---|---|---|---|---|---|---|---|---|
| $Y$ | 9 | 8 | 10 | 12 | 11 | 13 | 14 | 16 | 15 |

15. From the following regression equations: $8X - 10Y = -66, 40X - 18Y = 214$

    Find (a) average values of $X$ and $Y$.

    (b) correlation coefficient between the two variables $X$ and $Y$.

    (c ) standard deviation of $Y$.

16. Find the most likely price in Bombay ($X$), corresponding to the price of Rs.70 at Calcutta ($Y$) from the following data:

| | Bombay | Calcutta |
|---|---|---|
| Average price | 67 | 65 |
| Standard deviation | 3.5 | 2.5 |

17. The following data is based on the marks given for 450 students in statistics and economics at a certain examination:

| | Statistics | Economics |
|---|---|---|
| Means marks | 40 | 48 |
| Standard deviation | 12 | 16 |

Sum of the products of deviation of marks from their respective mean is 42.075. Give the equations to the two lines of regression. Estimate the average marks of regression.

18. From the following information, calculate the line of regression of $y$ on $x$.

| | $x$ | $y$ |
|---|---|---|
| Means | 40 | 60 |
| Standard deviation | 10 | 15 |
| Correlation coefficient | 0.7 | |

19. The correlation coefficient between two variables $x$ and $y$ is $r = 0.6$. If $\sigma_x = 1.5, \sigma_y = 2.0, \bar{x} = 10, \bar{y} = 20$, find the regression lines of (a) $y$ on $x$, (b) $x$ on $y$.

## Answers 10.2

1. When the father's height is 67.5 inches, the son's height is 68.19 inches.

2. $y = 11.9 - 0.65x$, $x = 16.4 - 1.3y$

3. $y = -1.739 + 0.8913x$, $x = 11.25 + 0.75y$

4. $y = 15.05 + 0.95x$, the value of $y$ when $x = 4$ is 18.85
   $x = -12.58 + 0.88y$, the value of $x$ when $y = 24$ is 8.73

5. $y = 16 + 1.6x$, $x = -1 + 0.4y$ 6. $y = 4.16 - 0.3x$

7. $y = 15.998 + 0.607x$, $x = 0.081 + 1.286y$

8. $y = 1.1 + 1.3x, x = 0.5 + 0.5y$

9. 12 kg

11. $x = 3.68 + 0.58y, r = 0.37$

12. $y = 15 + 2x$, demand = 40000 units

13. $\bar{x} = 130, \bar{y} = 90, r = 0.866$

14. $x = -6.4 + 0.95y$, $y = 7.25 + 0.95x$

15. (a) 13, 17 (b) $b_{yx} = 0.8$, $b_{xy} = 0.45$ (c) $\sigma_y = 4$

16. 72.6

17. $x = 22.24 + 0.37y$, $y = 22 + 0.65x, y = 54.5$

19. $y = 1.05x + 18$

20. $x = 0.45y + 1, y = 0.8x + 12$

CHAPTER ELEVEN

# Queueing theory

## 11.1 Introduction

In everyday life, there are many occasions like the theater or railway station, where we need to take tickets If many people arrive frequently, then they will have to wait for their tickets. Such problems arise at many places. Under such circumstances, the only alternative is to form a waiting line in order to get the service more effectively. The persons waiting to be served are called customers. The waiting line is called a queue. The person issuing the ticket is called a server. The server can only serve a single customer at a time, during which time the server is said to be busy. Otherwise, the server is said to be idle. Wherever the demand for service exceeds the capacity to render service, queues naturally form.

Queueing theory describes the behaviour of waiting lines. The essential components of queueing systems are customers (demands) who need to be served by a server (channel). If customers arrive too quickly, the server cannot handle them and the queue builds, in which case, the customers must wait. The arrival rate of customers is denoted by $\lambda$. If $\lambda$ is relatively small, then the queue will not grow over the long run, creating a stable system. If $\lambda$ is too large, the queue will grow without bounds and the system will go unstable.

Many queueing systems involve person-to-person service at fixed locations, such as bank teller services, barber shops, cinema theaters etc. Queueing theory was first applied to telephone engineering. It is now applicable to transportation service systems, commercial organisations, business–industrial service systems, social service systems and health care systems. A general queueing system can be represented as shown in Fig. 11.1

In a simple queueing model shown in Fig. 11.2, customers arrive from time to time and join a queue or waiting line, wait for service, and finally leave the system after being served.

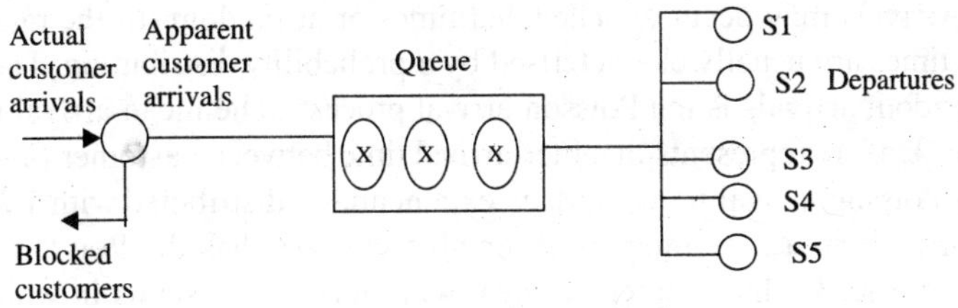


**Fig 11.1** *General queueing system*

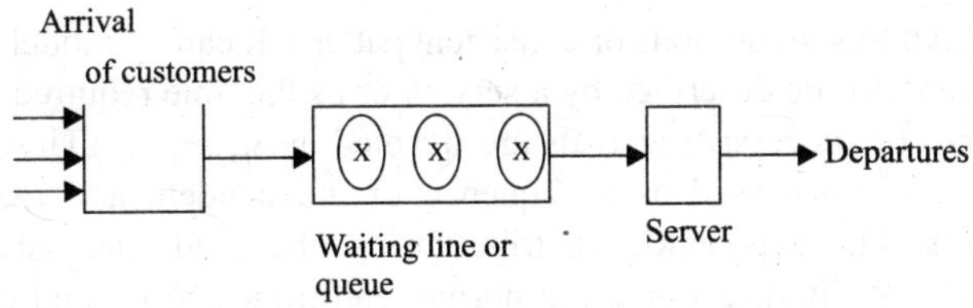


**Fig 11.2** *Queueing model*

Queueing system models solved mathematically provide the analyst with a powerful tool for designing and evaluating the performance of queueing systems. The members of system performance include server initialisation, length of waiting lines and delays of customers. Queueing theory is used to predict the members of system performance as a function of input parameters. The input parameters include line arrival rate of customers, line service demand of customers, the rate at which a server works and the member and arrangement of servers. Some of the input parameters are under the direct control of the management in the performance measures are under the indirect control of the management provide the relationship between the performance measures and the input parameters in adequately understood for the given system. We now discuss general characteristics of queues, the membership and the mathematical solution of some basic queueing models.

## 11.2 Queueing System

A queueing system can be completely described by

(i.) The arrival pattern of customers (or input)

(ii.) The service pattern (or the service mechanism) of servers

(iii.) The queue discipline

(iv.) Customer's behaviour

**(i) The arrival pattern**

The arrival pattern describes the way in which the customers arrive and join the system. The arrival process is usually characterised in terms of inter-arrival times of successive customers. Arrivals may occur at scheduled times or at random. In the random arrivals, the inter-arrival times are usually characterised by a probability distribution. The most important model for random arrivals is the Poisson arrival process. The mean arrival rate of customers is denoted by $\lambda$. If $A_n$ represents the inter-arrival time between customer $(n-1)$ and customer $n$, then for a Poisson arrival process, $A_n$ is exponentially distributed with $1/\lambda$ time units. The number of arrivals in a time interval of length $t$ say $N(t)$ has the Poisson distribution with mean $\lambda t$ customers. In this chapter, we deal with queueing systems in which the customers arrive in Poisson fashion.

**(ii) Service pattern (or service mechanism)**

Service pattern exhibits no obvious or consistent pattern. It can be modelled by using Poisson distribution. It can also be described by a service or as the time required to serve a customer. The service times of successive arrivals are denoted by $s_1, s_2, \ldots$. They may be of random duration and are characterised as a sequence of independent and identically distributed random variables. The exponential distribution can be used successively as a model of service times. The Weibull, gamma, log-normal and truncated normal distribution can also be used to model service times. In this chapter, we discuss queueing systems in which the service pattern follows a Poisson distribution with service rate $\mu$ or where the inter-service time has an exponential distribution with mean $\frac{1}{\mu}$.

**(iii) Queue discipline**

Queue refers to the logical ordering of customers in the queue and determines the manner in which a customer will be selected for service when a server becomes free. The simplest discipline is the first come, first served (FCFS) (or first-in-first out—FIFO) discipline according to which the customers are served in the order of their arrival.

Some of the other common queue disciplines are:

Last-in first out (LIFO)

Service-in-random order (SIRO)

Short processing time (SPT)

Service according to priority (PR)

**(iv) Customer's behaviour**

Various customers who are in a queue behave differently. The customers generally behave in four different ways.

**(a) Balking**

Some customers show reluctance for waiting in the queue. They do not join the queue. In some cases, a customer may decide to wait no matter how long the queue becomes. He may decide not to enter the waiting queue. The term balking is used when a customer refuses to join a waiting line (or queue).

**(b) Reneging**

Some customers may loose patience and leave the queue without getting service.

The term "reneging" is used to describe these customers that leave before their service is due to begin. When this occurs the rules for reneging must be specified.

**(c) Priorities**

Some members of a queue (waiting time) may have priority, meaning that they have a right to be served ahead of all other members of lower priority. There may be several members in the same priority class. A queueing discipline must be specified for the members within a class. The priority may be a pre-assigned attribute of the customer (entity). For example, women and senior citizens first or it may depend on the queue.

**(d) Jockeying**

A customer who has already joined a queue may leave it and join another queue. This happens if there are two or more parallel queues. Customers switching over from one queue to another queue due to personal reasons is known as jockeying.

## 11.3 Transient and Steady States

A queueing system is said to be in transient state when its opening characteristics are dependent on time.

A (queueing) system is said to be in steady state when the behaviour of the system is independent of time. Let $P_n(t)$ denote line probability that there are $n$ members within the system at time $t$. Then in steady state, we have

$$\operatorname*{lt}_{t\to\infty} P_n'(t) = 0 \qquad \text{(i.e., } \operatorname*{lt}_{t\to 0} P_n(t) = P_n \text{ i.e., independent of time } t\text{)}$$

## 11.4 Kendal's Notation

Traditionally, queueing model may be completely specified in the following symbolic form.

$(A|B|C)$: $(d|e)$

where $A =$ probability law of arrival (or arrival discipline)

$B =$ probability law according to which the customers are served (or service discipline)

$C =$ number of servers

$d =$ capacity of the system (maximum number in the system)

$e =$ queue discipline

If $d$ is omitted, "$d = \infty$" is implied. In this chapter, we treat only Markovian ($M$) arrival and service disciplines.

$M|M|1$ means the customers arrive in a Markovian manner (Poisson distributed with exponential inter-arrival times) manner and serviced in Markovian manner, and there is one server. Graphical representation of $M|M|1$ queue is shown in Fig. 11.3.

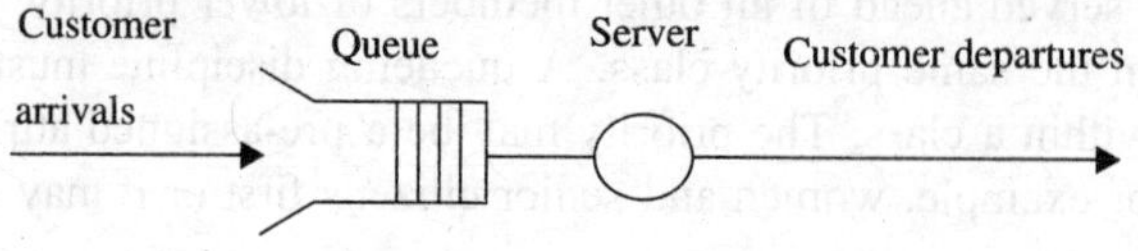


**Fig. 11.3** *A graphical representation or M|M|1 queue*

$M|M|2$ means that customers arrive in Markovian manner, are serviced in Markovian manner and there are two servers. A graphical representation of $M|M|2$ queue is shown in Fig. 11.4.

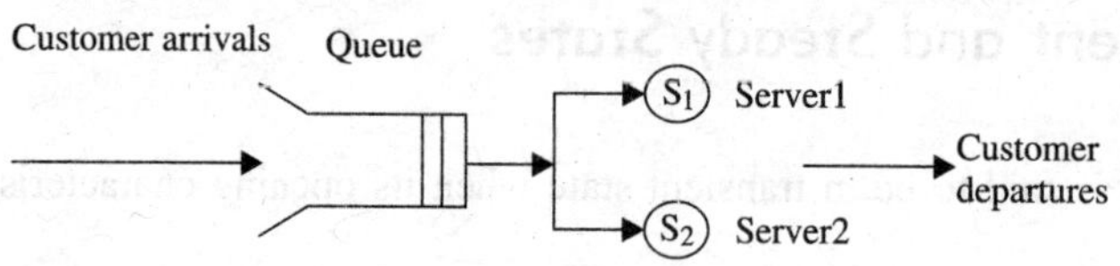


**Fig. 11.4** *Graphical representation of M|M|2 queue*

## 11.5 Classification of Queueing Models

Queueing models can be classified as follows

**(a) Probabilistic model**

When both arrival and service rates are unknown and assumed to be random variables, the queueing model is known as a probabilistic model. In this case, the probability distribution is obtained from past experience.

**(b) Deterministic model**

In this model, both arrival and service rates are exactly known (or fixed).

**(c) Mixed model**

When either arrival rate or service rate is exactly known and the other is not known, the queueing model is known as a mixed model.

In this chapter, we introduce only probabilistic models.

## 11.6 Notations

We use the following symbols and terminologies throughout the chapter.

$n =$ number of customers (units) in the system

Queue size = members of customers in line system (or line length)

$S_n =$ the state in which there are $n$ customers in the system

$P_n =$ steady state probability of having $n$ customers in the system

$P_n(t) =$ transient probability that exactly $n$ customers are in the system at time $t$

$\lambda_n =$ mean arrival rate when there are $n$ customers in the system

$\mu_n =$ mean service rate when there are $n$ customers in the system

$\lambda =$ mean arrival rate when $\lambda_n$ is constant for all $n$

$\mu =$ mean service rate when $\lambda_n$ is constant for all $n \geq 1$

$\rho = \dfrac{\lambda}{\mu} =$ traffic intensity (utilisation factor)

$L_s =$ average (expected) number of customers in the system (expedited queue size)

$L_q =$ average number of customers in the queue (average length of queue)

$L_n =$ average number of customers in non-empty queues.

$W_s =$ average (expected) waiting time of a customer in the system

$W_q =$ average (expected) waiting time of a customer in the queue

$W_w =$ average (expected) waiting time of a customer in the queue, if the customer has to wait.

Arrival of a customer is taken as a "birth" in the system. When an arrival occurs in the system which is in the state, $s_n$(consisting of $n$ customers), it is changed to the state $s_{n+1}$. On the other hand, a departure from the system is considered as a "death". If a system is in state $s_n$, a departure from state $s_n$ changes $s_n$ to $s_{n-1}$. This type of process is known as the birth–death process.

## 11.7 Arrival Distribution Theorem

**Theorem 11.1** If the arrivals are completely random, then the probability distribution of a number of arrivals in a fixed interval follows a Poisson distribution.

*Proof* To prove the theorem, we make the following assumptions.

(i) These are $n$ units in the system at time $t$, and the probability that exactly one arrival (birth) will occur during the small time interval $\Delta t$ (i.e., interval $t$ to $t + \Delta t$) is given by $\lambda\Delta t$.

(ii) $\Delta t$ is considered so small that the probability of more than one arrival in time $\Delta t$ is almost zero.

(iii) The process has independent increments. We wish to determine $P_n$ i.e., the probability of $n$ arrivals in a time interval of length $t$.

$P_n(t+\Delta t)$ is the probability of $n$ customers at time $t+\Delta t$.

**Case 1** When $n > 0$

In this case, there may be two mutually exclusive ways of having $n$ units (customers) at time $t+\Delta t$.

(a) There are $n$ units (customers) in the system at time $t$ and no arrival takes place during time interval $\Delta t$.

(b) There are $(n-1)$ units in the system at time $t$ and one arrival takes place during the time interval $\Delta t$.

Adding these two probabilities we get

$$P_n(t+\Delta t) = P_n(t)(1-\lambda\Delta t) + P_{n-1}(t)\lambda\Delta t$$

i.e, $P_n(t+\Delta t) = (1-\lambda\Delta t)P_n(t) + \lambda\Delta t P_{n-1}(t)$ (1)

**Case 2** When $n \geq 0$

We have $P_0(t+\Delta t) = P_0(t)(1-\lambda\Delta t)$ (2)

Re-arranging (1) and (2) and dividing by $\Delta t$, we get

$$\frac{P_n(t+\Delta t) - P_n(t)}{\Delta t} = -\lambda P_n(t) + \lambda P_{n-1}(t)$$

and

$$\frac{P_0(t+\Delta t) - P_0(t)}{\Delta t} = -\lambda P_0(t)$$

Applying limits as $\Delta t \to 0$, we obtain

$$P_n'(t) = -\lambda P_n(t) + \lambda P_{n-1}(t) \quad (n > 0) \tag{3}$$

$$P_0'(t) = -\lambda P_0(t) \quad \text{for } n = 0 \tag{4}$$

Equation (4) can be written as

$$\frac{P_0'(t)}{P_0(t)} = -\lambda$$

$$\text{or } \frac{d}{dt}(\log[P_0(t)]) = -\lambda$$

Integrating with respect to $t$, we obtain

$$\log P_0(t) = -\lambda t + A \tag{5}$$

The constant $A$ can be determined by using the boundary conditions

$$P_n(0) = \begin{cases} 1, & n = 0 \\ 0, & n > 0 \end{cases}$$

Since $P_0(0) = 1$ we get, $A = 0$, from equation (5).

Therefore, equation (5), can be written as

$$\log P_0(t) = -\lambda t$$

$$\text{or } P_0(t) = e^{-\lambda t}$$

Putting $n = 1$ in (3), we get

$$P_1'(t) = -\lambda P_1(t) + \lambda P_0(t)$$

$$\text{or } P_1'(t) = -\lambda P_1(t) + \lambda e^{-\lambda t}$$

$$\text{or } P'(t) + \lambda P_1(t) = \lambda e^{-\lambda t} \tag{6}$$

Equation (6) is a linear differential equation of order one. The integrating factor is

$$e^{\int \lambda dt} = e^{\lambda t}$$

Hence, the solution of (6) is

$$e^{\lambda t} P_1(t) = \int \lambda e^{-\lambda t} \cdot e^{\lambda t} dt + B$$

$$\text{i.e., } e^{\lambda t} P_1(t) = \int \lambda dt + B = \lambda t + B \tag{7}$$

Putting $t = 0$ in (7), we get $B = 0$.

Thus we obtain,

$$P_1(t) = \frac{\lambda t e^{-\lambda t}}{1!}$$

Similarly, we can show that

$$P_2(t) = \frac{(\lambda t)^2 e^{-\lambda t}}{2!}$$

$$P_3(t) = \frac{(\lambda t)^3 e^{-\lambda t}}{3!}$$

.

.

.

$$P_n(t) = \frac{(\lambda t)^n e^{-\lambda t}}{n!}$$

which is the Poisson distribution law.

### 11.7.1 Properties of Poisson arrivals

(i) Mean of arrivals $= \lambda t$

(ii) Expected number of arrivals in unit time $= \lambda$

(iii) $P_n(\Delta t) = 0,\ n \geq 2$

## 11.8 Model (M|M|1):(∞/FCFS)

Birth and death model: The probability distribution of queue length for a birth and death model $\rho^n(1-\rho)$, where $\rho = \frac{\lambda}{\mu} < 1, n > 0$.

*Proof*

Let $P_n(t+\delta\ t)$ denote the probability that there are $n$ units in the system at time $t+\Delta\ t$.

Four cases arise.

(i) There are $n-1$ units at time $t$ and one arrival occurs in time $\Delta\ t$ and nobody is served in time $\Delta\ t$.

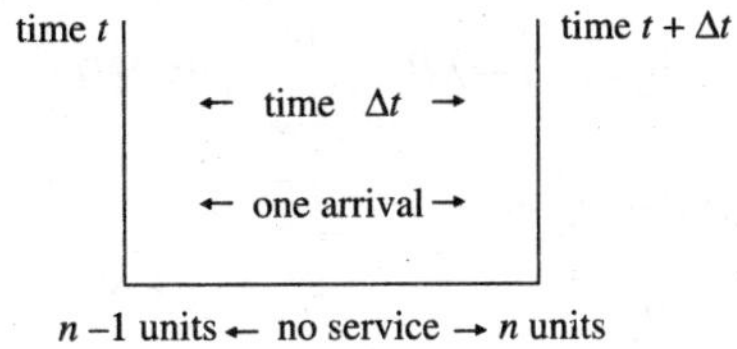


The probability in this case $P_{n-1}(t)\lambda\Delta t(1-\mu\Delta t)$

(ii) There are $n$ units in the system at time $t$, no arrival occurs in time $\Delta\, t$ and no unit is served in time $\Delta\, t$.

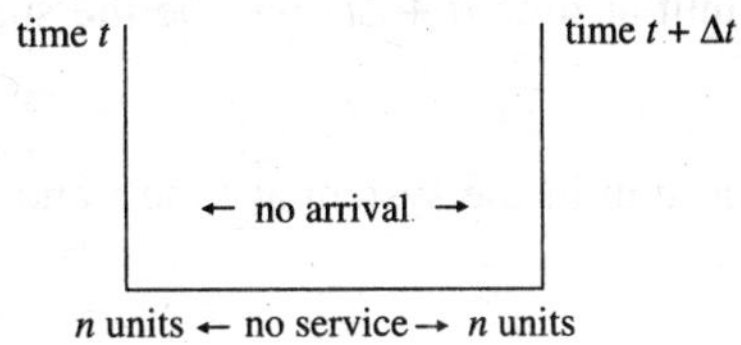


The probability in this case is

$$P_n(t)(1-\lambda\Delta t)(1-\mu\Delta t)$$

(iii) There are $(n+1)$ units in the system at time $t$. No unit arrives in time $\Delta t$ and one unit is served in time $\Delta t$.

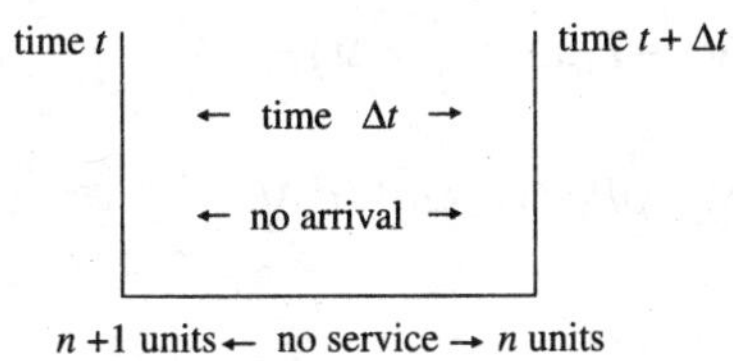


The probability in this case is

$$P_{n+1}(t)(1-\lambda\Delta t)\mu\Delta t$$

(iv) There are $n$ units at time $t$, one arrived in time $\Delta t$ and one is served.

The probability in this case in

$$P_n(t)\lambda\cdot\Delta t\cdot\mu(\Delta t)^2$$

where $\mu(\Delta t)^2$ is a negligible quantity. Since $(\Delta t)^2$ occurs, this case is not considered.

Therefore, the probability that there will be $n$ units in the system at time $(t+\Delta t)$ may be expressed as the sum of four independent compound probabilities by using the fundamental properties of probability, Poisson arrival and exponential service time.

We have

$$P_n(t+\Delta t) = P_n(t)(1-(\lambda+\mu)\Delta t) + P_{n-1}(t)\lambda\Delta t + P_{n+1}(t)\mu\Delta t + P_n(t)(\lambda\mu)(\Delta t)^2 \quad (1)$$

From (1), we get

$$\frac{P_n(t+\Delta t)-P_n(t)}{\Delta t}=-(\lambda+\mu)P_n(t)+\lambda P_{n-1}(t)+\mu P_{n-1}(t)$$

In the steady state, we have $P_n'(t)\to 0$

$$P_n(t)\to P_n$$

as $\Delta t\to 0$, we get

$$P_n'(t)=0=-(\lambda+\mu)P_n(t)+\lambda P_{n-1}(1)+\mu P_{n+1}(t) \tag{2}$$

Probability that there is no unit at time $(t+\Delta t)$ will be the sum of the following independent probabilities.

(i) Probability that there is no unit in the system at time $t$ and no arrived in time $\Delta t$
$=P_0(t)(1-\lambda\Delta t)$

(ii) Probability that there is one unit in the system at time $t$, one unit serviced in time $\Delta t$ and no arrival in $\Delta t$
$=P_1(t)\mu\Delta t(1-\lambda\Delta t)$

Adding these probabilites, we get

$$P_0(t+\Delta t)=P_0(t)(1-\lambda\Delta t)+P_1\mu\Delta t(1-\lambda\Delta t)$$

or $$\frac{P_0(t+\Delta t)-P_0(t)}{\Delta t}=-\lambda P_0(t)+\mu P_1(t)-\lambda\mu P_1(t)\Delta t$$

Applying limit, we obtain

$$\operatorname*{lt}_{\Delta t\to 0}\frac{P_0(t+\Delta t)-P_0(t)}{\Delta t}=-\lambda P_0(t)+\mu P_1(t)$$

or $P_0'(t)=-\lambda P_0(t)+\mu P_1(t)$

In steady state, $P_0'(t)=0$, therefore

$$0=-\lambda\, P_0(t)+\mu P_1(t)$$

or $P_1=\dfrac{\lambda}{\mu}P_0$

Taking $n=1$ in (2) we obtain

$$P_2=\frac{\lambda}{\mu}P_1$$

$$=\frac{\lambda}{\mu}\cdot\frac{\lambda}{\mu}P_0=\left(\frac{\lambda}{\mu}\right)^2 P_0$$

or $P_2 = \rho^2 P_0$

Taking $n = 2$ in (2), we obtain

$$\lambda P_1 - (\lambda + \mu)P_2 + \mu P_3 = 0$$

$$\text{or } \frac{-\lambda}{\mu} P_1 + \left(\frac{\lambda}{\mu} + 1\right) P_2 = P$$

$$-P_2 + \frac{\lambda}{\mu} P_2 + P_2 = P_3$$

$$\text{or } P_3 = \frac{\lambda}{\mu} P_2 = \frac{\lambda}{\mu} \cdot \left(\frac{\lambda}{\mu}\right)^2 P_0$$

$$\text{or } P_3 = \left(\frac{\lambda}{\mu}\right)^3 P_0$$

or $P_3 = \rho^3 P_0$

Proceeding as above, we get

$$P_n = \left(\frac{\lambda}{\mu}\right)^n P_0 = \rho^n P_0 \tag{3}$$

Since $\sum_{n=0}^{\infty} P_n = 1$, we get

$$P_0 + P_1 + P_2 + \ldots = 1$$

$$\text{or } P_0 + \frac{\lambda}{\mu} P_0 + \left(\frac{\lambda}{\mu}\right)^2 P_0 + \ldots = 1$$

$$\text{or } P_0 \left[1 + \frac{\lambda}{\mu} + \left(\frac{\lambda}{\mu}\right)^2 + \ldots\right] = 1 \tag{4}$$

Since LHS of (4) is an infinite geometric series with first term 1 and common ratio $\frac{\lambda}{\mu}(< 1)$, we have

$$P_0 \left(\frac{1}{1 - \frac{\lambda}{\mu}}\right) = 1$$

$$\text{or } P_0 = 1 - \frac{\lambda}{\mu} = 1 - \rho$$

Therefore, from (3) we have

$$P_n = \left(\frac{\lambda}{\mu}\right)^n P_0 = \left(\frac{\lambda}{\mu}\right)^n (1-\rho) = \rho^n(1-\rho)$$

Hence, proved.

## 11.9 Characteristics of a Model

### 11.9.1 Average (expected) number of customers in the system

By the definition of expected value we have

$$L_s = \sum_{n=0}^{\infty} nP_n$$

$$= \sum_{n=0}^{\infty} n\left(\frac{\lambda}{\mu}\right)^n \left(1-\frac{\lambda}{\mu}\right)$$

$$= \frac{\lambda}{\mu}\left(1-\frac{\lambda}{\mu}\right) \sum_{n=0}^{\infty} n\left(\frac{\lambda}{\mu}\right)^{n-1}$$

$$= \frac{\lambda}{\mu}\left(1-\frac{\lambda}{\mu}\right)\left[1+2\frac{\lambda}{\mu}+3\left(\frac{\lambda}{\mu}\right)^2+\ldots\right]$$

$$= \frac{\lambda}{\mu}\left(1-\frac{\lambda}{\mu}\right)\left(1-\frac{\lambda}{\mu}\right)^{-2}$$

$$= \frac{\lambda}{\mu}\left(1-\frac{\lambda}{\mu}\right)^{-1}$$

$$= \frac{\frac{\lambda}{\mu}}{1-\frac{\lambda}{\mu}} = \frac{\lambda}{\mu-\lambda}$$

$$\text{or } L_s = \frac{\rho}{1-\rho}$$

### 11.9.2 Average number of customers in the queue

Average length of the queue

$$= L_q = \text{(Queue length)}$$

$$= \text{Queue size} - \text{number of customers being served}$$

$$= \sum_{n=1}^{\infty} nP_n - \sum_{n=1}^{\infty} P_n$$

$$= \sum_{n=1}^{\infty} nP_n - \left\{ \sum_{n=0}^{\infty} P_n - P_0 \right\}$$

$$= \sum_{n=0}^{\infty} nP_n - (1 - \rho_0)$$

$$= L_s - (1 - \rho_0)$$

$$= \frac{\lambda}{\mu - \lambda} - \left[ 1 - \left( 1 - \frac{\lambda}{\mu} \right) \right]$$

$$= \frac{\lambda}{\mu - \lambda} - \frac{\lambda}{\mu}$$

$$= \frac{\lambda\mu - \lambda\mu + \lambda^2}{\mu(\mu - \lambda)} = \frac{\lambda^2}{\mu - (\mu - \lambda)}$$

$$\text{or } L_q = \frac{\rho^2}{1 - \rho}$$

### 11.9.3 Expected number of customers in a non-empty queue

Since the queue is non-empty, we have

$$L_n = \frac{E_{(n-1)}}{P_{(n-1>0)}} = \frac{E_{(n-1)}}{P_{(n>1)}} = \frac{E_{(n-1)}}{\sum\limits_{n=1}^{\infty} P_n}$$

$$= \frac{\lambda^2}{\mu(\mu - \lambda)} \cdot \frac{1}{\sum\limits_{n=2}^{\infty} \left( \frac{\lambda}{\mu} \right)^n \left( 1 - \frac{\lambda}{\mu} \right)}$$

$$= \frac{\lambda^2}{\mu(\mu-\lambda)} \cdot \frac{1}{\left(\frac{\lambda}{\mu}\right)^2 \left(1-\frac{\lambda}{\mu}\right) \sum_{n=0}^{\infty} \left(\frac{\lambda}{\mu}\right)^n}$$

$$= \frac{\mu}{\mu-\lambda} \frac{1}{\left(1-\frac{\lambda}{\mu}\right)\left(1+\frac{\lambda}{\mu}+\left(\frac{\lambda}{\mu}\right)^2+\ldots\right)}$$

$$= \frac{\mu}{\mu-\lambda} \frac{1}{\left(1-\frac{\lambda}{\mu}\right)\left(1-\frac{\lambda}{\mu}\right)^{-1}}$$

$$= \frac{\mu}{\mu-\lambda}$$

or $$L_n = \frac{1}{1-\rho}$$

### 11.9.4 Probability that the number of customers in the system is greater than or equal to *k*

We have

$$P(n \geq k) = \sum_{n=k}^{\infty} P_n$$

$$= \sum_{n=k}^{\infty} \left(\frac{\lambda}{\mu}\right)^n \left(1-\frac{\lambda}{\mu}\right)$$

$$= \left(\frac{\lambda}{\mu}\right)^k \left(1-\frac{\lambda}{\mu}\right) \sum_{n=k}^{\infty} \left(\frac{\lambda}{\mu}\right)^{n-k}$$

$$= \left(\frac{\lambda}{\mu}\right)^k \left(1-\frac{\lambda}{\mu}\right) \sum_{n=0}^{\infty} \left(\frac{\lambda}{\mu}\right)^n$$

$$= \left(\frac{\lambda}{\mu}\right)^k \left(1-\frac{\lambda}{\mu}\right) \left(1+\left(\frac{\lambda}{\mu}\right)+\left(\frac{\lambda}{\mu}\right)^2+\ldots\right)$$

$$= \left(\frac{\lambda}{\mu}\right)^k \left(1-\frac{\lambda}{\mu}\right)\left(1-\frac{\lambda}{\mu}\right)^{-1}$$

$$= \left(\frac{\lambda}{\mu}\right)^k$$

or $p[n > k] = \rho^k$

### 11.9.5 Probability density function of waiting time distributions

In the steady state, each customer has the same weighting time distribution. Let the probability that a customer begins to be served in the time interval $(w, w+dw)$ be denoted by $\psi(w)dw$, where $w$ is measured from the time of his arrival we suppose that a customer arrives at time $w = 0$ and service begins in the interval $(w, w+dw)$. Let A denote the customer.

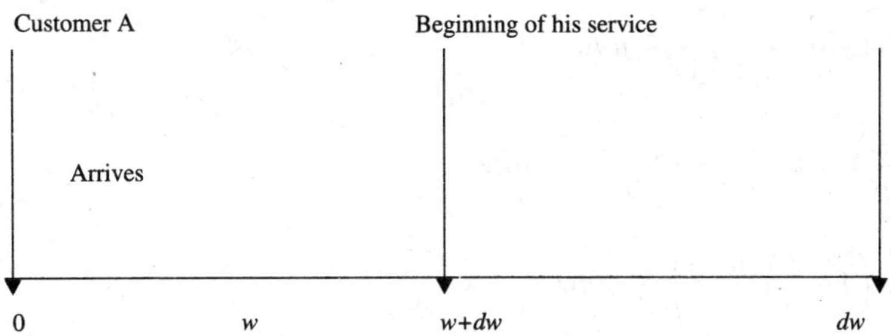


(i) There is a finite probability that the waiting time is zero.

(ii) There are $n$ customers already in the system. When customer A arrives, the $n$ customers must leave before the service of A begins. More precisely $(n-1)$ customers must leave during the time interval $(0, w)$ and the $n$th customer during $(w, w+dw)$

The mean rate of service is $\mu$ in unit time or $\mu w$ in time and the probability of $(n-1)$ departures in time $w$, during which the server is busy is the appropriate term of the Poisson distribution.

$$\frac{(\mu w)^{n-1} e^{-\mu w}}{(n-1)!}$$

Let there be $n$ units in the system. Then

$$\psi dw = \text{Prob}[\ (n-1) \text{ units are served at time } w] \times \text{Prob[one unit is served in time } dw]$$

$$\text{or } \psi(w)\, dw = \frac{(\mu w)^{n-1} e^{-\mu w}}{(n-1)!} \mu w$$

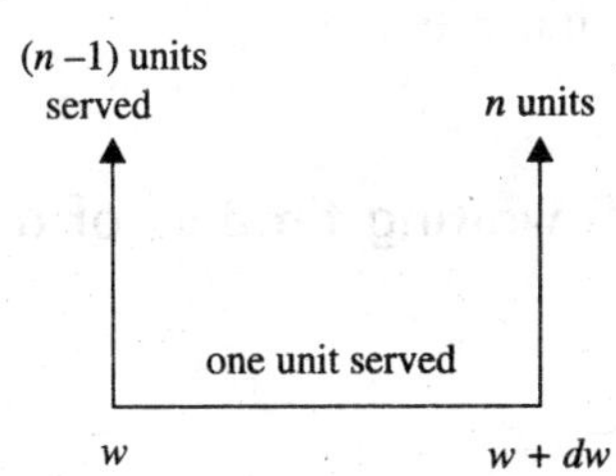


Let $W$ be the waiting time of a unit who has to wait such that $(w \le W \le w+dw)$.

Then the probability $\psi(w)dw$ is given by

$$\psi(w)dw = \text{Prob. } (w \leq w \leq w + dw)$$

= (Prob. that there are $n$ customers in the system when customer A arrives) × (Prob. that exactly $n-1$ customers leave in (0, $w$)) × (Prob. that the nth customer leaves in ($w$, $w+dw$)).

$$= \sum_{n=1}^{\infty} P_n \psi(w)\, dw$$

$$= \sum_{n=1}^{\infty} \rho^n (1-\rho) \frac{(\mu w)^{n-1} e^{-\mu w}}{(n-1)!} \mu\, dw$$

$$= \sum_{n=1}^{\infty} \left(\frac{\lambda}{\mu}\right)^n \left(1 - \frac{\lambda}{\mu}\right) e^{-\mu w} \frac{(\mu w)^{n-1}}{(n-1)!} \mu\, dw$$

$$= \sum_{n=1}^{\infty} \frac{\lambda^n}{\mu^n} \left(1 - \frac{\lambda}{\mu}\right) \frac{\mu^{n-1} w^{n-1}}{(n-1)!} \mu\, dw$$

$$= \sum_{n=1}^{\infty} \lambda \left(1 - \frac{\lambda}{\mu}\right) \frac{(\lambda w)^{n-1}}{(n-1)!} e^{-\mu w} dw$$

$$= \lambda \left(1 - \frac{\lambda}{\mu}\right) e^{-\mu w} e^{\lambda w} dw; (w > 0)$$

$$= \lambda \left(1 - \frac{\lambda}{\mu}\right) e^{-(\mu - \lambda)w} dw$$

Since $\sum_{n=1}^{\infty} \frac{(\lambda w)^{n-1}}{(n-1)!} = e^{\lambda w}$

Hence we define the probability density function as

$$\psi(w) = 1 - \frac{\lambda}{\mu}, \quad \text{if } w = 0$$

$$= \lambda \left(1 - \frac{\lambda}{\mu}\right) e^{-(\mu - \lambda)w} \quad \text{if } w > 0$$

### 11.9.6 Average (expected) waiting time $w_q$ of a customer in the queue

We have

$$w_q = \int_0^w w \frac{\lambda}{\mu} (\mu - \lambda) e^{-(\mu - \lambda)w} dw$$

$$= \frac{\lambda}{\mu}(\mu-\lambda)\int_0^{\infty} we^{-(\mu-\lambda)w}dw$$

$$= \frac{\lambda}{\mu}(\mu-\lambda)\left[\frac{we^{-(\mu-\lambda)w}}{-(\mu-w)} - \frac{e^{-(\mu-\lambda)w}}{-(\mu-w)^2}\right]_0^{\infty}$$

$$= -\frac{\lambda}{\mu}(\mu-\lambda)\left(0-\frac{1}{(\mu-\lambda)^2}\right) = \frac{\lambda}{\mu(\mu-\lambda)}$$

or $w_q = \dfrac{\rho}{\mu(1-\rho)}$

Probability that the waiting time of a customer in the queue exceeds $t$

$$= \frac{\lambda}{\mu}e^{-(\mu-\lambda)t}$$

Average waiting time of a customer in the queue, if he has to wait $= \dfrac{1}{\mu-\lambda}$

Average (expected) waiting time of a customer in the system $= w_s = \dfrac{1}{\mu-\lambda}$

Probability density function of the waiting time and waiting time distribution function of the system
$w_s(w) = 1 - e^{-\mu(1-\rho)}w, \; w > 0$

Probability that the waiting time of a customer in the system exceed $t = e^{-(\mu->)t}$

## 11.10 Little's Formula

This defines the relation between $L_s$, $L_q$, $w_s$ and $W_q$. From the above formula, we have

$$L_s = \lambda w_s \tag{1}$$

$$L_q = \lambda w_q \tag{2}$$

By definition, we have

$$w_s = w_q + \frac{1}{\mu} \tag{3}$$

Multiplying (3) by $\lambda$, we get

$$\lambda w_s = \lambda w_q + \frac{\lambda}{\mu}$$

i.e., $L_s = L_q + \dfrac{\lambda}{\mu}$

which is known as Little's formula.

## Solved examples

**Example 1** At a service centre, customers arrive at the rate of 10 per hour and are served at the rate of 15 per hour. Their arrival follows Poisson distribution and service is exponentially distributed. Find the average length and average waiting time in the system.

**Solution** We have

$\lambda$ = average arrival of customers = 10 per hour

$\mu$ = average service rate = 15 per hour

Traffic intensity

$$= \rho = \frac{\lambda}{\mu} = 10/15 = 0.66$$

Therefore,

Average queue length = $\lambda^2 / \mu(\mu - \lambda)$

$$L_q = \frac{10^2}{15(15-10)} = \frac{100}{(15)(5)} = \frac{100}{75} = \frac{4}{3}$$

Average waiting time in the system $= w_s = \dfrac{1}{(\mu - \lambda)} = \dfrac{1}{15-10} = \dfrac{1}{5}$ hours

**Example 2** In a public telephone booth, the arrivals are on the average 15 per hour. A call on the average takes 3 minutes. If there is just one phone, find

(a) expected number of callers in the booth at any time.

(b) the time the booth is expected to be idle.

**Solution** Here, we have

Mean arrival rate = $\lambda$ = 15 per hour

Mean service rate $= \mu = \dfrac{1}{3} \times 60 = 20$ per hour

(a) Expected length of the non-empty queue

$$\frac{\mu}{(\mu-\lambda)} = \frac{20}{(20-15)} = 4$$

$$\text{Traffic intensity} = \frac{\lambda}{\mu} = 15/20 = \frac{3}{4}$$

i.e., the service is busy 3/4 hours.

(b) Idle time i.e., the booth is expected to be idle

$$= 1-\rho = (1-3/4) \text{ hours} = 3/4 \text{ hour}$$

$$= 15 \text{ minutes}$$

**Example 3** Cars arrive at a petról pump having one petrol unit in Poisson fashion with an average of 10 cars per hour. The service time is distributed exponentially with a mean of 3 minutes. Find

(a) average number of cars in the system.

(b) average waiting time in the queue.

(c) average queue length.

(d) the probability that the number of cars in the system is 2.

**Solution** It is given that the mean arrival rate = $\lambda$ = 10 per hour

and, mean service rate = $\mu$ = (1/3) × 60 = 20 per hour

$\therefore \rho = \lambda/\mu = 10/20 = 1/2$

(a) Average number of cars in the system

$$L_s = \frac{\rho}{1-\rho} = \frac{1/2}{1-1/2} = 1 \text{ car}$$

(b) Average waiting time in the queue

$$= \frac{L_q}{\lambda} = \frac{\rho^2}{(1-\rho)\lambda} = \frac{(1/2)^2}{(1-1/2)^{10}} = \frac{(1/4)}{10} = \frac{1}{20} \text{ hrs}$$

$$= \frac{1}{20} \times 60 \text{ minutes} \equiv 3 \text{ minutes}$$

(c) average queue length $= L_q = \dfrac{\rho^2}{1-\rho} = \dfrac{(1/2)^2}{1-\frac{1}{2}} = \dfrac{1}{2} = 0.5$ cars

(d) Probability of $n$ units in the system is

$$P_n = \rho^n(1-\rho)$$

$$\therefore \quad P_2 = \rho^2(1-\rho) = \left(\frac{1}{2}\right)^2\left(1-\frac{1}{2}\right) = \left(\frac{1}{8}\right)$$

Hence, the probability that the number of cars in the system is $2 = \frac{1}{8}$

**Example 4** In a super market, the average arrival rate of customers is 10 in every 30 minutes following Poisson process. The average time taken by the cashier to list and calculate the customer's purchase is 2.5 minutes, following Poisson distribution. What is the probability that queue length exceeds 6? What is the expected time spent by customers in the system?

**Solution** Here, we have

$\lambda$ = mean arrival rate = 10/30 per hour

= 1/3 per hour

$\mu$ = mean service rate = 1/2.5 per minute

Therefore, $\rho = \lambda/\mu = (1/3)/(1/2.5) = 0.8333$

(a) Probability that queue size is greater than $n = 6$

$$= \rho^n = \rho^6 = (0.8333)^6 = 0.3348$$

(b) Expected waiting time in the system

$$= \frac{L_s}{\lambda} = \frac{\rho/(1-\rho)}{\lambda} = \frac{0.8333}{1-0.8333} \times 3$$

$$= \frac{0.8333}{0.1667} \times 3 = 14.996 \text{ minutes}$$

**Example 5** In a railway marshalling yard, goods trains arrive at a rate of 30 trains per day. Assuming that the inter-arrival time follows an exponential distribution and the service time is also exponential with an average of 36 mts, calculate the following:

(a) The mean queue size.

(b) If the input of the train increases to an average 33 per day, what will be the changes in (a) and (b).

**Solution** We have $\lambda$ = 30 trains per day = 30/60 $\times$24 =1/48 per minute

$\mu = 1/36$ trains per minute

$\therefore \rho = \lambda/\mu = 36/48 = 0.75$

(a) We have $L_s = \frac{\rho}{1-\rho} = \frac{0.75}{1-0.75} = 3$ , trains

Therefore, mean queue size is 3 trains

(b) $P(\geq 10) = \rho^n = (0.75)^{10} = 0.056$

i.e $P(\geq 10) = 0.56$

(c) When input of the trains increase to 33 trains per day, we get

$$\lambda = \frac{33}{60 \times 24} = \frac{11}{480} = 0.0229$$

$$\mu = \frac{1}{36} \text{ trains per minute}$$

$$\rho = \frac{\lambda}{\mu} = \frac{33 \times 36}{60 \times 24} = 0.825$$

$$L_s = \frac{\rho}{1-\rho} = \frac{0.825}{1-0.825} = \frac{0.825}{0.175} = 4.714 \cong 5$$

$L_s = 5$ trains

$P(\geq 10) = \rho^{10} = (0.825)^{10} = 0.1460$ trains

**Example 6** Patients arrive at a government hospital or emergency service at the rate of one every hour. Currently only one emergency can be handled at a time. Patients spend an average of 20 minutes receiving emergency care. The doctor wishes to have enough seats in the waiting room so that no more than about 1% of arriving patients will have to stand. Find

(a) the probabilities that a patient arriving at the hospital will have to wait.

(b) the average time a patient spends in the queue that forms.

(c) average time a patient spends in the queue.

(d) average time a patient spends in the system.

(e) probability that there will be five or more patients waiting for the service.

**Solution** We have $\lambda = 1$ hour

$$\mu = \frac{20}{60} = \frac{1}{3} \Rightarrow \mu = 3 \text{ per hour}$$

(a) The probability that a patient arriving at the hospital will have to wait

$$= \rho = \frac{\lambda}{\mu} = \frac{1}{3} = 0.333$$

(b) Average length of the queue

$$= L_q = \frac{\lambda^2}{\mu(\mu - \lambda)} = \frac{1}{3(3-1)} = \frac{1}{6} = 0.1667$$

(c) Average time a patient spends in the queue

$$w_q = \frac{\lambda}{\mu(\mu - \lambda)} = \frac{1}{3(3-1)} = \frac{1}{6} = 0.1667 \text{ hrs}$$

(d) Average time a patient spends in the system

$$w_s = \frac{\lambda}{\mu - \lambda} = \frac{1}{3-1} = 0.5 \text{ hrs}$$

(e) Probability that there will be five or more patients for the service is

$$P(n \geq 5) = \rho^5 = \left(\frac{1}{3^5}\right) = 0.0041$$

**Example 7** An airlines organisation has one reservation clerk duty in its local branch at any given time. The clerk handles information regarding passenger reservation and flight timings. Assume that the number of customers arriving during any given period is Poisson described with an arrival rate of eight per hour and that the reservation clerk can serve a customer in six minutes on an average with exponentially distributed service time.

(a) What is the probability that the system in busy?

(b) What is the average time a customer spends in the system?

(c) What is the average length of the queue and what is the number of customers in the system?

**Solution** It is given that the mean arrival rate is 8 customers/hr.

$\therefore \lambda =$ per hour

Mean service rate = $\mu$ = 60/6 =10 customers per hour

$$\therefore \qquad \rho = \frac{\lambda}{\mu} = \frac{8}{10} = \frac{4}{5} = 0.8$$

(a) The probability that the system is busy = traffic intensity = 0.8 = 80% of the time

(b) Average time spent by a customer in the system

$$= \frac{1}{\mu - \lambda} = \frac{1}{10-8} = \frac{1}{2} = 0.5$$

(c) Average length of the queue $= \dfrac{\lambda^2}{\mu(\mu-1)} = \dfrac{8^2}{10(10-8)}$

$$= \frac{64}{20} = 3.2$$

Number of the customers in the system = $\lambda/\mu - \lambda$

$$= \frac{8}{10-8} = \frac{8}{2} = 4 \text{ customers}$$

## 11.11 Model II: (M|M|S): (∞ |FCFS)

### Multi server model

Let there be $n$ units in the system. Let $S$ be the number of servers working independently of each other.

(i) If $n < S$

All the customers may be served simultaneously. There will be no queue. $(S - n)$ number of servers may remain idle and then we have

$$\mu_n = n\mu,\ n = 0, 1, 2, 3, \ldots, S$$

(ii) If $n \geq S$

All the servers are busy. Maximum number of customers waiting in the queue will be $n - S$.

Then, we have $\mu_n = S\mu$

and $\lambda_n = \lambda$

The steady state difference equations are

$$P_0(t + \Delta t) = P_0(t)(1 - \lambda\Delta t) + P_1(t)\mu\Delta t \text{ for } n = 0$$

$$P_n(t + \Delta t) = P_n(t)(1 - (\lambda + n\mu)\Delta t) + P_{n-1}(t)\lambda\Delta t + P_{n+1}(t)(n+1)\mu\Delta t$$

for $n = 1, 2, 3, \ldots S - 1$

and

$$P_n(t+\Delta t) = P_n(t)(1-(\lambda+S\mu)\Delta t) + P_{n-1}(t)\ \lambda\Delta t + P_{n+1}(t)\mu\Delta t$$

for $n = S,\ S+1,\ S+2,\ldots$

Dividing the above equation by $\Delta t$ and taking limits as $\Delta t \to 0$, we obtain the following difference equations

$$P_0'(t) = -\lambda P_0(t) + \mu P_1(t) \text{ for } n = 0 \tag{1}$$

$$P_n'(t) = -(\lambda+n\mu)P_n(t) + \lambda P_{n-1}(t) + (n+1)\mu P_{n+1}(t) \text{ for } n = 1, 2, 3, \ldots, S-1 \tag{2}$$

and $P_n'(t) = -(\lambda+S\mu)P_n(t) + \lambda P_{n-1}(t) + S\mu P_{n+1}(t)$ for $n =$ S, $S+1, \ldots$

Considering the case of steady state, i.e., when $t \to \infty$, $P_n(t) \to P_n$ $P_n'(t) \to 0$ for all $n$.

Hence, the equations (1), (2) and (3) can written as

$$0 = -\lambda P_0 + \mu P_1 \text{ for } n = 0$$

$$0 = -(\lambda+n\mu)P_n + \lambda P_{n-1} + (n+1)\mu P_{n+1} \text{ for } n \leq S-1$$

$$0 = -(\lambda+S\mu)P_n + \lambda P_{n-1} + S\mu P_{n+1} \text{ for } n \geq S$$

Here, we have $P_0 = P_0$ (initially)

$$P_1 = \frac{\lambda}{\mu}P_0$$

$$P_2 = \frac{\lambda}{\mu}P_1 = \frac{\lambda^2}{2!\mu^2}P_0 = \frac{1}{2!}\left(\frac{\lambda}{\mu}\right)^2 P_0$$

$$P_3 = \frac{\lambda}{3\mu}P_2 = \frac{1}{3!}\frac{\lambda^3}{\mu^3}P_0 = \frac{1}{3!}\left(\frac{\lambda}{\mu}\right)^3 P_0$$

.

.

.

$$P_n = \frac{\lambda}{n\mu}P_{n-1} = \frac{1}{n!}\left(\frac{\lambda}{\mu}\right)^n P_0 \text{ for } 1 \leq n \leq S$$

$$P_S = \frac{\lambda}{S\mu}P_{S-1} = \frac{1}{S!}\left(\frac{\lambda}{\mu}\right)^S P_0$$

$$P_{S+1} = \frac{1}{S} \cdot \frac{1}{S!} \left(\frac{\lambda}{\mu}\right)^{S+1} P_0$$

$$P_{S+2} = \frac{1}{S^2} \cdot \frac{1}{S!} \left(\frac{\lambda}{\mu}\right)^{S+2} P_0$$

In general, we get

$$P_n = P_{S+(n-S)} = \frac{1}{S^{n-S}} \cdot \frac{1}{S!} \left(\frac{\lambda}{\mu}\right)^n P_0 \text{ for } n \geq S$$

Since $\sum_{n=0}^{\infty} P_n = 1$, we can write

$$\sum_{n=0}^{S-1} P_n + \sum_{n=S}^{\infty} P_n = 1$$

$$\text{or } \sum_{n=0}^{S-1} \left[\frac{1}{n!} \left(\frac{\lambda}{\mu}\right)^n P_0\right] + \sum_{n=S}^{\infty} \left[\frac{1}{S^{n-S}} \cdot \frac{1}{S!} \left(\frac{\lambda}{\mu}\right)^n p_0\right] = 1$$

$$\text{or } P_0 \left[\sum_{n=0}^{S-1} \frac{S^n}{n!} \left(\frac{\lambda}{S\mu}\right)^n + \frac{1}{S!} \sum \frac{S^n}{S^{n-S}} \left(\frac{\lambda}{s\mu}\right)^n\right] = 1$$

$$\text{or } P_0 \left[\sum_{n=0}^{S-1} \frac{(S\rho)^n}{n!} + \frac{S^s}{S!} \sum_{n=S}^{\infty} \rho^n\right] = 1 \text{ (Since} \frac{\lambda}{S\mu} < \rho)$$

$$\text{or } P_0 = \left[\sum_{n=0}^{S-1} \frac{(S\rho)^n}{n!} + \frac{S^s}{S!} \sum_{n=s}^{\infty} \rho^n\right]^{-1} \quad (\frac{\lambda}{s\mu} = 1)$$

$$\text{or } P_0 = \left[\sum_{n=0}^{S-1} \frac{(S\rho)^n}{n!} + \frac{S^s}{S!} \left(\frac{\rho^s}{1-\rho}\right)\right]^{-1}$$

Hence, the steady distribution of n arrivals is

$$P_n = \begin{cases} \frac{1}{n!} \left(\frac{\lambda}{\mu}\right)^n P_0 & \text{if } n = 0, 1, \ldots S_{n-1} \\ \frac{1}{S!} \frac{1}{S^{n-s}} \left(\frac{\lambda}{\mu}\right)^n P_0 & \text{if } n = S, S+1 \end{cases}$$

### 11.11.1 Measures of model II

(a) Length of the queue

$$L_q = P_s \cdot \frac{\rho}{(1-\rho)^2}$$

where $P_s = \dfrac{\left(\dfrac{\lambda}{\mu}\right)^5 P_0}{S!}$

(b) Length of the system $L_s = \dfrac{\lambda}{\mu} + L_q$

(c) Waiting time in the queue $w_q = \dfrac{L_q}{\lambda}$

(d) Waiting time in the system $w_s = \dfrac{L_s}{\lambda}$

(e) The mean number of waiting individuals (customers) who actually wait in the queue = $(L|L > 0) = \dfrac{1}{1-\rho}$

(f) The mean waiting time in the queue for those who actually wait is $(w|w > 0) = \dfrac{1}{S\mu - \lambda}$

(g) Probability that there will be some one waiting $= \dfrac{P_s \cdot \rho}{1-\rho}$

## Solved examples

**Example 1** Four counters are being run on the frontier of a country to check passports and necessary papers of the tourists. The tourist chooses a counter at random. If the arrival at the frontier is Poisson at the rate of λ and the service time is exponential with parameter $\dfrac{\lambda}{2}$, what is the steady state average queue at each counter?

**Solution** We have

$$S = 4, \ \mu = \frac{1}{2} \text{ (given)}$$

$$\therefore \quad \rho = \frac{1}{S\mu} = \frac{1}{4\left(\frac{1}{2}\right)} = \frac{1}{2}$$

We get

$$P_0 = \left[\sum_{n=0}^{3} \frac{2^n}{n!} + \frac{4^4}{4!} \frac{\left(\frac{1}{2}\right)^4}{1-\frac{1}{2}}\right]^{-1}$$

$$= \left[1+2+2+\frac{4}{3}+\frac{4}{3}\right]^{-1} = \left[\frac{23}{3}\right]^{-1} = \frac{3}{23}$$

Hence, $L_q = \dfrac{\left(\frac{\lambda}{\mu}\right)^S}{S!} \cdot \dfrac{\rho}{(1-\rho)^2} \cdot P_0$

$$= \frac{2^4}{4!} \cdot \frac{\left(\frac{1}{2}\right)}{\left(1-\frac{1}{2}\right)^2} \cdot \frac{3}{23} = \frac{4}{23}$$

**Example 2** A super market has two girls serving at the counters. The customers arrive in a Poisson fashion at the rate of 12 per hour. The service time for each customer is exponential with mean 6 minutes. Find (i) the probability that an arriving customer has to wait for service, (ii) the average number of customers in the system, and (iii) the average time spent by a customer in the super market.

**Solutions** Since $\lambda = 12$ customers per hours (given), $\mu = 10$ customers per hours (given) and $s = 2$ (given), we get

(i) Probability that an arrived customer has to wait $= P(n \geq s)$

$$= \frac{\left(\frac{\lambda}{4}\right)^S}{S!\left(1-\frac{\lambda}{4s}\right)} P_0$$

where $P_0 = \left[1+\dfrac{12}{10}+\dfrac{1}{2!}\left(\dfrac{12}{10}\right)^2\left(\dfrac{20}{20-12}\right)\right]^{-1}$

$$= \left[1+\frac{6}{5}+\frac{9}{5}\right]^{-1} = \frac{1}{4}$$

$\therefore P(n \geq 2) = \dfrac{\left(\frac{12}{10}\right)^2 \cdot \frac{1}{4}}{2!\left(1-\frac{12}{10}\right)} = 0.45$ hrs $\therefore P(n \geq 2) = \dfrac{\left(\frac{12}{10}\right)^2 \cdot \frac{1}{4}}{2!\left(1-\frac{12}{10}\right)} = 0.45$ hrs

Probability that a customer arrived has to wait = 0.45

(ii) Average number of customers in the system $= L_s = \dfrac{1}{SS!}\dfrac{\left(\dfrac{\lambda}{\mu}\right)^{S+1}}{\left(1-\dfrac{\lambda}{\mu s}\right)^2}P_0 + \dfrac{\lambda}{\mu}$

$$= \frac{1}{2.2!}\frac{\left(\frac{12}{10}\right)^{2+1}}{\left(1-\frac{12}{20}\right)^2}\cdot\frac{1}{4}+\frac{12}{10} = 1.875$$

$$= 1.875 \simeq 2 \text{ customers}$$

(iii) Average time spent by a customer in the super market = average waiting time of a customer in the system

$$= \frac{1}{\lambda}\cdot L_s$$

$$= \frac{1.875}{12} \text{ hrs}$$

$$= \frac{1.875}{12} \times 60 \text{ minutes}$$

$$= 9.375 \text{ minutes}$$

**Example 3** A petrol pump station has 2 pumps. The service time follows the exponential distribution with a mean of 4 minutes and cars arrive for service in a Poisson process at the rate of 10 cars per hour. Find the probability that a customer has to wait for service. What proportion of time do the pumps remain idle?

**Solution** Here we have

$$\lambda = 10 \text{ cars/hr, service time} = \frac{1}{\mu} = \frac{4}{60} \text{ per hour}$$

$$\mu = \frac{60}{4} = 15 \text{ cars/hr}$$

and $S = 2$

Probability that a customer has to wait $P(n \geq 2) = P(n \geq 2)$

$$= \frac{\left(\frac{\lambda}{\mu}\right)^S P_0}{S!\left(1-\frac{\lambda}{\mu S}\right)} = \frac{\left(\frac{10}{15}\right)^2}{2!\left(1-\frac{10}{30}\right)}P_0$$

$$\text{where } P_0 = \left[\sum_{n=0}^{s-1} \frac{1}{n!}\left(\frac{\lambda}{\mu}\right)^n + \frac{1}{S!}\left(\frac{\lambda}{\mu}\right)^S \frac{\mu S}{\mu S - \lambda}\right]^{-1}$$

$$= \left[1 + \frac{1}{1!}\left(\frac{10}{15}\right) + \frac{1}{2!}\left(\frac{10}{15}\right)^2 \frac{30}{30-10}\right]^{-1}$$

$$= \left[1 + \frac{2}{3} + \frac{1}{2}\cdot\frac{100}{225}\cdot\frac{30}{20}\right]^{-1}$$

$$= \left[1 + \frac{2}{3} + \frac{1}{3}\right]^{-1} = \frac{1}{2}$$

$$\therefore P(n \geq 2) = \frac{\frac{100}{225}\cdot\frac{1}{2}}{2\left(\frac{30}{20}\right)} = \frac{1}{6} = 0.1667$$

The fraction of time when the pumps are busy = traffic intensity (ρ)

$$= \frac{\lambda}{\mu S} = \frac{10}{30} = \frac{1}{3}$$

Proportion of time the pumps remain idle

$$= 1 - \rho = 1 - \frac{1}{2} = \frac{2}{3} = 0.6667$$

i.e., 67%

**Example 4** Ships arrive at a port at the rate of one in every 4 hours with exponential distribution of inter arrival times. The time a ship occupies a berth for unloading has exponential distribution with an average of 10 hours. If the average delay of ships waiting for berths is to be kept below 14 hours how many berths should be provided at the port?

**Solution** Here, we have

$$\lambda = \frac{1}{4} \text{ per hour}$$

$$\mu = \frac{1}{10} \text{ per hour (given)}$$

Since $\frac{\lambda}{\mu S} < 1$, we get $\frac{10}{4s} < 1$ or $S > \frac{5}{2}$

Let $S = 3$, then we get

$$P_0 = \left[\sum_{n=0}^{S-1} \frac{1}{n!}\left(\frac{\lambda}{\mu}\right)^n + \frac{1}{S!}\left(\frac{\lambda}{\mu}\right)^S \frac{\mu S}{\mu S - \lambda}\right]^{-1}$$

$$= \left[1 + \left(\frac{5}{2}\right) + \frac{1}{2!}\left(\frac{5}{2}\right)^2 + \frac{1}{3!}\left(\frac{5}{2}\right)^3 \left(\frac{\frac{3}{10}}{\frac{3}{10} - \frac{1}{4}}\right)\right]^{-1}$$

$$= 0.0449$$

When $S = 3$,

$$w_q = \frac{1}{\mu} \cdot \frac{1}{SS!} \frac{\left(\frac{\lambda}{\mu}\right)^S}{\left(1 - \frac{\lambda}{\mu s}\right)^2} P_0$$

$$= 10 \cdot \frac{1}{3.3!} \frac{(5/2)^3}{\left(1 - \frac{5}{6}\right)^2} (0.0449)$$

$$= 14.0313 \text{ hrs}$$

which is greater than 14 hrs. Therefore, 3 berths are not enough (i.e., $S > 3$)

Hence, the number of berths to be provided at the port = 4.

## Exercise 11.1

1. A computer shop has a laser printer. The jobs for laser printing are randomly distributed approximately in a Poisson distribution with mean service rate of 10 jobs per hour, since job pages vary in length (pages to be printed). The jobs arrive at a rate of 6 per hour during the entire 8 hours work day. If the laser printer is valued at Rs 30/- per hour, determine

   (a) the percent time an arriving job has to wait.

   (b) average system time.

   (c) average idle time cost of the printer per day.

2. Trains arrive at the marshalling yard at a rate of 20 trains per day and the service time is 40 minutes. If the line capacity of the yard is limited to 6, find

   (a) the probability that the yard is empty.

   (b) the average number of trains in the system.

3. In a colour TV manufacturing plant, a loading unit takes exactly 10 minutes to load two TV sets into a wagon and again come back to the original position to load another set of TVs. If the arrival of TVs is a Poisson stream at an average of 2 TVs every 20 minutes, calculate the average waiting time of 2 TV sets in a stationary state. [Ans: 10 minutes]

4. Customers arrive at a one-man barber shop according to a Poisson process with a mean inter-arrival time of 20 minutes. They spend an average of 15 minutes in the barber chair. If an hour is used as a unit of time, then

   (a) what is the probability that a customer need not wait for a hair cut?

   (b) what is the expected number of customers in the barber shop and in the queue?

   (c) how much time can a customer expect to spend in the barber shop?

   (d) find the average time that the customer spends in the queue?

5 Assume that the mean duration of a telephone conversation is three minutes, and that no more than a three minute (average) wait for the phone may be tolerated.

   (i) What is the largest amount of incoming traffic that can be supported?

   (ii) Determine the maximum call rate that can be supported.

6. Assuming both queues are of Poisson type, taking an average rate of 20 per hour, is it better for a customer to get service at a single channel with mean service rate of 22 customers per hour or at one of two channels in parallel with mean service rate of 11 customers per hour for each of the two channels?

7. Customers arrive at a, 'one window drive' of a bank according to Poisson distribution with mean 10 per hour. Service time per customer is exponential with mean 5 minutes. The space in front of the window, including that for the service car can accommodate a maximum of three cars. Others can wait outside this space.

   (a) What is the probability that an arriving customer can drive directly to the space in front of the window?

   (b) What is the probability that an arriving customer will have to wait outside the indicated space?

   (c) How long is the customer expected to wait before starting service?

8. Consider a box office ticket window being manned by a single server. Customers arrive to purchase tickets according to a Poisson process with a mean rate of 30 per hour. The time required to serve a customer has an exponential distribution with a mean of 910 seconds. Determine the following:

   (a) fraction of the time the server is busy.

   (b) the average number of customers queueing for service.

   (c) the probability that there are more than 10 customers in the system.

9. Consider a self-service store with one cashier. Assume Poisson arrivals and exponential service times. Suppose that 9 customers arrive, on the average, every 5 minutes and the cashier can serve 9 in 5 minutes. Find
   (a) the average number of customers queueing for service.
   (b) the probability of having more than 10 customers in the system.
   (c) the probability that the customer has to queue for more than 2 minutes.

10. Patients arrive at a hospital at random with a mean arrival rate of 3 per hour. The department is served by one doctor, who spends on an average 15 minutes with each patient. Actual consulting times being exponentially distributed, find
    (a) the proportion of time that the doctor is idle.
    (b) the mean number of patients waiting to see the doctor.
    (c) the probability of being more than 3 patients waiting there.
    (d) the mean waiting time for patients.

11. The milk plant at a city distributed its products by trucks, loaded at the loading dock. It has its own fleet of trucks plus trucks of a private transport company. The trucks, are to wait in queue. The trucks' arrival is of 20 minutes. The service time is 4 minutes. Find
    (a) average number of trucks in the queue.
    (b) the probability that there are more than or equal to 4 trucks in the queue.
    (c) expected waiting time of a truck in the queue.
    (d) the variance of queue length.
    (e) the probability that the waiting time will exceed 10 minutes.

12. Weavers in a textile mill arrive at a department store room to obtain spare parts needed for keeping the looms running. The store is manned by one attendant. The average arrival rate of weavers per hour is 10 and service rate per hour is 12. Both arrival and service rates follow Poisson process. Determine
    (a) average length of waiting line
    (b) average time a machine spends in the system
    (c) percentage idle time of departmental store room

13. Find the probability that there are no customers in the system given that the number of channels in parallel is 3, mean arrival rate is 24 per hour and the mean service rate in each channel is 10 per hour. Compare the average time that a customer is in the system for the following 2 systems.
    (i) 5 channels in parallel with mean service rate of 10 per hour
    (ii) 1 channel with mean service rate of 30 per hour

14. A two channel waiting line with Poisson arrival has a mean rate of 50 per hour and exponential service with mean rate of 75 per hour for each channel. Find (i) the probability of an empty system and (ii) the probability that an arrival in the system will have to wait.

## Answers 11.1

1. (a) 60 (b) 15 mts (c) 96 Rs
2. (a) $\frac{4}{9}$ (b) $\frac{5}{4}$
3. 10 minutes
4. (a) 0.25 (b) 3 customers (c) 1 hour (d) 0.75 hr
5. (i) $\frac{1}{2}$ (ii) $\frac{1}{6}$ calls/minutes
6. 0.5 hr and 0.5238 hrs
7. (a) 0.42 (b) 0.58 (c) 0.4167
8. (a) 0.75 (b) 4.5 (c) 0.0563
9. (a) 8 (b) 0.3487 (c) 0.6033
10. (a) 15 minutes (b) $\frac{9}{4}$ (c) 0.421875 (d) 45 minutes
11. $\frac{1}{20}$ (b) 0.0016 (c) 1 (d) 0.3125 (e) 0.0271
12. (a) $\frac{25}{6}$ (b) 30 minutes (c) 10.67%
13. (i) 6 minutes (ii) 10 minutes
14. (i) 0.5 (ii) 0.111

## 11.12 Model III: (M|M|I):(*k*/FCFS)

In this system, the capacity is limited. The system can accommodate only $k$ customers, i.e., the maximum number of customers that can be accommodated is limited to $k$. If a customer arrives and finds that the queue is full, he has to go for his service elsewhere without joining the queue.

In this system, we assume

(i) The average arrival rate $\lambda_n = \lambda$ for $n = 0, 1, 2, 3, \ldots, k-1$

$$\lambda_n = 0 \text{ for } n \geq k$$

(ii) The average service rate $\mu_n = \mu$ for $n = 1, 2, 3, \ldots$

and obtain $P_n = \begin{cases} \left(\dfrac{\lambda}{\mu}\right)^n \left[\dfrac{1-\dfrac{\lambda}{\mu}}{1-\left(\dfrac{\lambda}{\mu}\right)^{k+1}}\right] & \text{if } \lambda \neq \mu \text{(i.e., } \rho \neq 1) \\ \dfrac{1}{k+1} & \text{if } \lambda = \mu \text{(i.e., } \rho = 1) \end{cases}$

The characteristics of the model are given below

(a) Average number of customers in the system

$$= L_s = P_0 \sum_{n=0}^{k} n\rho^n, \qquad \rho = \frac{\lambda}{\mu}$$

where $P_0 = \dfrac{1-\rho}{1-\rho^{k+1}}$

(b) Average number of customers in the queue

$$= L_q = L_s - \frac{\lambda}{\mu} \text{ when the arrival rate is } \lambda \text{ throughout.}$$

When the system is full

$$L_q = L_s - \frac{\lambda'}{\mu}, \text{ where } \lambda' = \mu(1-P_0)$$

where $\lambda'$ is called the effective arrival rate

## Solved examples

**Example 1** A barber shop has space to accommodate only 10 customers. He can serve only one person at a time. If customers comes to his shop and finds it full, they go to the next shop. If customers arrive randomly at an average rate of $\lambda = 10$ per hour and the barber's service time is a negative exponential with an average of $\dfrac{1}{\mu} = 5$ minutes per customers, find $P_0$ and $P_n$.

**Solution** It is given that $k = 10$

$$\lambda = \frac{10}{60} = \frac{1}{6}$$

$$\mu = \frac{1}{5}$$

$$\therefore \qquad \rho = \frac{\lambda}{\mu} = \frac{5}{6}$$

We obtain

$$\rho_0 = \frac{1-\rho}{1-\rho^{10+1}} = \frac{1-5/6}{1-(5/6)^{11}} = \frac{0.1667}{0.8665} = 0.1926$$

and $P_n = \left(\frac{1-\rho}{1-\rho^{k+1}}\right)\rho^n$

$$= \left(\frac{1-\left(\frac{5}{6}\right)}{1-\left(\frac{5}{6}\right)^{11}}\right)\cdot\left(\frac{5}{6}\right)^n$$

$$= 0.1926 \times \left(\frac{5}{6}\right)^n, \quad n = 0, 1, 2, \ldots, 10$$

**Example 2** If for a period of 2 hours in a day (9 AM to - 11 AM), trains arrive at the yard every 20 minutes but the service time continues to remain 36 minutes, calculate the following for the above period

(i) the probability that the yard is empty.

(ii) average queue length, assuming that the capacity of the yard is 4 trains only.

**Solution** Here $\lambda = \frac{1}{20}$ per minute

$$\mu = \frac{1}{36} \text{ per minute}$$

and $k = 4$

$$\therefore \quad \rho = \frac{\lambda}{\mu} = \frac{36}{20} = 1.8$$

(i) $P_0 = \frac{\rho - 1}{\rho^5 - 1} = \frac{1.8-1}{(1.8)^5 - 1} = 0.044$

$\therefore$ Probability that the yard is empty $= 0.044$

(ii) Average queue length

$$= L_s = P_0 \sum_{n=0}^{4} n\,\rho^n$$

$$= 0.04(0 + \rho + 2\rho^2 + 3\rho^3 + 4\rho^4)$$

$$= (0.04)(1.8 + 2(1.8)^2 + 3(1.8)^3 + 4(1.8)^4)$$

$$= 2.9 \simeq 3$$

∴ Average queue size = 3 trains

$$\text{or } L_s = \frac{\lambda}{\mu-\lambda} - \frac{(k+1)\left(\frac{\lambda}{\mu}\right)^{k+1}}{1-\left(\frac{\lambda}{\mu}\right)^{k+1}}, \quad (\lambda \neq \mu)$$

$$= \frac{\frac{1}{20}}{\frac{1}{36}-\frac{1}{20}} - \frac{5\left(\frac{26}{30}\right)^5}{1-\left(\frac{36}{20}\right)^5} = 3 \text{ (approx)}$$

## 11.13 Model IV (M|M|S): (*k*/FCFS)

### Multiple server with finite capacity

In this system, the maximum number of customers is limited, and more than one server services the customers.

We assume

$$\lambda_n = \begin{cases} \lambda & \text{for } n = 0, 1, 2, \ldots, k-1 \\ 0 & \text{for } n = k, k+1, \ldots \end{cases}$$

$$\mu_n = \begin{cases} n\mu & \text{for } n = 0, 1, 2, \ldots, S-1 \\ S\mu & \text{for } n = S, S+1, \ldots \end{cases}$$

also $1 < S < k$

$$\text{and obtain } P_n = \begin{cases} \frac{1}{n!}\left(\frac{\lambda}{\mu}\right)^n P_0, & \text{for } n < S \\ \frac{1}{S!S^{n-s}}\left(\frac{\lambda}{\mu}\right)^n P_0, & \text{for } S < nk \\ 0 & \text{for } n > k \end{cases}$$

$$\text{where } P_0 = \left[\sum_{n=0}^{S-1} \frac{1}{n!}\left(\frac{\lambda}{\mu}\right)^n + \sum_{n=S}^{k} \frac{1}{S^{n-S}S!}\left(\frac{\lambda}{\mu}\right)^n\right]^{-1}.$$

The characteristics of the model are:

(i) Average number of customers in the queue

= average queue length

$$= L_q = P_0 \left(\frac{\lambda}{\mu}\right)^S \frac{\rho}{S!(1-\rho)^2}\left[1-\rho^{k-S}-(t-S)(1-\rho)\rho^{k-S}\right]$$

where $\rho = \dfrac{\lambda}{\mu S}$

(ii) Average number of customers in the system

$$= L_s = L_q + s - \sum_{n=0}^{S-1}(S-n)P_n$$

Considering the effective arrival rate $\lambda'$, we have

$$L_s = L_q + \frac{\lambda'}{\mu}$$

where $\lambda' = \mu\left[S - \sum_{n=0}^{S-1}(S-n)P_n\right]$

(iii) Average waiting time of a customer in the queue:

$$w_q = \frac{L_q}{\lambda'}$$

(iv) Average waiting time of a customer in the system: $w_s = \dfrac{L_s}{\lambda'}$

**Example** A group in a computer browsing centre has 2 terminals. The average computing job requires 20 minutes of terminal time and each user requires some computation about once every half an hour. Assume that the arrival rate is Poisson and service rate is exponential and the group contains 6 users. Calculate the average number of users waiting to use one of the terminals in the computing job.

**Solution** Here we have

$$\frac{1}{\lambda} = \frac{30}{60} = \frac{1}{2}$$

$$\frac{1}{\mu} = \frac{20}{60} = \frac{1}{3}$$

Therefore $\lambda = 2$ per hour

$\mu = 3$ per hour

$s = 2$

and $k = 6$

$$P_0 = \left[\sum_{n=0}^{S-1} \frac{1}{n!}\left(\frac{\lambda}{\mu}\right)^n + \frac{1}{S!}\left(\frac{\lambda}{\mu}\right)^S \sum_{n=S}^{k}\left(\frac{\lambda}{\mu S}\right)^{n-S}\right]^{-1}$$

$$= 1 + \frac{2}{3} + \frac{1}{2!}\left(\frac{2}{3}\right)^2\left(1 + \frac{1}{3} + \frac{1}{3^2} + \frac{1}{3^3} + \frac{1}{3^4}\right)^{-1}$$

$$= 0.5004$$

$$\rho = \frac{\lambda}{\mu s} = \frac{2}{3.2} = \frac{1}{3}$$

$$L_q = P_0\left(\frac{\lambda}{\mu}\right)^S \frac{\rho}{S!(1-\rho)^2}\left[1 - \rho^{k-S} - (k-S)(1-\rho)\rho^{k-S}\right]$$

$$= (0.5004)\left(\frac{2}{3}\right)^2 \frac{\left(\frac{1}{3}\right)}{2!\left(\frac{2}{3}\right)^2}\left[1 - \left(\frac{1}{3}\right)^4 - 4\left(\frac{2}{3}\right)\left(\frac{1}{3}\right)^4\right]$$

$$= 0.0796$$

$$\therefore \quad L_s = L_q + S - \sum_{n=0}^{S-1}(S-n)p_n$$

$$= L_q + S - (2P_0 + P_1)$$

$$= 0.0796 + 2 - \left[2(0.5004) - \left(\frac{\lambda}{\mu}\right)P_0\right]$$

$$= 0.0796 + 2 - 2(0.5004) - \frac{2}{3}(0.5004)$$

$$= 0.7452$$

## Exercise 11.2

1. A car park contains 5 cars. The arrival of cars is Poisson distributed at a mean rate of 10 per hour. The length of time each car spends in the car park is a negative exponential distribution with a mean of 2 hours. How many cars are in the car park on average.
2. In a railway marshalling yard, goods trains arrive at a rate of 30 trains per day. Assume that the inter-arrival time follows an exponential distribution and the service time distribution is also an exponential with an average of 36 minutes. Calculate
   (a) probability that the yard is empty.
   (b) average queue length assuming that the line capacity of the yard in 9 trains.

3. A stenographer has 5 persons for whom she performs stenographic work. Arrival rate is Poisson distributed and service time is exponential. Average arrival rate is 4 per hour with an average service rate of 10 minutes. Find

   (a) the average waiting time of an arrival.

   (b) average length of waiting line

   (c) the average time an arrival spends in the system

4. A two-person barber shop has 5 chairs to accommodate waiting customers. Potential customers who arrive when all 5 chairs are full, leave without entering the barber shop. Customers arrive at the average rate of 3.76 per hour and spend an average 15 minutes in the barber chair.

   (i) What is the probability that a customer can get directly into the barber chair upon arrival?

   (ii) Compute $P_1$ and $P_5$.

   (iii) What is the expected number of customers waiting for a hair cut?

   (iv) What is the effective arrival rate?

5. A car servicing station has two bags where service can be offered simultaneously. Due to space limitation, only four cars are accepted for servicing.The arrival pattern is Poisson distributed with a mean of one car every minute during the peak hours.The service time is exponential with mean 6 minutes. Find

   (a) the average number of cars in the service station.

   (b) the average number of cars in the system during the peak hours.

   (c) the average waiting time a car spends in the system

6. A barber shop has two barbers and three chairs for customers. Assume that the customers arrive in Poisson fashion at a rate of 5 per hour and that each barber services customers according to an exponential distribution with mean of 15 minutes. Further if a customer arrivers and there are no empty chairs in the shop, he will leave. What is the probability that the shop is empty and what is the expected number of customers in the shop?

## Answers 11.2

1. 5
2. (a) 0.2649 (b) 3 trains
3. (a) 17.4 min (b) 1 (c) 22.4 min.
4. (i) 0.3671 (ii) 0.0166 (iii) 0.2450 (iv)3.7468
5. (a) 1.5687 (b) 3.535 (c) 0.8983 minutes.
6. 0.28; 1.226 customers $\approx$ 1 customer.

# CHAPTER TWELVE

# Stochastic Processes

## 12.1 Introduction

In dynamic systems, the input signals are rarely deterministic because input signals have an inherent fluctuation built around an underlying unpredictable processes. In such cases, we call them stochastic processes. In this chapter, we introduce stochastic processes; various types of stochastic processes, Markov chains and their properties. A stochastic process is also known as a random process. The concept of a stochastic process (i.e., a random process) is based on enlarging the random variable concept to include time.

Let $X$ be a random variable. Then, $X$ is a function of the possible outcomes of an experiment. If we assign, accordingly to some rule, a time function $x(t, s)$ to every outcome $s$, then the family of all such functions denoted by $X(t, s)$ is called a random process, which is also known as a stochastic process. The short form notation of a random process is $X(t)$. We often use the notation $X_1$ to denote the random variable associated with the process $X(t)$ at time $t_1$. When $t$ is fixed and $s$ is a random variable, a random process represents a random variable. The notation $x(t)$ is also used to represent a specific waveform of a random process denoted by $X(t)$. The set of all possible values of an individual member of the random process is called the state space.

The statistical properties of $X_1 = X(t_1)$ describe the statistical properties of the random process at time $t_1$. The expected value of $X_1$ is called the ensemble average. It is also called the mean value of the random process at time $t_1$.

## 12.2 Classification of Random Processes

A random process can be classified according to the characteristics of $t$ and the random variable $X = X(t)$. We shall now consider the following four cases based on $t$ and $X$ having values in the range $-\infty < t < \infty$ and $-\infty < x < \infty$.

### (i) Continuous random process

If $X$ is continuous and $t$ can have any of a continuum of values, then $X(t)$ is called a continuous random process (Fig. 12.1).

**Example:** Thermal agitation noise in conductors.

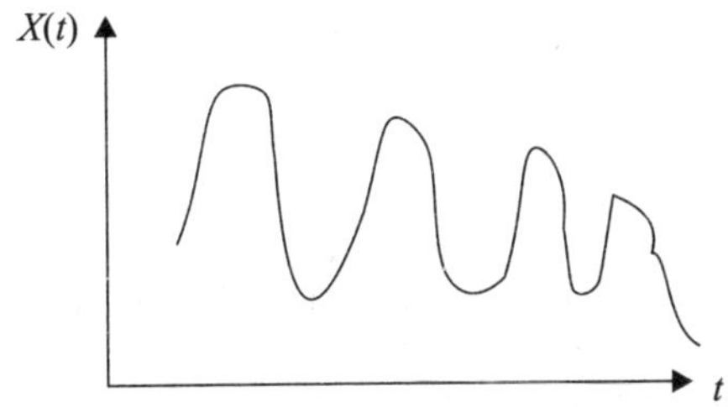


**Fig. 12.1** *Continuous random process*

### (ii) Discrete random process

A random process $X(t, s)$ which corresponds to the random variable $X$ having only discrete values while $t$ is continuous, is known as a discrete random process.

**Example:** The voltage when a switch is closed and open is an example of a discrete random process.

If $X(t)$ represents the number of telephone calls recorded into a switchboard of a company in $(0, t)$, then the set of all possible values of $X$, number of calls, is discrete, while $t$ is continuous.

### (iii) Continuous random sequence or discrete time random process

A random process for which $X$ is continuous but time has only discrete values is called a continuous random sequence.

It is also known as a discrete = time (DT) random process; its sample functions are referred to as a DT random signal.

A DT random process is a set of random variables denoted by $\{X(t_n)\}$ for sample times $t_n = n\ T_s\ ,\ n = 0, \pm 1, \pm 2, \ldots$ with $T_s$ called the sampling interval. The sampling rate is $\frac{1}{T_s}$ samples per second. A DT random process is also referred to as $X(nT_s)$.

### (iv) Discrete random sequence

A random process which corresponds to both time and the random variable being discrete is known as a discrete random sequence.

## Deterministic and non-deterministic processes

A random process can also be described by the form of its sample function.

If the future values of any sample function cannot be predicted exactly from the observed values, the process is called a non-deterministic process.

### Deterministic and non-deterministic random processes

A process is called deterministic if future values of any sample function can be predicted from past values.

A deterministic random process is also known as a predictable random process.

If the future values of any sample function cannot be predicted exactly from past values, then it is called a non-deterministic random process. It is also called a regular random process.

### Stationary and independent processes

We know that a random (stochastic) process becomes a random variable when time is fixed at some value. The random variables will possess statistical properties such as mean, variance etc, which are related to its density function. If two random variables are obtained from the process for two time instants, they will possess statistical properties related to their joint density function. In general, we can say that $N$ random variables will possess statistical properties related to their $N$-dimensional joint density function.

A random process is said to be stationary if all its statistical processes do not change with time. Other processes are known as non-stationary random processes. For a more concrete definition of stationary, we first define distribution and density functions.

Let $X(t)$ be a random process. For any particular time $t_1$, the distribution function associated with the random variable $X_1 = X(t_1)$ will be denoted by $F_X(x_1, t_1)$

It is defined as

$$F_X(x_1, t_1) = P\{X(t_1) \le x_1\} \tag{1}$$

for any real number $x_1$.

For two random variables $X_1 = X(t_1)$ and $X_2 = X(t_2)$, the second-order joint distribution function is the two-dimensional extension of (1)

$$F_X(x_1, x_2; t_1, t_2) = P\{X(t_1) \le x_1, X(t_2) \le x_2\} \tag{2}$$

Similarly, we can define the $N$th order joint distribution function for $N$ random variables.

$$X_i = X(t_i),\ i = 1, 2, 3, \ldots., N$$

$$F_X(x_1, x_2, \ldots, x_N; t_1, t_2, \ldots, t_N) = P\{X(t_1) \le x_1, \ldots, X(t_N) \le x_N\} \tag{3}$$

The joint functions corresponding to (1), (2) and (3) are

$$f_X(x_1, t_1) = \frac{\partial F_X(x_1, t_1)}{\partial x_1}$$

$$f_X(x_1, x_2; t_1, t_2) = \frac{\partial^2 F_X(x_1, x_2; t_1, t_2)}{\partial x_1 \partial x_2}$$

$$f_X(x_1,x_2,\dots,x_N;t_1,t_2,\dots,t_N) = \frac{\partial^N F_X(x_1,x_2,\dots,x_N;t_1,t_2,\dots,t_N)}{\partial x_1 \partial x_2 \dots \partial x_N}$$

## 12.3 Statistical Independence

Two random processes $X(t)$, $Y(t)$ are said to be statistically independent if the random variable group $X(t_1), X(t_2), \dots, X(t_N)$, is independent of the group $Y(t_1'), Y(t_2'), \dots, Y(t_M')$ for any choice of times $t_1, t_2, \dots, t_N, t_1', t_2', \dots, t_M'$.

We have

$$f_{XY}(x_1,x_2,\dots,x_N,y_1,y_2,\dots,y_M;t_1,t_2,\dots,t_N,t_1',t_2',\dots,t_M')$$

$$= f_X(x_1,x_2,\dots x_N;t_1,t_2,\dots,t_N)\, f_Y(y_1,y_2,\dots,y_M;t_1',t_2',\dots,t_M')$$

We now define a first-order and second-order stationery process.

## 12.4 First-order Stationary Process

A random process is called stationary to order one if its first-order density function does not change with shift in time origin

i.e., $f_X(x_1,t_1) = f_X(x_1,t_1+\Delta)$ (1)

Must be true for any $t_1$ and real number $\Delta$, if $X(t)$ is to be a first-order stationary process.

From (1) it is clear that $f_X(x_1,t_1)$ is independent of $t_1$ and the process mean value $E\{X(t)\}$ is a constant.

i.e., $E\{X(t)\} = \overline{X}$ = constant (2)

This can be proved as follows

For $X_1$, $E(X_1) = E[X(t_1)] = \int_{-\infty}^{\infty} x_1 f_X(x_1,t_1)\, dx_1$ (3)

For $X_2$, $E(X_2) = E[X(t_2)] = \int_{-\infty}^{\infty} x_2 f_X(x_1,t_2)\, dx_1$ (4)

(Note that $X_2$ has been replaced for convenience)

Let $t_2 = t_1 + \Delta$

Then using (1) and (3), we get

$$E(X_2) = E[X(t_1 + \Delta)] = \int_{-\infty}^{\infty} x_1 f_X(x_1, t_1)\, dx_1$$

Since $t_1$ and $\Delta$ are arbitrary

i.e., $E[X(t_1 + \Delta)] = E[X(t_1)]$ is a constant.

## 12.5 Second-order Stationary Process

A stochastic process is said to be stationary to order two if its second-order density function satisfies

$$f_X(x_1, x_2, t_1, t_2) = f_X(x_1, x_2; t_1 + \Delta, t_2 + \Delta) \tag{1}$$

for all $t_1, t_2$ and $\Delta$.

A second-order stationary process is also first-order stationary because the second-order density functions determines the lower, first-order density.

### Auto-correlation function

In general, the correlation $E[X_1 X_2] = E[X(t_1) X(t_2)]$ of a random process is a function of $t_1$ and $t_2$. This function is denoted by $R_{XY}(t_1, t_2)$ and is called the auto-correlation function of the random process $X(t)$.

We have

$$R_{XY}(t_1, t_2) = E[X(t_1) X(t_2)]$$

The auto-correlation function of a second-order stationary process is a function only of time differences and not absolute time.

If $\tau = t_2 - t_1$, we have

$$R_{XY}(t_1, t_2) = R_{XY}(t_1, t_1 + \tau) = E[X(t_1) X(t_1 + \tau)] = R_{XY}(\tau)$$

**Wide-sense stationary**

A random process $X(t)$ for which the conditions

$$E[X(t)] = \overline{X} = \text{Constant}$$

$$E[X(t)X(t+\tau)] = R_{XY}(\tau)$$

hold is called a wide-sense stationary process.

Let the process $X(t)$ and $Y(t)$ be wide-sense stationary. If their cross-correlation function, defined by

$$R_{XY}(t_1, t_2) = E[X(t_1)Y(t_2)]$$

is a function only of time difference and not absolute time, that is

$$R_{XY}(t_1, t_1+\tau) = E[X(t_1)Y(t_1+\tau)] = R_{XY}(\tau)$$

then $X(t)$ and $Y(t)$ are said to be jointly wide-sense stationary processes.

Let $X_1 = X(t_1), X_2 = X(t_2), \ldots, X_N = X(t_N)$ be $N$ random variables. A random process is stationary to order $N$, if its $N$th order density function is invariant to a time origin shift.

$$f_X(x_1, x_2, \ldots, x_N; t_1, t_2, \ldots, t_N) = f_X(x_1, x_2, \ldots, x_N; t_1+\Delta, \ldots, t_N+\Delta)$$

for all $t_1, t_2, \ldots .t_N$ and $\Delta$.

## 12.6 Time Average

Time average of a quantity is defined as

$$A[.] \lim_{T\to\infty} \frac{1}{2T} \int_{-T}^{T} [.]\, dt$$

$$E[X(t)] = \overline{X} = A[x(t_1)] = \overline{x}; \lim_{T\to\infty} \frac{1}{2T} \int_{-T}^{T} X^2(t)\, dt = R_X(0)$$

$$R_{xy}(\tau) = A[x(t)x(t+\tau)] \qquad \because A[.] \lim_{T\to\infty} \frac{1}{2T} \int_{-T}^{T} [.]\, dt$$

$$R_{xy}(\tau) = \underset{T\to\infty}{\text{lt}} \frac{1}{2T} \int_{-T}^{T} x(t)\, y(t+\tau)\, dt = R_{XY}(\tau)$$

$$t_2 = t_1 + \tau$$

where $A$ is used to denote the time average. Time average is taken over all time. Mean values of sample function and the time auto-correlation functions are defined by

$$\bar{x} = A[x(t)] = \underset{T\to\infty}{\text{lt}} \frac{1}{2T} \int_{-T}^{T} x(t)\, dt$$

$$R_{XX}(\tau) = A[x(t)\,x(t+\tau)]$$

$$= \underset{T\to\infty}{\text{lt}} \frac{1}{2T} \int_{-T}^{T} x(t)\, x(t+\tau)\, dt$$

## 12.7 Ergodicity

Ergodicity is a restrictive form of stationarity. Processes for which time and ensemble averages are interchangeable are called ergodic processes.

Thus, if $X(t)$ is an ergodic process, all time and corresponding ensemble averages are interchangeable.

The ergotic theorem states that the time averages $\bar{x}$ and $R_{xx}(\tau)$ are equal to the statistical averages $\overline{X}$ and $R_{XX}(\tau)$ respectively

i.e., $\bar{x} = \overline{X}$ and $R_{xx}(\tau) = R_{XX}(\tau)$

Processes that satisfy the ergodic theorem are known as ergodic processes.

Two random processes are called jointly ergodic if they are individually ergodic and also have a time cross-correlation function that equals the statistical cross-correlation function.

$$R_{xy}(\tau) = \underset{T\to\infty}{\text{lt}} \frac{1}{2T} \int_{-T}^{T} x(t)\, y(t+\tau)\, dt = R_{XY}(\tau)$$

## 12.8 Mean-Ergodic Process

A process $X(t)$ with a constant mean value $\overline{X}$ is called mean-ergodic if its statistical average $\overline{X}$ equals time average $\bar{x}$ of any sample function $x(t)$ with probability 1 for all sample functions, that is, if

$$E[X(t)] = \overline{X} = A[x(t_1)] = \overline{x}$$

with probability 1 for all $x(t)$.

### Mean-square ergodic process

A wide-sense stationary process $X(t)$ is ergodic in means square if

$$\lim_{T\to\infty} \frac{1}{2T} \int_{-T}^{T} X^2(t)\,dt = R_X(0)$$

## Solved examples

**Example 1** For a first-order stationary random process, show that the mean is constant.

**Solution** We have $E[X(t_1)] = \displaystyle\int_{-\infty}^{\infty} x_1 f(x_1, t_1)\,dx_1$

$$E[X(t_2)] = \int_{-\infty}^{\infty} x_2 f_X(x_2, t_2)\,dx_2$$

Let $t_2 = t_1 + \tau$

Then, $E[X(t_2)] = \displaystyle\int_{-\infty}^{\infty} x_2 f(x_2, t_1 + \tau)\,dx_1$

$$= \int_{-\infty}^{\infty} x_1 f(x_1, t_1 + \tau)\,dx_1 \quad (1)$$

(from the properties of definite integrals)

Since $X(t)$ is stationary, we have

$$f_X(x_1, t_1) = f_X(x_1, t_1 + \tau)$$

From (1), we get

$$E[X(t_2)] = \int_{-\infty}^{\infty} x_1 f(x_1, t_1)\,dx_1 = E[X(t_1)]$$

Therefore, $E[X(t)]$ is a constant

Hence, mean of a first-order stationary process is constant.

**Example 2** A stochastic process is described by $X(t) = A\sin t + B\cos t$, where $A$ and $B$ are independent random variables with zero means and equal standard deviations. Show that the process is stationary of the second order.

**Solution** We have $X(t) = A\sin t + B\cos t$ (given)

Therefore, $E[X(t)] = E[A\sin t + B\cos t]$

$$= \sin E(A) + \cos E(B) = 0$$

$$(\because E(A) = E(B) = 0)$$

and $E[X(t_1)X(t_2)] = E[[A\sin t_1 + B\cos t_1][A\sin t_2 + B\cos t_2]]$

$$= E\left[A^2 \sin t_1 \sin t_2 + AB\sin(t_1 + t_2) + B^2 \cos t_1 \cos t_2\right]$$

$$= \sin t_1 \sin t_2 E\left[A^2\right] + \sin(t_1 + t_2) E[AB] + \cos t_1 \cos t_2 E\left[B^2\right]$$

$$= \sigma^2 \cos(t_1 - t_2)$$

(since $E\left[A^2\right] = E\left[B^2\right] = \sigma^2$ and $E[AB] = 0$)

The second-order central moment depends only on $(t_1 - t_2) = \tau$. Hence, the given process is stationary.

**Example 3** Consider the random process $X(t) = \cos(\omega_0 t + \theta)$, where $\theta$ is uniformly distributed in the interval $-\pi$ to $\pi$. Show that $X(t)$ is stationary.

**Solution** Since $\theta$ is uniformly distributed on $[-\pi, \pi]$,

we have $f_\theta(\theta) = \dfrac{1}{2\pi}, \quad -\pi < \theta < \pi$

$= 0$, otherwise

The first moment of the random process at time $t_1$ is given by

$$E[X(t_1)] = \int_{-\pi}^{\pi} X(t_1) f_\theta(\theta)\, d\theta$$

$$= \int_{-\pi}^{\pi} \cos(\omega_0 t_1 + \theta) \frac{1}{2\pi} d\theta$$

$$= \frac{1}{2\pi}\int_{-\pi}^{\pi} \cos(\omega_0 t_1 + \theta)\, d\theta$$

$$= \frac{1}{2\pi}\left(\sin(\omega_0 t_1 + \theta)\right)_{-\pi}^{\pi}$$

$$= \frac{1}{2\pi}\left[\sin(\omega_0 t_1 + \pi) - \sin(\omega_0 t_1 - \pi)\right]$$

$$= \frac{1}{2\pi}\left[2\cos\omega_0 t_1 + \sin\pi\right] = 0$$

The second moment is

$$E\left[X^2(t_1)\right] = \int_{-\pi}^{\pi} \frac{1}{2\pi}\cos^2(\omega_0 t_1 + \theta)\, d\theta$$

$$= \frac{1}{2\pi}\int_{-\pi}^{\pi} \frac{1 - \cos(2\omega_0 t_1 + 2\theta)}{2}\, d\theta$$

$$= \frac{1}{4\pi}\left[\theta - \frac{\sin 2(\omega_0 t_1 + \theta)}{2}\right]_{-\pi}^{\pi}$$

$$= \frac{1}{4\pi}[2\pi - 0] = \frac{1}{2}$$

The first and second moments are independent of time.

Therefore, $X(t)$ is stationary.

**Example 4** Let $X(t) = A\cos\omega t + B\sin t$, where $A$ and $B$ are random variables with $E(A) = E(B) = E(B^2)$. Show that $X(t)$ is an ergodic processor.

**Solution** It is given that

$$X(t) = A\cos\omega t + B\sin\omega t$$

$$E(A) = E(B) = 0,\ E(A^2) = E(B^2)$$

By definition, the time average is

$$\bar{X} = \lim_{T\to\infty} \frac{1}{2T}\int_{-T}^{T} X(t)dt$$

$$= \underset{T\to\infty}{\text{lt}} \frac{1}{2T}\int_{-T}^{T}(A\cos\omega t + B\sin\omega t)dt$$

$$= \underset{T\to\infty}{\text{lt}} \frac{1}{2T}(2A)\int_{0}^{T}\cos\omega t\,dt + 0$$

Since $\cos\omega t$ is even and $\sin\omega t$

$$= \underset{T\to\infty}{\text{lt}} \frac{A}{T}\frac{\sin\omega t}{[\omega]}$$

$$= \underset{T\to\infty}{\text{lt}} \frac{A}{\omega T}[\sin\omega t] \tag{1}$$

Average is given by

$$E[X(t)] = E[A\cos\omega t + B\sin\omega t]$$

$$= E(A)\cos\omega t + E(B)\sin\omega t$$

$$= 0 + 0 = 0$$

From (1) and (2), we have

Time average = Ensemble average

Here the given process is ergodic.

## 12.9 Correlation Coefficient, Covariance and Auto-Correlation Function

If $X$ and $Y$ are two random variables, then the correlation coefficient $\rho$ is defined as

$$\rho = \frac{E(X-\bar{X})\cdot(Y-\bar{Y})}{\sqrt{E(X-\bar{X})^2 E(Y-\bar{Y})^2}} = E\left[\frac{(X-\bar{X})}{\sigma_X}\cdot\frac{(Y-\bar{Y})}{\sigma_Y}\right]$$

is called the covariance of the variables $X$ and $Y$. If the variables $X$ and $Y$ are independent, then their covariance is zero. If the correlation coefficient is zero, the variables are uncorrelated, but not necessary independent. If $\{X(t), Y(t)\}$ is a joint random process, then the covariance of the two sample random variables $X(t_1)$, $Y(t_k)$ is defined as

$$\text{Covariance } \{X(t_i), Y(t_k)\} = E\{[X(t_i) - \overline{X(t_i)}][Y(t_k) - \overline{(Y(t_k)}]\}$$

The concept can be extended to random processes.

When the two sample random variables are selected from a single process, the covariance coefficient is called auto covariance. In this case, we define

$$\text{Auto-covariance } \{X(t_i), X(t_k)\} = E\{[X(t_i) - \overline{X(t_i)}][X(t_k) - \overline{X(t_k)}]\}$$

If $\overline{X(t_i)} = 0$, $\overline{X(t_k)} = 0$, $Y(t_i i) = 0$, $(Y(t_i k)]) = 0$, we have

$$\text{Covariance of } \{X(t_i), Y(t_k)\} = E[X(t_i), Y(t_k)]$$

$$= E\{[X(t_i) - \overline{(X(t_i)}][Y(t_k) - \overline{Y(t_k)}]\}$$

$$\text{Auto-covariance of } \{X(t_i), Y(t_k)\} = E[X(t_i), X(t_k)]$$

$$= E\{[X(t_i) - \overline{X(t_i)}][X(t_k) - \overline{X(tk)}])$$

If the process is a stationary process of second order with a density function $f(x_i, X_i, t_i, t_k)$, the auto-covariance depends on the time lag $t_1 - t_k$ and we have

$$E\{[X(t_i) - \overline{X(t_i)}][X(t_k) - \overline{(X(t_k)}] = R_{XX}\{|t_i - t_k|\}$$

Auto-covariance of a stationary process is sometimes called the auto-correlation function, which was already defined as the correlation $E[X_1 X_2]$ of two random variables $X_1 = X(t_1)$ and $X_2 = X(t_2)$, where $t_1$ and $t_2$ are process time.

We have,

$$R_{XX}(t_1, t_2) = E[X(t_1), X(t_2)]$$

For assignments $t_1 = t$ and $t_2 = t_1 + \tau$ with $\tau$, a real number, we have

$$R_{XX}(t,\ t+\tau) = E[X(t)X(t+\tau)]$$

For wide-series stationary processes,

$$R_{XX}(\tau) = E[X(t)X(t+\tau)]$$

The auto-correlation of a stationary process function is given by

$$R_{XX}(\tau) = E[X(t_i)X(t_i+\tau)] = \int_{-\infty}^{\infty} x_1 x_2 f(x_1, x_2)\, dx$$

Properties of auto-correlation function: For a wide-sense stationary process, the auto-correlation function exhibits the following properties

(i) $|R_{XX}(\tau)| \le R_{XX}(0)$,

(ii) $R_{XX}(-\tau) = R_{XX}(\tau)$

(iii) $R_{XX}(0) = E(X^2(t))$

(iv) If $E[X(t) = \overline{X} \neq 0$, and $X(t)$ is ergodic with no periodic component then $\underset{(|\tau|)}{\text{lt}} R_{XX}(\tau) = \overline{X}^2$.

(v) If $X(t)$ is periodic, then $R_{XX}(\tau)$ also has the same period as its periodic component.

(vi) If $X(t)$ is ergodic, has zero mean and no period, then $\underset{|\tau|\to\infty}{\text{lt}} R_{XX}(\tau) = 0$.

(vii) $R_{XX}(\tau)$ cannot have an arbitrary shape.

## Solved examples

**Example 1** Consider an example consisting of four equally likely signals defined over the time interval [0,1] as follows

$$X(t) = \{2t+1, t+2, 3t+2, 4t+1\}$$

Find mean, second variance and auto-correlation of $X(t)$.

**Solution**

$$\text{Mean} = \mu = E[X(t)] = \frac{2t+1+t+2+3t+2+4t+1}{4}$$

$$\text{i.e., } \overline{X} = \frac{10t+6}{4} = \frac{5t+3}{2}$$

$$\text{Second moment} = E[X^2(t)] = \frac{(2t+1)^2+(t+2)^2+(3t+2)^2+(4t+1)^2}{4}$$

$$= \frac{4t^2+4t+1+t^2+4+4t+9t^2+4+12t+16t^2+8t+1}{4}$$

$$= \frac{30t^2+28t+10}{4} = \frac{15t^2+14t+5}{2}$$

$$\text{Variance} = \sigma_X^2 = E[X(t)^2] - \mu^2$$

$$= \frac{1}{2}[15t^2+14t+15] - \frac{1}{4}(5t+3)^2$$

$$= \frac{1}{4(5t^2 - 2t + 1)}$$

Auto-correlation $R_{XX(\tau,\tau)} = E[X(t)X(t+\tau)]$

$$= \frac{1}{4}[(4t+1)(4t+4\tau+1) + (2t+1)(2t+2\tau+1) + (t+2)(t+\tau+2) + (3t+2)(3t+3\tau+2)]$$

$$= \frac{1}{2}[15t^2 + 14t + 5 + 15t\tau + 7\tau]$$

**Example 2** Show that $R_{XX}(0) = \overline{X^2}$.

**Solution**

We have

$$R_{XX}(\tau) = E[X(t)X(t+\tau)]$$

$$= R_{XX}(0) = E[X(t)X(t)]$$

$$= E[X(t)^2] = \overline{X^2}$$

**Example 3** Show that the random process $X(t) = A\cos(\omega_0 + \theta)$ is a wide-sense stationary process, where $A$, $\omega_0$ are constants and $\theta$ is a uniformly distributed random variable on the interval $(0, 2\pi)$.

**Solution:**

Since $X(t) = A\cos(\omega_0 + \theta)$, we set

$$E[X(t)] = \int_0^{2\pi} A\cos(\omega_0 + \theta) f(\theta) d\theta$$

where $f(\theta) = \frac{1}{2\pi}, 0 < \theta < 2\pi$

$= 0$ elsewhere

$$E[X(t)] = \frac{1}{2\pi}\int_0^{2\pi} A\cos(\omega_0 t + \theta) d\theta$$

$$= \frac{A}{2\pi}[\sin(\omega_0 t + \theta)]_0^{2\pi}$$

The auto-correlation function is

$$R_{XX}(t,t+\tau) = E[A\cos(\omega_0 t+\theta)+\cos(\omega_0 t+\theta)]$$

$$= \frac{A^2}{2}[E(\cos(\omega_0 t)+\cos 2(\omega_0 t+\omega_0\tau+2\theta)]$$

$$= \frac{A^2}{2}E(\cos(\omega_0 t)+\frac{A^2}{2}E\cos 2(\omega_0 t+\omega_0\tau+2\theta)$$

$$= \frac{A^2}{2}\cos(\omega_0 t)+0 = \frac{A^2}{2}\cos(\omega_0 t)$$

Since the auto-correlation depends on the value $\tau$ and the mean value is constant, $X(t)$ is wide stationary.

**Example 4** Show that the function $A\cos(\omega_0\tau)$ is an auto-correlation function.

**Solution**

We have

$$R_{XX}(\tau) = A\cos(\omega_0\tau)$$

$$R_{XX}(-\tau) = A\cos(-\omega_0\tau) = A\cos(\omega_0\tau)$$

$$= R_{XX}(\tau)$$

$|R_{XX}(\tau)| = A$ and $R_{XX}(0) = A$

Therefore, $|R_{XX}(\tau)| = R_{XX}(0)$

Both the conditions satisfied. Hence, $A\cos(\omega_0\tau)$ is an auto-correlation function.

**Example 5** Given that the auto-correlation function for a stationary periodic process is

$$R_{XX}(\tau) = 25+\frac{4}{1+t^2}$$

find the mean and variance of the process $X(t)$.

**Solution:**

$$[E(X^2)] = \mu^2 = X^2 = \underset{|T|\to\infty}{\text{lt}} R_{XX}(\tau)$$

$$= \underset{|T|\to\infty}{\text{lt}} 25+\frac{4}{1+t^2}$$

$$= 25 + 0 = 25$$

$$\text{Mean} = \pm 5$$

$$E(X^2) = \sigma_{X^2} - [E(X^2)]$$

$$= 29 - 25 = 4$$

Function $R_{XX}(\tau)$ is called auto-correlation ergodic if and only if $\underset{T\to\infty}{\text{lt}} \int_{-T}^{T} X(t)X(t+\tau)dt = R_{XX}(\tau)$.

**Example 6** Given a stationary continuous process $X(t)$ with auto-correlation

$$X(t) = 10 \cos(100t + \theta)$$

where $\theta$ is a random process with a uniform probability distribution in the interval, show that $X(t)$ is ergodic in the auto-correlation function.

**Solution**

$$R_{XX}(\tau) = E[X(t)X(t+\tau)]$$

$$= E[100 \cos(100t + \tau) \cos(100t + 100\tau + \theta)]$$

$$= 50 \cos 100\tau$$

The time auto-correlation function is $E(10^2 \cos(200t + 100 + 2\theta))$

$$= \frac{1}{T} \int_{-\frac{T}{2}}^{\frac{T}{2}} X(t)X(t+\tau)d\tau$$

$$= \frac{1}{T} \int_{-\frac{T}{2}}^{\frac{T}{2}} 100 \cos(100t + \theta) \cos(100t + 100t + \theta)d\tau$$

$$= \frac{1}{T} \int_{-\frac{T}{2}}^{\frac{T}{2}} 50 \cos(100\tau)d\tau + \frac{1}{T} \int_{-\frac{T}{2}}^{\frac{T}{2}} 50 \cos(100t + 100\tau + 2\theta)d\tau$$

$$50 \cos(100\tau) = R_{XX}(\tau) = 0$$

Variance $R_{XX}(\tau) = 0$

From this, the process is an ergodic auto-correlation function.

**Example 7** Determine the mean and variance of

$$R_{XX}(\tau) = \frac{4\tau^2 + 100}{t^2}$$

**Solution**

We have $(E(x(t)))^2 = \underset{\tau\to\infty}{\text{lt}}\ R_{XX}$

$$= \underset{\tau\to\infty}{\text{lt}} \frac{4\tau^2 + 100}{\tau^2 + 4}$$

$$= \underset{\tau\to\infty}{\text{lt}} \frac{4 + \dfrac{100}{\tau^2}}{1 + \dfrac{4}{\tau^2}}$$

$$= \frac{4}{1} = 4$$

$$\therefore \bar{x}^2 = 4$$

Hence $E(x(t)) = \bar{x} = 2$

Variance $= E(X^2(t)) - E(X(t))^2$

We have $E(X^2(t) - R_{XX}(0)$

Since $R_{XX}(\tau) = \dfrac{4\tau^2 + 100}{\tau^2 + 4}$, we get

$$R_{XX}(0) = \frac{100}{4} = 25$$

$$\therefore \text{var}[X(t)] = 25 - 2^2 = 21$$

Hence, mean = 2, and variance = 21

## 12.10 Cross-Correlation

The cross-correlation function of the two random variables $X(t)$ and $Y(t)$ is defined by

$$R_{XY}(t_1, t_2) = E(X(t_1), Y(t_2)) \tag{1}$$

where $X(t_1)$ and $Y(t_2)$ are random variables.

Setting $t_1 = t$ and $\tau = t_2 - t_1$, we may write (1) as

$$R_{XY}(t,t+\tau) = E(X(t)Y(t+\tau))$$

If $X(t)$ and $Y(t)$ are at least jointly wide-sense stationary, $R_{XY}(t,t+\tau)$ is independent of absolute time and we can write

$$R_{XY}(t,t+\tau) = E(X(t)Y(t+\tau))$$

## 12.11 Orthogonal Process

If $R_{XY}(t,t+\tau) = 0$, then the process $X(t)$ and $Y(t)$ are called orthogonal processes.

### Properties of cross-correlation function

(i) $R_{XY}(-\tau) = R_{YX}(\tau)$

(ii) $|R_{XY}(\tau)| \le \sqrt{R_{XX}(0)R_{YY}(0)}$

(iii) $|R_{XY}(\tau)| \le \frac{1}{2}[R_{XX}(0) + R_{YY}(0)]$

### Auto-covariance and cross-covariance functions

The auto-covariance functions is defined by

$$C_{XY}(t,t+\tau) = R_{XY}(t,t+\tau) - E(X(t))E(Y(t+\tau))$$

It can also be written as

$$C_{XY}(t,t+\tau) = R_{XY}(t,t+\tau) - E(X(t))E(Y(t+\tau))$$

The cross-covariance function of the process $X(t)$ and $Y(t)$ is defined by

$$C_{XY}(t,t+\tau) = R_{XY}(t,t+\tau) - E(X(t))E(Y(t+\tau))$$

i.e., $C_{XY}(t,t+\tau) = R_{XY}(t,t+\tau) - E(X(t))E(Y(t+\tau))$

The random process $X(t)$ and $Y(t)$ are uncorrelated if

$$C_{XY}(t,t+\tau) = 0$$

i.e., if $R_{XY}(t,t+\tau) = E(X(t))E(Y(t+\tau))$

If the processes are at least jointly wide-sense stationary, we have

$$C_{XY}(\tau) = R_{XY}(\tau) - \bar{\lambda}^2$$

$$C_{XY}(\tau) = R_{XY}(\tau) - \overline{XY}$$

If $C_{XY}(t, t+\tau) = 0$, then the random process $X(t)$ and $Y(t)$ are uncorrelated.

## 12.12 The Power Density Spectrum

Consider a random process $x(t)$. Let $x(t)$ be the sample function and $x_T(t)$ be defined as follows

$$x_T(t) = \begin{cases} x(t), & -T < t < T \\ 0, & \text{elsewhere.} \end{cases} \tag{1}$$

for a finite $T$. Let us assume that $x_T(\omega)$ has bounded variation such that

$$\int_{-T}^{T} X_T(t)dt \tag{2}$$

and will have a fourier transform, which we denote by $X_T(\omega)$. It is defined by

$$x_T(\omega) = \int_{-T}^{T} x_T(t)e^{-j\omega t}dt$$

$$= \int_{-T}^{T} x(t)e^{-j\omega t}dt$$

The energy contained in the $x(t)$ in the interval $(-T, T)$ is

$$E(T) = \int_{-T}^{T} x_T^2(t)dt = \int_{-T}^{T} x^2(t)dt \tag{3}$$

By Parseval's theorem, the energy is related to $x_T(\omega)$.

Hence, we have, dividing (3) by $2T$, the average power $P(T)$ in $x(t)$ over the interval $(-T, T)$, where

$$P(t) = \frac{1}{2\pi}\int_{-T}^{T} x^2(t)dt = \frac{1}{2\pi}\int_{\infty}^{\infty} \frac{|x_T(\omega)|^2}{2\pi}d\omega \tag{4}$$

Applying the limit as $T \to \infty$ and taking the expected value of (4), we obtain a power density spectrum for the random process $x(t)$. It is important to note that the limiting operation is performed last. After performing these operations we get,

$$P_{XX} = \underset{T\to\infty}{\text{lt}} \frac{1}{2T}\int_{-T}^{T} E[x^2(t)]dt \tag{5}$$

Equation (5) establishes two facts.

(i) The average power $P_{XX}$ is a random process $x(t)$, given by the time interval of its second moment, i.e.

$$P_{XX} = \underset{T\to\infty}{\text{lt}} \frac{1}{2T}\int_{-T}^{T} E[x^2(t)]dt$$

$$= A\{E[x^2(t)]\}$$

For a process, that is at least wide-sense stationary $E[x^2(t)] = \overline{x^2}$ is constant and $P_{XX} = \overline{x^2}$.

(ii) $P_{XX}$ can be obtained by a frequency domain integration. If we obtain the power density spectrum for the random process by

$$S_{XX}(\omega) = \underset{T\to\infty}{\text{lt}} \frac{E[|X_T(\omega)|^2]}{2T} \tag{7}$$

The power formula is given by

$$P_{XX} = \frac{1}{2\pi}\int_{\infty}^{\infty} \underset{T\to\infty}{\text{lt}} \frac{E|X_T(\omega)|^2}{2T} d\omega$$

$$P_{XX} = \frac{1}{2\pi}\int_{-\infty}^{\infty} S_{XX}(\omega)d\omega \tag{8}$$

The power density spectrum is also called the power specified density or power spectrum. We occasionally use the name 'power density'.

### Properties of power density spectrum

(i) $S_{XX}(\omega) \geq 0$

(ii) $S_{XX}(-\omega) = S_{XX}(\omega)$ (where $x(t)$ is real)

(iii) $S_{XX}(\omega)$ is real

(iv) $\frac{1}{2\pi}\int_{-\infty}^{\infty} S_{XX}(\omega)d\omega = A\{E[x^2(t)]\}$

(v) $S_{XX}(\omega) = \omega^2 S_{XX}(\omega)$ (where $\dot{x}(t) = \frac{dx(t)}{dt}$)

For a wide-sense stationary process, we have

$$S_{XX} = \int_{-\infty}^{\infty} R_{XX}(\tau)e^{(-j\omega\tau)}d\tau$$

and $R_{XX} = \frac{1}{2\pi}\int_{-\infty}^{\infty} S_{XX}(\omega)e^{(-j\omega\tau)}d\omega$

**Bandwidth of the power density spectrum**

If $x(t)$ is a lowpass process, then it is called the base band. $S_{XX}(\omega)$ has characteristics similar to probability density function, expect for the fact that the area of $S_{XX}$ is not necessarily unity. By dividing $S_{XX}(\omega)$ with its area, we can form a new function with area unity.

For a real process, we know that $S_{XX}(\omega)$ is an even function, its mean value is zero and standard deviation is the square root of the second moment.

For density function, the standard deviation is a measure of its spread. The analogous quantity for the normalised power spectrum is a measure of its spread. It is called rms bandwidth and is denoted by $w_{\text{rms}}$ (rad/s). Thc rms bandwith is given by

$$w_{\text{rms}}^2 = \frac{\int_{-\infty}^{\infty} \omega^2 S_{XX}(\omega)d\omega}{\int_{-\infty}^{\infty} S_{XX}(\omega)d\omega}$$

## Solved examples

**Example 1** Find the power density spectrum for $R_{XX}(\tau) = k^2 e^{-2\alpha|\tau|}$.

**Solution**

Using the definition of power density spectrum, we have

$$S_{XX}(\omega) = \int_{-\infty}^{\infty} R_{XX}(\tau)e^{-i\omega\tau}dt$$

$$= \int_{-\infty}^{\infty} k^2 e^{-2\alpha|\tau|} e^{-i\omega\tau} d\tau + \int_{0}^{\infty} k^2 e^{-2\alpha|\tau|} e^{-i\omega\tau} d\tau$$

$$= \int_{-\infty}^{0} k^2 e^{2\alpha|\tau|} e^{-i\omega\tau} dt + \int_{0}^{\infty} k^2 e^{-2\alpha|\tau|} e^{-i\omega\tau} d\tau$$

$$= k^2 \int_{-\infty}^{0} e^{(2\alpha - i\omega)\tau} d\tau + k^2 \int_{0}^{\infty} e^{-(2\alpha + i\omega)\tau} d\tau$$

$$= k^2 \left[\frac{e^{(2\alpha - i\omega)\tau}}{(2\alpha - i\omega)}\right]_{-\infty}^{0} + k^2 \left[\frac{e^{-(2\alpha - i\omega)\tau}}{-(2\alpha + i\omega)}\right]_{0}^{\infty}$$

$$= k^2 \left[ \frac{1}{2\alpha - i\omega} + \frac{1}{2\alpha + i\omega} \right]$$

$$= k^2 \left[ \frac{2\alpha + i\omega + 2\alpha - i\omega}{(2\alpha)^2 - i^2\omega^2} \right]$$

$$= k^2 \left[ \frac{4\alpha}{4\alpha^2 + \omega^2} \right] = \frac{4k^2\alpha}{4\alpha^2 + \omega^2}$$

**Example 2** An ergodic process is known to have an auto-correlation function of the form $R_{XX}(\tau) = 1 - |\tau|, \quad |\tau| \le 1$

$$= 0, \qquad |\tau| > 1$$

Find the spectral analysis.

**Solution**

$$S_{XX}(\tau) = \int_{-\infty}^{\infty} R_{XX}(\tau) e^{-i\omega\tau} d\tau$$

$$= \int_{-1}^{1} (1 - |\tau|) e^{-i\omega\tau} d\tau$$

$$= \int_{-1}^{0} (1+\tau) e^{-i\omega\tau} d\tau + \int_{0}^{1} (1-\tau) e^{-i\omega\tau} d\tau$$

$$= \left[ (1+\tau) \frac{e^{-i\omega\tau}}{-i\omega} \right]_{-1}^{0} - \int_{0}^{1} \frac{e^{-i\omega\tau}}{-i\omega} \frac{d}{d\tau}(1+\tau) d\tau + \left[ \frac{(1-\tau)e^{-\omega\tau}}{-i\omega} \right]_{0}^{1} - \int_{0}^{1} \frac{e^{-i\omega\tau}}{-i\omega} \frac{d}{d\tau}(1-\tau) d\tau$$

$$= \frac{-1}{i\omega} - \int_{-1}^{0} \frac{e^{-i\omega\tau}}{-i\omega} d\tau + \frac{1}{i\omega} - \int_{0}^{1} \frac{e^{-i\omega\tau}}{-i\omega} d\tau$$

$$= \frac{1}{\omega^2}(1 - e^{i\omega}) + \frac{1}{\omega^2}(1 - e^{-i\omega})$$

$$= \frac{1}{\omega^2}(2 - e^{i\omega} - e^{-i\omega})$$

$$= \frac{1}{\omega^2}(2 - 2\cos\omega)$$

$$= \frac{2}{\omega^2}(1-\cos\omega) = \frac{2}{\omega^2}(1-(1-2\sin^2\omega/2)]$$

$$= \frac{2}{\omega^2}(1-1+2\sin^2\omega/2) = \frac{4}{\omega^2}\sin^2\omega/2$$

$$= \left[\frac{\sin\omega/2}{\omega/2}\right]^2$$

**Example 3** Find $\omega_{rms}$ for the power spectrum given by

$$S_{XX}(\omega) = \frac{10}{\left[\left(1+\left(\frac{\omega}{10}\right)^2\right)\right]^2}$$

where the 2-db band width is 10 rad/s.

**Solution**

We have

$$\int_{-\infty}^{\infty} S_{XX}(\omega) = \int_{-\infty}^{\infty} \frac{10d\omega}{\left[\left(1+\left(\frac{\omega}{10}\right)^2\right)\right]^2}$$

$$= 10^5 \int_{-\infty}^{\infty} \frac{d\omega}{[(100+\omega^2)]^2}$$

$$= 10^5 \left\{\left[\frac{\omega}{200(100+\omega^2)}\right]_{-\infty}^{\infty} + \frac{1}{2000}\left(\tan^{-1}\frac{\omega}{10}\right)_{-\infty}^{\infty}\right\}$$

$$= 10^5 \left\{0 + \frac{1\left(\frac{\pi}{2}+\frac{\pi}{2}\right)}{2000}\right\}$$

$$= \frac{10^5}{20000}\pi = \frac{10^2}{2}\pi = 50\pi$$

and $\displaystyle\int_{-\infty}^{\infty} S_{XX}(\omega)\,\omega^2\,d\omega = 10^5 \int_{-\infty}^{\infty} \frac{10\omega^2 d\omega}{\left[\left(1+\left(\frac{\omega}{10}\right)^2\right)\right]^2}$

$$= 10^5 \left\{ \left[ \frac{-\omega}{2(100+\omega^2)} \right]_{-\infty}^{\infty} + \frac{1}{20} \left[ \tan^{-1} \frac{\omega}{10} \right]_{-\infty}^{\infty} \right\}$$

$$= \frac{10^5 \pi}{(0+20)} = 5000\ \pi$$

$$= \frac{10^5 \times \pi}{20} = 500\ \pi$$

$$\omega_{rms}^2 = \frac{500\pi}{50\pi} = 100$$

Hence $\omega_{rms} = \sqrt{100} = 10$ rad/s

## 12.13 Wiener–Khinchine Relations

If $x(t)$ is a wide-sense stationary random process, the power density spectrum can be obtained by the Fourier transform of the auto-correlation function. We get

$$S_{XX}(\omega) = \int_{-\infty}^{\infty} R_{XX}(\tau) e^{-i\omega\tau} d\tau \tag{1}$$

Conversely we have,

$$R_{XX}(\tau) = \frac{1}{2\pi} \int_{-\infty}^{\infty} S_{XX}(\omega) e^{i\omega\tau} d\omega \tag{2}$$

The relations (1) and (2) are known as the Wiener–Khinchine relations.

### Cross-power density spectrum

Consider two real random processes $X(t)$ and $Y(t)$. Let $x(t)$ and $y(t)$ denote the sample function of $x(t)$ and $y(t)$ respectively. Define $x_T(t)$ and $y_T(t)$ as follows

$$x_T(t) = \begin{cases} x(t), & -T < t < T \\ 0, & \text{elsewhere} \end{cases}$$

$$\text{and } y_T(t) = \begin{cases} y(t), & -T < t < T \\ 0, & \text{elsewhere} \end{cases}$$

where $x_T(t)$ and $y_T(t)$ are to have bounded variation and be magnitude integrable over $(-T, T)$. Since, $x_T(t)$ and $y_T(t)$ both possess Fourier transforms, we denote them by $X_T(w)$ and $Y_T(w)$ respectively.

Let the cross-power $P_{xy}(T)$ in the two processes within $(-T, T)$ be defined by

$$P_{XY}(T) = \frac{1}{2T}\int_{-T}^{T} x(t)y(t)dt$$

$$= \frac{1}{2T}\int_{-T}^{T} x(t)y(t)dt$$

By Parseval's theorem we have,

$$P_{XY}(T) = \frac{1}{2T}\int_{-T}^{T} x(t)y(t)dt = \frac{1}{2T}\int_{-T}^{T} x_T(t)y_T(t)dt \tag{1}$$

which is a random quantity forming the average cross-power $\overline{P}_{xy}(T)$. By taking the expected value in (1), we get

$$\bar{P}_{XY}(T) = \frac{1}{2T}\int_{-T}^{T} R_{XY}(t,\ t)dt \tag{2}$$

$$= \frac{1}{2\pi}\int_{-\infty}^{\infty} \frac{[X_T^*(\omega)Y_T(\omega)]}{2T}d\omega$$

Applying the limit as T→ ∞, we form the total average cross-power $P_{XY}$ as follows

$$P_{XY} = \operatorname*{lt}_{T\to\infty} \frac{1}{2T}\int_{-T}^{T} R_{XY}(t,\ t) \tag{3}$$

$$= \frac{1}{2\pi}\int_{-\infty}^{\infty} \frac{E[X_T^*(\omega)Y_T(\omega)]}{2T}d\omega$$

The integrand involving ω can be defined as a cross-power density spectrum. It is a function of ω which is given by

$$S_{XY}(\omega) = \operatorname*{lt}_{T\to\infty} \frac{E[X_T^*(\omega)\ Y_T(\omega)]}{2T} \tag{4}$$

Thus, we get the cross-power formula

$$S_{YX}(\omega) = \operatorname*{lt}_{T\to\infty} \frac{E(Y_T^*(\omega)X_T(\omega))}{2T} \tag{5}$$

Similarly, we can define

$$S_{XY}(\omega) = \underset{T\to\infty}{\text{lt}} \frac{E(Y_T(\omega)X_T(\omega))}{2T} \tag{6}$$

and $P_{YX} = \dfrac{1}{2\pi}\displaystyle\int_{-\infty}^{\infty} S_{YX}(\omega) = P_{xy}^*$

The cross-power density spectrum and the time average of cross-correlation function form a Fourier transform pair. For jointly wide-sense stationary processes we have,

$$S_{XY}(\omega) = \int_{-\infty}^{\infty} R_{XY}(\tau)e^{-i\omega\tau}d\tau$$

$$S_{YX}(\omega) = \int_{-\infty}^{\infty} R_{YX}(\tau)e^{-i\omega\tau}d\tau$$

$$R_{XY}(\tau) = \frac{1}{2\pi}\int_{-\infty}^{\infty} S_{XY}(\omega)e^{-i\omega\tau}d\omega$$

$$R_{XY}(\tau) = \frac{1}{2\pi}\int_{-\infty}^{\infty} S_{XY}(\omega)e^{-i\omega\tau}d\omega$$

The cross-power spectrum of real random processes $x(t)$ and $y(t)$ have the following properties

(i) $S_{YX}(\omega) = S_{YX}(-\omega) = S_{YX}^*(\omega)$.

(ii) Re$[S_{XY}(\omega)]$ and Re $[S_{YX}(\omega)]$ are even functions of $\omega$.

(iii) Im $[S_{XY}(\omega)]$ and Im $[S_{YX}(\omega)]$ are odd functions of $\omega$.

(iv) If $x(t)$ and $y(t)$ are orthogonal, $S_{XY}(\omega) = 0$ and $S_{YX}(\omega) = 0$.

(v) If $x(t)$ and $y(t)$ are uncorrelated and have constant means $\bar{x}$ and $\bar{y}$, then $S_{XY}(\omega) = S_{YX}(\omega) = 2\pi\bar{x}\bar{y}S(\omega)$.

**Example** A cross power is defined by

$$S_{XY}(\omega) = \begin{cases} a+ib\omega, & -w < \omega < w \\ 0, & \text{elsewhere} \end{cases}$$

where $\omega > 0$, $a$ and $b$ are real constants. Find the cross-correlation function.

**Solution** We have

$$R_{XY} = \frac{1}{2\pi}\int_{-\infty}^{\infty} S_{XY}(\omega)e^{i\omega\tau}d\omega = \frac{1}{2\pi}\int_{-w}^{w}\left(a+i\frac{b\omega}{w}\right)e^{i\omega\tau}d\omega$$

$$= \frac{a}{2\pi}\int_{-w}^{w} e^{-i\omega\tau}d\omega + i\frac{b\omega}{2\pi w}\int_{-w}^{w} we^{-i\omega\tau}d\omega$$

$$= \frac{a}{2\pi}\left[\frac{e^{-i\omega\tau}}{i\tau}\right]\left[\frac{e^{-i\omega\tau}}{i\tau}\right]_{-w}^{w} + i\frac{b}{2\pi w}\left[e^{-i\omega\tau}\left(\frac{\omega}{i\tau}-\frac{1}{(i\tau)^2}\right)\right]_{-w}^{w}$$

$$= \frac{1}{\pi w\tau^2}[(aw\tau - b)\sin w\tau + bw\tau\cos w\tau]$$

## Exercise 12.2

1. Define
   (i) Random process
   (ii) Time average
   (iii) Discrete random process
   (iv) Continuous random process
   (v) Discrete random sequence
   (vi) Continuous random sequence

2. Define the terms
   (a) Stationary random process
   (b) First-order random process
   (c) Second-order stationary process
   (d) Wide-sense stationary process

3. A random process is defined by $x(t) = A$, where $A$ is a continuous random variable uniformly distributed on (0,1). Show that $x(t)$ is a wide-sense stationary process.

4. Given the random process $x(t) = A\cos\omega t + B\sin\omega t$, where $\omega$ is a constant, $A$ and $B$ are uncorrelated zero mean random variables having different density functions but the same variance $\sigma^2$, show that $x(t)$ is wide-sense stationary.

5. A random process is defined by $x(t) = At$, where $A$ is a continuous random variable uniformly distributed on (0,1). Show that the mean of the process is $\frac{1}{2}$ and $R_{xx}(\tau) = \frac{1}{2}[t(t+\pi)]$.

6. If $x(t) = \sin(\omega t + y)$, where $y$ is uniformly distributed in $(0, 2\pi)$, show that $x(t)$ is wide-sense stationary.

7. Define

   (i) Ergodic random process
   (ii) Joint ergodicity
   (iii) Mean ergodic random process

8. Define the terms

   (i) Auto-correlation
   (ii) Cross-correlation

9. Which of the following are suitable auto-correlation functions:

   (a) $A\cos\omega_0\tau$
   (b) $A\sin\omega_0\tau$
   (c) $A\pi\left(\frac{\tau}{\tau_0}\right)$, where $\pi(x)$ is a unit area rectangular function
   (d) $A^n\left(\frac{\tau}{\tau_0}\right)$, where $n(x)$ is the unit triangular function. [Ans: (a) (c) and (d)]

## 12.14 Markov Process

A stochastic process $\{X(t) : t \in T\}$ is called a Markov process if for any $t_0 < t_1 < t_2 < .... < t_n < t$, the conditional distribution of $X(t)$ for given values of $X(t_0), X(t_1), X(t_2), ...., X(t_n)$ depends only on $X(t_n)$.

i.e., $PX(t) \leq n | X(t_n) = X_n ; X(t_{n-1}) = X_{n-1}$,

$$X(t_0) = X$$

$$P[X(t) \leq n | X(t_n) = X_n]$$

In stochastic processes, we have two types of Markov processes.

(i) Markov processes with discrete state space

(ii) Markov processes with continuous state space

## 12.15 Markov Chain

Let $X(t)$ be a Markov process. If $X(t)$ possesses Markov property and takes only discrete values, whether $t$ is discrete or continuous, then $X(t)$ is called a Markov chain.

For example, consider the number of heads obtained in the first $n$ tosses of a coin. The possible values are 0, 1, 2, 3 ... $n$.

$$\text{Let } S_{n+1} = \begin{cases} x+1, & \text{if}\,(n+1)\text{th trial results in a head} \\ x, & \text{if}\,(n+1)\text{th trial result in a tail} \end{cases}$$

From this we have

$$S_n = X_0 + X_1 + ..... + X_n$$

i.e., the conditional probability of $S_{n+1}$ given $S_n$ depends on $S_n$ but not on the manner how the value of $S_n$ is derived.

If we let

$$P[X_n = a_n/X_{n-1} = a_{n-1}/X_{n-2} = a_{n-2}, ....., X_0 = a_0]$$

such that

$$P[X_n = a_n/X_{n-1} = a_{n-1}]\,\forall n$$

then this process $x_n \forall n,\ n = 0, 1, 2, ....$ is called a Markov chain.

Markov chains are finite. Therefore, it is possible to characterise the chain's state totality by a vector function of time. Assuming that there are $n$ states in the Markov chain, we can classify the chain as given in the next section.

## 12.16 Classification of States

1. State $j$ is said to be accessible from state $i$ if $P_{ij}^n > 0$ for some $n \geq 0$.
2. Two states $i$ and $j$ which are accessible to each other are said to communicate
   i. State $i$ communicates with itself for all $i \geq 0$.
   ii. If state $i$ communicates with state $j$ and state $j$ communicates with state $k$, then state $i$ communicates with state $k$.
3. Two states that communicate are in the same class. Two classes of a state are identical or disjoint.
4. The Markov chain is irreducible if there is only one class.
5. A state is said to be an absorbing state if no other state is accessible from it, i.e., from absorbing state $I$, $p_{ij} = 1$.

   **Markov chain is absorbing if**

   i. it has at least one absorbing state

ii. it is possible to go from every non-absorbing state to at least one absorbing state.

6. State $i$ is said to be recurrent if starting in state $i$, the process will re-enter state $i$ with probability $p_i = 1$.

   A state $i$ is recurrent if $\sum_{n=1}^{\infty} P_{ij}^{n} = \infty$.

   A state $i$ is positive recurrent, if starting in $i$, the expected time until the process returns to state $i$ is infinite. For a process, a recurrent state need not be positive recurrent. Such states are called null recurrent.

7. State $i$ is said to be transient, if starting from state $i$, the process will re-enter state $i$ with probability $i < 1$.

8. A state is called an essential state if it communicates with every state it leads to.

9. A state is said have a period $d$ if $P_{ij}^{n}$, whenever $n$ is not divisible by $d$ and $d$ is the largest integer with this property.

   If a state $i$ is periodic with period $d$, the g.c.d of the set $\{n|P_{ij}^{n} \geq 0, n \geq 1\}$ is $d > 1$.

   A state is said to be a periodic if its period is 1.

10. Positive recurrent, aperiodic states are called ergodic.

## 12.17 Examples of Markov Chains

(a) A random walk on the line is a Markov chain.

(b) Random placement of balls. Consider a sequence of independent trails each consisting of placing a ball at random in one of the $P$ given cells.

We say that the system is in state $\in_n$, if exactly $k$ cells are occupied. This determines a Markov chain with states $\in_0, \in_1, \ldots, \in_p$ and transition probabilities such that

$$P_{ij} = \frac{j}{P}, \qquad P_{i,i+j} = \frac{P-j}{P}$$

$P_{ik} = 0$ for all other combinations of $j$ and $k$

If initially all cells are empty, the distribution $\{a_k\}$ is determined by $a_0 = 1$ and $a_k = 0$ for $k > 0$.

## 12.18 Probability Vector

A vector $v = (v_1, v_2, \ldots, v_n)$ is called a probability vector if the components are non-negative and their sum is equal to 1.

**Stochastic matrix**

A square matrix $p = (p_{ij})$ is called a stochastic matrix if each of the rows is a probability vector, if each entry $P$ is non-negative and the sum of entries in each row is 1.

## Solved Examples

**Example 1** The matrix

$$P = \begin{bmatrix} 0 & 1 & 0 \\ 1/2 & 1/6 & 1/3 \\ 1/3 & 2/3 & 0 \end{bmatrix} \text{ is a stochastic matrix.}$$

**Example 2**

The matrix $A = \begin{bmatrix} 0 & 1 \\ 1/2 & 1/2 \end{bmatrix}$ is a stochastic matrix.

## 12.19 Limiting Distribution

The limiting distribution $\{v_k\}$ of an aperiodic irreducible Markov chain, whether it is discrete or a continuous parameter, with a finite number of states is given by

$$V_k = \frac{C_k}{\Sigma_j C_j}$$

where $C_j = \Sigma w_i(T_j)$.

## 12.20 Regular Stochastic Matrix

A stochastic matrix $P$ is said to be regular if all the entries of some power $P_n$ are positive.

## 12.21 Probability Distribution

Consider the Markov chain $\{X_1, X_2, ....\}$ for which $X_1$ is the first member. $X_1$ is called the initial random variable of the process.

Let $P_1^{(0)}$ denote the probability that $X_1$ takes the value $a_1, P_2^0$, denote the probability that $X_1$ takes the value $a_2$, and so on. Then the probability distribution of $X_1$ may be represented by the row matrix

$$A^{(0)} = [P_1^{(0)}.P_2^{(0)}...P_m^{(0)}] \tag{1}$$

This is called the initial probability distribution of the chain. If the probability distribution of $X_1$ is known, then the probability distribution of $X_{n+1}$ referred to as the $n$th step probability distribution of the system is represented by the row matrix

$$A^{(n)} \tag{2}$$

where $A^{(n)} = A^0 P^n$

Since $A^{(0)}$ is a row matrix of order $m$ and $P^n$ is an $m \times m$ matrix, we observe that $A^{(n)}$ is a row matrix of order $m$, of the form

$$A^{(n)} = A^{(0)} = \left[P_1^{(0)}, P_2^{(0)}, \ldots P_m^{(0)}\right]$$

$$\text{Suppose } \underset{n\to\infty}{\text{lt}} P_i^{(n)} = V_i, i = 1, 2, \ldots, m \tag{3}$$

$$\text{where } v_i \text{ are non-negative, independent of } P_i^{(0)} \text{ and } \sum_{i=1}^{\infty} V_i = 1, \tag{4}$$

then the row matrix

$$A^{\infty} = (v_1, v_2, \ldots, v_m) = V \tag{5}$$

represents what is called a steady-state distribution of the process. The row matrix $V$ is called a stationary probability vector of the chain.

For a given chain, the steady-state distribution is determined by the formulae

$$V = vP \tag{6}$$

## Solved examples

**Example 1** Show that the matrix

$$A = \begin{bmatrix} 0 & 1 \\ 1/2 & 1/2 \end{bmatrix}$$

is a stochastic matrix. Check whether it is regular.

**Solution** Consider $A = \begin{bmatrix} 0 & 1/2 \\ 1/2 & 1/2 \end{bmatrix}$.

All the entries of A are positive and sum of all entries in each row is equal to 1.

$\therefore A$ is a stochastic matrix.

$$\text{Now } A^2 = \begin{bmatrix} 0 & 1 \\ 1/2 & 1/2 \end{bmatrix}\begin{bmatrix} 0 & 1 \\ 1/2 & 1/2 \end{bmatrix}$$

$$= \begin{bmatrix} 0+1/2 & 0+1/2 \\ 0+1/4 & 1/2+1/2 \end{bmatrix}$$

$$= \begin{bmatrix} 1/2 & 1/2 \\ 1/4 & 1/4 \end{bmatrix}$$

Each entry in $A^2$ is positive and the sum of all entries in each row is equal to 1

i.e., $\sum_{i=1}^{\infty} a_i = 1$

Hence, $A$ is regular.

**Example 2** Three boys A, B and C are throwing a ball to each other. A always throws the ball to B and B always throws the ball to C. But C is just as likely to throw the ball to B as to A. If C was the first person to throw the ball, find the probabilities that

(i) A has the ball

(ii) B has the ball

(iii) C has the ball after three throws.

**Solution** We have

$P_{11}$ = Probability that the ball is thrown from A to A $= \infty$

$P_{12}$ = Probability that the ball is thrown from A to B $= 1$

$P_{13}$ = Probability that the ball is thrown from A to C $= 0$

$P_{21}$ = Probability that the ball is thrown from B to A $= 0$

$P_{22}$ = Probability that the ball is thrown from B to B $= 0$

$P_{23}$ = Probability that the ball is thrown from B to C $= 1$

$P_{31}$ = Probability that the ball is thrown from C to A $= \frac{1}{2}$

$P_{32}$ = Probability that the ball is thrown from C to B $= \frac{1}{2}$

$P_{33}$ = Probability that the ball is thrown from C to C $= 0$

The transition matrix is

$$P = \begin{bmatrix} 0 & 1 & 0 \\ 0 & 0 & 1 \\ 1/2 & 1/2 & 0 \end{bmatrix}$$

Hence, the 3-step transition matrix is

$$P^3 = \begin{bmatrix} 0 & 1 & 0 \\ 0 & 0 & 1 \\ 1/2 & 1/2 & 0 \end{bmatrix} \begin{bmatrix} 0 & 1 & 0 \\ 0 & 0 & 1 \\ 1/2 & 1/2 & 0 \end{bmatrix} \begin{bmatrix} 0 & 1 & 0 \\ 0 & 0 & 1 \\ 1/2 & 1/2 & 0 \end{bmatrix}$$

$$= \begin{bmatrix} 1/2 & 1/2 & 0 \\ 0 & 1/2 & 1/2 \\ 1/4 & 1/4 & 1/2 \end{bmatrix}$$

Now if the ball was first thrown by C, the initial probability distribution is given by the row matrix

$$A^{(0)} = \left[P_1^{(0)}, P_2^{(0)}, P_3^{(0)}\right] = [0, 0, 1]$$

After three rows, the probability is given by

$$\left[P_1^{(0)}, P_2^{(0)}, P(0)_3\right] = [A^{(0)} P^3]$$

$$= [0, 0, 1] \begin{bmatrix} 1/2 & 1/2 & 0 \\ 0 & 1/2 & 1/2 \\ 1/4 & 1/4 & 1/2 \end{bmatrix}$$

$$= \begin{bmatrix} 1/4 & 1/4 & 1/4 \end{bmatrix}$$

Thus after three rows, the probability that the ball is with

(i) A is 1/4

(ii) B is 1/4

(iii) C is 1/4

**Example 3** In a cascade of binary communication channels, the symbols 1 and 0 arc transmitted in successive stages. In any stage, the probability that a transmitted 1 is received as a 1 is 0.75 and the probability that 0 is received as a 0 is 0.5. Find the probability that

(i) 1 transmitted in the first stage is received correctly, and

(ii) 0 transmitted in the first stage is received as 1 after the third stage.

**Solution:** We have

$P_{11}$ = Probability that a transmitted 1 is received as 1 = 3/4

$P_{12}$ = Probability that a transmitted 1 is received as 0 = 1/4

$P_{21}$ = Probability that a transmitted 0 is received as 1 = 1/2

$P_{22}$ = Probability that a transmitted 0 is received as 0 = 1/2

The transition matrix is

$$P = \begin{bmatrix} 3/4 & 1/ \\ 1/2 & 1/2 \end{bmatrix}$$

Hence, we get

$$P^3 = \begin{bmatrix} 43/64 & 21/64 \\ 21/32 & 11/32 \end{bmatrix}$$

$$= \begin{bmatrix} 0.6719 & 0.3281 \\ 0.6563 & 0.3437 \end{bmatrix}$$

Therefore,

(i) The probability that 1 is transmitted in the first stage and is received correctly as 1 after the third states is 43/64 = 0.6719.

(ii) The probability that 0 is transmitted in the first stage and is received as a 1 after the third stage is 21/ 32 = 0.6563.

**Example 4** Consider the transition probability matrix of a Markov chain as

$$\begin{bmatrix} 0.3 & 0.7 & 0 \\ 0.1 & 0.4 & 0.5 \\ 0 & 0.2 & 0.8 \end{bmatrix}$$

Check whether the chain is irreducible.

**Solution** If we assume the states as 0, 1 and 2, it is possible to go from 0 to 1 with a probability 0.7 and from state 1 to 2 with probability 0.5. Hence it is possible to go from 0 to 2. In this way, all the states communicate with each other in the given Markov chain and thus, the given chain is irreducible.

**Example 5** A person owning a scooter has the opinion to switch over to a scooter, bike or a car next time with the probability of (0.3, 0.5, 0.2). If the transition probability matrix is

$$\begin{bmatrix} 0.4 & 0.3 & 0.3 \\ 0.2 & 0.5 & 0.3 \\ 0.25 & 0.25 & 0.5 \end{bmatrix}$$

what are the probabilities of vehicles to his fourth purchase?

**Solution** We have

$$P(1) = \begin{bmatrix} 0.4 & 0.3 & 0.3 \\ 0.2 & 0.5 & 0.3 \\ 0.25 & 0.25 & 0.5 \end{bmatrix}$$

$$\therefore P^3 = [P(1)]^3 = \begin{bmatrix} 0.4 & 0.3 & 0.3 \\ 0.2 & 0.5 & 0.3 \\ 0.25 & 0.25 & 0.5 \end{bmatrix} \begin{bmatrix} 0.4 & 0.3 & 0.3 \\ 0.2 & 0.5 & 0.3 \\ 0.25 & 0.25 & 0.5 \end{bmatrix} \begin{bmatrix} 0.4 & 0.3 & 0.3 \\ 0.2 & 0.5 & 0.3 \\ 0.25 & 0.25 & 0.5 \end{bmatrix}$$

$$= \begin{bmatrix} 0.277 & 0.351 & 0.372 \\ 0.269 & 0.359 & 0.372 \\ 0.275 & 0.345 & 0.380 \end{bmatrix}$$

$\therefore$ Probabilities of his fourth purchase

$$= \begin{pmatrix} 0.3 & 0.5 & 0.2 \end{pmatrix} P^3 = (0.2726, 0.3538, 0.3736)$$

**Example 6** Is the transition matrix $P = \begin{bmatrix} 1 & 0 & 0 \\ 0 & 1/2 & 1/2 \\ 0 & 1/3 & 2/3 \end{bmatrix}$

an absorbing Markov chain?

**Solution** We have

$$P = \begin{bmatrix} 1 & 0 & 0 \\ 0 & 1/2 & 1/2 \\ 0 & 1/3 & 2/3 \end{bmatrix}$$

$S_1$ is the only absorbing state and this cannot be reached from $S_2$ or $S_3$. Hence $P$ is not an absorbing Markov chain.

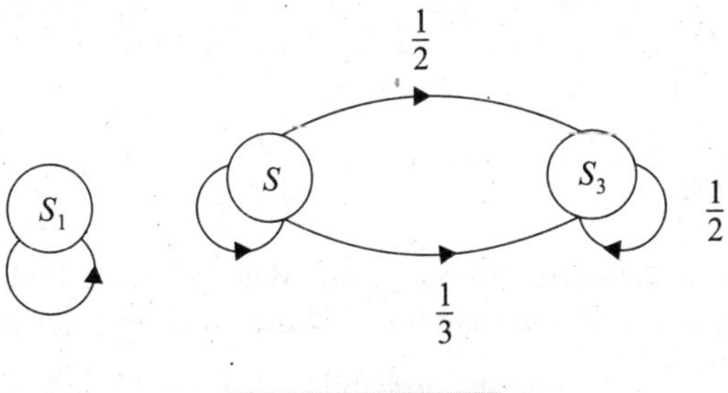


## Exercise 12.3

1. Consider the Markov chain with transient probability matrix $\begin{bmatrix} 0.4 & 0.6 & 0 & 0 \\ 0.3 & 0.7 & 0 & 0 \\ 0.2 & 0.4 & 0.1 & 0.3 \\ 0 & 0 & 0 & 1 \end{bmatrix}$. Is

it irreducible? If no, find the classes. Find the nature of states.

2. Find $P(n)$, for the homogeneous Markov chain with the following transition probability matrix $P = \begin{bmatrix} 1-a & a \\ b & 1-b \end{bmatrix}$, where $0 < a < 1, 0 < b < 1$

3. Check whether the following Markov chain is regular $\begin{bmatrix} 0 & 1/2 & 1/2 & 0 \\ 1/2 & 0 & 0 & 1/2 \\ 1/2 & 0 & 0 & 1/2 \\ 0 & 1/2 & 1/2 & 0 \end{bmatrix}$.

4. A student's study habits are as follows:

   If he studies one night, he is 60% sure not to study next night. On the other hand, if he does not study one night, he is 80% sure to study the next night. In the long run, how often does he study?

5. Define (i) stochastic process (ii) Markov chain

6. Define the terms: (i) absorbing state (ii) irreducible Markov chain (iii) recurrent state

7. Check whether the following matrices are stochastic matrices. (i) $A = \begin{bmatrix} 2/3 & 0 & 1/3 \\ -1/2 & 1/2 & 3/4 \\ 1/3 & 1/3 & 1/3 \end{bmatrix}$.

   (ii) $A = \begin{bmatrix} 3/4 & 1/4 \\ 1/3 & 1/3 \end{bmatrix}$.

8. Show that the matrix $A = \begin{bmatrix} 0 & 1 \\ 1/2 & 1/2 \end{bmatrix}$ is regular.

9. Consider the Markov chain with TPM

   $$P = \begin{bmatrix} 0 & 1 & 0 \\ 1/2 & 0 & 1/2 \\ 0 & 1 & 0 \end{bmatrix}$$

   Show that $P^2 n = P^2$.

10. A drilling process is considered to be a state Markov chain. If the process is there in the production company, it is in the state 0 and if the process is not there, it is in state 1. The TPM is $P = \begin{bmatrix} 0.1 & 0.2 \\ 0.3 & 0.4 \end{bmatrix}$. Find the probability that the drilling process will be there for 3 days from 1 day.

## Answers 12.3

1. Not irreducible. The classes are [0 1], [2], [3]. State 3 is an absorbing state.

2. The limiting transition matrix is $\underset{n\to\infty}{\text{lt}} P(n) = \begin{bmatrix} \frac{b}{a+b} & \frac{a}{a+b} \\ \frac{b}{a+b} & \frac{a}{a+b} \end{bmatrix}$.

4. [ 57% of the time]

7. (a) No (b) No

10. $P(3) = \begin{bmatrix} 0.037 & 0.054 \\ 0.081 & 0.218 \end{bmatrix}$

# APPENDIX

## Table I Logarithms

| | 0 | 1 | 2 | 3 | 4 | 5 | 6 | 7 | 8 | 9 | 1 | 2 | 3 | 4 | 5 | 6 | 7 | 8 | 9 |
|---|---|---|---|---|---|---|---|---|---|---|---|---|---|---|---|---|---|---|---|
| 10 | 0000 | 0043 | 0086 | 0128 | 0170 | 0212 | 0253 | 0294 | 0334 | 0374 | 4 | 8 | 12 | 17 | 21 | 25 | 29 | 33 | 37 |
| 11 | 0414 | 0453 | 0492 | 0531 | 0569 | 0607 | 0645 | 0682 | 0719 | 0755 | 4 | 8 | 11 | 15 | 19 | 23 | 26 | 30 | 34 |
| 12 | 0792 | 0828 | 0864 | 0899 | 0934 | 0969 | 1004 | 1038 | 1072 | 1106 | 3 | 7 | 10 | 14 | 17 | 21 | 24 | 28 | 31 |
| 13 | 1139 | 1173 | 1206 | 1239 | 1271 | 1303 | 1335 | 1367 | 1399 | 1430 | 3 | 6 | 10 | 13 | 16 | 19 | 23 | 26 | 29 |
| 14 | 1461 | 1492 | 1523 | 1553 | 1584 | 1614 | 1644 | 1673 | 1703 | 1732 | 3 | 6 | 9 | 12 | 15 | 18 | 21 | 24 | 27 |
| 15 | 1761 | 1790 | 1818 | 1847 | 1875 | 1903 | 1931 | 1959 | 1987 | 2014 | 3 | 6 | 8 | 11 | 14 | 17 | 20 | 22 | 25 |
| 16 | 2041 | 2068 | 2095 | 2122 | 2148 | 2175 | 2201 | 2227 | 2253 | 2279 | 3 | 5 | 8 | 11 | 13 | 16 | 18 | 21 | 24 |
| 17 | 2304 | 2330 | 2355 | 2380 | 2405 | 2430 | 2455 | 2480 | 2504 | 2529 | 2 | 5 | 7 | 10 | 12 | 15 | 17 | 20 | 22 |
| 18 | 2553 | 2577 | 2601 | 2625 | 2648 | 2672 | 2695 | 2718 | 2742 | 2765 | 2 | 5 | 7 | 9 | 12 | 14 | 16 | 19 | 21 |
| 19 | 2788 | 2810 | 2833 | 2856 | 2878 | 2900 | 2923 | 2945 | 2967 | 2989 | 2 | 4 | 7 | 9 | 11 | 13 | 16 | 18 | 20 |
| 20 | 3010 | 3032 | 3054 | 3075 | 3096 | 3118 | 3139 | 3160 | 3181 | 3201 | 2 | 4 | 6 | 8 | 11 | 13 | 15 | 17 | 19 |
| 21 | 3222 | 3243 | 3263 | 3284 | 3304 | 3324 | 3345 | 3365 | 3385 | 3404 | 2 | 4 | 6 | 8 | 10 | 12 | 14 | 16 | 18 |
| 22 | 3424 | 3444 | 3464 | 3483 | 3502 | 3522 | 3541 | 3560 | 3579 | 3598 | 2 | 4 | 6 | 8 | 10 | 12 | 14 | 15 | 17 |
| 23 | 3617 | 3636 | 3655 | 3674 | 3692 | 3711 | 3729 | 3747 | 3766 | 3784 | 2 | 4 | 6 | 7 | 9 | 11 | 13 | 15 | 17 |
| 24 | 3802 | 3820 | 3838 | 3856 | 3874 | 3892 | 3909 | 3927 | 3945 | 3962 | 2 | 4 | 5 | 7 | 9 | 11 | 12 | 14 | 16 |
| 25 | 3979 | 3997 | 4014 | 4031 | 4048 | 4065 | 4082 | 4099 | 4116 | 4133 | 2 | 3 | 5 | 7 | 9 | 10 | 12 | 14 | 15 |
| 26 | 4150 | 4166 | 4183 | 4200 | 4216 | 4232 | 4249 | 4265 | 4281 | 4298 | 2 | 3 | 5 | 7 | 8 | 10 | 11 | 13 | 15 |
| 27 | 4314 | 4330 | 4346 | 4362 | 4378 | 4393 | 4409 | 4425 | 4440 | 4456 | 2 | 3 | 5 | 6 | 8 | 9 | 11 | 13 | 14 |
| 28 | 4472 | 4487 | 4502 | 4518 | 4533 | 4548 | 4564 | 4579 | 4594 | 4609 | 2 | 3 | 5 | 6 | 8 | 9 | 11 | 12 | 14 |
| 29 | 4624 | 4639 | 4654 | 4669 | 4683 | 4698 | 4713 | 4728 | 4742 | 4757 | 1 | 3 | 4 | 6 | 7 | 9 | 10 | 12 | 13 |
| 30 | 4771 | 4736 | 4800 | 4814 | 4829 | 4843 | 4857 | 4871 | 4886 | 4900 | 1 | 3 | 4 | 6 | 7 | 9 | 10 | 11 | 13 |
| 31 | 4914 | 4928 | 4942 | 4955 | 4969 | 4983 | 4997 | 5011 | 5024 | 5038 | 1 | 3 | 4 | 6 | 7 | 8 | 10 | 11 | 12 |
| 32 | 5051 | 5065 | 5079 | 5092 | 5105 | 5119 | 5132 | 5145 | 5159 | 5172 | 1 | 3 | 4 | 5 | 7 | 8 | 9 | 11 | 12 |
| 33 | 5185 | 5198 | 5211 | 5224 | 5237 | 5250 | 5263 | 5276 | 5289 | 5302 | 1 | 3 | 4 | 5 | 6 | 8 | 9 | 10 | 12 |
| 34 | 5315 | 5328 | 5340 | 5353 | 5366 | 5378 | 5391 | 5403 | 5416 | 5428 | 1 | 3 | 4 | 5 | 6 | 8 | 9 | 10 | 11 |
| 35 | 5441 | 5453 | 5465 | 5478 | 5490 | 5502 | 5514 | 5527 | 5539 | 5551 | 1 | 2 | 4 | 6 | 8 | 7 | 9 | 10 | 11 |
| 36 | 5563 | 5575 | 5587 | 5599 | 5611 | 5623 | 5635 | 5647 | 5658 | 5670 | 1 | 2 | 4 | 5 | 6 | 7 | 8 | 10 | 11 |
| 37 | 5682 | 5694 | 5705 | 5717 | 5729 | 5740 | 5752 | 5763 | 5775 | 5786 | 1 | 2 | 3 | 5 | 6 | 7 | 8 | 9 | 10 |
| 38 | 5798 | 5809 | 5821 | 5832 | 5843 | 5855 | 5866 | 5877 | 5888 | 5899 | 1 | 2 | 3 | 5 | 6 | 7 | 8 | 9 | 10 |
| 39 | 5911 | 5922 | 5933 | 5944 | 5955 | 5966 | 5977 | 5988 | 5999 | 6010 | 1 | 2 | 3 | 4 | 5 | 7 | 8 | 9 | 10 |
| 40 | 6021 | 6031 | 6042 | 6053 | 6064 | 6075 | 6085 | 6096 | 6107 | 6117 | 1 | 2 | 3 | 4 | 5 | 6 | 8 | 9 | 10 |
| 41 | 6128 | 6133 | 6149 | 6160 | 6170 | 6180 | 6191 | 6201 | 6212 | 6222 | 1 | 2 | 3 | 4 | 5 | 6 | 7 | 8 | 9 |
| 42 | 6232 | 6243 | 6253 | 6263 | 6274 | 6284 | 6294 | 6304 | 6314 | 6325 | 1 | 2 | 3 | 4 | 5 | 6 | 7 | 8 | 9 |
| 43 | 6335 | 6345 | 6355 | 6365 | 6375 | 6385 | 6395 | 6405 | 6415 | 6425 | 1 | 2 | 3 | 4 | 5 | 6 | 7 | 8 | 9 |
| 44 | 6435 | 6444 | 6454 | 6464 | 6474 | 6484 | 6493 | 6503 | 6513 | 6522 | 1 | 2 | 3 | 4 | 5 | 6 | 7 | 8 | 9 |
| 45 | 6532 | 6542 | 6551 | 6561 | 6571 | 6580 | 6590 | 6599 | 6609 | 6618 | 1 | 2 | 3 | 4 | 5 | 6 | 7 | 8 | 9 |
| 46 | 6628 | 6637 | 6646 | 6656 | 6665 | 6675 | 6684 | 6693 | 6702 | 6712 | 1 | 2 | 3 | 4 | 5 | 6 | 7 | 7 | 8 |
| 47 | 6721 | 6730 | 6739 | 6749 | 6758 | 6767 | 6776 | 6785 | 6794 | 6803 | 1 | 2 | 3 | 4 | 5 | 5 | 6 | 7 | 8 |
| 48 | 6812 | 6821 | 6830 | 6839 | 6848 | 6857 | 6866 | 6875 | 6884 | 6893 | 1 | 2 | 3 | 4 | 4 | 5 | 6 | 7 | 8 |
| 49 | 6902 | 6911 | 6920 | 6928 | 6937 | 6946 | 6955 | 6964 | 6972 | 6981 | 1 | 2 | 3 | 4 | 4 | 5 | 6 | 7 | 8 |
| 50 | 6990 | 6998 | 7007 | 7016 | 7024 | 7033 | 7042 | 7050 | 7059 | 7067 | 1 | 2 | 3 | 3 | 4 | 5 | 6 | 7 | 8 |
| 51 | 7076 | 7084 | 7093 | 7101 | 7110 | 7118 | 7126 | 7135 | 7143 | 7152 | 1 | 2 | 3 | 3 | 4 | 5 | 6 | 7 | 8 |
| 52 | 7160 | 7168 | 7177 | 7185 | 7193 | 7202 | 7210 | 7218 | 7226 | 7235 | 1 | 2 | 2 | 3 | 4 | 5 | 6 | 7 | 7 |
| 53 | 7243 | 7251 | 7259 | 7267 | 7275 | 7284 | 7292 | 7300 | 7308 | 7316 | 1 | 2 | 2 | 3 | 4 | 5 | 6 | 6 | 7 |
| 54 | 7324 | 7332 | 7340 | 7348 | 7356 | 7364 | 7372 | 7380 | 7388 | 7396 | 1 | 2 | 2 | 3 | 4 | 5 | 6 | 6 | 7 |

| 55 | 7404 | 7412 | 7419 | 7427 | 7435 | 7443 | 7451 | 7459 | 7466 | 7474 | 1 | 2 | 2 | 3 | 4 | 5 | 5 | 6 | 7 |
|---|---|---|---|---|---|---|---|---|---|---|---|---|---|---|---|---|---|---|---|
| 56 | 7482 | 7490 | 7497 | 7505 | 7513 | 7520 | 7528 | 7536 | 7543 | 7551 | 1 | 2 | 2 | 3 | 4 | 5 | 5 | 6 | 7 |
| 57 | 7559 | 7566 | 7574 | 7582 | 7589 | 7597 | 7604 | 7612 | 7619 | 7627 | 1 | 2 | 2 | 3 | 4 | 5 | 5 | 6 | 7 |
| 58 | 7634 | 7642 | 7649 | 7657 | 7664 | 7672 | 7679 | 7686 | 7694 | 7701 | 1 | 1 | 2 | 3 | 4 | 4 | 5 | 6 | 7 |
| 59 | 7709 | 7716 | 7723 | 7731 | 7738 | 7745 | 7752 | 7760 | 7767 | 7774 | 1 | 1 | 2 | 3 | 4 | 4 | 5 | 6 | 7 |
| 60 | 7782 | 7789 | 7796 | 7803 | 7810 | 7818 | 7825 | 7832 | 7839 | 7846 | 1 | 1 | 2 | 3 | 4 | 4 | 5 | 6 | 6 |
| 61 | 7853 | 7860 | 7868 | 7875 | 7882 | 7889 | 7896 | 7903 | 7910 | 7917 | 1 | 1 | 2 | 3 | 4 | 4 | 5 | 6 | 6 |
| 62 | 7924 | 7931 | 7938 | 7945 | 7952 | 7959 | 7966 | 7973 | 7980 | 7987 | 1 | 1 | 2 | 3 | 3 | 4 | 5 | 6 | 6 |
| 63 | 7993 | 8000 | 8007 | 8014 | 8021 | 8028 | 8035 | 8041 | 8048 | 8055 | 1 | 1 | 2 | 3 | 3 | 4 | 5 | 5 | 6 |
| 64 | 8062 | 8069 | 8075 | 8082 | 8089 | 8096 | 8102 | 8109 | 8116 | 8122 | 1 | 1 | 2 | 3 | 3 | 4 | 5 | 5 | 6 |
| 65 | 8129 | 8136 | 8142 | 8149 | 8156 | 8162 | 8169 | 8176 | 8182 | 8189 | 1 | 1 | 2 | 3 | 3 | 4 | 5 | 5 | 6 |
| 66 | 8195 | 8202 | 8209 | 8215 | 8222 | 8228 | 8235 | 8241 | 8248 | 8254 | 1 | 1 | 2 | 3 | 3 | 4 | 5 | 5 | 6 |
| 67 | 8261 | 8267 | 8274 | 8280 | 8287 | 3293 | 8299 | 8306 | 8312 | 8319 | 1 | 1 | 2 | 3 | 3 | 4 | 5 | 5 | 6 |
| 68 | 8325 | 8331 | 8338 | 8344 | 8351 | 8357 | 8363 | 8370 | 8376 | 8382 | 1 | 1 | 2 | 3 | 3 | 4 | 4 | 5 | 6 |
| 69 | 8388 | 8395 | 8401 | 8407 | 8414 | 8420 | 8426 | 8432 | 8439 | 8445 | 1 | 1 | 2 | 2 | 3 | 4 | 4 | 5 | 6 |
| 70 | 8451 | 8457 | 8463 | 8470 | 8476 | 8482 | 8488 | 8494 | 8500 | 8506 | 1 | 1 | 2 | 2 | 3 | 4 | 4 | 5 | 6 |
| 71 | 8513 | 8519 | 8525 | 8531 | 8537 | 8543 | 8549 | 8555 | 8561 | 8567 | 1 | 1 | 2 | 2 | 3 | 4 | 4 | 5 | 5 |
| 72 | 8573 | 8579 | 8585 | 8591 | 8597 | 3603 | 8609 | 8615 | 8621 | 8627 | 1 | 1 | 2 | 2 | 3 | 4 | 4 | 5 | 5 |
| 73 | 8633 | 8639 | 8645 | 8651 | 8657 | 8663 | 8669 | 8675 | 8681 | 8686 | 1 | 1 | 2 | 2 | 3 | 4 | 4 | 5 | 5 |
| 74 | 8692 | 8697 | 8704 | 8710 | 8716 | 8722 | 8727 | 8733 | 8739 | 8745 | 1 | 1 | 2 | 2 | 3 | 4 | 4 | 5 | 5 |
| 75 | 8751 | 8756 | 8762 | 8768 | 8774 | 8779 | 8785 | 8791 | 8797 | 8802 | 1 | 1 | 2 | 2 | 3 | 3 | 4 | 5 | 5 |
| 76 | 8808 | 8814 | 8820 | 8825 | 8831 | 8837 | 8842 | 8848 | 8854 | 8859 | 1 | 1 | 2 | 2 | 3 | 3 | 4 | 5 | 5 |
| 77 | 8865 | 8871 | 8876 | 8882 | 8887 | 8893 | 8899 | 8904 | 8910 | 8915 | 1 | 1 | 2 | 2 | 3 | 3 | 4 | 4 | 5 |
| 78 | 8921 | 8927 | 8932 | 8938 | 8943 | 8949 | 8954 | 8960 | 8965 | 8971 | 1 | 1 | 2 | 2 | 3 | 3 | 4 | 4 | 5 |
| 79 | 8976 | 8982 | 8987 | 8993 | 8998 | 9004 | 9009 | 9015 | 9020 | 9025 | 1 | 1 | 2 | 2 | 3 | 3 | 4 | 4 | 5 |
| 80 | 9031 | 9036 | 9042 | 9047 | 9053 | 9058 | 9063 | 9069 | 9074 | 9079 | 1 | 1 | 2 | 2 | 3 | 3 | 4 | 4 | 5 |
| 81 | 9085 | 9090 | 9096 | 9101 | 9106 | 9112 | 9117 | 9122 | 9128 | 9133 | 1 | 1 | 2 | 2 | 3 | 3 | 4 | 4 | 5 |
| 82 | 9138 | 9143 | 9149 | 9157 | 9159 | 9165 | 9170 | 9175 | 9180 | 9186 | 1 | 1 | 2 | 2 | 3 | 3 | 4 | 4 | 5 |
| 83 | 9191 | 9196 | 9201 | 9206 | 9212 | 9217 | 9222 | 9227 | 9232 | 9238 | 1 | 1 | 2 | 2 | 3 | 3 | 4 | 4 | 5 |
| 84 | 9243 | 9248 | 9253 | 9258 | 9263 | 9269 | 9274 | 9279 | 9284 | 9289 | 1 | 1 | 2 | 2 | 3 | 3 | 4 | 4 | 5 |
| 85 | 9294 | 9299 | 9304 | 9309 | 9315 | 9320 | 9325 | 9330 | 9335 | 9340 | 1 | 1 | 2 | 2 | 3 | 3 | 4 | 4 | 5 |
| 86 | 9345 | 9350 | 9355 | 9360 | 9365 | 9370 | 9375 | 9380 | 9385 | 9390 | 1 | 1 | 2 | 2 | 3 | 3 | 4 | 4 | 5 |
| 87 | 9395 | 9400 | 9405 | 9410 | 9415 | 9420 | 9425 | 9430 | 9435 | 9440 | 0 | 1 | 1 | 2 | 2 | 3 | 3 | 4 | 4 |
| 88 | 9445 | 9450 | 9455 | 9460 | 9465 | 9469 | 9474 | 9479 | 9484 | 9489 | 0 | 1 | 1 | 2 | 2 | 3 | 3 | 4 | 4 |
| 89 | 9494 | 9499 | 9504 | 9509 | 9513 | 9518 | 9523 | 9528 | 9533 | 9538 | 0 | 1 | 1 | 2 | 2 | 3 | 3 | 4 | 4 |
| 90 | 9542 | 9547 | 9552 | 9557 | 9562 | 9566 | 9571 | 9576 | 9581 | 9586 | 0 | 1 | 1 | 2 | 2 | 3 | 3 | 4 | 4 |
| 91 | 9590 | 9595 | 9600 | 9605 | 9609 | 9614 | 9619 | 9624 | 9628 | 9633 | 0 | 1 | 1 | 2 | 2 | 3 | 3 | 4 | 4 |
| 92 | 9638 | 9643 | 9647 | 9652 | 9657 | 9661 | 9666 | 9671 | 9675 | 9680 | 0 | 1 | 1 | 2 | 2 | 3 | 3 | 4 | 4 |
| 93 | 9685 | 9689 | 9694 | 9699 | 9703 | 9708 | 9713 | 9717 | 9722 | 9727 | 0 | 1 | 1 | 2 | 2 | 3 | 3 | 4 | 4 |
| 94 | 9731 | 9736 | 9741 | 9745 | 9750 | 9754 | 9759 | 9763 | 9768 | 9773 | 0 | 1 | 1 | 2 | 2 | 3 | 3 | 4 | 4 |
| 95 | 9777 | 9782 | 9786 | 9791 | 9795 | 9800 | 9805 | 9809 | 9814 | 9818 | 0 | 1 | 1 | 2 | 2 | 3 | 3 | 4 | 4 |
| 96 | 9823 | 9827 | 9632 | 9836 | 9841 | 9845 | 9850 | 9854 | 9859 | 9863 | 0 | 1 | 1 | 2 | 2 | 3 | 3 | 4 | 4 |
| 97 | 9868 | 9872 | 9871 | 9887 | 9886 | 9890 | 9894 | 9899 | 9903 | 9908 | 0 | 1 | 1 | 2 | 2 | 3 | 3 | 4 | 4 |
| 98 | 9912 | 9917 | 9921 | 9926 | 9930 | 9934 | 9939 | 9943 | 9948 | 9952 | 0 | 1 | 1 | 2 | 2 | 3 | 3 | 4 | 4 |
| 99 | 9956 | 9961 | 9965 | 9969 | 9974 | 9978 | 9983 | 9987 | 9991 | 9996 | 0 | 1 | 1 | 2 | 2 | 3 | 3 | 3 | 4 |

**Table II Anti-logarithms**

| | 0 | 1 | 2 | 3 | 4 | 5 | 6 | 7 | 8 | 9 | 1 | 2 | 3 | 4 | 5 | 6 | 7 | 8 | 9 |
|---|---|---|---|---|---|---|---|---|---|---|---|---|---|---|---|---|---|---|---|
| .50 | 3162 | 3170 | 3177 | 3184 | 3192 | 3199 | 3206 | 3214 | 3221 | 3228 | 1 | 1 | 2 | 3 | 4 | 4 | 5 | 6 | 7 |
| .51 | 3236 | 3243 | 3251 | 3258 | 3266 | 3273 | 3281 | 3289 | 3296 | 3304 | 1 | 2 | 2 | 3 | 4 | 5 | 5 | 6 | 7 |
| .52 | 3211 | 3319 | 3327 | 3334 | 3342 | 3350 | 3357 | 3365 | 3373 | 3381 | 1 | 2 | 2 | 3 | 4 | 5 | 5 | 6 | 7 |
| .53 | 3388 | 3396 | 3404 | 3412 | 3420 | 3428 | 3436 | 3443 | 3451 | 3459 | 1 | 2 | 2 | 3 | 4 | 5 | 6 | 6 | 7 |
| .54 | 3467 | 3475 | 3483 | 3491 | 3499 | 3508 | 3516 | 3527 | 3532 | 3540 | 1 | 2 | 2 | 3 | 4 | 5 | 6 | 6 | 7 |
| .55 | 3548 | 3556 | 3565 | 3573 | 3581 | 3589 | 3597 | 3606 | 3614 | 3622 | 1 | 2 | 2 | 3 | 4 | 5 | 6 | 7 | 7 |
| .56 | 3631 | 3639 | 3648 | 3656 | 3664 | 3673 | 3681 | 3690 | 3698 | 3707 | 1 | 2 | 3 | 3 | 4 | 5 | 6 | 7 | 8 |
| .57 | 3715 | 3724 | 3733 | 3741 | 3750 | 3758 | 3767 | 3776 | 3784 | 3793 | 1 | 2 | 3 | 3 | 4 | 5 | 6 | 7 | 8 |
| .58 | 3802 | 3811 | 3819 | 3828 | 3837 | 3846 | 3855 | 3864 | 3873 | 3882 | 1 | 2 | 3 | 4 | 4 | 5 | 6 | 7 | 8 |
| .59 | 3890 | 3899 | 3908 | 3917 | 3926 | 3936 | 3945 | 3954 | 3963 | 3972 | 1 | 2 | 3 | 4 | 5 | 5 | 6 | 7 | 8 |
| .60 | 3981 | 3990 | 3999 | 4009 | 4018 | 4027 | 4036 | 4046 | 4055 | 4064 | 1 | 2 | 3 | 4 | 5 | 6 | 6 | 7 | 8 |
| .61 | 4074 | 4083 | 4093 | 4102 | 4111 | 4121 | 4130 | 4140 | 4150 | 4159 | 1 | 2 | 3 | 4 | 5 | 6 | 7 | 8 | 9 |
| .62 | 4169 | 4178 | 4188 | 4198 | 4207 | 4217 | 4227 | 4236 | 4246 | 4256 | 1 | 2 | 3 | 4 | 5 | 6 | 7 | 8 | 9 |
| .63 | 4266 | 4276 | 4285 | 4295 | 4305 | 4315 | 4325 | 4335 | 4345 | 4355 | 1 | 2 | 3 | 4 | 5 | 6 | 7 | 8 | 9 |
| .64 | 4365 | 4375 | 4385 | 4395 | 4406 | 4416 | 4426 | 4436 | 4446 | 4457 | 1 | 2 | 3 | 4 | 5 | 6 | 7 | 8 | 9 |
| .65 | 4467 | 4477 | 4487 | 4498 | 4508 | 4519 | 4529 | 4539 | 4550 | 4560 | 1 | 2 | 3 | 4 | 5 | 6 | 7 | 8 | 9 |
| .66 | 4571 | 4581 | 4592 | 4603 | 4613 | 4624 | 4634 | 4645 | 4656 | 4667 | 1 | 2 | 3 | 4 | 5 | 6 | 7 | 9 | 10 |
| .67 | 4677 | 4688 | 4699 | 4710 | 4721 | 4732 | 4742 | 4753 | 4764 | 4775 | 1 | 2 | 3 | 4 | 5 | 7 | 8 | 9 | 10 |
| .68 | 4786 | 4797 | 4808 | 4819 | 4831 | 4842 | 4853 | 4864 | 4875 | 4887 | 1 | 2 | 3 | 4 | 6 | 7 | 8 | 9 | 10 |
| .69 | 4898 | 4909 | 4920 | 4932 | 4943 | 4955 | 4966 | 4977 | 4989 | 5000 | 1 | 2 | 3 | 5 | 6 | 7 | 8 | 9 | 10 |
| .70 | 5012 | 5023 | 5035 | 5047 | 5058 | 5070 | 5082 | 5093 | 5105 | 5117 | 1 | 2 | 4 | 5 | 6 | 7 | 8 | 9 | 11 |
| .71 | 5129 | 5140 | 5152 | 5164 | 5176 | 5188 | 5200 | 5212 | 5224 | 5236 | 1 | 2 | 4 | 5 | 6 | 7 | 8 | 10 | 11 |
| .72 | 5248 | 5260 | 5272 | 5284 | 5297 | 5309 | 5321 | 5333 | 5346 | 5358 | 1 | 2 | 4 | 5 | 6 | 7 | 9 | 10 | 11 |
| .73 | 5370 | 5383 | 5395 | 5408 | 5420 | 5433 | 5445 | 5458 | 5470 | 5483 | 1 | 3 | 4 | 5 | 6 | 8 | 9 | 10 | 11 |
| .74 | 5495 | 5508 | 5521 | 5534 | 5546 | 5559 | 5572 | 5585 | 5598 | 5610 | 1 | 3 | 4 | 5 | 6 | 8 | 9 | 10 | 12 |
| .75 | 5623 | 5636 | 5649 | 5662 | 5675 | 5689 | 5702 | 5715 | 5728 | 5741 | 1 | 3 | 4 | 5 | 7 | 8 | 9 | 10 | 12 |
| .76 | 5754 | 5768 | 5781 | 5794 | 5808 | 5821 | 5834 | 5848 | 5861 | 5875 | 1 | 3 | 4 | 5 | 7 | 8 | 9 | 11 | 12 |
| .77 | 5888 | 5902 | 5916 | 5929 | 5943 | 5957 | 5970 | 5984 | 5998 | 6012 | 1 | 3 | 4 | 5 | 7 | 8 | 10 | 11 | 12 |
| .78 | 6026 | 6039 | 6053 | 6067 | 6081 | 6095 | 6109 | 6124 | 6138 | 6152 | 1 | 3 | 4 | 6 | 7 | 8 | 10 | 11 | 13 |
| .79 | 6166 | 6180 | 6194 | 6209 | 6223 | 6237 | 6252 | 6266 | 6281 | 6295 | 1 | 3 | 4 | 6 | 7 | 9 | 10 | 11 | 13 |
| .80 | 6310 | 6324 | 6339 | 6353 | 6368 | 6383 | 6397 | 6412 | 6427 | 6442 | 1 | 3 | 4 | 6 | 7 | 9 | 10 | 12 | 13 |
| .81 | 6457 | 6471 | 6486 | 6501 | 6516 | 6531 | 6546 | 6561 | 6577 | 6592 | 2 | 3 | 5 | 6 | 8 | 9 | 11 | 12 | 14 |
| .82 | 6607 | 6622 | 6637 | 6653 | 6668 | 6683 | 6699 | 6714 | 6730 | 6745 | 2 | 3 | 5 | 6 | 8 | 9 | 11 | 12 | 14 |
| .83 | 6761 | 6776 | 6792 | 6808 | 6823 | 6839 | 6855 | 6871 | 6887 | 6902 | 2 | 3 | 5 | 6 | 8 | 9 | 11 | 13 | 14 |
| .84 | 6918 | 6934 | 6950 | 6966 | 6982 | 6998 | 7015 | 7031 | 7047 | 7063 | 2 | 3 | 5 | 6 | 8 | 10 | 11 | 13 | 15 |
| .85 | 7079 | 7096 | 7112 | 7129 | 7145 | 7161 | 7178 | 7194 | 7211 | 7228 | 2 | 3 | 5 | 7 | 8 | 10 | 12 | 13 | 15 |
| .86 | 7244 | 7261 | 7278 | 7295 | 7311 | 7328 | 7345 | 7362 | 7379 | 7396 | 2 | 3 | 5 | 7 | 8 | 10 | 12 | 13 | 15 |
| .87 | 7413 | 7430 | 7447 | 7464 | 7482 | 7499 | 7516 | 7534 | 7551 | 7568 | 2 | 3 | 5 | 7 | 9 | 10 | 12 | 14 | 16 |
| .88 | 7586 | 7603 | 7621 | 7638 | 7656 | 7674 | 7691 | 7709 | 7727 | 7745 | 2 | 4 | 5 | 7 | 9 | 11 | 12 | 14 | 16 |
| .89 | 7762 | 7780 | 7798 | 7816 | 7834 | 7852 | 7870 | 7889 | 7907 | 7925 | 2 | 4 | 5 | 7 | 9 | 11 | 13 | 14 | 16 |
| .90 | 7943 | 7962 | 7980 | 7998 | 8017 | 8035 | 8054 | 8072 | 8091 | 8110 | 2 | 4 | 6 | 7 | 9 | 11 | 13 | 15 | 17 |
| .91 | 8128 | 8147 | 8166 | 8185 | 8204 | 8222 | 8241 | 8260 | 8279 | 8299 | 2 | 4 | 6 | 8 | 9 | 11 | 13 | 15 | 17 |
| .92 | 8318 | 8337 | 8356 | 8375 | 8395 | 8414 | 8433 | 8453 | 8472 | 8492 | 2 | 4 | 6 | 8 | 10 | 12 | 14 | 15 | 17 |
| .93 | 8511 | 8531 | 8551 | 8570 | 8590 | 8610 | 8630 | 8650 | 8670 | 8690 | 2 | 4 | 6 | 8 | 10 | 12 | 14 | 16 | 18 |
| .94 | 8710 | 8730 | 8750 | 8770 | 8790 | 8810 | 8831 | 8851 | 8872 | 8892 | 2 | 4 | 6 | 8 | 10 | 12 | 14 | 16 | 18 |
| .95 | 8913 | 8933 | 8954 | 8974 | 8995 | 9016 | 9036 | 9057 | 9078 | 9099 | 2 | 4 | 6 | 8 | 10 | 12 | 15 | 17 | 19 |
| .96 | 9120 | 9141 | 9162 | 9183 | 9204 | 9226 | 9247 | 9268 | 9290 | 9311 | 2 | 4 | 6 | 8 | 11 | 13 | 15 | 17 | 19 |
| .97 | 9333 | 9354 | 9376 | 9397 | 9419 | 9441 | 9462 | 9484 | 9506 | 9528 | 2 | 4 | 7 | 9 | 11 | 13 | 15 | 17 | 20 |
| .98 | 9550 | 9572 | 9594 | 9616 | 9638 | 9661 | 9683 | 9705 | 9727 | 9750 | 2 | 4 | 7 | 9 | 11 | 13 | 16 | 18 | 20 |
| .99 | 9772 | 9795 | 9817 | 9840 | 9863 | 9886 | 9908 | 9931 | 9954 | 9977 | 2 | 5 | 7 | 9 | 11 | 14 | 16 | 18 | 20 |

**Table III Binomal distribution function**

| | | θ | | | | | | | | | |
|---|---|---|---|---|---|---|---|---|---|---|---|
| *n* | *x* | **.05** | **.10** | **.15** | **.20** | **.25** | **.30** | **.35** | **.40** | **.45** | **.50** |
| 1 | 0 | .9500 | .9000 | .8500 | .8000 | .7500 | .7000 | .6500 | .6000 | .5500 | .5000 |
| | 1 | .0500 | .1000 | .1500 | .2000 | .2500 | .3000 | .3500 | .4000 | .4500 | .5000 |
| 2 | 0 | .9025 | .8100 | .7225 | .6400 | .5625 | .4900 | .4225 | .3600 | .3025 | .2500 |
| | 1 | .0950 | .1800 | .2550 | .3200 | .3750 | .4200 | .4550 | .4800 | .4950 | .5000 |
| | 2 | .0025 | .0100 | .0225 | .0400 | .0625 | .0900 | .1225 | .1600 | .2025 | .2500 |
| 3 | 0 | .8574 | .7290 | .6141 | .5120 | .4219 | .3430 | .2746 | .2160 | .1664 | .1250 |
| | 1 | .1354 | .2430 | .3251 | .3840 | .4219 | .4410 | .4436 | .4320 | .4084 | .3750 |
| | 2 | .0071 | .0270 | .0574 | .0960 | .1406 | .1890 | .2389 | .2880 | .3341 | .3750 |
| | 3 | .0001 | .0010 | .0034 | .0080 | .0156 | .0270 | .0429 | .0640 | .0911 | .1250 |
| 4 | 0 | .8145 | .6561 | .5220 | .4096 | .3164 | .2401 | .1785 | .1296 | .0915 | .0625 |
| | 1 | .1715 | .2916 | .3685 | .4096 | .4219 | .4116 | .3845 | .3456 | .2995 | .2500 |
| | 2 | .0315 | .0486 | .0975 | .1536 | .2109 | .2646 | .3105 | .3456 | .3675 | .3750 |
| | 3 | .0005 | .0036 | .0115 | .0256 | .0469 | .0756 | .1115 | .1536 | .2005 | .2500 |
| | 4 | .0000 | .0001 | .0005 | .0016 | .0039 | .0081 | .0150 | .0256 | .0410 | .0625 |
| 5 | 0 | .7738 | .5905 | .4437 | .3277 | .2373 | .1681 | .1160 | .0778 | .0503 | .0312 |
| | 1 | .2036 | .3280 | .3915 | .4096 | .3955 | .3602 | .3124 | .2592 | .2059 | .1562 |
| | 2 | .0214 | .0729 | .1382 | .2048 | .2637 | .3087 | .3364 | .3456 | .3369 | .3125 |
| | 3 | .0011 | .0081 | .0244 | .0512 | .0879 | .1323 | .1811 | .2304 | .2757 | .3125 |
| | 4 | .0000 | .0004 | .0022 | .0064 | .0146 | .0284 | .0488 | .0768 | .1128 | .1562 |
| | 5 | .0000 | .0000 | .0001 | .0003 | .0010 | .0024 | .0053 | .0102 | .0185 | .0312 |
| 6 | 0 | .7351 | .5314 | .3771 | .2621 | .1780 | .1176 | .0754 | .0467 | .0277 | .0156 |
| | 1 | .2321 | .3543 | .3993 | .3932 | .3560 | .3025 | .2437 | .1866 | .1359 | .0938 |
| | 2 | .0305 | .0984 | .1762 | .2458 | .2966 | .3241 | .3280 | .3110 | .2780 | .2344 |
| | 3 | .0021 | .0146 | .0415 | .0819 | .1318 | .1852 | .2355 | .2765 | .3032 | .3125 |
| | 4 | .0001 | .0012 | .0055 | .0154 | .0330 | .0595 | .0951 | .1382 | .1861 | .2344 |
| | 5 | .0000 | .0001 | .0004 | .0015 | .0044 | .0102 | .0205 | .0369 | .0609 | .0938 |
| | 6 | .0000 | .0000 | .0000 | .0001 | .0002 | .0007 | .0018 | .0041 | .0083 | .0156 |
| 7 | 0 | .6983 | .4783 | .3206 | .2097 | .1335 | .0824 | .0490 | .0280 | .0152 | .0078 |
| | 1 | .2573 | .3720 | .3960 | .3670 | .3115 | .2471 | .1848 | .1306 | .0872 | .0547 |
| | 2 | .0406 | .1240 | .2097 | .2753. | .3115 | .3177 | .2985 | .2613 | .2140 | .1641 |
| | 3 | .0036 | .0230 | .0617 | .1147 | .1730 | .2269 | .2679 | .2903 | .2918 | .2734 |
| | 4 | .0002 | .0026 | .0109 | .0287 | .0577 | .0972 | .1442 | .1935 | .2388 | .2734 |
| | 5 | .0000 | .0002 | .0012 | .0043 | .0115 | .0250 | .0466 | .0774 | .1172 | .1641 |
| | 6 | .0000 | .0000 | .0001 | .0004 | .0013 | .0036 | .0084 | .0172 | .0320 | .0547 |
| | 7 | .0000 | .0000 | .0000 | .0000 | .0001 | .0002 | .0006 | .0016 | .0037 | .0078 |
| 8 | 0 | .6634 | .4305 | .2725 | .1678 | .1001 | .0576 | .0319 | .0168 | .0084 | .0039 |
| | 1 | .2793 | .3826 | .3847 | .3355 | .2670 | .1977 | .1373 | .0896 | .0548 | .0312 |
| | 2 | .0515 | .1488 | .2376 | .2936 | .3115 | .2965 | .2587 | .2090 | .1569 | .1094 |
| | 3 | .0054 | .0331 | .0839 | .1468 | .2076 | .2541 | .2786 | .2787 | .2568 | .2188 |
| | 4 | .0004 | .0046 | .0185 | .0459 | .0865 | .1361 | .1875 | .2322 | .2627 | .2734 |

**Table III (continued)**

| | | θ | | | | | | | | | |
|---|---|---|---|---|---|---|---|---|---|---|---|
| *n* | *x* | **.05** | **.10** | **.15** | **.20** | **.25** | **.30** | **.35** | **.40** | **.45** | **.50** |
| 8 | 5 | .0000 | .0004 | .0026 | .0092 | .0231 | .0467 | .0808 | .1239 | .1719 | .2188 |
| | 6 | .0000 | .0000 | .0002 | .0011 | .0038 | .0100 | .0217 | .0413 | .0703 | .1094 |
| | 7 | .0000 | .0000 | .0000 | .0001 | .0004 | .0012 | .0033 | .0079 | .0164 | .0312 |
| | 8 | .0000 | .0000 | .0000 | .0000 | .0000 | .0001 | .0002 | .0007 | .0017 | .0039 |
| 9 | 0 | .6302 | .3874 | .2316 | .1342 | .0751 | .0404 | .0207 | .0101 | .0046 | .0020 |
| | 1 | .2685 | .3874 | .3679 | .3020 | .2253 | .1556 | .1004 | .0605 | .0339 | .0176 |
| | 2 | .0629 | .1722 | .2597 | .3020 | .3003 | .2668 | .2162 | .1612 | .1110 | .0703 |
| | 3 | .0077 | .0446 | .1069 | .1762 | .2336 | .2668 | .2716 | .2508 | .2119 | .1641 |
| | 4 | .0006 | .0074 | .0283 | .0661 | .1168 | .1715 | .2194 | .2508 | .2600 | .2461 |
| | 5 | .0000 | .0008 | .0050 | .0165 | .0389 | .0735 | .1181 | .1672 | .2128 | .2461 |
| | 6 | .0000 | .0001 | .0006 | .0028 | .0087 | .0210 | .0424 | .0743 | .1160 | .1641 |
| | 7 | .0000 | .0000 | .0000 | .0003 | .0012 | .0039 | .0098 | .0212 | .0407 | .0703 |
| | 8 | .0000 | .0000 | .0000 | .0000 | .0001 | .0004 | .0013 | .0035 | .0083 | .0176 |
| | 9 | .0000 | .0000 | .0000 | .0000 | .0000 | .0000 | .0001 | .0003 | .0008 | .0020 |
| 10 | 0 | .5987 | .3487 | .1969 | .1074 | .0563 | .0282 | .0135 | .0060 | .0025 | .0010 |
| | 1 | .3151 | .3874 | .3474 | .2684 | .1877 | .1211 | .0725 | .0403 | .0207 | .00098 |
| | 2 | .0746 | .1937 | .2759 | .3020 | .2816 | .2335 | .1757 | .1209 | .0763 | .0439 |
| | 3 | .0105 | .0574 | .01295 | .2013 | .2503 | .2668 | .2522 | .2150 | .1665 | .1172 |
| | 4 | .0010 | .0112 | .0401 | .0881 | .1460 | .2001 | .2377 | .2508 | .2384 | .2051 |
| | 5 | .0001 | .0015 | .0085 | .0264 | .0584 | .1029 | .1536 | .2007 | .2340 | .2461 |
| | 6 | .0000 | .0001 | .0012 | .0055 | .0162 | .0368 | .0689 | .1115 | .1596 | .2051 |
| | 7 | .0000 | .0000 | .0001 | .0008 | .0031 | .0090 | .0212 | .0425 | .0746 | .1172 |
| | 8 | .0000 | .0000 | .0000 | .0001 | .0004 | .0014 | .0043 | .0106 | .0229 | .0439 |
| | 9 | .0000 | .0000 | .0000 | .0000 | .0000 | .0001 | .0005 | .0016 | .0042 | .0098 |
| | 10 | .0000 | .0000 | .0000 | .0000 | .0000 | .0000 | .0000 | .0001 | .0003 | .0006 |
| 11 | 0 | .5688 | .3138 | .1673 | .0859 | .0422 | .0198 | .0088 | .0036 | .0014 | .0005 |
| | 1 | .3293 | .3835 | .3248 | .2362 | .1549 | .0932 | .0518 | .0266 | .0125 | .0054 |
| | 2 | .0867 | .2131 | .2866 | .2953 | .2581 | .1998 | .1395 | .0887 | .0513 | .0269 |
| | 3 | .0137 | .0710 | .1517 | .2215 | .2581 | .2568 | .2254 | .1774 | .1259 | .0806 |
| | 4 | .0014 | .0158 | .0536 | .1107 | .1721 | .2201 | .2428 | .2365 | .2060 | .1611 |
| | 5 | .0001 | .0025 | .0132 | .0388 | .0803 | .1321 | .1830 | .2207 | .2360 | .2256 |
| | 6 | .0000 | .0003 | .0023 | .0097 | .0268 | .0566 | .0985 | .1471 | .1931 | .2253 |
| | 7 | .0000 | .0000 | .0003 | .0017 | .0064 | .0173 | .0379 | .0701 | .1128 | .1611 |
| | 8 | .0000 | .0000 | .0000 | .0002 | .0011 | .0037 | .0102 | .0234 | .0462 | .0806 |
| | 9 | .0000 | .0000 | .0000 | .0000 | .0001 | .0005 | .0018 | .0052 | .0126 | .0269 |
| | 10 | .0000 | .0000 | .0000 | .0000 | .0000 | .0000 | .0002 | .0007 | .0021 | .0054 |
| | 11 | .0000 | .0000 | .0000 | .0000 | .0000 | .0000 | .0000 | .0000 | .0002 | .0005 |
| 12 | 0 | .5404 | .2824 | .1422 | .0687 | .0317 | .0138 | .0057 | .0022 | .0008 | .0002 |
| | 1 | .3413 | .3766 | .3012 | .2062 | .1267 | .0712 | .0368 | .0174 | .0075 | .0029 |
| | 2 | .0988 | .2301 | .2924 | .2835 | .2323 | .1678 | .1088 | .6039 | .0339 | .0131 |
| | 3 | .0173 | .0852 | .1720 | .2362 | .2581 | .2397 | .1954 | .1419 | .0923 | .0537 |
| | 4 | .0021 | .0213 | .0683 | .1329 | .1936 | .2311 | .2367 | .2128 | .1700 | .1208 |
| | 5 | .0002 | .0038 | .0193 | .0532 | .1062 | .1585 | .2039 | .2270 | .2225 | .1934 |
| | 6 | .0000 | .0005 | .0040 | .0155 | .0401 | .0792 | .1281 | .1766 | .2124 | .2256 |
| | 7 | .0000 | .0000 | .0006 | .0033 | .0115 | .0291 | .0591 | .1009 | .1489 | .1934 |
| | 8 | .0000 | .0000 | .0001 | .0005 | .0024 | .0078 | .0199 | .0420 | .0762 | .1208 |
| | 9 | .0000 | .0000 | .0000 | .0001 | .0004 | .0015 | .0048 | .0125 | .0277 | .0537 |

**Table III (continued)**

| | | θ | | | | | | | | | |
|---|---|---|---|---|---|---|---|---|---|---|---|
| *n* | *x* | **.05** | **.10** | **.15** | **.20** | **.25** | **.30** | **.35** | **.40** | **.45** | **.50** |
| 12 | 10 | .0000 | .0000 | .0000 | .0000 | .0000 | .0002 | .0008 | .0025 | .0068 | .0161 |
| | 11 | .0000 | .0000 | .0000 | .0000 | .0000 | .0000 | .0001 | .0003 | .0010 | .0029 |
| | 12 | .0000 | .0000 | .0000 | .0000 | .0000 | .0000 | .0000 | .0000 | .0001 | .0002 |
| 13 | 0 | .5133 | .2542 | .1209 | .0550 | .0238 | .0097 | .0037 | .0013 | .0004 | .0001 |
| | 1 | .3512 | .3672 | .2774 | .1787 | .1029 | .0540 | .0259 | .0113 | .0045 | .0016 |
| | 2 | .1109 | .2448 | .2937 | .2680 | .2059 | .1388 | .0836 | .0453 | .0220 | .0095 |
| | 3 | .0214 | .0997 | .1900 | .2457 | .2517 | .2181 | .1651 | .1107 | .0660 | .0349 |
| | 4 | .0028 | .0277 | .0838 | .1535 | .2097 | .2337 | .2222 | .1845 | .1350 | .0873 |
| | 5 | .0003 | .0055 | .0266 | .0691 | .1258 | .1803 | .2154 | .2214 | .1989 | .1571 |
| | 6 | .0000 | .0008 | .0063 | .0230 | .0559 | .1030 | .1546 | .1968 | .2169 | .2095 |
| | 7 | .0000 | .0001 | .0011 | .0058 | .0186 | .0442 | .0833 | .1312 | .1775 | .2095 |
| | 8 | .0000 | .0000 | .0001 | .0011 | .0047 | .0142 | .0336 | .0656 | .1089 | .1571 |
| | 9 | .0000 | .0000 | .0000 | .0001 | .0009 | .0034 | .0101 | .0243 | .0495 | .0873 |
| | 10 | .0000 | .0000 | .0000 | .0000 | .0001 | .0006 | .0022 | .0065 | .0162 | .0349 |
| | 11 | .0000 | .0000 | .0000 | .0000 | .0000 | .0001 | .0003 | .0012 | .0036 | .0095 |
| | 12 | .0000 | .0000 | .0000 | .0000 | .0000 | .0000 | .0000 | .0001 | .0005 | .0016 |
| | 13 | .0000 | .0000 | .0000 | .0000 | .0000 | .0000 | .0000 | .0000 | .0000 | .0001 |
| 14 | 0 | .4877 | .2288 | .1028 | .0440 | .0178 | .0068 | .0024 | .0008 | .0002 | .0001 |
| | 1 | .3593 | .3559 | .2539 | .1539 | .0832 | .0407 | .0181 | .0073 | .0027 | .0009 |
| | 2 | .1229 | .2570 | .2912 | .2501 | .1802 | .1134 | .0634 | .0317 | .0141 | .0056 |
| | 3 | .0259 | .1142 | .2056 | .2501 | .2402 | .1943 | .1366 | .0845 | .0462 | .0222 |
| | 4 | .0037 | .0349 | .0998 | .1720 | .2202 | .2290 | .2022 | .1549 | .1040 | .0611 |
| | 5 | .0004 | .0078 | .0352 | .0860 | .1468 | .1963 | .2178 | .2066 | .1701 | .1222 |
| | 6 | .0000 | .0013 | .0093 | .0322 | .0734 | .1262 | .1759 | .2066 | .2088 | .1833 |
| | 7 | .0000 | .0002 | .0019 | .0092 | .0280 | .0618 | .1082 | .1574 | .1952 | .2095 |
| | 8 | .0000 | .0000 | .0003 | .0020 | .0082 | .0232 | .0510 | .0918 | .1398 | .1833 |
| | 9 | .0000 | .0000 | .0000 | .0003 | .0018 | .0066 | .0183 | .0408 | .0762 | .1222 |
| | 10 | .0000 | .0000 | .0000 | .0000 | .0003 | .0014 | .0049 | .0136 | .0312 | .0611 |
| | 11 | .0000 | .0000 | .0000 | .0000 | .0000 | .0002 | .0010 | .0033 | .0093 | .0222 |
| | 12 | .0000 | .0000 | .0000 | .0000 | .0000 | .0000 | .0001 | .0005 | .0019 | .0056 |
| | 13 | .0000 | .0000 | .0000 | .0000 | .0000 | .0000 | .0000 | .0001 | .0002 | .0009 |
| | 14 | .0000 | .0000 | .0000 | .0000 | .0000 | .0000 | .0000 | .0000 | .0000 | .0001 |
| 15 | 0 | .4633 | .2059 | .0874 | .0352 | .0134 | .0047 | .0016 | .0005 | .0001 | .0000 |
| | 1 | .3658 | .3432 | .2312 | .1319 | .0668 | .0305 | .0126 | .0047 | .0016 | .0005 |
| | 2 | .1348 | .2669 | .2856 | .2309 | .1559 | .0916 | .0476 | .0219 | .0090 | .0032 |
| | 3 | .0307 | .1285 | .2184 | .2501 | .2252 | .1700 | .1110 | .0634 | .0318 | .0139 |
| | 4 | .0049 | .0428 | .1156 | .1876 | .2252 | .2186 | .1792 | .1268 | .0780 | .0417 |
| | 5 | .0006 | .0105 | .0449 | .1032 | .1651 | .2061 | .2123 | .1859 | .1404 | .0916 |
| | 6 | .0000 | .0019 | .0132 | .0430 | .0917 | .1472 | .1906 | .2066 | .1914 | .1527 |
| | 7 | .0000 | .0003 | .0030 | .0138 | .0393 | .0811 | .1319 | .1771 | .2013 | .1964 |
| | 8 | .0000 | .0000 | .0005 | .0035 | .0131 | .0348 | .0710 | .1181 | .1647 | .1964 |
| | 9 | .0000 | .0000 | .0001 | .0007 | .0034 | .0116 | .0298 | .0612 | .1048 | .1527 |
| | 10 | .0000 | .0000 | .0000 | .0001 | .0007 | .0030 | .0096 | .0245 | .0515 | .0916 |
| | 11 | .0000 | .0000 | .0000 | .0000 | .0001 | .0006 | .0024 | .0074 | .0191 | .0417 |
| | 12 | .0000 | .0000 | .0000 | .0000 | .0000 | .0001 | .0004 | .0016 | .0052 | .0139 |
| | 13 | .0000 | .0000 | .0000 | .0000 | .0000 | .0000 | .0001 | .0003 | .0010 | .0032 |
| | 14 | .0000 | .0000 | .0000 | .0000 | .0000 | .0000 | .0000 | .0000 | .0001 | .0005 |
| | 15 | .0000 | .0000 | .0000 | .0000 | .0000 | .0000 | .0000 | .0000 | .0000 | .0000 |

**Table III (continued)**

| | | θ | | | | | | | | | |
|---|---|---|---|---|---|---|---|---|---|---|---|
| *n* | *x* | **.05** | **.10** | **.15** | **.20** | **.25** | **.30** | **.35** | **.40** | **.45** | **.50** |
| 16 | 0 | .4401 | .1853 | .0743 | .0281 | .0100 | .0033 | .0010 | .0003 | .0001 | .0000 |
| | 1 | .3706 | .3294 | .2097 | .1126 | .0535 | .0228 | .0087 | .0030 | .0009 | .0002 |
| | 2 | .1463 | .2745 | .2775 | .2111 | .1336 | .0732 | .0353 | .0150 | .0056 | .0018 |
| | 3 | .0359 | .1423 | .2285 | .2463 | .2979 | .1465 | .0888 | .0468 | .0215 | .0085 |
| | 4 | .0061 | .0514 | .1311 | .2001 | .2252 | .2040 | .1553 | .1014 | .0572 | .0278 |
| | 5 | .0008 | .0137 | .0555 | .1201 | .1802 | .2099 | .2008 | .1623 | .1123 | .0667 |
| | 6 | .0001 | .0028 | .0180 | .0550 | .1101 | .1649 | .1982 | .1983 | .1684 | .1222 |
| | 7 | .0000 | .0004 | .0045 | .0197 | .0524 | .1010 | .1524 | .1889 | .1969 | .1746 |
| | 8 | .0000 | .0001 | .0009 | .0055 | .0197 | .0487 | .0923 | .1417 | .1812 | .1964 |
| | 9 | .0000 | .0000 | .0001 | .0012 | .0058 | .0185 | .0442 | .0840 | .1318 | .1746 |
| | 10 | .0000 | .0000 | .0000 | .0002 | .0014 | .0056 | .0167 | .0392 | .0755 | .1222 |
| | 11 | .0000 | .0000 | .0000 | .0000 | .0002 | .0013 | .0049 | .0142 | .0337 | .0667 |
| | 12 | .0000 | .0000 | .0000 | .0000 | .0000 | .0002 | .0011 | .0040 | .0115 | .0278 |
| | 13 | .0000 | .0000 | .0000 | .0000 | .0000 | .0000 | .0002 | .0008 | .0029 | .0085 |
| | 14 | .0000 | .0000 | .0000 | .0000 | .0000 | .0000 | .0000 | .0001 | .0005 | .0018 |
| | 15 | .0000 | .0000 | .0000 | .0000 | .0000 | .0000 | .0000 | .0000 | .0001 | .0002 |
| | 16 | .0000 | .0000 | .0000 | .0000 | .0000 | .0000 | .0000 | .0000 | .0000 | .0000 |
| 17 | 0 | .4181 | .1668 | .0631 | .0225 | .0075 | .0023 | .0007 | .0002 | .0000 | .0000 |
| | 1 | .3741 | .3150 | .1893 | .0957 | .0426 | .0169 | .0060 | .0019 | .0005 | .0001 |
| | 2 | .1575 | .2800 | .2673 | .1914 | .1136 | .0581 | .0260 | .0102 | .0035 | .0010 |
| | 3 | .0415 | .1556 | .2359 | .2393 | .1893 | .1245 | .0701 | .0341 | .0144 | .0052 |
| | 4 | .0076 | .0605 | .1457 | .2093 | .2209 | .1868 | .1320 | .0796 | .0411 | .0182 |
| | 5 | .0010 | .0175 | .0668 | .1361 | .1914 | .2081 | .1849 | .1379 | .0875 | .0472 |
| | 6 | .0001 | .0039 | .0236 | .0680 | .1276 | .1784 | .1091 | .1839 | .1432 | .0944 |
| | 7 | .0000 | .0007 | .0065 | .0267 | .0668 | .1201 | .1685 | .1927 | .1841 | .1484 |
| | 8 | .0000 | .0001 | .0014 | .0084 | .0279 | .0644 | .1134 | .1606 | .1883 | .1855 |
| | 9 | .0000 | .0000 | .0003 | .0021 | .0093 | .0276 | .0611 | .1070 | .1540 | .1855 |
| | 10 | .0000 | .0000 | .0000 | .0004 | .0025 | .0095 | .0263 | .0571 | .1008 | .1484 |
| | 11 | .0000 | .0000 | .0000 | .0001 | .0005 | .0026 | .0090 | .0242 | .0525 | .0944 |
| | 12 | .0000 | .0000 | .0000 | .0000 | .0001 | .0006 | .0024 | .0081 | .0215 | .0472 |
| | 13 | .0000 | .0000 | .0000 | .0000 | .0000 | .0001 | .0005 | .0021 | .0068 | .0182 |
| | 14 | .0000 | .0000 | .0000 | .0000 | .0000 | .0000 | .0001 | .0004 | .0016 | .0052 |
| | 15 | .0000 | .0000 | .0000 | .0000 | .0000 | .0000 | .0000 | .0001 | .0003 | .0010 |
| | 16 | .0000 | .0000 | .0000 | .0000 | .0000 | .0000 | .0000 | .0000 | .0000 | .0001 |
| | 17 | .0000 | .0000 | .0000 | .0000 | .0000 | .0000 | .0000 | .0000 | .0000 | .0000 |
| 18 | 0 | .3972 | .1501 | .0536 | .0180 | .0056 | .0016 | .0004 | .0001 | .0000 | .0000 |
| | 1 | .3763 | .3002 | .1704 | .0811 | .0338 | .0126 | .0042 | .0012 | .0003 | .0001 |
| | 2 | .1683 | .2835 | .2556 | .1723 | .0958 | .0458 | .0100 | .0069 | .0022 | .0006 |
| | 3 | .0473 | .1680 | .2406 | .2297 | .1704 | .1046 | .0547 | .0246 | .0095 | .0031 |
| | 4 | .0093 | .0700 | .1592 | .2153 | .2130 | .1681 | .1104 | .0614 | .0291 | .0117 |
| | 5 | .0014 | .0218 | .0787 | .1507 | .1988 | .2017 | .1664 | .1146 | .0666 | .0327 |
| | 6 | .0002 | .0052 | .0301 | .0816 | .1436 | .1873 | .1941 | .1655 | .1181 | .0708 |
| | 7 | .0000 | .0010 | .0091 | .0350 | .0820 | .1376 | .1792 | .1892 | .1657 | .1214 |
| | 8 | .0000 | .0002 | .0022 | .0120 | .0376 | .0811 | .1327 | .1734 | .1864 | .1669 |
| | 9 | .0000 | .0000 | .0001 | .0033 | .0139 | .0386 | .0794 | .1284 | .1694 | .1855 |

**Table III (continued)**

| | | θ | | | | | | | | | |
|---|---|---|---|---|---|---|---|---|---|---|---|
| *n* | *x* | .05 | .10 | .15 | .20 | .25 | .30 | .35 | .40 | .45 | .50 |
| 18 | 10 | .0000 | 0000 | .0001 | .0008 | .0042 | .0149 | .0385 | .0771 | .1248 | .1669 |
| | 11 | .0000 | .0000 | .0000 | .0001 | .0010 | .0046 | .0151 | .0374 | .0742 | .1214 |
| | 12 | .0000 | .0000 | .0000 | .0000 | .0002 | .0012 | .0047 | .0145 | .0354 | .0708 |
| | 13 | .0000 | .0000 | .0000 | .0000 | .0000 | .0002 | .0012 | .0045 | .0134 | .0327 |
| | 14 | .0000 | .0000 | .0000 | .0000 | .0000 | .0000 | .0002 | .0011 | .0039 | .0117 |
| | 15 | .0000 | .0000 | .0000 | .0000 | .0000 | .0000 | .0000 | .0002 | .0009 | .0031 |
| | 16 | .0000 | .0000 | .0000 | .0000 | .0000 | .0000 | .0000 | .0000 | .0001 | .0006 |
| | 17 | .0000 | .0000 | .0000 | .0000 | .0000 | .0000 | .0000 | .0000 | .0000 | .0001 |
| | 18 | .0000 | .0000 | .0000 | .0000 | .0000 | .0000 | .0000 | .0000 | .0000 | .0000 |
| 19 | 0 | .3774 | .1351 | .0456 | .0144 | .0042 | .0011 | .0003 | .0001 | .0000 | .0000 |
| | 1 | .3774 | .2852 | .1529 | .0685 | .0268 | .0093 | .0029 | .0008 | .0002 | .0000 |
| | 2 | .1787 | .2852 | .2428 | .1540 | .0803 | .0358 | .0138 | .0046 | .0013 | .0003 |
| | 3 | .0533 | .1796 | .2428 | .2182 | .1517 | .0869 | .0422 | .0175 | .0062 | .0018 |
| | 4 | .0112 | .0798 | .1714 | .2182 | .2023 | .1491 | .0909 | .0467 | .0203 | .0074 |
| | 5 | .0018 | .0266 | .0907 | .1636 | .2023 | .1916 | .1468 | .0933 | .0497 | .0222 |
| | 6 | .0002 | .0069 | .0374 | .0955 | .1574 | .1916 | .1844 | .1451 | .0949 | .0518 |
| | 7 | .0000 | .0014 | .0122 | .0443 | .0974 | .1525 | .1844 | .1797 | .1443 | .0691 |
| | 8 | .0000 | .0002 | .0032 | .0166 | .0487 | .0981 | .1489 | .1797 | .1771 | .1442 |
| | 9 | .0000 | .0000 | .0007 | .0051 | .0198 | .0514 | .0980 | .1464 | .1771 | .1762 |
| | 10 | .0000 | .0000 | .0001 | .0013 | .0066 | .0220 | .0528 | .0976 | .1449 | .1762 |
| | 11 | .0000 | .0000 | .0000 | .0003 | .0018 | .0077 | .0233 | .0532 | .0970 | .1442 |
| | 12 | .0000 | .0000 | .0000 | .0000 | .0004 | .0022 | .0083 | .0237 | .0529 | .0961 |
| | 13 | .0000 | .0000 | .0000 | .0000 | .0001 | .0005 | .0024 | .0085 | .0233 | .0518 |
| | 14 | .0000 | .0000 | .0000 | .0000 | .0000 | .0001 | .0006 | .0024 | .0082 | .0222 |
| | 15 | .0000 | .0000 | .0000 | .0000 | .0000 | .0000 | .0001 | .0005 | .0022 | .0074 |
| | 16 | .0000 | .0000 | .0000 | .0000 | .0000 | .0000 | .0000 | .0001 | .0005 | .0018 |
| | 17 | .0000 | .0000 | .0000 | .0000 | .0000 | .0000 | .0000 | .0000 | .0001 | .0003 |
| | 18 | .0000 | .0000 | .0000 | .0000 | .0000 | .0000 | .0000 | .0000 | .0000 | .0000 |
| | 19 | .0000 | .0000 | .0000 | .0000 | .0000 | .0000 | .0000 | .0000 | .0000 | .0000 |
| 20 | 0 | .3585 | .1216 | .0388 | .0115 | .0032 | .0008 | .0002 | .0000 | .0000 | .0000 |
| | 1 | .3774 | .2702 | .1368 | .0576 | .0211 | .0068 | .0020 | .0005 | .0001 | .0000 |
| | 2 | .1887 | .2852 | .2293 | .1369 | .0669 | .0278 | .0100 | .0031 | .0008 | .0002 |
| | 3 | .0596 | .1901 | .2428 | .2054 | .1339 | .0716 | .0323 | .0123 | .0040 | .0011 |
| | 4 | .0133 | .0898 | .1821 | .2182 | .1897 | .1304 | .0738 | .0350 | .0139 | .0046 |
| | 5 | .0022 | .0319 | .1028 | .1746 | .2023 | .1789 | .1272 | .0746 | .0365 | .0148 |
| | 6 | .0003 | .0089 | .0454 | .1091 | .1686 | .1916 | .1712 | .1244 | .0746 | .0370 |
| | 7 | .0000 | .0020 | .0160 | .0545 | .1124 | .1643 | .1844 | .1659 | .1221 | .0739 |
| | 8 | .0000 | .0004 | .0046 | .0222 | .0609 | .1144 | .1614 | .1797 | .1623 | .1201 |
| | 9 | .0000 | .0001 | .0011 | .0074 | .0271 | .0654 | .1158 | .1597 | .1771 | .1602 |
| | 10 | .0000 | .0000 | .0002 | .0020 | .0099 | .0308 | .0686 | .1171 | .1593 | .1762 |
| | 11 | .0000 | .0000 | .0000 | .0005 | .0030 | .0120 | .0336 | .0710 | .1185 | .1602 |
| | 12 | .0000 | .0000 | .0000 | .0001 | .0008 | .0039 | .0136 | .0355 | .0727 | .1201 |
| | 13 | .0000 | .0000 | .0000 | .0000 | .0002 | .0010 | .0045 | .0146 | .0366 | .0739 |
| | 14 | .0000 | .0000 | .0000 | .0000 | .0000 | .0002 | .0012 | .0049 | .0150 | .0370 |
| | 15 | .0000 | .0000 | .0000 | .0000 | .0000 | .0000 | .0003 | .0013 | .0049 | .0148 |
| | 16 | .0000 | .0000 | .0000 | .0000 | .0000 | .0000 | .0000 | .0003 | .0013 | .0046 |
| | 17 | .0000 | .0000 | .0000 | .0000 | .0000 | .0000 | .0000 | .0000 | .0002 | .0011 |
| | 18 | .0000 | .0000 | .0000 | .0000 | .0000 | .0000 | .0000 | .0000 | .0000 | .0002 |
| | 19 | .0000 | .0000 | .0000 | .0000 | .0000 | .0000 | .0000 | .0000 | .0000 | .0000 |
| | 20 | .0000 | .0000 | .0000 | .0000 | .0000 | .0000 | .0000 | .0000 | .0000 | .0000 |

**Table IV Poisson distribution function**

| | λ | | | | | | | | | |
|---|---|---|---|---|---|---|---|---|---|---|
| $x$ | **0.1** | **0.2** | **0.3** | **0.4** | **0.5** | **0.6** | **0.7** | **0.8** | **0.9** | **1.0** |
| 0 | .9048 | .8187 | .7408 | .6703 | .6065 | .5488 | .4966 | .4493 | .4066 | .3679 |
| 1 | .0905 | .1637 | .2222 | .2681 | .3033 | .3293 | .3476 | .3595 | .3659 | .3679 |
| 2 | .0045 | .0164 | .0333 | .0536 | .0758 | .0988 | .1217 | .1438 | .1647 | .1839 |
| 3 | .0002 | .0011 | .0033 | .0072 | .0126 | .0198 | .0284 | .0383 | .0494 | .0613 |
| 4 | .0000 | .0001 | .0002 | .0007 | .0016 | .0030 | .0050 | .0077 | .0111 | .0153 |
| 5 | .0000 | .0000 | .0000 | .0001 | .0002 | .0004 | .0007 | .0012 | .0020 | .0031 |
| 6 | .0000 | .0000 | .0000 | .0000 | .0000 | .0000 | .0001 | .0002 | .0003 | .0005 |
| 7 | .0000 | .0000 | .0000 | .0000 | .0000 | .0000 | .0000 | .0000 | .0000 | .0001 |

| | λ | | | | | | | | | |
|---|---|---|---|---|---|---|---|---|---|---|
| $x$ | **1.1** | **1.2** | **1.3** | **1.4** | **1.5** | **1.6** | **1.7** | **1.8** | **1.9** | **2.0** |
| 0 | .3329 | .3012 | .2725 | .2466 | .2231 | .2019 | .1827 | .1653 | .1496 | .1353 |
| 1 | .3662 | .3614 | .3543 | .3452 | .3347 | .3230 | .3106 | .2975 | .2842 | .2707 |
| 2 | .2014 | .2169 | .2303 | .2417 | .2510 | .2584 | .2640 | .2678 | 2700 | .2707 |
| 3 | .0738 | .0867 | .0998 | .1128 | .1255 | .1378 | .1496 | .1607 | .1710 | .1804 |
| 4 | .0203 | .0260 | .0324 | .0395 | .0471 | .0551 | .0636 | .0723 | .0812 | .0902 |
| 5 | .0045 | .0062 | .0084 | .0111 | .0141 | .0176 | .0216 | .0260 | .0309 | .0361 |
| 6 | .0008 | .0012 | .0018 | .0026 | .0035 | .0047 | .0061 | .0078 | .0098 | .0120 |
| 7 | .0001 | .0002 | .0003 | .0005 | .0008 | .0011 | .0015 | .0020 | .0027 | .0034 |
| 8 | .0000 | .0000 | .0001 | .0001 | .0001 | .0002 | .0003 | .0005 | .0006 | .0009 |
| 9 | .0000 | .0000 | .0000 | .0000 | .0000 | .0000 | .0001 | .0001 | .0001 | .0002 |

| | λ | | | | | | | | | |
|---|---|---|---|---|---|---|---|---|---|---|
| $x$ | **2.1** | **2.2** | **2.3** | **2.4** | **2.5** | **2.6** | **2.7** | **2.8** | **2.9** | **3.0** |
| 0 | .1225 | .1108 | .1003 | .0907 | .0821 | .0743 | .0672 | .0608 | .0550 | .0498 |
| 1 | .2572 | .2438 | .2306 | .2177 | .2052 | .1931 | .1815 | .1703 | .1596 | .1494 |
| 2 | .2700 | .2681 | .2652 | .2613 | .2565 | .2510 | .2450 | .2384 | .2314 | .2240 |
| 3 | .1890 | .1966 | .2033 | .2090 | .2138 | .2176 | .2205 | .2225 | .2237 | .2240 |
| 4 | .0992 | .1082 | .1169 | .1254 | .1336 | .1414 | .1488 | .1557 | .1622 | .1680 |
| 5 | .0417 | .0476 | .0538 | .0602 | .0668 | .0735 | .0804 | .0872 | .0940 | .1008 |
| 6 | .0146 | .0174 | .0206 | .0241 | .0278 | .0319 | .0362 | .0407 | .0455 | .0504 |
| 7 | .0044 | .0055 | .0068 | .0083 | .0099 | .0118 | .0139 | .0163 | .0188 | .0216 |
| 8 | .0011 | .0015 | .0019 | .0025 | .0031 | .0038 | .0047 | .0057 | .0068 | .0081 |
| 9 | .0003 | .0004 | .0005 | .0007 | .0009 | .0011 | .0014 | .0018 | .0022 | .0027 |
| 10 | .0001 | .0001 | .0001 | .0002 | .0002 | .0003 | .0004 | .0005 | .0006 | .0008 |
| 11 | .0000 | .0000 | .0000 | .0000 | .0000 | .0001 | .0001 | .0001 | .0002 | .0002 |
| 12 | .0000 | .0000 | .0000 | .0000 | .0000 | .0000 | .0000 | .0000 | .0000 | .0001 |

**Table IV (continued)**

| | λ | | | | | | | | | |
|---|---|---|---|---|---|---|---|---|---|---|
| *x* | **3.1** | **3.2** | **3.3** | **3.4** | **3.5** | **3.6** | **3.7** | **3.8** | **3.9** | **4.0** |
| 0 | .0450 | .0408 | .0369 | .0334 | .0302 | .0273 | .0247 | .0224 | .0202 | .0183 |
| 1 | .1397 | .1304 | .1217 | .1135 | .1057 | .0984 | .0915 | .0850 | .0789 | .0733 |
| 2 | .2165 | .2087 | .2008 | .1929 | .1850 | .1771 | .1692 | .1615 | .1539 | .1465 |
| 3 | .2237 | .2226 | .2209 | .2186 | .2158 | .2125 | .2087 | .2046 | .2001 | .1954 |
| 4 | .1734 | .1781 | .1823 | .1858 | .1888 | .1912 | .1931 | .1944 | .1951 | .1954 |
| 5 | .1075 | .1140 | .1203 | .1264 | .1322 | .1377 | .1429 | .1477 | .1522 | .1563 |
| 6 | .0555 | .0608 | .0662 | .0716 | .0771 | .0826 | .0881 | .0936 | .0989 | .1042 |
| 7 | .0246 | .0278 | .0312 | .0348 | .0385 | .0425 | .0466 | .0508 | .0551 | .0595 |
| 8 | .0095 | .0111 | .0129 | .0148 | .0169 | .0191 | .0215 | .0241 | .0269 | .0298 |
| 9 | .0033 | .0040 | .0047 | .0056 | .0066 | .0076 | .0089 | .0102 | .0116 | .0132 |
| 10 | .0010 | .0013 | .0016 | .0019 | .0023 | .0028 | .0033 | .0039 | .0045 | .0053 |
| 11 | .0003 | .0004 | .0005 | .0006 | .0007 | .0009 | .0011 | .0013 | .0016 | .0019 |
| 12 | .0001 | .0001 | .0001 | .0002 | .0002 | .0003 | .0003 | .0004 | .0005 | .0006 |
| 13 | .0000 | .0000 | .0000 | .0000 | .0001 | .0001 | .0001 | .0001 | .0002 | .0002 |
| 14 | .0000 | .0000 | .0000 | .0000 | .0000 | .0000 | .0000 | .0000 | .0000 | .0001 |

| | λ | | | | | | | | | |
|---|---|---|---|---|---|---|---|---|---|---|
| *x* | **4.1** | **4.2** | **4.3** | **4.4** | **4.5** | **4.6** | **4.7** | **4.8** | **4.9** | **5.0** |
| 0 | .0166 | .0150 | .0136 | .0123 | .0111 | .0101 | .0091 | .0082 | .0074 | .0067 |
| 1 | .0679 | .0630 | .0583 | .0540 | .0500 | .0462 | .0427 | .0395 | .0365 | .0337 |
| 2 | .1393 | .1323 | .1254 | .1188 | .1125 | .1063 | .1005 | .0948 | .0894 | .0842 |
| 3 | .1904 | .1852 | .1798 | .1743 | .1687 | .1631 | .1574 | .1517 | .1460 | .1404 |
| 4 | .1951 | .1944 | .1933 | .1917 | .1898 | .1875 | .1849 | .1820 | .1789 | .1755 |
| 5 | .1600 | .1633 | .1662 | .1687 | .1708 | .1725 | .1738 | .1747 | .1753 | .1755 |
| 6 | .1093 | .1143 | .1191 | .1237 | .1281 | .1323 | .1362 | .1398 | .1432 | .1462 |
| 7 | .0640 | .0686 | .0732 | .0778 | .0824 | .0869 | .0914 | .0959 | .1002 | .1044 |
| 8 | .0328 | .0360 | .0393 | .0428 | .0463 | .0500 | .0537 | .0575 | .0614 | .0653 |
| 9 | .0150 | .0168 | .0188 | .0209 | .0232 | .0255 | .0280 | .0307 | .0334 | .0363 |
| 10 | .0061 | .0071 | .0081 | .0092 | .0104 | .0118 | .0132 | .0147 | .0164 | .0181 |
| 11 | .0023 | .0027 | .0032 | .0037 | .0043 | .0049 | .0056 | .0064 | .0073 | .0082 |
| 12 | .0008 | .0009 | .0011 | .0014 | .0016 | .0019 | .0022 | .0026 | .0030 | .0034 |
| 13 | .0002 | .0003 | .0004 | .0005 | .0006 | .0007 | .0008 | .0009 | .0011 | .0013 |
| 14 | .0001 | .0001 | .0001 | .0001 | .0002 | .0002 | .0003 | .0003 | .0004 | .0005 |
| 15 | .0000 | .0000 | .0000 | .0000 | .0001 | .0001 | .0001 | .0001 | .0001 | .0002 |

| | λ | | | | | | | | | |
|---|---|---|---|---|---|---|---|---|---|---|
| *x* | **5.1** | **5.2** | **5.3** | **5.4** | **5.5** | **5.6** | **5.7** | **5.8** | **5.9** | **6.0** |
| 0 | .0061 | .0055 | .0050 | .0045 | .0041 | .0037 | .0033 | .0030 | .0027 | .0025 |
| 1 | .0311 | .0287 | .0265 | .0244 | .0225 | .0207 | .0191 | .0176 | .0162 | .0149 |
| 2 | .0793 | .0746 | .0701 | .0659 | .0618 | .0580 | .0544 | .0509 | .0477 | .0446 |
| 3 | .1348 | .1293 | .1239 | .1185 | .1133 | .1082 | .1033 | .0985 | .0938 | .0892 |
| 4 | .1719 | .1681 | .1641 | .1600 | .1558 | .1515 | .1472 | .1428 | .1383 | .1339 |

**Table IV (continued)**

| | λ | | | | | | | | | |
|---|---|---|---|---|---|---|---|---|---|---|
| $x$ | **5.1** | **5.2** | **5.3** | **5.4** | **5.5** | **5.6** | **5.7** | **5.8** | **5.9** | **6.0** |
| 5 | .1753 | .1748 | .1740 | .1728 | .1714 | .1697 | .1678 | .1656 | .1632 | .1606 |
| 6 | .1490 | .1515 | .1537 | .1555 | .1571 | .1584 | .1594 | .1601 | .1605 | .1606 |
| 7 | .1086 | .1125 | .1163 | .1200 | .1234 | .1267 | .1298 | .1326 | .1353 | .1377 |
| 8 | .0692 | .0731 | .0771 | .0810 | .0849 | .0887 | .0925 | .0962 | .0998 | .1033 |
| 9 | .0392 | .0423 | .0454 | .0486 | .0519 | .0552 | .0586 | .0620 | .0654 | .0688 |
| 10 | .0200 | .0220 | .0241 | .0262 | .0285 | .0309 | .0334 | .0359 | .0386 | .0413 |
| 11 | .0093 | .0104 | .0116 | .0129 | .0143 | .0157 | .0173 | .0190 | .0207 | .0225 |
| 12 | .0039 | .0045 | .0051 | .0058 | .0065 | .0073 | .0082 | .0092 | .0102 | .0113 |
| 13 | .0015 | .0018 | .0021 | .0024 | .0028 | .0032 | .0036 | .0041 | .0046 | .0052 |
| 14 | .0006 | .0007 | .0008 | .0009 | .0011 | .0013 | .0015 | .0017 | .0019 | .0022 |
| 15 | .0002 | .0002 | .0003 | .0003 | .0004 | .0005 | .0006 | .0007 | .0008 | .0009 |
| 16 | .0001 | .0001 | .0001 | .0001 | .0001 | .0002 | .0002 | .0002 | .0003 | .0003 |
| 17 | .0000 | .0000 | .0000 | .0000 | .0000 | .0001 | .0001 | .0001 | .0001 | .0001 |

| | λ | | | | | | | | | |
|---|---|---|---|---|---|---|---|---|---|---|
| $x$ | **6.1** | **6.2** | **6.3** | **6.4** | **6.5** | **6.6** | **6.7** | **6.8** | **6.9** | **7.0** |
| 0 | .0022 | .0020 | .0018 | .0017 | .0015 | .0014 | .0012 | .0011 | .0010 | .0009 |
| 1 | .0137 | .0126 | .0116 | .0106 | .0098 | .0090 | .0082 | .0076 | .0070 | .0064 |
| 2 | .0417 | .0390 | .0364 | .0340 | .0318 | .0296 | .0276 | .0258 | .0240 | .0223 |
| 3 | .0848 | .0806 | .0765 | .0726 | .0688 | .0652 | .0617 | .0584 | .0552 | .0521 |
| 4 | .1294 | .1249 | .1205 | .1162 | .1118 | .1076 | .1034 | .0992 | .0952 | .0912 |
| 5 | .1579 | .1549 | .1519 | .1487 | .1454 | .1420 | .1385 | .1349 | .1314 | .1277 |
| 6 | .1605 | .1601 | .1595 | .1586 | .1575 | .1562 | .1546 | .1529 | .1511 | .1490 |
| 7 | .1399 | .1418 | .1435 | .1450 | .1462 | .1472 | .1480 | .1486 | .1489 | .1490 |
| 8 | .1066 | .1099 | .1130 | .1160 | .1188 | .1215 | .1240 | .1263 | .1284 | .1304 |
| 9 | .0723 | .0757 | .0791 | .0825 | .0858 | .0891 | .0923 | .0954 | .0985 | .1014 |
| 10 | .0441 | .0469 | .0498 | .0528 | .0558 | .0588 | .0618 | .0649 | .0679 | .0710 |
| 11 | .0245 | .0265 | .0285 | .0307 | .0330 | .0353 | .0377 | .0401 | .0426 | .0452 |
| 12 | .0124 | .0137 | .0150 | .0164 | .0179 | .0194 | .0210 | .0227 | .0245 | .0264 |
| 13 | .0058 | .0065 | .0073 | .0081 | .0089 | .0098 | .0108 | .0119 | .0130 | .0142 |
| 14 | .0025 | .0029 | .0033 | .0037 | .0041 | .0046 | .0052 | .0058 | .0064 | .0071 |
| 15 | .0010 | .0012 | .0014 | .0016 | .0018 | .0020 | .0023 | .0026 | .0029 | .0033 |
| 16 | .0004 | .0005 | .0005 | .0006 | .0007 | .0008 | .0010 | .0011 | .0013 | .0014 |
| 17 | .0001 | .0002 | .0002 | .0002 | .0003 | .0003 | .0004 | .0004 | .0005 | .0006 |
| 18 | .0000 | .0001 | .0001 | .0001 | .0001 | .0001 | .0001 | .0002 | .0002 | .0002 |
| 19 | .0000 | .0000 | .0000 | .0000 | .0000 | .0000 | .0000 | .0001 | .0001 | .0001 |

| | λ | | | | | | | | | |
|---|---|---|---|---|---|---|---|---|---|---|
| $x$ | **7.1** | **7.2** | **7.3** | **7.4** | **7.5** | **7.6** | **7.7** | **7.8** | **7.9** | **8.0** |
| 0 | .0008 | .0007 | .0007 | .0006 | .0006 | .0005 | .0005 | .0004 | .0004 | .0003 |
| 1 | .0059 | .0054 | .0049 | .0045 | .0041 | .0038 | .0035 | .0032 | .0029 | .0027 |
| 2 | .0208 | .0194 | .0180 | .0167 | .0156 | .0145 | .0134 | .0125 | .0116 | .0107 |
| 3 | .0492 | .0464 | .0438 | .0413 | .0389 | .0366 | .0345 | .0324 | .0305 | .0286 |
| 4 | .0874 | .0836 | .0799 | .0764 | .0729 | .0696 | .0663 | .0632 | .0602 | .0573 |
| 5 | .1241 | .1204 | .1167 | .1130 | .1094 | .1057 | .1021 | .0986 | .0951 | .0916 |
| 6 | .1468 | .1445 | .1420 | .1394 | .1367 | .1339 | .1311 | .1282 | .1252 | .1221 |
| 7 | .1489 | .1486 | .1481 | .1474 | .1465 | .1454 | .1442 | .1428 | .1413 | .1396 |
| 8 | .1321 | .1337 | .1351 | .1363 | .1373 | .1382 | .1388 | .1392 | .1395 | .1396 |
| 9 | .1042 | .1070 | .1096 | .1121 | .1144 | .1167 | .1187 | .1207 | .1224 | .1241 |

**Table IV (continued)**

| x | 7.1 | 7.2 | 7.3 | 7.4 | 7.5 | 7.6 | 7.7 | 7.8 | 7.9 | 8.0 |
|---|---|---|---|---|---|---|---|---|---|---|
| 10 | .0740 | .0770 | .0800 | .0829 | .0858 | .0887 | .0914 | .0941 | .0967 | .0993 |
| 11 | .0478 | .0504 | .0531 | .0558 | .0585 | .0613 | .0640 | .0667 | .0695 | .0722 |
| 12 | .0283 | .0303 | .0323 | .0344 | .0366 | .0388 | .0411 | .0434 | .0457 | .0481 |
| 13 | .0154 | .0168 | .0181 | .0196 | .0211 | .0227 | .0213 | .0260 | .0278 | .0296 |
| 14 | .0078 | .0086 | .0095 | .0104 | .0113 | .0123 | .0134 | .0145 | .0157 | .0169 |
| 15 | .0037 | .0041 | .0046 | .0051 | .0057 | .0062 | .0069 | .0075 | .0083 | .0090 |
| 16 | .0016 | .0019 | .0021 | .0024 | .0026 | .0030 | .0033 | .0037 | .0041 | .0045 |
| 17 | .0007 | .0008 | .0009 | .0010 | .0012 | .0013 | .0015 | .0017 | .0019 | .0021 |
| 18 | .0003 | .0003 | .0004 | .0004 | .0005 | .0006 | .0006 | .0007 | .0008 | .0009 |
| 19 | .0001 | .0001 | .0001 | .0002 | .0002 | .0002 | .0003 | .0003 | .0003 | .0004 |
| 20 | .0000 | .0000 | .0001 | .0001 | .0001 | .0001 | .0001 | .0001 | .0001 | .0002 |
| 21 | .0000 | .0000 | .0000 | .0000 | .0000 | .0000 | .0000 | .0000 | .0001 | .0001 |

| x | 8.1 | 8.2 | 8.3 | 8.4 | 8.5 | 8.6 | 8.7 | 8.8 | 8.9 | 9.0 |
|---|---|---|---|---|---|---|---|---|---|---|
| 0 | .0003 | .0003 | .0002 | .0002 | .0002 | .0002 | .0002 | .0002 | .0001 | .0001 |
| 1 | .0025 | .0023 | .0021 | .0019 | .0017 | .0016 | .0014 | .0013 | .0012 | .0011 |
| 2 | .0100 | .0092 | .0086 | .0079 | .0074 | .0068 | .0063 | .0058 | .0054 | .0050 |
| 3 | .0269 | .0252 | .0237 | .0222 | .0208 | .0195 | .0183 | .0171 | .0160 | .0150 |
| 4 | .0544 | .0517 | .0491 | .0466 | .0443 | .0420 | .0398 | .0377 | .0357 | .0337 |
| 5 | .0882 | .0849 | .0816 | .0784 | .0752 | .0722 | .0692 | .0663 | .0635 | .0607 |
| 6 | .1191 | .1160 | .1128 | .1097 | .1066 | .1034 | .1003 | .0972 | .0941 | .0911 |
| 7 | .1378 | .1358 | .1338 | .1317 | .1294 | .1271 | .1247 | .1222 | .1197 | .1171 |
| 8 | .1395 | .1392 | .1388 | .1382 | .1375 | .1366 | .1356 | .1344 | .1332 | .1318 |
| 9 | .1256 | .1269 | .1280 | .1290 | .1299 | .1306 | .1311 | .1315 | .1317 | .1318 |
| 10 | .1017 | .1040 | .1063 | .1084 | .1104 | .1123 | .1140 | .1157 | .1172 | .1186 |
| 11 | .0749 | .0776 | .0802 | .0828 | .0853 | .0878 | .0902 | .0925 | .0948 | .0910 |
| 12 | .0505 | .0530 | .0555 | .0579 | .0604 | .0629 | .0654 | .0679 | .0703 | .0728 |
| 13 | .0315 | .0334 | .0354 | .0374 | .0395 | .0416 | .0438 | .0459 | .0481 | .0504 |
| 14 | .0182 | .O196 | .0210 | .0225 | .0240 | .0256 | .0272 | .0289 | .0306 | .0324 |
| 15 | .0098 | .0107 | .0116 | .0126 | .0136 | .0147 | .0158 | .0169 | .0182 | .0194 |
| 16 | .0050 | .0055 | .0060 | .0066 | .0072 | .0079 | .0086 | .0093 | .0101 | .0109 |
| 17 | .0024 | .0026 | .0029 | .0033 | .0036 | .0040 | .0044 | .0048 | .0053 | .0058 |
| 18 | .0011 | .0012 | .0014 | .0015 | .0017 | .0019 | .0021 | .0024 | .0026 | .0029 |
| 19 | .0005 | .0005 | .0006 | .0007 | .0008 | .0009 | .0010 | .0011 | .0012 | .0014 |
| 20 | .0002 | .0002 | .0002 | .0003 | .0003 | .0004 | .0004 | .0005 | .0005 | .0006 |
| 21 | .0001 | .0001 | .0001 | .0001 | .0001 | .0002 | .0002 | .0002 | .0002 | .0003 |
| 22 | .0000 | .0000 | .0000 | .0000 | .0001 | .0001 | .0001 | .0001 | .0001 | .0001 |

| x | 9.1 | 9.2 | 9.3 | 9.4 | 9.5 | 9.6 | 9.7 | 9.8 | 9.9 | 10 |
|---|---|---|---|---|---|---|---|---|---|---|
| 0 | .0001 | .0001 | .0001 | .0001 | .0001 | .0001 | .0001 | .0001 | .0001 | .0000 |
| 1 | .0010 | .0009 | .0009 | .0008 | .0007 | .0007 | .0006 | .0005 | .0005 | .0005 |
| 2 | .0046 | .0043 | .0040 | .0037 | .0034 | .0031 | .0029 | .0027 | .0025 | .0023 |
| 3 | .0140 | .0131 | .0123 | .0115 | .0107 | .0100 | .0093 | .0087 | .0081 | .0076 |
| 4 | .0319 | .0302 | .0285 | .0269 | .0254 | .0240 | .0226 | .0213 | .0201 | .0189 |

**Table IV (continued)**

| | λ | | | | | | | | | |
|---|---|---|---|---|---|---|---|---|---|---|
| *x* | **9.1** | **9.2** | **9.3** | **9.4** | **9.5** | **9.6** | **9.7** | **9.8** | **9.9** | **10** |
| 5 | .0581 | .0555 | .0530 | .0506 | .0483 | .0460 | .0439 | .0418 | .0398 | .0378 |
| 6 | .0881 | .0851 | .0822 | .0793 | .0764 | .0736 | .0709 | .0682 | .0656 | .0631 |
| 7 | .1145 | .1118 | .1091 | .1064 | .1037 | .1010 | .0982 | .0955 | .0928 | .0901 |
| 8 | .1302 | .1286 | .1269 | .1251 | .1232 | .1212 | .1191 | .1170 | .1148 | .1126 |
| 9 | .1317 | .1315 | .1311 | .1306 | .1300 | .1293 | .1284 | .1274 | .1263 | .1251 |
| 10 | .1198 | .1210 | .1219 | .1228 | .1235 | .1241 | .1245 | .1249 | .1250 | .1251 |
| 11 | .0991 | .1012 | .1031 | .1049 | .1067 | .1083 | .1098 | .1112 | .1125 | .1137 |
| 12 | .0752 | .0776 | .0799 | .0822 | .0844 | .0866 | .0888 | .0908 | .0928 | .0948 |
| 13 | .0526 | .0549 | .0572 | .0594 | .0617 | .0640 | .0662 | .0685 | .0707 | .0729 |
| 14 | .0342 | .0361 | .0380 | .0399 | .0419 | .0439 | .0459 | .0479 | .0500 | .0521 |
| 15 | .0208 | .0221 | .0235 | .0250 | .0265 | .0281 | .0297 | .0313 | .0330 | .0347 |
| 16 | .0118 | .0127 | .0137 | .0147 | .0157 | .0168 | .0180 | .0192 | .0204 | .0217 |
| 17 | .0063 | .0069 | .0075 | .0081 | .0088 | .0095 | .0103 | .0111 | .0119 | .0128 |
| 18 | .0032 | .0035 | .0039 | .0042 | .0046 | .0051 | .0055 | .0060 | .0065 | .0071 |
| 19 | .0015 | .0017 | .0019 | .0021 | .0023 | .0026 | .0028 | .0031 | .0034 | .0037 |
| 20 | .0007 | .0008 | .0009 | .0010 | .0011 | .0012 | .0014 | .0015 | .0017 | .0019 |
| 21 | .0003 | .0003 | .0004 | .0004 | .0005 | .0006 | .0006 | .0007 | .0008 | .0009 |
| 22 | .0001 | .0001 | .0002 | .0002 | .0002 | .0002 | .0003 | .0003 | .0004 | .0004 |
| 23 | .0000 | .0001 | .0001 | .0001 | .0001 | .0001 | .0001 | .0001 | .0002 | .0002 |
| 24 | .0000 | .0000 | .0000 | .0000 | .0000 | .0000 | .0000 | .0001 | .0001 | .0001 |

| | λ | | | | | | | | | |
|---|---|---|---|---|---|---|---|---|---|---|
| *x* | **11** | **12** | **13** | **14** | **15** | **16** | **17** | **18** | **19** | **20** |
| 0 | .0000 | .0000 | .0000 | .0000 | .0000 | .0000 | .0000 | .0000 | .0000 | .0000 |
| 1 | .0002 | .0001 | .0000 | .0000 | .0000 | .0000 | .0000 | .0000 | .0000 | .0000 |
| 2 | .0010 | .0004 | .0002 | .0001 | .0000 | .0000 | .0000 | .0000 | .0000 | .0000 |
| 3 | .0037 | .0018 | .0008 | .0004 | .0002 | .0001 | .0000 | .0000 | .0000 | .0000 |
| 4 | .0102 | .0053 | .0027 | .0013 | .0006 | .0003 | .0001 | .0001 | .0000 | .0000 |
| 5 | .0224 | .0127 | .0070 | .0037 | .0019 | .0010 | .0005 | .0002 | .0001 | .0001 |
| 6 | .0411 | .0255 | .0152 | .0087 | .0048 | .0026 | .0014 | .0007 | .0004 | .0002 |
| 7 | .0646 | .0437 | .0281 | .0174 | .0104 | .0060 | .0034 | .0018 | .0010 | .0005 |
| 8 | .0888 | .0655 | .0457 | .0304 | .0194 | .0120 | .0072 | .0042 | .0024 | .0013 |
| 9 | .1085 | .0874 | .0661 | .0473 | .0324 | .0213 | .0135 | .0083 | .0050 | .0029 |
| 10 | .1194 | .1048 | .0859 | .0663 | .0486 | .0341 | .0230 | .0150 | .0095 | .0058 |
| 11 | .1194 | .1144 | .1015 | .0844 | .0663 | .0496 | .0355 | .0245 | .0164 | .0106 |
| 12 | .1094 | .1144 | .1099 | .0984 | .0829 | .0661 | .0504 | .0368 | .0259 | .0176 |
| 13 | .0926 | .1056 | .1099 | .1060 | .0956 | .0814 | .0658 | .0509 | .0378 | .0271 |
| 14 | .0728 | .0905 | .1021 | .1000 | .1024 | .0930 | .0800 | .0655 | .0514 | .0387 |
| 15 | .0534 | .0724 | .0885 | .0989 | .1024 | .0992 | .0906 | .0786 | .0650 | .0516 |
| 16 | .0367 | .0543 | .0719 | .0866 | .0960 | .0992 | .0963 | .0884 | .0772 | .0646 |
| 17 | .0237 | .0383 | .0550 | .0713 | .0847 | .0934 | .0963 | .0936 | .0863 | .0760 |
| 18 | .0145 | .0256 | .0397 | .0554 | .0706 | .0830 | .0909 | .0936 | .0911 | .0844 |
| 19 | .0084 | .0161 | .0272 | .0409 | .0557 | .0699 | .0814 | .0887 | .0911 | .0888 |
| 20 | .0046 | .0097 | .0177 | .0288 | .0418 | .0559 | .0692 | .0798 | .0866 | .0888 |
| 21 | .0024 | .0055 | .0109 | .0191 | .0299 | .0426 | .0560 | .0684 | .0783 | .0846 |
| 22 | .0012 | .0030 | .0065 | .0121 | .0204 | .0310 | .0433 | .0560 | .0676 | .0769 |
| 23 | .0006 | .0016 | .0037 | .0074 | .0133 | .0216 | .0320 | .0438 | .0559 | .0669 |
| 24 | .0003 | .0008 | .0020 | .0043 | .0083 | .0144 | .0226 | .0328 | .0442 | .0557 |

**Table IV (continued)**

| | | | | | λ | | | | | |
|---|---|---|---|---|---|---|---|---|---|---|
| *x* | **11** | **12** | **13** | **14** | **15** | **16** | **17** | **18** | **19** | **20** |
| 25 | .0001 | .0004 | .0010 | .0024 | .0050 | .0092 | .0154 | .0237 | .0336 | .0446 |
| 26 | .0000 | .0002 | .0005 | .0013 | .0029 | .0057 | .0101 | .0164 | .0246 | .0343 |
| 27 | .0000 | .0001 | .0002 | .0001 | .0016 | .0034 | .0063 | .0109 | .0173 | .0254 |
| 28 | .0000 | .0000 | .0001 | .0003 | .0009 | .0019 | .0038 | .0010 | .0111 | .0181 |
| 29 | .0000 | .0000 | .0001 | .0002 | .0004 | .0011 | .0023 | .0044 | .0071 | .0125 |
| 30 | .0000 | .0000 | .0000 | .0001 | .0002 | .0006 | .0013 | .0026 | .0049 | .0083 |
| 31 | .0000 | .0000 | .0000 | .0000 | .0001 | .0003 | .0001 | .0015 | .0030 | .0054 |
| 32 | .0000 | .0000 | .0000 | .0000 | .0001 | .0001 | .0004 | .0009 | .0018 | .0034 |
| 33 | .0000 | .0000 | .0000 | .0000 | .0000 | .0001 | .0002 | .0005 | .0010 | .0020 |
| 34 | .0000 | .0000 | .0000 | .0000 | .0000 | .0000 | .0001 | .0002 | .0006 | .0012 |
| 35 | .0000 | .0000 | .0000 | .0000 | .0000 | .0000 | .0000 | .0001 | .0003 | .0001 |
| 36 | .0000 | .0000 | .0000 | .0000 | .0000 | .0000 | .0000 | .0001 | .0002 | .0004 |
| 37 | .0000 | .0000 | .0000 | .0000 | .0000 | .0000 | .0000 | .0000 | .0001 | .0002 |
| 38 | .0000 | .0000 | .0000 | .0000 | .0000 | .0000 | .0000 | .0000 | .0000 | .0001 |
| 39 | .0000 | .0000 | .0000 | .0000 | .0000 | .0000 | .0000 | .0000 | .0000 | .0001 |

**Table V Standard normal distribution**

| $z$ | .00 | .01 | .02 | .03 | .04 | .05 | .06 | .07 | .08 | .09 |
|---|---|---|---|---|---|---|---|---|---|---|
| 0.0 | .0000 | .0040 | .0080 | .0120 | .0160 | .0199 | .0239 | .0279 | .0319 | .0359 |
| 0.1 | .0398 | .0438 | .0478 | .0517 | .0557 | .0596 | .0636 | .0675 | .0714 | .0753 |
| 0.2 | .0793 | .0832 | .0871 | .0910 | .0948 | .0987 | .1026 | .1064 | .1103 | .1141 |
| 0.3 | .1179 | .1217 | .1255 | .1293 | .1331 | .1368 | .1406 | .1443 | .1480 | .1517 |
| 0.4 | .1554 | .1591 | .1628 | .1664 | .1700 | .1736 | .1772 | .1808 | .1844 | .1879 |
| 0.5 | .1915 | .1950 | .1985 | .2019 | .2054 | .2088 | .2123 | .2157 | .2190 | .2224 |
| 0.6 | .2257 | .2291 | .2324 | .2357 | .2389 | .2422 | .2454 | .2486 | .2517 | .2549 |
| 0.7 | .2580 | .2611 | .2642 | .2673 | .2704 | .2734 | .2764 | .2794 | .2823 | .2852 |
| 0.8 | .2881 | .2910 | .2939 | .2967 | .2995 | .3023 | .3051 | .3078 | .3106 | .3133 |
| 0.9 | .3159 | .3186 | .3212 | .3238 | .3264 | .3289 | .3315 | .3340 | .3365 | .3389 |
| 1.0 | .3413 | .3438 | .3461 | .3485 | .3508 | .3531 | .3554 | .3577 | .3599 | .3621 |
| 1.1 | .3643 | .3665 | .3686 | .3708 | .3729 | .3749 | .3770 | .3790 | .3810 | .3830 |
| 1.2 | .3849 | .3869 | .3888 | .3907 | .3925 | .3944 | .3962 | .3980 | .3997 | .4015 |
| 1.3 | .4032 | .4049 | .4066 | .4082 | .4099 | .4115 | .4131 | .4147 | .4162 | .4177 |
| 1.4 | .4192 | .4207 | .4222 | .4236 | .4251 | .4265 | .4279 | .4292 | .4306 | .4319 |
| 1.5 | .4332 | .4345 | .4357 | .4370 | .4382 | .4394 | .4406 | .4418 | .4429 | .4441 |
| 1.6 | .4452 | .4463 | .4474 | .4484 | .4495 | .4505 | .4515 | .4525 | .4535 | .4545 |
| 1.7 | .4554 | .4564 | .4573 | .4582 | .4591 | .4599 | .4608 | .4616 | .4625 | .4633 |
| 1.8 | .4641 | .4649 | .4656 | .4664 | .4671 | .4678 | .4686 | .4693 | .4699 | .4706 |
| 1.9 | .4713 | .4719 | .4726 | .4732 | .4738 | .4744 | .4750 | .4756 | .4761 | .4767 |
| 2.0 | .4772 | .4778 | .4783 | .4788 | .4793 | .4798 | .4803 | .4808 | .4812 | .4817 |
| 2.1 | .4821 | .4826 | .4830 | .4834 | .4838 | .4842 | .4846 | .4850 | .4854 | .4857 |
| 2.2 | .4861 | .4864 | .4868 | .4871 | .4875 | .4878 | .4881 | .4884 | .4887 | .4890 |
| 2.3 | .4893 | .4896 | .4898 | .4901 | .4904 | .4906 | .4909 | .4911 | .4913 | .4916 |
| 2.4 | .4918 | .4920 | .4922 | .4925 | .4927 | .4929 | .4931 | .4932 | .4934 | .4936 |
| 2.5 | .4938 | .4940 | .4941 | .4943 | .4945 | .4946 | .4948 | .4949 | .4951 | .4952 |
| 2.6 | .4953 | .4955 | .4956 | .4957 | .4959 | .4960 | .4961 | .4962 | .4963 | .4964 |
| 2.7 | .4965 | .4966 | .4967 | .4968 | .4969 | .4970 | .4971 | .4972 | .4973 | .4974 |
| 2.8 | .4974 | .4975 | .4976 | .4977 | .4977 | .4978 | .4979 | .4979 | .4980 | .4981 |
| 2.9 | .4981 | .4982 | .4982 | .4983 | .4984 | .4984 | .4985 | .4985 | .4986 | .4986 |
| 3.0 | .4987 | .4987 | .4987 | .4988 | .4988 | .4989 | .4989 | .4989 | .4990 | .4990 |

Also, for $z = 4.0$, 5.0 and 6.0, the probabilities are 0.49997, 0.4999997 and 0.499999999.

**Table VI Student's $t$-Distribution**
**Values of $t_{\alpha,\nu}$**

| $\nu$ | $\alpha$ = .10 | $\alpha$ = 0.5 | $\alpha$ = 0.25 | $\alpha$ = .01 | $\alpha$ = .005 |
|---|---|---|---|---|---|
| 1 | 3.078 | 6.314 | 12.706 | 31. 821 | 63.657 |
| 2 | 1.886 | 2.920 | 4.303 | 6.965 | 9.925 |
| 3 | 1.638 | 2.353 | 3.182 | 4.541 | 5.841 |
| 4 | 1.533 | 2.132 | 2.776 | 3.747 | 4.604 |
| 5 | 1.476 | 2.015 | 2.571 | 3.365 | 4.032 |
| 6 | 1.440 | 1.943 | 2.447 | 3.143 | 3.707 |
| 7 | 1.415 | 1.895 | 2.365 | 2.998 | 3.499 |
| 8 | 1.397 | 1.860 | 2.306 | 2.896 | 3.355 |
| 9 | 1.383 | 1.833 | 2.262 | 2.821 | 3.250 |
| 10 | 1.372 | 1.812 | 2.228 | 2.764 | 3.169 |
| 11 | 1.363 | 1.796 | 2.201 | 2.718 | 3.106 |
| 12 | 1.356 | 1.782 | 2.179 | 2.681 | 3.055 |
| 13 | 1.350 | 1.771 | 2.160 | 2.650 | 3.012 |
| 14 | 1.345 | 1.761 | 2.145 | 2.624 | 2.977 |
| 15 | 1.341 | 1.753 | 2.131 | 2.602 | 2.947 |
| 16 | 1.337 | 1.746 | 2.120 | 2.583 | 2.921 |
| 17 | 1.333 | 1.740 | 2.110 | 2.567 | 2.898 |
| 18 | 1.330 | 1.734 | 2.101 | 2.552 | 2.878 |
| 19 | 1.328 | 1.729 | 2.093 | 2.539 | 2.861 |
| 20 | 1.325 | 1.725 | 2.086 | 2.528 | 2.845 |
| 21 | 1.323 | 1.721 | 2.080 | 2.518 | 2.831 |
| 22 | 1.321 | 1.717 | 2.074 | 2.508 | 2.819 |
| 23 | 1.319 | 1.714 | 2.069 | 2.500 | 2.807 |
| 24 | 1.318 | 1.711 | 2.064 | 2.492 | 2.797 |
| 25 | 1.316 | 1.708 | 2.060 | 2.485 | 2.787 |
| 26 | 1.315 | 1.706 | 2.056 | 2.479 | 2.779 |
| 27 | 1.314 | 1.703 | 2.052 | 2.473 | 2.771 |
| 28 | 1.313 | 1.701 | 2.048 | 2.467 | 2.763 |
| 29 | 1.311 | 1.699 | 2.045 | 2.462 | 2.756 |
| inf. | 1.282 | 1.645 | 1.960 | 2.326 | 2.576 |

**Table VII $\chi^2$ distribution**

**Values of $\chi^2_{\alpha,\nu}$**

| $\nu$ | $\alpha = .995$ | $\alpha = .99$ | $\alpha = .975$ | $\alpha = .95$ | $\alpha = .05$ | $\alpha = .025$ | $\alpha = .01$ | $\alpha = .005$ | $\nu$ |
|---|---|---|---|---|---|---|---|---|---|
| 1 | .0000393 | .000157 | .000982 | .00393 | 3.841 | 5.024 | 6.635 | 7.879 | 1 |
| 2 | .0100 | .0201 | .0506 | .103 | 5.991 | 7.378 | 9.210 | 10.597 | 2 |
| 3 | .0717 | .115 | .216 | .352 | 7.815 | 9.348 | 11.345 | 12.838 | 3 |
| 4 | .207 | .297 | .484 | .711 | 9.488 | 11.143 | 13.277 | 14.860 | 4 |
| 5 | .412 | .554 | .831 | 1.145 | 11.070 | 12.832 | 15.086 | 16.750 | 5 |
| 6 | .676 | .872 | 1.237 | 1.635 | 12.592 | 14.449 | 16.812 | 18.548 | 6 |
| 7 | .989 | 1.239 | 1.690 | 2.167 | 14.067 | 16.013 | 18.475 | 20.278 | 7 |
| 8 | 1.344 | 1.646 | 2.180 | 2.733 | 15.507 | 17.535 | 20.090 | 21.955 | 8 |
| 9 | 1.735 | 2.088 | 2.700 | 3.325 | 16.919 | 19.023 | 21. 666 | 23.589 | 9 |
| 10 | 2.156 | 2.558 | 3.247 | 3.940 | 18.307 | 20.483 | 23.209 | 25.188 | 10 |
| 11 | 2.603 | 3.053 | 3.816 | 4.575 | 19.675 | 21.920 | 24.725 | 26.757 | 11 |
| 12 | 3.074 | 3.571 | 4.404 | 5.226 | 21.026 | 23.337 | 26.217 | 28.300 | 12 |
| 13 | 3.565 | 4.107 | 5.009 | 5.892 | 22.362 | 24.736 | 27.688 | 29.819 | 13 |
| 14 | 4.075 | 4.660 | 5.629 | 6.571 | 23.685 | 26.119 | 29.141 | 31.319 | 14 |
| 15 | 4.601 | 5.229 | 6.262 | 7.261 | 24.996 | 27.488 | 30.578 | 32.801 | 15 |
| 16 | 5.142 | 5.812 | 6.908 | 7.962 | 26.296 | 28.845 | 32.000 | 34.267 | 16 |
| 17 | 5.697 | 6.408 | 7.564 | 8.672 | 27.587 | 30.191 | 33.409 | 35.718 | 17 |
| 18 | 6.265 | 7.015 | 8.231 | 9.390 | 28.869 | 31.526 | 34.805 | 37.156 | 18 |
| 19 | 6.844 | 7.633 | 8.907 | 10.117 | 30.144 | 32.852 | 36.191 | 38.582 | 19 |
| 20 | 7.434 | 8.260 | 9.591 | 10.851 | 31.410 | 34.170 | 37.566 | 39.997 | 20 |
| 21 | 8.034 | 8.897 | 10.283 | 11.591 | 32.671 | 35.479 | 38.932 | 41.401 | 21 |
| 22 | 8.643 | 9.542 | 10.982 | 12.338 | 33.924 | 36.781 | 40.298 | 42.796 | 22 |
| 23 | 9.260 | 10.196 | 11.689 | 13.091 | 35.172 | 38.076 | 41.638 | 44.181 | 23 |
| 24 | 9.886 | 10.856 | 12.401 | 13.848 | 36.415 | 39.364 | 42.980 | 45.558 | 24 |
| 25 | 10.520 | 11.524 | 13.120 | 14.611 | 37.652 | 40.646 | 44.314 | 46.928 | 25 |
| 26 | 11.160 | 12.198 | 13.844 | 15.379 | 38.885 | 41.923 | 45.642 | 48.290 | 26 |
| 27 | 11.808 | 12.879 | 14.573 | 16.151 | 40.113 | 43.194 | 46.963 | 49.645 | 27 |
| 28 | 12.461 | 13.565 | 15.308 | 16.928 | 41.337 | 44.461 | 48.278 | 50.993 | 28 |
| 29 | 13.121 | 14.256 | 16.047 | 17.708 | 42.557 | 45.722 | 49.588 | 52.336 | 29 |
| 30 | 13.787 | 14.953 | 16.791 | 18.493 | 43.773 | 46.979 | 50.892 | 53.672 | 30 |

**Table VIII** *F*-Distribution

**Values of** $F_{0.5,\nu1,\nu2}$

$\nu_1$ = Degrees of freedom for numerator; $\nu_2$ = Degrees of freedom for denominator

| $\nu_2$ | 1 | 2 | 3 | 4 | 5 | 6 | 7 | 8 | 9 | 10 | 12 | 15 | 20 | 24 | 30 | 40 | 60 | 120 | ∞ |
|---|---|---|---|---|---|---|---|---|---|---|---|---|---|---|---|---|---|---|---|
| 1 | 161 | 200 | 216 | 225 | 230 | 234 | 237 | 239 | 241 | 242 | 244 | 246 | 248 | 249 | 250 | 251 | 252 | 253 | 254 |
| 2 | 18.5 | 19.0 | 19.2 | 19.2 | 19.3 | 19.3 | 19.4 | 19.4 | 19.4 | 19.4 | 19.4 | 19.4 | 19.4 | 19.5 | 19.5 | 19.5 | 19.5 | 19.5 | 19.5 |
| 3 | 10.1 | 9.55 | 9.28 | 9.12 | 9.01 | 8.94 | 8.89 | 8.85 | 8.81 | 8.79 | 8.74 | 8.70 | 8.66 | 8.64 | 8.62 | 8.59 | 8.57 | 8.55 | 8.53 |
| 4 | 7.71 | 6.94 | 6.59 | 6.39 | 6.26 | 6.16 | 6.09 | 6.04 | 6.00 | 5.96 | 5.91 | 5.86 | 5.80 | 5.77 | 5.75 | 5.72 | 5.69 | 5.66 | 5.63 |
| 5 | 6.61 | 5.79 | 5.41 | 5.19 | 5.05 | 4.95 | 4.88 | 4.82 | 4.77 | 4.74 | 4.68 | 4.62 | 4.56 | 4.53 | 4.50 | 4.46 | 4.43 | 4.40 | 4.37 |
| 6 | 5.99 | 5.14 | 4.76 | 4.53 | 4.39 | 4.28 | 4.21 | 4.15 | 4.10 | 4.06 | 4.00 | 3.94 | 3.87 | 3.84 | 3.81 | 3.77 | 3.74 | 3.70 | 3.67 |
| 7 | 5.59 | 4.74 | 4.35 | 4.12 | 3.97 | 3.87 | 3.79 | 3.73 | 3.68 | 3.64 | 3.57 | 3.51 | 3.44 | 3.41 | 3.38 | 3.34 | 3.30 | 3.27 | 3.23 |
| 8 | 5.32 | 4.46 | 4.07 | 3.84 | 3.69 | 3.58 | 3.50 | 3.44 | 3.39 | 3.35 | 3.28 | 3.22 | 3.15 | 3.12 | 3.08 | 3.04 | 3.01 | 2.97 | 2.93 |
| 9 | 5.12 | 4.26 | 3.86 | 3.63 | 3.48 | 3.37 | 3.29 | 3.23 | 3.18 | 3.14 | 3.07 | 3.01 | 2.94 | 2.90 | 2.86 | 2.83 | 2.79 | 2.75 | 2.71 |
| 10 | 4.96 | 4.10 | 3.71 | 3.48 | 3.33 | 3.22 | 3.14 | 3.07 | 3.02 | 2.98 | 2.91 | 2.85 | 2.77 | 2.74 | 2.70 | 2.66 | 2.62 | 2.58 | 2.54 |
| 11 | 4.84 | 3.98 | 3.59 | 3.36 | 3.20 | 3.09 | 3.01 | 2.95 | 2.90 | 2.85 | 2.79 | 2.72 | 2.65 | 2.61 | 2.57 | 2.53 | 2.49 | 2.45 | 2.40 |
| 12 | 4.75 | 3.89 | 3.49 | 3.26 | 3.11 | 3.00 | 2.91 | 2.85 | 2.80 | 2.75 | 2.69 | 2.62 | 2.54 | 2.51 | 2.47 | 2.43 | 2.38 | 2.34 | 2.30 |
| 13 | 4.67 | 3.81 | 3.41 | 3.18 | 3.03 | 2.92 | 2.83 | 2.77 | 2.71 | 2.67 | 2.60 | 2.53 | 2.46 | 2.42 | 2.38 | 2.34 | 2.30 | 2.25 | 2.21 |
| 14 | 4.60 | 3.74 | 3.34 | 3.11 | 2.96 | 2.85 | 2.76 | 2.70 | 2.65 | 2.60 | 2.53 | 2.46 | 2.39 | 2.35 | 2.31 | 2.27 | 2.22 | 2.18 | 2.13 |
| 15 | 4.54 | 3.68 | 3.29 | 3.06 | 2.90 | 2.79 | 2.71 | 2.64 | 2.59 | 2.54 | 2.48 | 2.40 | 2.33 | 2.29 | 2.25 | 2.20 | 2.16 | 2.11 | 2.07 |
| 16 | 4.49 | 3.63 | 3.24 | 3.01 | 2.85 | 2.74 | 2.66 | 2.59 | 2.54 | 2.49 | 2.42 | 2.35 | 2.28 | 2.24 | 2.19 | 2.15 | 2.11 | 2.06 | 2.01 |
| 17 | 4.45 | 3.59 | 3.20 | 2.96 | 2.81 | 2.70 | 2.61 | 2.55 | 2.49 | 2.45 | 2.38 | 2.31 | 2.23 | 2.19 | 2.15 | 2.10 | 2.06 | 2.01 | 1.96 |
| 18 | 4.41 | 3.55 | 3.16 | 2.93 | 2.77 | 2.66 | 2.58 | 2.51 | 2.46 | 2.41 | 2.34 | 2.27 | 2.19 | 2.15 | 2.11 | 2.06 | 2.02 | 1.97 | 1.92 |
| 19 | 4.38 | 3.52 | 3.13 | 2.90 | 2.74 | 2.63 | 2.54 | 2.48 | 2.42 | 2.38 | 2.31 | 2.23 | 2.16 | 2.11 | 2.07 | 2.03 | 1.98 | 1.93 | 1.88 |
| 20 | 4.35 | 3.49 | 3.10 | 2.87 | 2.71 | 2.60 | 2.51 | 2.45 | 2.39 | 2.35 | 2.28 | 2.20 | 2.12 | 2.08 | 2.04 | 1.99 | 1.95 | 1.90 | 1.84 |
| 21 | 4.32 | 3.47 | 3.07 | 2.84 | 2.68 | 2.57 | 2.49 | 2.42 | 2.37 | 2.32 | 2.25 | 2.18 | 2.10 | 2.05 | 2.01 | 1.96 | 1.92 | 1.87 | 1.81 |
| 22 | 4.30 | 3.44 | 3.05 | 2.82 | 2.66 | 2.55 | 2.46 | 2.40 | 2.34 | 2.30 | 2.23 | 2.15 | 2.07 | 2.03 | 1.98 | 1.94 | 1.89 | 1.84 | 1.78 |
| 23 | 4.28 | 3.42 | 3.03 | 2.80 | 2.64 | 2.53 | 2.44 | 2.37 | 2.32 | 2.27 | 2.20 | 2.13 | 2.05 | 2.01 | 1.96 | 1.91 | 1.86 | 1.81 | 1.76 |
| 24 | 4.26 | 3.40 | 3.01 | 2.78 | 2.62 | 2.51 | 2.42 | 2.36 | 2.30 | 2.25 | 2.18 | 2.11 | 2.03 | 1.98 | 1.94 | 1.89 | 1.84 | 1.79 | 1.73 |
| 25 | 4.24 | 3.39 | 2.99 | 2.76 | 2.60 | 2.49 | 2.40 | 2.34 | 2.28 | 2.24 | 2.16 | 2.09 | 2.01 | 1.96 | 1.92 | 1.87 | 1.82 | 1.77 | 1.71 |
| 30 | 4.17 | 3.32 | 2.92 | 2.69 | 2.53 | 2.42 | 2.33 | 2.27 | 2.21 | 2.16 | 2.09 | 2.01 | 1.93 | 1.89 | 1.84 | 1.79 | 1.74 | 1.68 | 1.62 |
| 40 | 4.08 | 3.23 | 2.84 | 2.61 | 2.45 | 2.34 | 2.25 | 2.18 | 2.12 | 2.08 | 2.00 | 1.92 | 1.84 | 1.79 | 1.74 | 1.69 | 1.64 | 1.58 | 1.51 |
| 60 | 4.00 | 3.15 | 2.76 | 2.53 | 2.37 | 2.25 | 2.17 | 2.10 | 2.04 | 1.99 | 1.92 | 1.84 | 1.75 | 1.70 | 1.65 | 1.59 | 1.53 | 1.47 | 1.39 |
| 120 | 3.92 | 3.07 | 2.68 | 2.45 | 2.29 | 2.18 | 2.09 | 2.02 | 1.96 | 1.91 | 1.83 | 1.75 | 1.66 | 1.61 | 1.55 | 1.50 | 1.43 | 1.35 | 1.25 |
| ∞ | 3.84 | 3.00 | 2.60 | 2.37 | 2.21 | 2.10 | 2.01 | 1.94 | 1.88 | 1.83 | 1.75 | 1.67 | 1.57 | 1.52 | 1.46 | 1.39 | 1.32 | 1.22 | 1.00 |

**Table VIII (continued)**

**Values of $F_{0.1,v1,v2}$**

$v_2$ = Degrees of freedom for numerator

$v_1$ = Degrees of freedom for denominator

| | 1 | 2 | 3 | 4 | 5 | 6 | 7 | 8 | 9 | 10 | 12 | 15 | 20 | 24 | 30 | 40 | 60 | 120 | ∞ |
|---|---|---|---|---|---|---|---|---|---|---|---|---|---|---|---|---|---|---|---|
| 1 | 4,052 | 5,000 | 5,403 | 5,625 | 5,764 | 5,859 | 5,928 | 5,982 | 6,023 | 6,056 | 6,106 | 6,157 | 6,209 | 6,235 | 6,261 | 6,287 | 6,313 | 6,339 | 6,366 |
| 2 | 98.5 | 99.0 | 99.2 | 99.2 | 99.3 | 99.3 | 99.4 | 99.4 | 99.4 | 99.4 | 99.4 | 99.4 | 99.4 | 99.5 | 99.5 | 99.5 | 99.5 | 99.5 | 99.5 |
| 3 | 34.1 | 30.8 | 29.5 | 28.7 | 28.2 | 27.9 | 27.7 | 27:5 | 27.3 | 27.2 | 27.1 | 26.9 | 26.7 | 26.6 | 26.5 | 26.4 | 26.3 | 26.2 | 26.1 |
| 4 | 21.2 | 18.0 | 16.7 | 16.0 | 15.5 | 15.2 | 15.0 | 14.8 | 14.7 | 14.5 | 14.4 | 14.2 | 14.0 | 13.9 | 13.8 | 13.7 | 13.7 | 13.6 | 13.5 |
| 5 | 16.3 | 13.3 | 12.1 | 11.4 | 11.0 | 10.7 | 10.5 | 10.3 | 10.2 | 10.1 | 9.89 | 9.72 | 9.55 | 9.47 | 9.38 | 9.29 | 9.20 | 9.11 | 9.02 |
| 6 | 13.7 | 10.9 | 9.78 | 9.15 | 8.75 | 8.47 | 8.26 | 8.10 | 7.98 | 7.87 | 7.72 | 7.56 | 7.40 | 7.31 | 7.23 | 7.14 | 7.06 | 6.97 | 6.88 |
| 7 | 12.2 | 9.55 | 8.45 | 7.85 | 7.46 | 7.19 | 6.99 | 6.84 | 6.72 | 6.62 | 6.47 | 6.31 | 6.16 | 6.07 | 5.99 | 5.91 | 5.82 | 5.74 | 5.65 |
| 8 | 11.3 | 8.65 | 7.59 | 7.01 | 6.63 | 6.37 | 6.18 | 6.03 | 5.91 | 5.81 | 5.67 | 5.52 | 5.36 | 5.28 | 5.20 | 5.12 | 5.03 | 4.95 | 4.86 |
| 9 | 10.6 | 8.02 | 6.99 | 6.42 | 6.06 | 5.80 | 5.61 | 5.47 | 5.35 | 5.26 | 5.11 | 4.96 | 4.81 | 4.73 | 4.65 | 4.57 | 4.48 | 4.40 | 4.31 |
| 10 | 10.0 | 7.56 | 6.55 | 5.99 | 5.64 | 5.39 | 5.20 | 5.06 | 4.94 | 4.85 | 4.71 | 4.56 | 4.41 | 4.33 | 4.25 | 4.17 | 4.08 | 4.00 | 3.91 |
| 11 | 9.65 | 7.21 | 6.22 | 5.67 | 5.32 | 5.07 | 4.89 | 4.74 | 4.63 | 4.84 | 4.40 | 4.25 | 4.10 | 4.02 | 3.94 | 3.86 | 3.78 | 3.69 | 3.60 |
| 12 | 9.33 | 6.93 | 5.95 | 5.41 | 5.06 | 4.82 | 4.64 | 4.50 | 4.39 | 4.30 | 4.16 | 4.01 | 3.86 | 3.78 | 3.70 | 3.62 | 3.54 | 3.45 | 3.36 |
| 13 | 9.07 | 6.70 | 5.74 | 5.21 | 4.86 | 4.62 | 4.44 | 4.30 | 4.19 | 4.10 | 3.96 | 3.82 | 3.66 | 3.59 | 3.51 | 3.43 | 3.34 | 3.25 | 3.17 |
| 14 | 8.86 | 6.51 | 5.56 | 5.04 | 4.70 | 4.46 | 4.28 | 4.14 | 4.03 | 3.94 | 3.80 | 3.66 | 3.51 | 3.43 | 3.35 | 3.27 | 3.18 | 3.09 | 3.00 |
| 15 | 8.68 | 6.36 | 5.42 | 4.89 | 4.56 | 4.32 | 4.14 | 4.00 | 3.89 | 3.80 | 3.67 | 3.52 | 3.37 | 3.29 | 3.21 | 3.13 | 3.05 | 2.96 | 2.87 |
| 16 | 8.53 | 6.23 | 5.29 | 4.77 | 4.44 | 4.20 | 4.03 | 3.89 | 3.78 | 3.69 | 3.55 | 3.41 | 3.26 | 3.18 | 3.10 | 3.02 | 2.93 | 2.84 | 2.75 |
| 17 | 8.40 | 6.11 | 5.19 | 4.67 | 4.34 | 4.10 | 3.93 | 3.79 | 3.68 | 3.59 | 3.46 | 3.31 | 3.16 | 3.08 | 3.00 | 2.92 | 2.83 | 2.75 | 2.65 |
| 18 | 8.29 | 6.01 | 5.09 | 4.58 | 4.25 | 4.01 | 3.84 | 3.71 | 3.60 | 3.51 | 3.37 | 3.23 | 3.08 | 3.00 | 2.92 | 2.84 | 2.75 | 2.66 | 2.57 |
| 19 | 8.19 | 5.93 | 5.01 | 4.50 | 4.17 | 3.94 | 3.77 | 3.63 | 3.52 | 3.43 | 3.30 | 3.15 | 3.00 | 2.92 | 2.84 | 2.76 | 2.67 | 2.58 | 2.49 |
| 20 | 8.10 | 5.85 | 4.94 | 4.43 | 4.10 | 3.87 | 3.70 | 3.56 | 3.46 | 3.37 | 3.23 | 3.09 | 2.94 | 2.86 | 2.78 | 2.69 | 2.61 | 2.52 | 2.42 |
| 21 | 8.02 | 5.78 | 4.87 | 4.37 | 4.04 | 3.81 | 3.64 | 3.51 | 3.40 | 3.31 | 3.17 | 3.03 | 2.88 | 2.80 | 2.72 | 2.64 | 2.55 | 2.46 | 2.36 |
| 22 | 7.95 | 5.72 | 4.82 | 4.31 | 3.99 | 3.76 | 3.59 | 3.45 | 3.35 | 3.26 | 3.12 | 2.98 | 2.83 | 2.75 | 2.67 | 2.58 | 2.50 | 2.40 | 2.31 |
| 23 | 7.88 | 5.66 | 4.76 | 4.26 | 3.94 | 3.71 | 3.54 | 3.41 | 3.30 | 3.21 | 3.07 | 2.93 | 2.78 | 2.70 | 2.62 | 2.54 | 2.45 | 2.35 | 2.26 |
| 24 | 7.82 | 5.61 | 4.72 | 4.22 | 3.90 | 3.67 | 3.50 | 3.36 | 3.26 | 3.17 | 3.03 | 2.89 | 2.74 | 2.66 | 2.58 | 2.49 | 2.40 | 2.31 | 2.21 |
| 25 | 7.77 | 5.57 | 4.68 | 4.18 | 3.86 | 3.63 | 3.46 | 3.32 | 3.22 | 3.13 | 2.99 | 2.85 | 2.70 | 2.62 | 2.53 | 2.45 | 2.36 | 2.27 | 2.17 |
| 30 | 7.56 | 5.39 | 4.51 | 4.02 | 3.70 | 3.47 | 3.30 | 3.17 | 3.07 | 2.98 | 2.84 | 2.70 | 2.55 | 2.47 | 2.39 | 2.30 | 2.21 | 2.11 | 2.01 |
| 40 | 7.31 | 5.18 | 4.31 | 3.83 | 3.51 | 3.29 | 3.12 | 2.99 | 2.89 | 2.80 | 2.66 | 2.52 | 2.37 | 2.29 | 2.20 | 2.11 | 2.02 | 1.92 | 1.80 |
| 60 | 7.08 | 4.98 | 4.13 | 3.65 | 3.34 | 3.12 | 2.95 | 2.82 | 2.72 | 2.63 | 2.50 | 2.35 | 2.20 | 2.12 | 2.03 | 1.94 | 1.84 | 1.73 | 1.60 |
| 120 | 6.85 | 4.79 | 3.95 | 3.48 | 3.17 | 2.96 | 2.79 | 2.66 | 2.56 | 2.47 | 2.34 | 2.19 | 2.03 | 1.95 | 1.86 | 1.76 | 1.66 | 1.53 | 1.38 |
| ∞ | 6.63 | 4.61 | 3.78 | 3.32 | 3.02 | 2.80 | 2.64 | 2.51 | 2.41 | 2.32 | 2.18 | 2.04 | 1.88 | 1.79 | 1.70 | 1.59 | 1.47 | 1.32 | 1.00 |

# BIBLIOGRAPHY


Childers, D.G. *Probability and Random Processes.* Chicago: Irwin. 1997.

Clarke , A.B and R.L. Disney. *Probability and Random Processes for Engineers and Scientists.* York: Wiley. 1970.

Davenport, W.B. Jr. *Probability and Random processes, An Introduction for Scientists and Engineers.* New York: McGraw-Hill. 1970.

Feller,W. *An Introduction to Probability Theory and its Applications.* New York: John Wiley and Sons. 1966.

Helstorm, R.M. *Probability and Stochastic Processes for Engineers.* New York: Macmillan. 1984.

Hoel, Paul G. *Introduction to Mathematical Statistics.* India: Asia Publishing House. 1960.

Johnson, Richard A . *Miller and Freund's Probability and Statistics for Engineers.* India: Prentice-Hall of India Private limited. 2000.

Kanda Swamy , P., K. Thilagavathi and K. Gunavathi. *Probability and Statistics and Queueing Theory.* New Delhi: S.Chand. 2007.

Kapur, J. N. and H.C. Saxena. *Mathematical Statistics.* India: S.Chand and Company. 2006.

Khinchine,A . *Works on the Theory of Queues.* New York: John Wiley and Sons. 1961.

Medhi, J. *Stochastic Processes.* New Delhi: New Age International (P) Limited Publishers. 1993.

Miller, Irwin and Marylees Miller. *John E. Freund's Mathematical Statistics with Applications.* Pearson Prentice Hall. 2007.

Peebles Jr, Peyton. Z. *Probability, Random Variables and Random Signals Basic.* New York: TMH. 2002.

Snedecor, G.W. and W. G. Cochran. Statistical Methods. Iowa: Iowa State College Press. 1967.

Takacs, K. *An Introduction to the Theory of Queues.* New York: Oxford University Press. 1962.

# INDEX